Neuroprotective Signal Transduction

Contemporary Neuroscience

Gene Therapy for Neurological Disorders and Brain Tumors, edited by **E. Antonio Chiocca** and **Xandra Breakefield,** 1998

Highly Selective Neurotoxins: Basic and Clinical Applications, edited by **Richard M. Kostrzewa,** 1998

Neuroinflammation: Mechanisms and Management, edited by **Paul L. Wood,** 1998

Neuroprotective Signal Transduction, edited by **Mark P. Mattson,** 1998

Clinical Pharmacology of Cerebral Ischemia, edited by **Gert J. Ter Horst** and **Jakob Korf,** 1997

Molecular Mechanisms of Dementia, edited by **Wilma Wasco** and **Rudolph E. Tanzi,** 1997

Neurotransmitter Transporters: Structure, Function, and Regulation, edited by **Maarten E. A. Reith,** 1997

Motor Activity and Movement Disorders: Research Issues and Applications, edited by **Paul R. Sanberg, Klaus-Peter Ossenkopp,** and **Martin Kavaliers,** 1996

Neurotherapeutics: Emerging Strategies, edited by **Linda M. Pullan** and **Jitendra Patel,** 1996

Neuron–Glia Interrelations During Phylogeny: II. Plasticity and Regeneration, edited by **Antonia Vernadakis** and **Betty I. Roots,** 1995

Neuron–Glia Interrelations During Phylogeny: I. Phylogeny and Ontogeny of Glial Cells, edited by **Antonia Vernadakis** and **Betty I. Roots,** 1995

The Biology of Neuropeptide Y and Related Peptides, edited by **William F. Colmers** and **Claes Wahlestedt,** 1993

Psychoactive Drugs: Tolerance and Sensitization, edited by **A. J. Goudie** and **M. W. Emmett-Oglesby,** 1989

Experimental Psychopharmacology, edited by **Andrew J. Greenshaw** and **Colin T. Dourish,** 1987

Developmental Neurobiology of the Autonomic Nervous System, edited by **Phyllis M. Gootman,** 1986

The Auditory Midbrain, edited by **Lindsay Aitkin,** 1985

Neurobiology of the Trace Elements, edited by **Ivor E. Dreosti** and **Richard M. Smith**

Vol. 1: *Trace Element Neurobiology and Deficiencies,* 1983

Vol. 2: *Neurotoxicology and Neuropharmacology,* 1983

Neuroprotective Signal Transduction

Edited by

Mark P. Mattson

University of Kentucky,
Lexington, KY

Humana Press **Totowa, New Jersey**

999 Riverview Drive, Suite 208
Totowa, New Jersey 07512

For additional copies, pricing for bulk purchases, and/or information about other Humana titles, contact Humana at the above address or at any of the following numbers: Tel.: 973-256-1699; Fax: 973-256-8341; E-mail: humana@mindspring.com, or visit our Website: http://humanapress.com

This publication is printed on acid-free paper. ∞
ANSI Z39.48-1984 (American Standards Institute) Permanence of Paper for Printed Library Materials.

Cover design by Mark P. Mattson and Patricia F. Cleary.

Printed in the United States of America. 10 9 8 7 6 5 4 3 2 1

Library of Congress Cataloging-in-Publication Data

Neuroprotective signal transduction/edited by Mark P. Mattson.
p. cm.—(Contemporary neuroscience)
Includes index.
ISBN 0-89603-473-9 (alk. paper)
1. Nervous system—Degeneration—Chemoprevention. 2. Cellular signal transduction. 3. Nervous system—Degeneration—Molecular aspects. 4. Molecular neurobiology. 5. Neurotrophic functions. 6. Growth factors. I. Mattson, Mark Paul. II. Series.
[DNLM: 1. Growth Substances—physiology. 2. Growth Substances—pharmacology. 3. Signal Transduction—physiology. 4. Nervous System Diseases—physiopathology. 5. Neurons—physiology. 6. Neuroprotective Agents—pharmacology. QU 107 N494 1998]
RC365.N479 1998
616.8—dc21
DNLM/DLC
for Library of Congress 97-37497
CIP

Preface

The primordial soup from which cellular life as we know it arose was an oxidizing environment, and also consisted of high concentrations of various ions leached from the earth's crust. From those beginnings, cells had to develop mechanisms to prevent free radicals and certain ions from accumulating within themselves. Genes that encode antioxidant enzymes evolved to provide cells with the ability to survive and flourish in oxidizing environments. Ion-motive ATPases and calcium-binding proteins were early players in the ion removal game. During the same time period that cells were developing defense systems, they took advantage of the chemical properties of free radicals and ions, such as Ca^{2+}, and used them to good effect in molecular signal transduction pathways that allowed them to respond to environmental signals. Although many of the cellular systems that regulate free radicals and Ca^{2+} in cells therefore have their origins hundreds of millions of years ago and are present in essentially all cell types, more elaborate mechanisms have recently evolved. Nerve cells, in particular, possess elegant molecular machineries for the control of ionic movements, as well as for the generation and utilization of free radicals.

In order to prevent their demise, cells evolved injury response mechanisms designed to squelch free radical accumulation and suppress ionic overloads. Many of these injury response pathways are activated by free radicals and calcium themselves, and involve either delayed transcription-dependent cascades or rapid posttranslational events. For example, free radicals can induce activation of the transcription factor NFκB, which may then induce the expression of such antioxidant enzymes as manganese superoxide dismutase. An example of a rapid response is Ca^{2+}-induced generation of cyclic GMP, which then activates a kinase that, in turn, leads to activation of K^+ channels, membrane hyperpolarization, and suppression of Ca^{2+} influx.

As multicellular organisms emerged approximately 700 million years ago, pressure was placed on them to develop intercellular signaling mechanisms to alert their neighbors of impending doom and allow them to respond in ways that prevent their damage and death. In the nervous system a variety of neurotrophic factors and cytokines are induced by injury and released from neurons and glial cells. These proteins bind to specific cell surface receptors on neurons, where they initiate various signaling cascades that usually involve activation of multiple kinases and, ultimately, transcription factors. A major focus of *Neuroprotective Signal Transduction* concerns the cellular and molecular underpinnings of such neuroprotective signal transduction cascades. Knowledge of such cascades, and the ways in which they can be manipulated (physiologically and pharmacologically), has led to the identification of novel strategies for preventing neuron degeneration in a variety of human neurodegenerative conditions, including stroke, traumatic brain injury, Alzheimer's disease, Parkinson's disease, Huntington's disease, and amyotrophic lateral sclerosis.

Neuroprotective Signal Transduction covers signaling pathways of an array neuroprotective factors, ranging from neurotrophins (NGF, BDNF, NT-3, and NT-4/5) to growth factors (bFGF, IGF-1, GDNF) to cytokines (TNF, IL-1β, and TGFβ) to secreted amyloid precursor proteins to protease nexin-1. Additional consideration is given to cytoprotective signaling events that occur within injured neurons independently of intercellular signals. Throughout the chapters, fundamental information on mechanisms of neuron cell death is provided. It is hoped that this novel focus on cellular and molecular signal transduction pathways relevant to human neurodegenerative conditions will spur further basic research in this emerging field. Ultimately, detailed knowledge of the molecular cascades the nervous system has evolved to protect itself from injury will lead to effective means of stimulating such cascades to reduce neuron cell damage in human patients.

ACKNOWLEDGMENTS

I wish to thank first and foremost, Grace Sears for her meticulous editing of each chapter, for efficiently completing revisions in consultation with the authors, and for her patience in dealing with the details of organizing the book. I am also indebted to the young scientists who have passed through my laboratory; they generated much data and many ideas relevant to the topic of the present book. They include: Bin Cheng, Ginger Smith-Swintosky, Steve Barger, Bob Mark, Katsutoshi Furukawa, Susie Goodman, Anna Bruce, Emmanuelle Blanc, Jeff Keller, Qing Guo, and Inna Kruman. Finally, I thank those in my life whose understanding provided me with the time and encouragement to pursue my research—Joanne, Elliot, and Emma.

Mark P. Mattson

Contents

Contributors

ILKNUR AY, *CNS Growth Factor Lab, Massachusetts General Hospital—East, Charlestown, MA*
STEVEN W. BARGER, *Department of Internal Medicine and Anatomy, College of Medicine, University of Arkansas, Little Rock, AK*
WAYNE A. CASS, *Department of Anatomy and Neurobiology, University of Kentucky, Lexington, KY*
WAYNE M. CLARK, *Department of Neurology, Oregon Health Sciences University, Portland, OR*
THOMAS L. DECKWERTH, *Department of Molecular Biology and Pharmacology, Washington University School of Medicine, St. Louis, MO*
STEVE ESTUS, *Center on Aging, University of Kentucky, Lexington, KY*
BARRY W. FESTOFF, *Department of Neurobiological Research, VA Medical Center, University of Kansas, Kansas City, MO*
SETH P. FINKLESTEIN, *CNS Growth Factor Lab, Massachusetts General Hospital—East, Charlestown, MA*
NORMAN FISCHER, *Department of Inflammation, Amgen, Inc., Boulder, CO*
KATSUTOSHI FURUKAWA, *Center on Aging, University of Kentucky, Lexington, KY*
SHOEI FURUKAWA, *Laboratory of Molecular Biology, Gifu Pharmaceutical University, Gifu, Japan*
DON M. GASH, *Department of Anatomy and Neurobiology, University of Kentucky, Lexington, KY*
CECILIA M. KEARNS, *University of Kentucky Medical Center, Lexington, KY*
PATRICK H. KITZMAN, *Department of Anatomy and Neurobiology, University of Kentucky Medical Center, Lexington, KY*
J. H. KORDOWER, *Department of Neurological Sciences, Rush Presbyterian—St. Luke's Medical Center, Chicago, IL*
JOSEF KRIEGLSTEIN, *Institut für Pharmakologie und Toxicologie, Marburg, Germany*
KERSTIN KRIEGLSTEIN, *Institut für Pharmakologie und Toxicologie, Marburg, Germany*
DAVID MARTIN, *Department of Inflammation, Amgen, Inc., Boulder, CO*
MARK P. MATTSON, *Center on Aging, University of Kentucky, Lexington, KY*
GERALD MILLER, *Department of Inflammation, Amgen, Inc., Boulder, CO*
ELLIOTT J. MUFSON, *Department of Neurological Sciences, Rush Presbyterian—St. Luke's Medical Center, Chicago, IL*
TOSHITAKA NABESHIMA, *Department of Neuropsychopharmacology and Hospital Pharmacy, Nagoya, Japan*

ATSUMI NITTA, *Laboratory of Molecular Biology, Gifu Pharmaceutical University, Gifu, Japan*
ROBERT M. SAPOLSKY, *Department of Biological Sciences, Stanford University, Stanford, CA*
VIRGINIA L. SMITH-SWINTOSKY, *The R. W. Johnson Pharmaceutical Research Institute, Springhouse, PA*
JOE E. SPRINGER, *Department of Anatomy and Neurobiology, University of Kentucky Medical Center, Lexington, KY*
JUSTIN A. ZIVIN, *Department of Neurology, Oregon Health Sciences University, Portland, OR*

1

Neuroprotective Strategies Involving the Neurotrophins and Their Signaling Pathways

Joe E. Springer and Patrick H. Kitzman

1. THE NEUROTROPHIC FACTOR HYPOTHESIS

Nerve growth factor (NGF) was the first neurotrophic factor isolated and sequenced, and much of the knowledge in the field has been based on studies related to this classic growth factor. It is now evident that NGF is a member of a larger family termed the neurotrophins, which consists of brain-derived neurotrophic factor (BDNF), neurotrophin-3 (NT-3), NT-4/5, and NT-6 (Bothwell, 1995). The neurotrophins affect a number of developing and adult neuronal populations in both the peripheral nervous system (PNS) and central nervous system (CNS), and can provide neuroprotection to neurons axotomized by mechanical injury or damaged by neurotoxins (Table 1). These observations have led to the hypothesis that the neurotrophins have therapeutic potential in the treatment of a wide range of insults to the nervous system, including neurodegenerative disorders, peripheral neuropathies, and traumatic injury.

Early studies on NGF provided the first evidence that neurotrophic factors are released by target structures and bind to membrane-associated receptors located on responsive neurons. Originally, one criterion of the neurotrophic factor hypothesis maintained that, following release and binding to membrane receptors, the factor was internalized and retrogradely transported to the cell body. In this early scheme, it was thought that the retrograde transport of the neurotrophin/receptor complex was necessary for initiation of the trophic message to the cell body. There are several lines of evidence that support the necessity of retrograde transport for neurotrophin-mediated neuronal responses. First, there is a high correlation between neuronal survival and the level of neurotrophin expression in the peripheral target area. Specifically, changes in neurotrophin levels in the target during development correlate with changes in neuronal survival (Johnson et al., 1980; Anderson et al., 1995). Second, disruption of retrograde transport by axotomy (Springer et al., 1987) or by the administration of colchicine (Purves, 1976) leads to the death of neurotrophin-responsive neurons and the increased accumulation of the corresponding neurotrophin in the target (Grahm et al., 1983). Third, neurons can extend axons a relatively long distance to their target area. Thus, signals generated in the axon terminal would have to diffuse a great distance to the cell body in order to induce neuronal survival.

From: Neuroprotective Signal Transduction *edited by M. P. Mattson*
Humana Press Inc., Totowa, NJ

Table 1
Neurotrophin-Responsive Neuronal Populations in the PNS and CNS

Neurotrophin	Neuronal population	Reference
NGF	Cholinergic basal forebrain	Martinez et al., 1985; Mobley et al., 1985; Hefti, 1986; Williams et al., 1986; Gage et al., 1988
	Cerebellar Purkinje	Cohen-Cory, 1991
	Sympathetic and sensory	Levi-Montalcini, 1982; Johnson et al., 1980; Ruit et al., 1992
BDNF	Midbrain dopamine	Hyman et al., 1994
	Cholinergic basal forebrain	Anderson, et al., 1995; Morse et al., 1993
	Motor neurons	Sendtner et al., 1992; Henderson et al., 1993; Koliatsos et al., 1993; Yan et al., 1992
	Cerebellar granule	Lindholm et al., 1993; Segal et al., 1992
	Retinal ganglia	Rodriquez-Tebar, 1989
	Sensory and nodose ganglia	Lindsay et al, 1985; Hohn et al., 1990; Maisonpierre et al., 1990
NT-3	Locus ceruleus	Friedman et al., 1993
	Cerebellar Purkinje	Mount et al., 1994
	Motor neurons	Henderson et al., 1993
	Sympathetic and sensory	Hohn et al., 1990; Rosenthal et al., 1990
NT-4/5	Motor neurons	Henderson et al., 1993; Friedman et al., 1995
	Sensory	Davies et al., 1993b; Ibanez et al., 1993
	Cerebellar granule	Gao et al., 1995
	Striatal	Widmer and Hefti, 1994; Ardelt et al., 1994

The importance of retrograde transport on NGF-mediated neuronal responses was demonstrated in a classic study in which NGF was introduced directly into the cytoplasm of PC12 cells and was shown to have no effect on NGF-mediated neurite outgrowth (Heumann et al., 1981). Additionally, introduction of antibodies against NGF into the cytoplasm did not inhibit neurite outgrowth induced by the addition of NGF to the culture medium (Heumann et al., 1981). Thus, NGF-mediated responses did not occur because of direct binding of NGF to cytoplasmic elements or regulatory portions on genes.

More contemporary studies have led to the hypothesis that retrograde transport of the neurotrophin is not necessarily obligatory for expression of trophic properties. Specifically, the current dogma contends that the neurotrophins bind to specific high-affinity receptors containing intracellular tyrosine kinase domains (Barbacid, 1994; Kaplan and Stephens, 1994), which in turn activate a cascade of intracellular events resulting in immediate (altered synaptic efficacy) and long-term (altered gene expression) cellular responses. To date, three receptors (Trks) have been identified that serve as selective ligands for the neurotrophins. NGF binds primarily to Trk A, BDNF and NT-4/5 bind preferentially to Trk B, and NT-3 binds to Trk C. NT-3 is somewhat promiscuous, in that it exhibits the ability to bind with relatively high affinity to all three Trk receptors.

The early studies on receptor-mediated events were based in large part on findings involving the low-affinity p75 neurotrophin receptor (Springer, 1988). However, as the discovery of new members of the neurotrophin family and the Trks emerged, it became evident that the p75 alone was not sufficient to account for how all members of the neurotrophins function. For example, although p75 does bind to all of the neurotrophins examined to date, this receptor is not found in most adult CNS neuronal populations that respond to neurotrophins other than NGF. In the PNS, however, there is evidence that p75 is associated with neurons that contain Trk A and Trk B, implying some functional role for p75 in the actions of NGF, BDNF, and NT-4/5. One of the major drawbacks related to the role of p75 is that the trophic effects of the neurotrophins are mediated by high-affinity receptors, but p75 binds with only low affinity. This has raised several questions about the function of p75. Of particular note is whether p75 is even necessary for neurotrophin signaling. In a recent study, Gage and coworkers found that the activation of the Trk A receptor alone is all that is required to rescue axotomized basal forebrain cholinergic neurons in adult rats (Lucidi-Phillipi et al., 1996). These cholinergic neurons also exhibited a robust sprouting response following Trk A receptor activation, which is similar to that seen following NGF infusions (Gage et al., 1988). However, when p75 activation was blocked, these same cholinergic neurons exhibited significantly less axonal sprouting (Lucidi-Phillipi et al., 1996). In another recent study analyzing the role of p75, Bothwell and coworkers demonstrated that the transport of BDNF from the developing retina to the isthmo-optic nucleus can be significantly reduced in the presence of antibodies blocking Trk B or p75 (von Bartheld et al., 1996b). However, the blockade of p75 had no effect on cell survival in the isthmo-optic nucleus; inactivation of Trk B was found to block the trophic effect of BDNF. These findings suggest that the transport of BDNF alone is not sufficient for trophic expression, but that the activation of Trk B is required (von Bartheld et al., 1996b).

Taken together, these studies provide in vivo evidence that p75 does play a role in certain neurotrophin-related events, such as the facilitation of transport and phenotypic expression. Other recent studies have revealed a role by which p75 can mediate neuronal responses independent of Trk interaction.

The present evidence related to this family of growth factors, and to the Trks in particular, has resulted in a modified definition of the neurotrophic factor hypothesis (Korsching, 1993). Neurotrophic factors, and the neurotrophins in particular, can be target-derived or supplied locally, implying an autocrine and/or paracrine function (Acheson et al., 1995). The actions of the neurotrophins are receptor-mediated, and involve an intracellular cascade that can account for both immediate and long-term cellular responses. The task of elucidating the exact role of internalization and retrograde transport of the neurotrophin remains, although this may simply reflect the process by which these peptides are degraded once they reach the lysosomal pool. It should be noted, however, that each member of the neurotrophin family has been shown to be retrogradely transported (Hendry et al., 1974; Palmateir et al., 1984; Seiler and Schwab, 1984; Johnson et al., 1987; DiStefano et al., 1992; Curtis et al., 1995; Ehlers et al., 1995; von Bartheld et al., 1996a,b), and that the high-affinity Trk A receptor (in addition to the low-affinity p75 receptor) has also been shown to be transported in a retrograde fashion (Loy et al., 1994; Ehlers et al., 1995).

Regardless of the role of retrograde transport, a new understanding of how the trophic signals of the neurotrophins are transmitted has and will continue to guide the development of therapeutic tools aimed at mimicking the signal transduction pathways. In addition, understanding events that regulate endogenous neurotrophin synthesis and release may also provide insight into ways in which neuroprotection can be afforded to compromised or injured nerve cells. Implicit in these strategies is the understanding of how the neurotrophins function, and several recent studies utilizing transgenic technology have provided new clues to the way in which these proteins function in the PNS and CNS.

2. MECHANISMS OF ACTION OF NEUROTROPHINS

2.1. *Transgenic Mice*

If one of the tenets of the neurotrophic hypothesis is correct, namely, that the number of neurons that survive the period of naturally occurring cell death is dependent on the availability of neurotrophins expressed in the peripheral target area, then altering the amount of the various neurotrophins, or their receptors, should elicit a change in the number of surviving neurons. Two strategies employed to test this hypothesis in vivo require the development of several lines of transgenic mice. In one strategy, the gene encoding for a specific neurotrophin or its receptor is disrupted, thus resulting in lines of knockout (KO) mice (Table 2). In the second strategy, the gene for a specific neurotrophin is linked to a tissue-specific promoter that can induce the expression of the neurotrophin at a specific time during development.

2.1.1. *NGF and Trk A Knockouts*

NGF was the first of the neurotrophins shown to be essential for the survival of sympathetic and neural crest-derived sensory neurons in the PNS. Neurotrophin dependence of dorsal root ganglia (DRG) neurons is segregated according to sensory modality with DRG neurons mediating nociception, requiring NGF, and expressing Trk A receptor mRNA and protein (Johnson et al., 1980; Ruit et al., 1992; Mu et al., 1993; Crowley et al., 1994; Smeyne et al., 1994).

Studies utilizing either NGF KO or Trk A KO mice have confirmed most of the original findings by demonstrating dramatic decreases in the number of neurons remaining in the spinal cord ganglia (SCG), DRG, and trigeminal ganglia (Crowley et al., 1994; Smeyne et al., 1994; Silos-Santiago et al., 1995). In addition, a preferential loss of small neurons and unmyelinated and small-diameter myelinated axons was demonstrated (Crowley et al., 1994; Silos-Santiago et al., 1995). In the CNS, there was an elimination of CGRP (a peptide marker of nociceptive function) immunoreactivity within the dorsal horn of the spinal cord (Crowley et al., 1994; Silos-Santiago et al., 1995).

Behaviorally, these mice display a decreased responsiveness to painful stimuli. These results help to strengthen the hypothesis that NGF and its receptor are required for the survival of neurons involved in nociceptive function. Also in the CNS, a decrease in AChE staining in cholinergic fibers that project from the medium septum to the hippocampus was observed (Smeyne et al., 1994). However, there was no significant loss in basal forebrain neurons, suggesting Trk A may be involved in the outgrowth of cholinergic axons or in the maintenance of the cholinergic phenotype. As noted above, a recent study has demonstrated that activation of Trk A alone will afford neuroprotection and elicit sprouting of axotomized basal forebrain cholinergic neurons in adult rats (Lucidi-Phillipi et al., 1996).

2.1.2. BDNF and Trk B Knockouts

BDNF has been shown to promote the survival of cultured peripheral sensory neurons, and enhances the survival of DRG and nodose ganglia neurons following infusion into embryos (Hofer and Barde, 1988). In addition, BDNF can promote the survival of embryonic midbrain dopaminergic neurons and motor neurons (Oppenheim et al., 1992; Yan et al., 1992; Hyman et al., 1994). Disruption of the BDNF gene confirmed the survival-promoting effects of BDNF on trigeminal, DRG, nodose, and vestibular ganglia neurons (Ernfors et al., 1994a; Jones et al., 1994; Conover et al., 1995; Liu et al., 1995). In these studies, there was no effect of the BDNF gene KO on the number of SCG neurons. In previous studies, BDNF was found to provide neuroprotection for developing axotomized motor neurons (Oppenheim et al., 1992; Yan et al., 1992), leading to the hypothesis that these neurons normally require this neurotrophin for survival. However, there was no effect on motor neuron survival in mice containing the disrupted BDNF gene, indicating that other members of the neurotrophin family that function through the Trk B receptor may also be important.

In contrast to the results seen with the BDNF KO mice, mice in which the Trk B gene was disrupted exhibited a 35% decrease in the number of neurons in the lumbar spinal cord and a 70% decrease in the number of facial motor neurons (Klein et al., 1993). This findings provide evidence that another ligand (e.g., NT-3) may be acting through the Trk B receptor to promote the survival of motor neurons.

2.1.3. NT-4 Knockouts

Studies show that NT-4 is neuroprotective in at least three contexts: It promotes the survival of cultured embryonic sensory neurons (Davies et al., 1993b; Ibanez et al., 1993; Hyman et al. 1994) and striatal neurons (Ardelt et al., 1994); it rescues axotomized motor neurons (Koliatsos et al., 1994; Friedman et al., 1995); and it supports the survival of nodose-petrosal neurons and central dopaminergic neurons grown in culture (Hyman et al., 1994). To date, two studies have looked at the effects of disruption of the NT-4 gene. NT-4 KO mice display a significant loss in nodose-petrosal and geniculate ganglia neurons (Conover et al., 1995; Liu et al., 1995). Yet, in contrast to previous reports, the loss of NT-4 expression had no effect on motor neurons of the facial nucleus or dopaminergic neurons in the substantia nigra. Since a decrease in motor neuron survival was seen in the Trk B KO mice, but not in the BDNF KO or the NT-4 KO mice, this further suggests that another ligand is acting through the Trk B receptor to promote motor neuron survival.

Table 2
Effects of Neurotrophin and Neurotrophin Receptor Knockouts on Peripheral and Central Nervous System Neuronal Populations

Knockout	Phenotype behavior	PNS (% Neuronal loss)	CNS (% Neuron loss and changes in Immunoreactivity)	References
NGF and Trk A	Display severe sensory and sympathetic neuropathies; ↓ Response to painful stimuli; most die by 3 wk	SCG (81%) [3]; DRG (70–80%) [1,3]; Nodose (no loss) [3]; Trigeminal (70%) [3]; DRG-small neurons are preferentially lost [1–3]; Small non-Trk A expressing NGF-dependent DRG neurons (–100%) [2];	Motor neurons (no loss) [1,3]; Basal forebrain-neuronalatrophy [1]; ↓ ChAT [1], and ↓ Axonal branching [3]; BS-1 immunoreactivity in dorsal horn (complete loss) [2]; 95% loss of myelinated and 50% loss of small-diameter myelinated axons [2]; Loss of CGRP immunoreactive fibers in spinal cord [1];	1. Crowley et. al., 1994 (NGF KO); 2. Silos-Santiago et. al. 1995 (Trk A KO); 3. Smeyne et. al., 1994 (Trk A KO)
p75	Loss of heat sensitivity and development of ulcers in the distal extremities	Footpad skin—↓ in CGRP and Sub. P immunoreactive fibers [1]; DRGRs appear smaller [1]; Cutaneous sensory trigeminal neurons display a ↓ response to NGF [2]; SCG neurons display ↓ response to NGF [3]; ↓ sympathetic innervation to pineal gland and sweat glands [3];		1. Lee et. al., 1992; 2. Davies et. al., 1993a; 3. Lee et. al., 1994a,b

BDNF and Trk B	Feed poorly; In Trk B and BDNF most die in 48 hs; In BDNF, head turning, spinning in wk 2	SCG (no loss) [2–3]; DRG (30–50%) [2–4]; Vestibular Ganglia (>80%) [1–3,5]; Nodose Ganglia (45–65%) [1–3,5]; Trigeminal ganglia (25–55%) [1,2,4,5]; Geniculate ganglia (48%) [3,5];	Facial motor neurons; Trk B (70%) [4]; BDNF (no loss) [2,3]; Vmes (40–50%) [2,3]; Cortical, ↓ NPY [3]; Facial Motor Nucleus (no loss) [1–2]; Lumbar Spinal motor neurons (no loss) [1,2]—Trk B –35% at L2–L5, but not at sacral levels [4];	1. Conover et al., 1995 (BDNF KO); 2. Ernfors et al., 1994a (BDNF KO); 3. Jones et al., 1994 (BDNF KO); 4. Klein et al., 1993 (Trk B KO); 5. Liu et al., 1995 (BDNF KO)
NT-4	No real behavioral changes in these animals	Trigeminal ganglia (no loss) [2]; Vesitbular ganglia—(21%) [2]; SCG (no loss) [2]; Nodose ganglia (59%) [1,2]; Geniculate ganglia (50%) [2]; DRG (no loss) [1,2]; Cochlear ganglia (no loss) [1];	Facial motor nucleus (no loss) [1,2]; Central dopaminergic neurons (no loss) [2];	1. Conover et al., 1995 (NT-4 KO); 2. Liu et al., 1995 (NT-4 KO)
NT-3 and Trk C	Abnormal limb postures; Reduced activity; Loss of proprioception; Die within 30 d	SCG (55%) (NT-3) [1,3]; DRG Trk C (20%) [4]; NT-3 (55–78%) [1,3]; Nodose 47% [1,3]; Trigeminal ganglia (65%) (NT-3) [1,3]; 47% ↓ in myelinated axons [4]; Geniculate ganglia (25%) [3]; Vestibular ganglia (23–34%) [2–3]; Cochlear ganglia (85%) [2–3];	Motor neurons: Trk C (30%) [4]; NT-3 (no loss) [1,3]; Vmes (50%) [1,3]; la muscle afferents absent in cervical and thoracic spinal cord [1]; ↓ area of dorsal columns [4]; Enteric ganglia (no changes) [3];	1. Ernfors et al., 1994b (NT-3 KO); 2. Ernfors et al., 1995 (NT-3 KO); 3. Farinas et al., 1994 (NT-3 KO); 4. Klein et al., 1994 (Trk C KO);

2.1.4. NT-3 and Trk C Knockouts

In vitro studies have shown that populations of neurons responsive to NT-3 include primary sensory, sympathetic, motor, enteric, locus ceruleus, and cerebellar neurons (Hohn et al., 1990; Maisonpierre et al., 1990; Rosenthal et al., 1990; Henderson et al., 1993). In mice lacking the NT-3 gene, a significant decrease in the number of neurons was observed in the SCG, DRG, mescencephalic nucleus of V, nodose, trigeminal, geniculate, vestibular, and cochlear ganglia (Ernfors et al., 1994b, 1995; Farinas et al., 1994).

No loss of motor neurons was observed in the NT-3 KO mice. However, spinal proprioceptive afferents, along with muscle spindles and Golgi tendon organs, were absent. This lack explains the severity of the movement defects observed in these animals. These animals also exhibit a loss of slowly adapting mechanoreceptors and their target structures, the Merkel cells (Airaksinen et al., 1996).

In Trk C KO mice, there is a 20% decrease in the number of DRG neurons (Klein et al., 1994). These mice (like the NT-3 KO mice) also display a lack of 1a afferents, as well as a significant decrease in myelinated axons. As would be predicted, these animals display abnormal movements and postures, which strongly suggests that NT-3/Trk C plays a primary role in proprioception. However, since the abnormalities seen in the Trk C KO mice are less severe than those seen in the NT-3 KO mice, NT-3 may utilize other Trk receptors, in addition to the Trk C receptor, to mediate its effects. Such a hypothesis is consistent with the finding that NT-3 binds to all three Trk receptor subtypes.

2.1.5. p75 Knockouts

The function of the p75 receptor has yet to be completely elucidated. Although mice with a null mutation to the p75 gene grow to maturity and do not display the dramatic abnormalities seen in the Trk A KOs, several alterations are evident in these mice. There is a decrease in the CGRP and Sub P (another marker for nociceptive function) immunoreactive fibers in the footpads of these mice. This correlates with an observed loss of heat sensitivity (Lee et al., 1992). In addition, trigeminal ganglia and SCG neurons display a decreased sensitivity to NGF (Lee et al., 1992; Davies et al., 1993a). These results suggest that one function of the p75 receptor is to modulate neuronal responsiveness to NGF, which is consistent with the finding that p75 potentiates the action of NGF in the presence of Trk A (Hantzopoulos et al., 1994).

2.1.6. Mice Overexpressing NGF, BDNF, and NT-3 in Skin

Mice in which NGF has been overexpressed in the epidermis exhibit an increase in the number of CGRP and tyrosine hydroxylase (TH) immunoreactive fibers innervating the skin, suggesting an increase in sensory and sympathetic innervation. The increase of immunoreactive fibers is further evidence that NGF is important in the development of both systems. The trigeminal ganglia and SCG in these mice hypertrophied, with a significant increase in the total number of neurons found in the trigeminal ganglia (Albers et al., 1994). As might be predicted, the number of trigeminal ganglia neurons expressing Trk A increased fivefold (Goodness et al., 1996); there was also two-fold increase in the number of trigeminal neurons that expressed Trk C mRNA.

One interpretation under investigation is that overexpression of NGF reduces or prevents the naturally occurring cell death in at least one population of sensory neurons. Overexpression of NGF also induces a novel sympathetic projection to the cell bodies of sensory neurons (Davis et al., 1994). Davis and coworkers further indicate that the overexpression of NGF may be linked to the development of sympathetic hyperplasia, which can occur following peripheral nerve injury.

As described in Section 2.1., a subpopulation of trigeminal, DRG, nodose, and vestibular ganglia neurons are lost in BDNF knockout animals (Ernfors et al., 1994a; Jones et al., 1994; Conover et al., 1995; Liu et al., 1995). To examine the effects of BDNF on the development of sensory neurons innervating the skin, BDNF was overexpressed in the dermis using a tissue-specific keratin promoter (Albers et al., 1994). High levels of BDNF protein expression in the skin resulted in an increased innervation of regions underlying the epidermis, and was composed primarily of myelinated and unmyelinated small-diameter fibers (LeMaster et al., 1995).

No change occurred in the number of neurons containing Trk B mRNA in the DRG; the trigeminal ganglia was found to contain a threefold higher number of neurons expressing Trk B mRNA and 1.5-fold higher number of Trk A mRNA expressing neurons (LeMaster et al., 1996). A twofold increase in the number of neurons containing p75 was also seen in the trigeminal ganglia. The increases in Trk A- and Trk B-expressing neurons in the trigeminal ganglia correlate with the changes observed in whisker pad innervation (LeMaster et al., 1995), a primary target of the trigeminal neurons.

Overexpression of NT-3 in the epidermis induced a 70% increase in the number of trigeminal ganglia neurons, a 42% increase in DRG neurons, and a 32% increase in SCG neurons (Albers et al., 1996). The increase in DRG neurons was restricted primarily to the population of large- and small-diameter neurons. Although a 60% increase in myelinated fibers in the cutaneous saphenous nerve was also shown in these mice, the ratio of myelinated to unmyelinated fibers remained the same, since there was also a 65% increase in the number of unmyelinated fibers in these animals. Mice having high levels of NT-3 in skin also exhibited an increase in Merkel cells and touch domes, structures thought to be involved in the relay of tactile sensory stimulation (Albers et al., 1996). This observation corroborates the findings of a recent study demonstrating a loss of Merkel cells in the NT-3 KO mice (Airaksinen et al., 1996). Both studies, in conjunction with results observed in culture and the knockout studies, support the hypothesis that NT-3 enhances the survival of sensory and sympathetic neurons.

2.1.7. *Summary of Transgenic Studies*

Transgenic animal studies have provided significant advances toward understanding the role of the neurotrophins in the development of CNS and PNS neuronal populations. Of particular note is the finding that certain Trk-containing neurons do not require the corresponding neurotrophin ligand for survival during development. For example, although basal forebrain cholinergic neurons can be rescued from degeneration following axotomy with exogenous NGF treatment in adults (Hefti, 1986; Williams et al., 1986), they do not degenerate in either the Trk A or NGF KO mice (Smeyne et al., 1994).

The numerous studies utilizing transgenic technology, lesion models, and in vitro paradigms have provided substantial evidence for treating certain nervous system diseases with the neurotrophins. However, delivery of these charged proteins into the CNS and potential peripheral effects has hampered the use of the neurotrophins as therapeutic agents. Two approaches to be discussed that circumvent the delivery issue involve the regulation of neurotrophin synthesis in vivo, and the use of agents that mimic neurotrophin-selective signal transduction pathways.

3. EVENTS CONTROLLING NEUROTROPHIN SYNTHESIS

The primary goal of using neurotrophins as therapeutic tools is to enhance their action on damaged or compromised neurons. Several strategies have been envisioned, including the use of exogenous replacement either by direct infusions into the CNS (Olson, 1993) or the transplantation of neurotrophin-producing cells (Gage et al., 1991), and the administration of neurotrophins conjugated to molecules that allow delivery across the blood–brain barrier (Friden et al., 1993). Two approaches to be discussed in this chapter involve the administration of compounds that regulate neurotrophin synthesis and release (*see* Chapter 5), and ones that mimic Trk receptor activation (Hempstead, 1993) and/or the intracellular signaling cascade (*see* Chapter 16).

Among the first researchers to observe that neurotrophin synthesis can be regulated in the CNS were Gall and coworkers. They demonstrated that events that modify neuronal activity result in changes in neurotrophin mRNA expression in a number of forebrain structures. Initially, it was shown that stimuli intense enough to result in seizures would result in robust changes in neurotrophin mRNA expression in such areas as the hippocampal formation and neocortex (Gall and Isackson, 1989). Subsequent studies showed that less intense stimuli, such as elicitation of a single after-discharge by electrical stimulation or activation of long-term potentiation (LTP), would also increase NGF and BDNF mRNA expression in the hippocampal formation (Gall, 1993; Springer et al., 1994). The significant role of the excitatory neurotransmitter glutamate (Zafra et al., 1991), a common theme throughout these investigations, is supported by experiments showing that selective activation of the *N*-methyl-D-aspartate (NMDA) glutamate receptor subtype increases NGF and BDNF mRNA expression in the adult rat hippocampal formation in the absence of seizures or excitotoxicity (Gwag et al., 1993; Gwag and Springer, 1993). Moreover, the induction of NGF and BDNF mRNA following LTP can be blocked by NMDA-receptor antagonists, but not by antagonists that block the AMPA/kainate glutamate receptor subtypes (Springer et al., 1994).

These findings have led to speculation that events associated with neuronal activity and plasticity in the hippocampal formation may be under the control of glutamate-mediated neurotrophin expression. In addition to increasing neurotrophins and their receptor messages, neuronal activity could also alter expression of signaling pathway proteins. Thus, neurotrophins may work in concert with neuronal activity to promote neuronal survival.

It can be hypothesized that a loss of the glutamatergic regulatory process would have a profound impact on the function and survival of neurotrophin-responsive neuronal populations that project to the hippocampal formation. For example, neurons located in the entorhinal cortex, which give rise to the primary glutamatergic input to the hippocampal formation (angular bundle-perforant path), have been shown to degenerate in patients with Alzheimer's disease (Gomez-Isla et al., 1996). It is conceivable that, over time, the continued loss of this regulatory pathway would compromise the ability of hippocampal neurons to synthesize required levels of certain neurotrophins. One consequence would be a loss of function and ultimately death of neurons that project to the hippocampal formation and require neurotrophic support.

Although it is not practical to imagine the use of glutamatergic agonists as therapeutic agents for stimulating neurotrophin synthesis, these observations do provide substantial evidence that endogenous neurotrophin levels can be modified. Taking advantage of this finding, several strategies are being considered for modifying endogenous neurotrophin synthesis in the CNS, and the effects of several pharmacological agents on neurotrophin expression have been summarized elsewhere (Carswell, 1993).

A second approach to be discussed in this book involves the examination of compounds that could mimic or activate neurotrophin-specific pathways. Implicit in this strategy is that there is a good understanding of the neurotrophin signaling pathways, and several advances have been made toward this end. The final section of this chapter will focus on what is known about the different downstream events that occur following neurotrophin binding (*see also* Chapter 16).

4. NEUROTROPHIN SIGNALING PATHWAYS

4.1. *The p75 Receptor Revisited*

In addition to binding to members of the Trk receptor family (Hantzopoulos et al., 1994), all of the neurotrophins bind to the low-affinity neurotrophin receptor (or p75). There are several lines of evidence that p75 may serve numerous functions in mediating cellular responses to the neurotrophins. For example, the p75 receptor has been shown to be important in the retrograde transport of neurotrophins (Curtis et al., 1995; von Bartheld et al., 1996b). In addition, the p75 receptor may play a role in apoptosis (Rabizadeh et al., 1993; Barrett and Bartlett et al., 1994; Cortazzo et al., 1996) and Schwann cell migration (Anton et al., 1994). Of particular note is the growing body of evidence demonstrating that p75 may modulate the function of Trk receptors. This modulation can be in the form of increasing ligand-binding specificity of Trk A (Benedetti et al., 1993; Lee et al., 1994b), increased rate of association of NGF with Trk (Mahadeo et al., 1994), and changes in NGF-mediated Trk tyrosine kinase activity (Barker et al., 1994; Kahle et al., 1994; Verdi et al., 1994). This modulatory function of p75 may involve physical interaction between p75 and Trk (Huber and Chao, 1995; Wolf et al., 1995; Ross et al., 1996). p75 may also influence activation of Trk B and Trk C receptors (Hantzopolos et al., 1994).

More recent studies have shown that p75 may activate cellular responses without the participation of Trk receptors. Signaling through the p75 receptor has been shown to activate portions of the sphingomyelin pathway (Dobrowsky et al., 1994, 1995; Carter et al., 1996). Specifically, p75 has been shown to activate ceramide and NF-κB; p75-dependent sphingomyelin hydrolysis is modulated by Trk tyrosine kinase activity (Dobrowsky et al., 1995). One recent hypothesis proposes that activation of p75 leads to the activation of two separate pathways. One pathway leads to the activation of ceramide and the second pathway leads to NF-κB activation (Siebenlist et al., 1994; Bothwell, 1996). NGF binding to p75 activates both of the pathways; binding of BDNF or NT-3 to p75 activates only the ceramide pathway. The two-pathway hypothesis is supported by a recent study that demonstrated that activation of NF-κB occurred only when NGF was bound to p75, but not when BDNF or NT-3 were bound to p75. In addition, using an antibody against p75 inhibits NGF binding more effectively than the binding of BDNF and NT-3 (von Bartheld et al., 1996b). This suggests that the neurotrophins bind to separate sites on the p75 receptor.

4.2. *Neurotrophin–Trk Interaction*

The neurotrophins induce their biological activities by binding to members of the tyrosine kinase family of receptors, which in turn activate signaling pathways within the cell. Each neurotrophin has a primary receptor in which it interacts: NGF with Trk A, BDNF and NT4/5 with Trk B, and NT-3 with Trk C. However, there is some promiscuity among the receptors (e.g., NT-3 can also bind to Trk A and Trk B). Most of the studies that have tried to elucidate the signaling mechanism(s) of the neurotrophins have concentrated on the Trk A receptor and have utilized PC12 cell cultures. Autophosphorylation of the Trk A receptor, induced by NGF binding, has been shown to activate multiple pathways, including the Ras/ERK, PLCδ, PI-3, and SNT pathways (*see* Fig. 1. Vetter et al., 1991; Soltoff et al., 1992; Obermeier et al., 1993a,b; Yao and Cooper, 1995).

4.2.1. *Ras/ERK Pathway*

Activation of the Trk receptor induces tyrosine phosphorylation, which links the receptor with the adapter proteins Shc, Grb2, and Sos (a protein that promotes the exchange of GDP for GTP). This exchange activates the Ras protein, which in turn activates a cascade of protein kinases, including Raf and MEK (*see* review by Segal and Greenberg, 1996). MEK activates the mitogen-activated protein kinase (MAPK), an extracellular signal-regulated kinase (ERK), which then translocates into the nucleus, where it phosphorylates such transcription factors as Elk1. Elk1 binds to the serum response factor (SRF), which is bound to the serum response element (SRE). Ras can also activate CREB kinase, which is translocated into the nucleus, where it phosphorylates the transcription factor cAMP response-element binding protein (CREB). CREB binds to the cAMP-Ca response element (Ca-CRE) and activates transcription of immediate early genes (IEGs) in cooperation with the SER complex. The activation of IEGs can lead to the activation of delayed response genes (*see* review by Segal and Greenberg, 1996). This pathway is proposed to be involved in NGF-mediated cellular differentiation (Kaplan and Stephens, 1994), and is mediated in part by autophosphorylation that occurs at two tyrosine residues outside the kinase domain of Trk (Obermeier et al., 1993a; Stephens et al., 1994).

4.2.2. *PLCg Pathway*

Autophosphorylation of the Trk receptor has also been shown to activate PLCγ (Vetter et al., 1991; Obermeier et al., 1993a). It is hypothesized that PLCγ interacts with the Ras pathway to help activate ERKs, and contributes to NGF-mediated differentiation in PC12 cells (Obermeier et al., 1993a; Loeb et al., 1994; Stephens et al., 1994). The point at which PLCγ can affect the Ras pathway has not been fully demonstrated, although it can be postulated that the activation of, and/or interaction with, MEKs is related to IP_3-mediated increases in intracellular calcium, or DAG-induced activation of PKC. As in the SHC/Ras pathway, it appears that there are two tyrosine residues on Trk that mediate the interaction and activation of PLCγ (Obermeier et al., 1993a; Loeb et al., 1994; Stephens et al., 1994).

4.2.3. *Suc-Associated Neurotrophic Factor-Induced Tyrosine-Phosphorylated Target (SNT) Pathway*

The SNT signaling cascade is believed to be involved in the removal of the cell from the cell cycle in response to NGF–Trk interaction (Rabin et al., 1993). Specifically, the

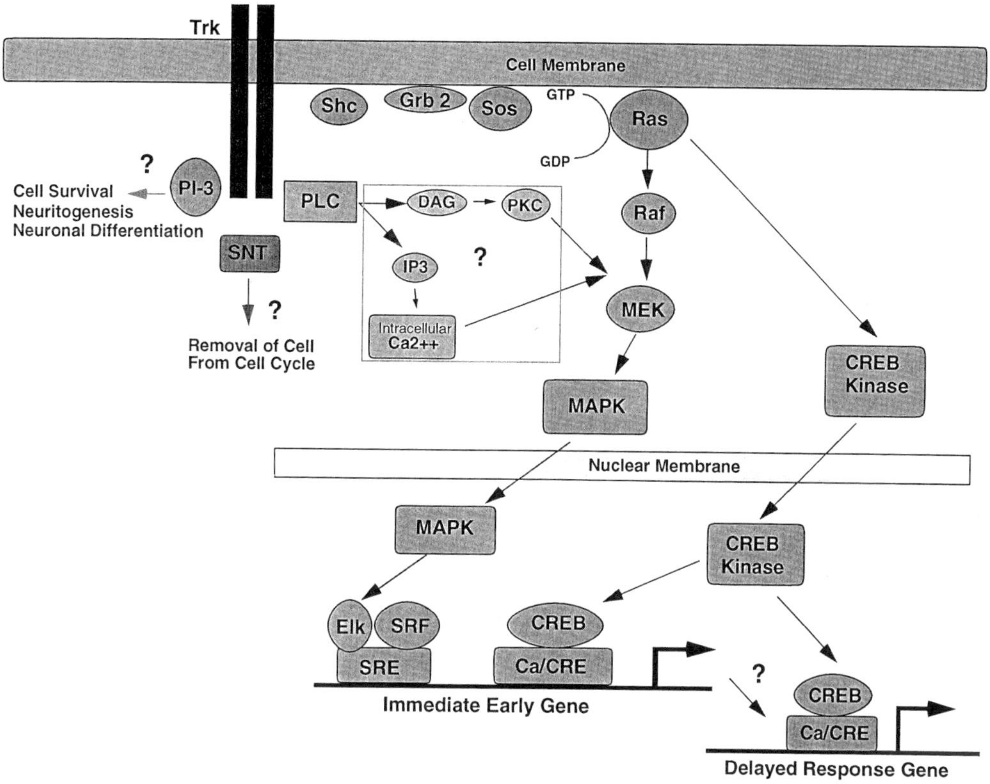

Fig. 1. Schematic diagram depicting several known and proposed intracellular signaling pathways activated following Trk receptor stimulation.

SNT pathway is activated when cells are treated with a differentiation factor (e.g., NGF), but not a mitogenic factor (e.g., EGF). The SNT pathway appears to work independently of the Ras pathway, but is similar to Ras in that it is highly dependent on tyrosine kinase activity associated with Trk activation.

4.2.4. Phosphatidylinositol-3′ (PI-3) Kinase Pathway

The PI-3 pathway is a fourth signaling mechanism that has been linked to Trk autophosphorylation and is believed to be involved in NGF-mediated cellular survival (Soltoff et al., 1992; Obermeier et al., 1993b; Yao and Cooper, 1995). PI-3 can initiate several intracellular pathways, but the precise downstream neurotrophin signaling mechanism(s) related to this kinase have not been determined.

4.2.5. Trk-Mediated Multiple Signaling Pathways

Two of the main questions that have been the foundation of many studies looking at signal transduction are: Why would a single receptor type activate multiple signaling pathways; and how can a single pathway or combination of pathways be differentially activated?

As to the question of why multiple pathways are linked to an individual receptor, the utilization of multiple signaling pathways by the Trk receptors would allow a single receptor type to induce multiple cellular responses, such as cell survival, cytoskeletal rearrangement, cell migration, and chemotaxis. In addition, the activity of certain promoters may require the simultaneous presence of multiple transcription factors. However, the problem arising from having multiple possible signaling pathways is how to activate each of the pathways individually or in various combinations. Several hypothetical mechanisms have been proposed to deal with this dilemma.

1. Different cellular responses can be induced by changes in the duration of ERK activation (i.e., transient vs prolonged activation). For example, NGF has been shown to promote the sustained activation of MAP/ERK kinases, whereas the activation of the Ras/ERK system by epidermal growth factor (EGF) is transient (Muroya et al., 1992; Traverse et al., 1992). Studies have shown that sustained ERK activation is associated with translocation of ERKs into the nucleus, but transient activation is not (Traverse et al., 1992). Since translocated ERKs activate transcription factors, alteration in the amount of translocated ERKs within the nucleus will lead to an alteration of transcription factor activation.

   It is evident, therefore, that the kinetics of the system can contribute to the activation of several parallel pathways following the binding of a neurotrophin to its Trk receptor. Indeed, small changes in neurotrophin binding has been shown to produce qualitative differences in downstream responses, including gene expression (Green et al., 1992). According to this concept, one downstream pathway would be favored when neurotrophin binding levels are relatively low or transient; additional pathways would be recruited when ligand-binding levels are higher or sustained. A corollary to the level of activation concept is that the regulation of different signaling pathways may also be controlled by the rate of internalization, receptor turnover, and retrograde transport. For example, the SNT and PI-3 pathways may be important when cells leave the cell cycle and undergo differentiation, and these pathways may be favored, depending on the levels of available neurotrophin and subsequent receptor activation. Such mechanisms of pathway control are supported by the well-known fact that the levels of NGF are high in several target structures during early developmental stages, and that the levels decline once critical phases of development are accomplished.
2. Some factors can directly inhibit transcription factor activity or indirectly downregulate certain components within the signal pathway itself. Such a regulatory process would allow for selective activation of individual signaling mechanisms. For example, IκB binds to and inhibits NFκB activity (*see* review by Siebenlist et al., 1994) and ICER competes with CREBs for target sites (Molina et al., 1993).
3. The amount of signal necessary to activate a transcription factor might be regulated by phosphorylation of multiple pathway proteins. In this scenario, activation of multiple pathways leading to the phosphorylation of a single critical protein would increase specificity at the bottom of the pathway by setting a threshold for factor activation and would also allow tolerance of signal-independent noise or low-level cross-talk between pathways.
4. Many factors bind DNA as homodimers, or as heterodimers with related proteins, which may differ in DNA-binding specificity. Therefore, the interaction of one intracellular factor with another would enhance the ability of either one to interact more efficiently with binding sites on the DNA. Thus, the activity of many promoters may require the simultaneous presence of multiple transcription factors.

These hypothetical mechanisms are contingent on which transcription factors or other pathway components are present and expressed in the cell. It is conceivable that many of the different component proteins are developmentally regulated and are thus only expressed at certain times during development. Therefore, only certain pathways could be activated at specific times over the developmental time window.

5. CONCLUSIONS

Neurotrophins affect a wide range of neuronal cell types in the CNS and PNS, making it difficult to target one neuronal population with systemic treatment strategies. Nevertheless, great strides are being made in understanding how the neurotrophins are regulated and what intracellular pathways are important for trophic signaling. Knowledge of these neuroprotective events are necessary for the development of compounds aimed at facilitating neurotrophin synthesis or activation of the signal transduction pathways. Finally, additional issues that need to be addressed include the neurotrophin-mediated neuroprotective signal pathways involved in necrotic and apoptotic forms of cell death; these are considered elsewhere in this book.

REFERENCES

Acheson, A., Conover, J. C., Fandl, J. P., DeChiara, T. M., Russell, M., Thadani, A., et al. (1995) A BDNF autocrine loop in adult sensory neurons prevents cell death. *Nature* **374,** 450–453.

Airaksinen, M. S., Koltzenburg, M., Lewin, G. R., Masu. Y., Helbig, C., Wolf, E., et al. (1996) Specific subtypes of cutaneous mechanoreceptors require neurotrophin-3 following peripheral nerve innervation. *Neuron* **16,** 287–295.

Albers, K. M., Wright, D. E., and Davis, B. M. (1994) Overexpression of nerve growth factor in epidermis of transgenic mice causes hypertrophy of the peripheral nervous system. *J. Neurosci.* **14,** 1422–1432.

Albers, K. M., Perrone, T. N., Goodness, T. P., Jones, M. E., Green, M. A., and Davis, B. M. (1996) Cutaneous overexpression of NT-3 increases sensory and sympathetic neuron number and enhances touch dome and hair follicle innervation. *J. Cell Biol.* **134,** 1–11.

Alderson, R. F., Alterman, A. L., Barde, Y.-A., and Lindsay, R. M. (1990) Brain-derived neurotrophic factor increases survival and differentiated functions of rat septal cholinergic neurons in culture. *Neuron* **5,** 297–306.

Anderson, K. D., Alderson, R. F., Altar, C. A., DiStefano, P. S., Corcoran, T. L., Lindsay, R. M., and Wiegand, S. J. (1995) Differential distribution of exogenous BDNF, NGF, and NT-3 in the brain corresponds to the relative abundance and distribution of high-affinity and low-affinity neurotrophin receptors. *J. Comp. Neurol.* **357,** 296–317.

Anton, E. S., Weskamp, G., Reichardt, L. F., and Matthew, W. D. (1994) Nerve growth factor and its low-affinity receptor promote Schwann cell migration. *Proc. Natl. Acad. Sci. USA* **91,** 2795–2799.

Ardelt, A. A., Flaris, N. A., and Roth, K. A. (1994) Neurotrophin-4 selectively promotes survival of striatal neurons in organotypic slice culture. *Brain Res.* **647,** 340–344.

Barbacid, M. (1994) The Trk family of neurotrophin receptors. *J. Neurobiol.* **25,** 1386–1403.

Barker, P. A. and Shooter, E. M. (1994) Disruption of NGF binding to the low-affinity neurotrophin receptor $p75_{LNTR}$ reduces NGF binding to Trk A on PC12 cells. *Neuron* **13,** 203–215.

Barrett, G. L. and Bartlett, P. F. (1994) The p75 nerve growth factor receptor mediates survival or death depending on the stage of sensory neuron development. *Proc. Natl. Acad. Sci. USA* **91,** 6501–6505.

Benedetti, M., Levi, A., and Chao, M. V. (1993) Differential expression of nerve growth factor receptors leads to altered binding affinity and neurotrophin responsiveness. *Proc. Natl. Acad. Sci. USA* **90,** 7859–7863.

Bothwell, M. (1996) $p75^{NTR}$: a receptor after all. *Science* **272,** 506–507.

Bothwell, M. (1995) Functional interactions of neurotrophins and neurotrophin receptors. *Annu. Rev. Neurosci.* **18,** 223–253.

Carswell, S. (1993) The potential for treating Alzheimer's disease with NGF-inducing compounds. *Exp. Neurol.* **124,** 36–42.

Carter, B. D., Kaltschmidt, C., Kaltschmidt, B., Offenhauser, N., Bohm-Matthaei, R., Baeuerle, P. A., Barde, Y. (1996) Selection of activation of NF-κB by nerve growth factor through the neurotrophin receptor p75. *Science* **272,** 542–545.

Cohen-Cory, S., Dreyfus, C. F., and Black, I. B. (1991) NGF and excitatory neurotransmitters regulate survival and morphogenesis of cultured cerebellar Purkinje cells. *J. Neurosci.* **11,** 462–471.

Conover, J. C., Erickson, J. T., Katz, D. M., Bianchi, L. M., Poueymirou, W. T., McClain, J., et al. (1995) Neuronal deficits, not involving motor neurons, in mice lacking BDNF and/or NT4. *Nature* **375,** 235–238.

Cortazzo, M. H., Kassis, E. S., Sproul, K. A., and Schor, N. F. (1996) Nerve growth factor (NGF)-mediated protection of neural crest cells from antimitotic agent-induced apoptosis: a role of the low-affinity NGF receptor. *J. Neurosci.* **16,** 3895–3899.

Crowley, C., Spencer, S. D., Nishimura, M. C., Chen, K. S., Pitts-Meek, S., Armanini, M. P., et al. (1994) Mice lacking nerve growth factor display perinatal loss of sensory and sympathetic neurons yet develop basal forebrain cholinergic neurons. *Cell* **76,** 1001–1011.

Curtis, R., Adryan, K. M., Strak, J. L., Park, J. S., Campton, D. L., Wescamp, G., et al. (1995) Differential role of the low-affinity neurotrophin receptor (p75) in retrograde axonal transport of the neurotrophins. *Neuron* **14,** 1201–1211.

Davies, A. M., Lee, K., and Jaenisch, R. (1993a) p75-deficient trigeminal sensory neurons have altered response to NGF but not to other neurotrophins. *Neuron* **11,** 565–574.

Davies, A. M., Horton, A., Burton, L. E., Schmelzer, C., Vandlen, R., and Rosenthal, A. (1993b) Neurotrophin-4/5 is a mammalian-specific survival factor for distinct populations of sensory neurons. *J. Neurosci.* **13,** 4961–4967.

Davis, B. M., Katz, D. M., Seroogy, K. M., Albers, K. M. (1994) Overexpression of NGF in transgenic mice induces novel sympathetic projections to primary sensory neurons. *J. Comp. Neurol.* **349,** 464–474.

DiStefano, P. S., Friedman, B., Radziejewski, C., Alexander, C., Boland, P., Schick, C. M., Lindsay, R. M., and Wiegand, S. J. (1992) The neurotrophins BDNF, NT-3, and NGF display distinct patterns of rerograde axonal transport in peripheral and central neurons. *Neuron* **8,** 983–993.

Dobrowsky, R. T., Werner, M. H., Castellino, A. M., Chao, M. V., and Hannun, Y. A. (1994) Activation of the sphingomyelin cycle through the low-affinity neurotrophin receptor. *Science* **265,** 1596–1599.

Dobrowsky, R. T. Jenkins, G. M., and Hannun, Y. A. (1995) Neurotrophins induce sphingomyelin hydrolysis. *J. Biol. Chem.* **270,** 22,135–22,142.

Ehlers, M. D., Kaplan, D. R., Price, D. L., and Koliatsos, V. E. (1995) NGF-stimulated retrograde transport of Trk A in the mammalian nervous system. *J. Cell Biol.* **130,** 149–156.

Ernfors, P., Lee, K., and Jaenisch, R. (1994a) Mice lacking brain-derived neurotrophic factor develop with sensory deficits. *Nature* **368,** 147–150.

Ernfors, P., Lee, K., Kucera, J., Jaenisch, R. (1994b) Lack of neurotrophin-3 leads to deficiencies in the peripheral nervous system and loss of limb proprioceptive afferents. *Cell* **77,** 503–512.

Ernfors, P., Van De Water, T., Loring, J., and Jaenisch, R. (1995) Complementary roles of BDNF and NT-3 in vestibular and auditory development. *Neuron* **14,** 1153–1164.

Farinas, I., Jones, K. R., Backus, C., Wang, X., and Reichardt, L. F. (1994) Severe sensory and sympathetic deficits in mace lacking neurotrophin-3. *Nature* **369,** 658–661.

Friden, P. M., Walus, L. R., Watson, P., Doctrow, S. R., Kozarich, J. W., Backman, C. (1993) Blood–brain barrier penetration and in vivo activity of an NGF conjugate. *Science* **259,** 373–377.

Friedman, W. J., Ibanez, C. F., Hallbook, F., Persson, H., Cain, L. D., Dreyfus, C. F., and Black, I. B. (1993) Differential actions of neurotrophins in the locus coeruleus and basal forebrain. *Exp. Neurol.* **119,** 72–78.

Friedman, B., Kleinfeld, D., Ip, N. Y., Verge, V. M., Moulton, R., Boland, P. (1995) BDNF and NT-4/5 exert neurotrophic influences on injured adult spinal motor neurons. *J. Neurosci.* **15,** 1044–1056.

Gage, F. H., Armstrong, D. M., Williams, L. R., and Varon, S. (1988) Morphological response of axotomized septal neurons to nerve growth factor. *J. Comp. Neurol.* **289,** 147–155.

Gage, F. H., Kawaja, M. D., and Fisher, L. J. (1991) Genetically-modified cells: applications for intracerebral grafting. *Trends Neurosci.* **14,** 328–333.

Gall, C. M. and Isackson, P. J. (1989) Limbic seizures increase neuronal production of messenger RNA for nerve growth factor. *Science* **245,** 758–761.

Gall, C. M. (1993) Seizure-induced changes in neurotrophin expression: implications for epilepsy. *Exp. Neurol.* **124,** 150–162.

Gao, W. Q., Zheng, J. L., and Karihaloo, M. (1995) Neurotrophin 4/5 (NT-4/5) and brain-derived neurotrophic factor (BDNF) act at later stages of cerebellar granule cell devleopment. *J. Neurosci.* **15,** 2656–2667.

Gnahn, H., Hefti, F., Heumann, R., Schwab, M. E., and Thoenen, H. (1983) Time course of the elevation of nerve growth factor (NGF) content in the hippocampus and septum following lesions of the septohippocampal pathway in rats. *Brain Res.* 45–52.

Gomez-Isla, T., Price, J. L., McKeel, D. W., Morris, J. C., Growdon, J. H., and Hyman, B. T. (1996) Profound loss of layer II entorhinal cortex neurons occurs in very mild Alzheimer's disease. *J. Neurosci.* **16,** 4491–4500.

Goodness, T., Albers, K. M., Davis, F. E., and Davis, B. M. (1997) Overexpression of nerve growth factor increases sensory neuron size and modulates Trk expression. *Eur. J. Neurosci.*, in press.

Green, J. B. A., New, H. V., and Smith, J. C. (1992) Responses of embryonic xenopus cells to activin and FGF are separated by multiple dose thresholds and correspond to distinct axes of the mesoderm. *Cell* **71,** 731–739.

Gwag, B. J., Sessler, F. M., Waterhouse, B. D., and Springer, J. E. (1993) Regulation of nerve growth factor mRNA in the hippocampal formation: effects of N-methyl-d-aspartate receptor activation. *Exp. Neurol.* **121,** 160–171.

Gwag, B. J. and Springer, J. E. (1993) Activation of NMDA receptors increases brain-derived neurotrophic factor (BDNF) mRNA expression in the hippocampal formation. *Neuroreport* **5,** 125–128.

Hantzopoulos, P. A., Suri, C., Glass, D. J., Goldfarb, M. P., and Yancopoulos, G. D. (1994) The low-affinity NGF receptor, p75, can collaborate with each of the Trks to potentiate functional responses to the neurotrophins. *Neuron* **13,** 187–201.

Hefti, F. (1986) Nerve growth factor promotes survival of septal cholinergic neurons after fimbrial transections. *J. Neurosci.* **6,** 2155–2162.

Hempstead, B. L. (1993) Strategies for modulating Trk receptor activity. *Exp. Neurol.* **124,** 31–35.

Henderson, C. E., Camu, W., Mettling, C., Gouin, A., Poulsen, K., Karhaloo, M. (1993) Neurotrophins promote motor neuron survival and are present in embryonic limb bud. *Nature* **363,** 266–270.

Hendry, I. A., Stockel, K., Thoenen, H., and Iversen, L. L. (1974) The retrograde axonal transport of nerve growth factor. *Brain Res.* **68,** 103–121.

Heumann, R., Schwab, M., and Theonen, H. (1981) A second messenger required for nerve growth factor biological activity? *Nature* **292,** 838–840.

Hofer, M. M. and Barde, Y.-A. (1988) Brain-derived neurotrophic factor prevents neuronal death in vivo. *Nature* **331,** 261–262.

Hohn, A., Leibrock, J., Bailey, K., and Barde, Y.-A. (1990) Identification and characterization of a novel member of the nerve growth/brain-derived neurotrophic factor family. *Nature* **344,** 339–341.

Huber, J. L. and Chao, M. V. (1995) Mesenchymal and neuronal cell expression of the p75 neurotrophin receptor gene occur by different mechanisms. *Dev. Biol.* **167,** 227–238.

Hyman, C., Juhasz, M., Jackson, C., Wright, P., Ip, N. Y., and Lindsay, R. M. (1994) Overlapping and distinct actions of the neurotrophins BDNF, NT-3, and NT-4/5 on cultured dopaminergic and GABAergic neurons of the ventral mesecephalon. *J. Neurosci.* **14,** 335–347.

Ibanez, C. F., Ernfors, P., Timmusk, T., Ip, N. Y., Arenas, E., Yancopoulos, G. D., and Persson, H. (1993) Neurotrophin-4 is a target-derived neurotrophic factor for neurons of the trigeminal ganglion. *Development* **117,** 1345–1353.

Johnson, E. M., Gorin, P. D., Bradeis, L. D., and Pearson, J. (1980) Dorsal root ganglion neurons are destroyed by exposure in utero to maternal antibody to nerve growth factor. *Science* **210,** 916–918.

Johnson Jr., E. M., Taniuchi, M., Clark, B. H., Springer, J. E., Koh, S., Tayrien, M., and Loy, R. (1987) Demonstration of the retrograde transport of nerve growth factor receptor in the peripheral and central nervous system. *J. Neurosci.* **7,** 923–929.

Jones, K. R., Farinas, I., Backus, C., and Reichardt, L. F. (1994) Targeted disruption of the BDNF gene perturbs brain and sensory neuron development but not motor neuron development. *Cell* **76,** 989–999.

Kahle, P., Barker, P. A., Shooter, E. M., and Hertel, C. (1994) p75 nerve growth factor receptor modulates p140-Trk A kinase activity, but not ligand internalization, in PC12 cells. *J. Neurosci. Res.* **38,** 599–606.

Kaplan, D. R. and Stephens, R. M. (1994) Neurotrophin signal transduction by the Trk receptor. *J. Neurobiol.* **25,** 1404–1417.

Klein, R., Smeyne, R. J., Wurst, W., Long, L. K., Auerbach, A. B., Joyner, A. L., Barbacid, M. (1993) Targeted disruption of the Trk B neurotrophin receptor gene results in nervous system lesions and neonatal death. *Cell* **75,** 113–122.

Klein, K., Silos-Santiago, I., Smeyne, R. J., Lira, S. A., Brambilla, R., Bryant, S., et al. (1994) Disruption of the neurotrophin-3 receptor gene Trk C eliminates Ia muscle afferents and results in abnormal movements. *Nature* **368,** 249–251.

Koliatsos, V. E., Cayouette, M. H., Berkemeier, L. R., Clatterbuck, R. E., Price, D. L., and Rosenthal, A. (1994) Neurotrophin 4/5 is a trophic factor for mammalian facial motor neurons. *Proc. Natl. Acad. Sci. USA* **91,** 3304–3308.

Koliatsos, V. E., Clatterbuck, R. E., Winslow, J. W., Cayouette, M. H., and Price, D. L. (1993) Evidence that brain-derived neurotrophic factor is a trophic factor for motor neurons in vivo. *Neuron* **10,** 359–367.

Korsching, S. (1993) The neurotrophic factor concept: a reexamination. *J. Neurosci.* **13,** 2739–2748.

Lee, K., Li, E., Huber, J., Landis, S. C., Sharpe, A. H., Chao, M. V., Jaenisch, R. (1992) Targeted mutation of the gene encoding the low-affinity NGF receptor p75 leads to deficits in the peripheral sensory nervous system. *Cell* **69,** 737–749.

Lee, K., Bachman, K., Landis, S., and Jaenisch, R. (1994a) Dependence on p75 for innervation of some sympathetic targets. *Science* **263,** 1447–1449.

Lee, K., Davies, A. M., and Jaenisch, R. (1994b) p75-deficient embryonic dorsal sensory and neonatal sympathetic neurons display a decreased sensitivity to NGF. *Development* **120,** 1027–1033.

LeMaster, A. M., Davis B. M., and Albers, K. M. (1995) Overexpression of BDNF in epidermis of transgenic mice causes changes in trk receptor expression in sensory ganglia and alterations in skin innervation. *Soc. Neurosci. Abs.* **21,** 1054.

LeMaster, A. M., Johnson, J. E., Kitzman, P. H., Davis, B. M., and Albers, K. M. (1996) Examination of DRG neurons and behavioral testing in mice overexpressing BDNF in epidermis. *Soc. Neurosci. Abs.* **22,** 549.

Levi-Montalcini, R. (1982) Developmental neurobiology and the natural history of nerve growth factor. *Annu. Rev. Neurosci.* **5,** 341–362.

Lindholm, D., Dechant, G., Heisenberg, C. P., and Thoenen, H. (1993) Brain-derived neurotrophic factor is a survival factor for cultured cerebellar granule neurons and protects them against glutamate-induced neurotoxicity. *Eur. J. Neurosci.* **5,** 1455–1464.

Lindsay, R. M., Thoenen, H., and Barde, Y.-A. (1985) Placode and neural crest-derived sensory neurons are responsive at early developmental stages to brain-derived neurotrophic factor. *Dev. Biol.* **112,** 319–328.
Liu, X., Ernfors, P., Wu, H., and Jaenisch, R. (1995) Sensory but not motor neuron deficits in mice lacking NT4 and BDNF. *Nature* **375,** 238–240.
Loeb, D. M., Stephens, R. M., Copeland, T., Kaplan, D. R., and Greene, L. A. (1994) A Trk NGF receptor point mutation affecting interaction with PLCγ abolishes NGF-promoted peripherin induction, but not neurite outgrowth. *J. Biol. Chem.* **269,** 8901–8910.
Loy, R., Lachyankar, M. B., Condon, P. J., Poluha, D. K., and Ross, A. H. (1994) Retrograde axonal transport and lesion-induced upregulation of the Trk A NGF receptor. *Exp. Neurol.* **130,** 377–386.
Lucidi-Phillipi, C. A., Clary, D. O., Reichardt, L. F., and Gage, F. H. (1996) Trk A activation is sufficient to rescue axotomized cholinergic neurons. *Neuron* **16,** 653–663.
Mahadeo, D., Kaplan, L., Chao, M. V., and Hempstead, B. L. (1994) High-affinity nerve growth factor binding displays a faster rate of association than p140-Trk binding. *J. Biol. Chem.* **269,** 6884–6891.
Maisonpierre, P. C., Belluscia, L., Squinto, S., Ip, N. Y., Furth, M. E., Linsay, R. M., and Yancopoulos, G. D. (1990) Neurotrophin-3: a neurotrophic factor related to NGF and BDNF. *Science* **247,** 1446–1451.
Martinez, H. J., Dreyfus, C. E., Jonakait, G. M., and Black, I. A. (1985) Nerve growth factor promotes cholinergic development in brain striatal cultures. *Proc. Natl. Acad. Sci. USA* **82,** 7777–7781.
Mobley, W. C., Rutkowoski, J. L., Tennekoon, G. I., Buckanan, K., and Johnston, M. V. (1985) Choline acetyltransferase activity in striatum of neonatal rats increased by nerve growth factor. *Science* **229,** 284–287.
Molina, C. A., Foulkey, N. S., Lalli, E., and Sussone-Corsi, P. (1993) Inducibility and negative autoregulation of CREM: an alternative promoter directs the expression of ICER, and early response repressor. *Cell* **75,** 875–886.
Morse, J. K., Weigand, S. J., Anderson, K., You, Y., Cai, N., Carnahan, J., et al. (1993) Brain-derived neurotrophic factor (BDNF) prevents the degeneration of medial septal cholinergic neurons following fimbria transection. *J. Neurosci.* **13,** 4146–4156.
Mount, H. T., Dreyfus, C. F., and Black, I. B. (1994) Neurotrophin-3 selectively increases Purkinje cell survival. *Neuroreport* **5,** 2497–2500.
Mu, X., Silos-Santiago, I., Carroll, S. L., and Snider, W. D. (1993) Neurotrophin receptor genes are expressed in distinct patterns in developing dorsal root ganglia. *J. Neurosci.* **13,** 4029–4041.
Muroya, K., Hattori, S., and Nakamura, S. (1992) Nerve growth factor induces rapid accumulation of the GTP-bound form of p21(ras) in rat pheochromocytoma PC12 cells. *Oncogene* **7,** 277–281.
Obermeier, A., Lammers, R., Wiesmuller, K., Jung, G., Schlessinger, J., and Ullrich, A. (1993a) Identification of Trk binding sites for SHC and phosphatidylinositol 3′-kinase and formation of a multimeric signaling complex. *J. Biol. Chem.* **268,** 22,963–22,966.
Obermeier, A., Halfter, H., Weismuller, K., Jung, G., Schlessinger, J., and Ullrich, A. (1993b) Tyrosine 785 is a major determinant of Trk-substrate interaction. *EMBO J.* **12,** 933–941.
Olson, L. (1993) NGF and the treatment of Alzheimer's disease. *Exp. Neurol.* **124,** 5–15.
Oppenheim, R. W., Qin-Wei, Y., Prevette, D., and Yan, Q. (1992) Brain-derived neurotrophic factor rescues developing avian motoneurons from cell death. *Nature* **360,** 755–757.
Palmatier, M. A., Hartman, B. K., and Johnson, E. M. (1984) Demonstration of retrogradely transported endogenous nerve growth factor in axons of sympathetic neurons. *J. Neurosci.* **4,** 751–756.
Purves, D. (1976) Functional and structural changes in mammalian sympathetic neurons following colchicine application to post-ganglionic nerves. *J. Physiol.* **259,** 159–175.
Rabin, S. J., Cleghon, V., and Kaplan, D. R. (1993) SNT, a differentiation specific target of neurotrophic factor-induced tyrosine kinase activity in neurons and PC12 cells. *Mol. Cell Biol.* **13,** 2203–2213.

Rabizadeh, S., Oh, J., Zhong, L., Yang, J., Bitler, C. M., Butcher, L. L., Bredesen, D. E. (1993) Induction of apoptosis by the low-affinity NGF receptor. *Science* **261,** 345–348.

Rodriguez-Tebar, A., Jeffrey, P. L., Thonen, H., and Barde, Y. A. (1989) The survival of chick-retinal ganglion cells in response to brain-derived neurotrophic factor depends on their embryonic age. *Dev. Biol.* **136,** 296–303.

Rosenthal, A., Goeddel, D. V., Nguyen, T., Lewis, M., Shih, A., Laramee, G. R., Nikolics, K., and Winslow, J. W. (1990) Primary structure and biological activity of a novel human neurotrophic factor. *Neuron* **i,** 767–773.

Ross, A. H., Daou, M., McKinnon, C. A., Condon, P. J., Lachyankar, M. B., Stephens, R. M., Kaplan, D. R., and Wolf, D. E. (1996) The neurotrophin receptor, gp75, forms a complex with the receptor tyrosine kinase Trk A. *J. Cell Biol.* **132,** 945–953.

Ruit, K. G., Elliott, J. L., Osborne, P. A., Yan, Q., and Snider, W. D. (1992) Selective dependence of mammalian dorsal root ganglion neurons on nerve growth factor during embryonic development. *Neuron* **8,** 573–587.

Segal, R. A. and Greenberg, M. E. (1996) Intracellular signaling pathways activated by neurotrophic factors. *Ann. Rev. Neurosci.* **19,** 463–489.

Segal, R. A., Takahashi, H., and McKay, R. D. (1992) Changes in neurotrophin responsiveness during the development of cerebellar granule neurons. *Neuron* **9,** 1041–1052.

Seiler, M. and Schwab, M. E. (1984) Specific retrograde transport of nerve growth factor (NGF) from neocortex to nucleus basalis in the rat. *Brain Res.* **300,** 33–39.

Sendtner, M., Holtmann, B., Kolbeck, R., Thoenen, H., and Barde, Y.-A. (1992) Brain-derived neurotrophic factor prevents the death of motoneurons in newborn rats after nerve section. *Nature* **360,** 757–759.

Siebenlist, U., Franzoso, G., and Brown, K. (1994) Structure, regulation and function of NF-κB. *Annu. Rev. Cell Biol.* **10,** 405–455.

Silos-Santiago, I., Molliver, D. C., Ozaki, S., Smeyne, R. J., Fagan, A. M. Barbacid, M., and Snider, W. D. (1995) Non-Trk A-expressing small DRG neurons are lost in Trk A deficient mice. *J. Neurosci.* **15,** 5929–5942.

Smeyne, R. J., Klein, R., Schnapp, A., Long, L. K., Bryant, S., Lewin, A., Lira, S. A., and Barbacid, M. (1994) Severe sensory and sympathetic neuropathies in mice carrying a disrupted Trk/NGF receptor gene. *Science* **368,** 246–248.

Soltoff, S. P., Rabin, S. L., Cantley, L. C., and Kaplan, D. R. (1992) Nerve growth factor promotes the activation of phosphatidylinositol 3-kinase and its association with the Trk tyrosine kinase. *J. Biol. Chem.* **267,** 17472–17477.

Springer, J. E., Koh, S., Tayrien, M. W. and Loy, R. (1987) Basal forebrain magnocellular neurons stain for nerve growth factor receptor: correlation with cholinergic cell bodies and effects of axotomy. *J. Neurosci. Res.* **17,** 111–118.

Springer, J. E. (1988) Nerve growth factor receptors in the central nervous system. *Exp. Neurol.* **102,** 354–365.

Springer, J. E., Gwag, B. J. and Sessler, F. M. (1994) Neurotrophic factor mRNA expression in dentate gyrus is increased following in vivo stimulation of the angular bundle. *Brain Res.* **23,** 135–143.

Stephens, R. M., Loeb, D. M., Copeland, T. D., Pawson, T., Greene, L. A., and Kaplan, D. R. (1994) Trk receptor uses redundant signal transduction pathways involving SHC and PLCγl to mediate NGF response. *Neuron* **12,** 691–705.

Traverse, S., Gomez, N., Paterson, H., Marshall, C., and Cohen, P. (1992) Sustained activation of the mitogen-activated protein (MAP) kinase cascade may be required for differentiation of PC12 cells. Comparison of the effects of nerve growth factor and epidermal growth factor. *Biochem. J.* **288,** 351–355.

Verdi, J. M., Birren, S. J., Ibanez, C. F., Persson, H., Kaplan, D. R., Beneditti, M., Chao, M. V., and Anderson, D. J. (1994) p75LNGFR regulates Trk signal transduction and NGF-induced neuronal differentiation in MAH cells. *Neuron* **12,** 733–745.

Vetter, M. L., Martin-Zanca, D., Parada, L. F., Bishop, M. J., and Kaplan, D. R. (1991) Nerve growth factor rapidly stimulates tyrosine phosphorylation of phospholipase C-γ1 by a kinase associated with the product of the Trk protooncogene. *Proc. Natl. Acad. Sci. USA* **88,** 5650–5654.

von Bartheld, C. S., Byers, M. R., Williams, R., and Bothwell, M. (1996a) Anterograde transport and axo-dendritic transfer of neurotrophins in the developing visual system. *Nature* **379,** 830–833.

von Bartheld, C. S., Williams, R., Lefcort, F., Clary, D. O., Reichardt, L. F., and Bothwell, M. (1996b) Retrograde transport of neurotrophins form the eye to the brain in chick embryos: role of the p75NTR and Trk B receptors. *J. Neurosci.* **16,** 2995–3008.

Widmar, H. R. and Hefti, F. (1994) Neurotrophin-4/5 promotes survival and differentiation of rat striatal neurons devleoping in culture. *Eur. J. Neurosci.* **6,** 1669–1679.

Williams, L. R., Varon, S., Peterson, G. M., Wictorin, K., Fischer, W., Bjorklund, A. and Gage, F. H. (1986) Continuous infusion of nerve growth factor prevents basal forebrain neuronal death after fimbria fornix transection. *Proc. Natl. Acad. Sci. USA* **83,** 9231–9235.

Wolf, D. E., McKinnon, C. A., Daou, M., Stephens, R. M., Kaplan, D. R., and Ross, A. H. (1995) interaction with Trk A immobilizes gp75 in the high affinity nerve growth factor receptor complex. *J. Biol. Chem.* **270,** 2133–2138.

Yan, Q., Elliott, J., and Snider, W. D. (1992) Brain-derived neurotrophic factor rescues spinal motor neurons from axotomy-induced cell death. *Nature* **360,** 753–755.

Yao, R. and Cooper, G. M. (1995) Requirement for phosphatidylinositol-3 kinase in the prevention of apoptosis by nerve growth factor. *Science* **267,** 2003–2005.

Zafra, F., Castren, E., Thoenen, H. and Lindholm, D. (1991) Interplay between glutamate and gamma-aminobutyric acid transmitter systems in the physiological regulation of brain-derived neurotrophic factor and nerve growth factor synthesis in hippocampal neurons. *Proc. Natl. Acad. Sci. USA* **88,** 10,037–10,041.

2

Nerve Growth Factor and Its Receptors in the Primate Forebrain

Alterations in Alzheimer's Disease and Potential Use in Experimental Therapeutics

Elliott J. Mufson and J. H. Kordower

1. INTRODUCTION

The selective vulnerability of specific neuronal populations in the brain is a feature of various neurological disorders, including Alzheimer's disease (AD). A consistent and early pathological feature of AD is the degeneration of cholinergic neurons in the basal forebrain (Bartus, 1982; Whitehouse et al., 1983; Mufson et al., 1989). Cortical cholinergic deficits resulting from atrophy and degeneration of basal forebrain neurons correlate better than any other neurochemical change with the memory impairment seen in patients with AD (Bierer et al., 1995). It is now well documented that neurotrophic molecules play a pivotal role in the survival and maintenance of some adult neurons, including those of the cholinergic basal forebrain (CBF) (*see* Hefti et al., 1989, for review). Nerve growth factor (NGF) is the prototype neurotrophic molecule, which was first characterized by Levi-Montalcini and coworkers as a protein supporting the survival and phenotypic differentiation of catecholamine-containing peripheral sympathetic ganglia (*see* Levi-Montalcini and Angelleti, 1968; Thoenen and Barde, 1980). It is now well established that NGF is one member of a superfamily of neurotrophic substances that includes brain-derived neurotrophic factor (BDNF), neurotrophin-3 (NT-3), neurotrophin-4 (NT-4), neurotrophin-5 (NT-5), and NT-6 (Maisonpierre et al., 1990; Bothwell, 1991; Ibanez et al., 1992; Ip et al., 1992; Jing et al., 1992; Barbacid, 1994; Koster et al., 1994). Six regions of amino-acid nonidentity confer specificity for each neurotrophin (Ip et al., 1993). Initially, a potential role for NGF as a trophic molecule for the CBF neurons was inferred from the seminal retrograde transport studies of Schwaab et al. (1979) and Seiler and Schwaab (1984). In these studies, retrogradely labeled neurons were observed within the septal–diagonal band complex or the nucleus basalis magnocellularis following injection of iodinated NGF into cortex and hippocampus, respectively. Autoradiographic investigations revealed that neurons that bind radiolabeled NGF codistribute with CBF neurons (Richardson et al., 1986). Converging lines of evidence then indicated that NGF modulates the development, survival, and maintenance of CBF neurons (Hefti et al., 1989). NGF is found in highest concentrations within target sites of CBF perikarya, such

From: Neuroprotective Signal Transduction *edited by M. P. Mattson*
Humana Press Inc., Totowa, NJ

as the cerebral cortex, hippocampus, and olfactory bulb (Whittemore and Seiger, 1987). NGF increases choline acetyltransferase (ChAT) levels in cultured septal and striatal cholinergic perikarya (Hefti et al., 1985; Martinez et al., 1985). Intraventricular injections of NGF in neonatal rats increases ChAT activity in the basal forebrain, hippocampus, cortex, and striatum (Gnahn et al., 1983; Mobley et al., 1985). Evidence also has been accumulating that NGF may modulate regenerative events within the CBF. For example, infusion of NGF, transplantation of NGF-containing mouse salivary gland, and grafting of capsules containing cells modified to secrete NGF rescue medial septal cholinergic neurons that undergo degeneration after transection of the fimbria–fornix system (e.g., Hefti et al., 1986; Williams et al., 1986; Springer et al., 1987; Emerich et al., 1994; Kordower et al., 1994c). Indeed, some of these approaches attenuate cognitive deficits resulting from septohippocampal axotomy (Will and Hefti, 1985; Springer et al., 1987). Aged rats exhibit cognitive impairments, degeneration of the CBF neurons, and changes in NGF, similar to that seen in animals with experimental lesions of this system (*see* reviews in Collier and Coleman, 1991; Scott and Crutcher, 1994; Scott et al., 1994). For example, aged rats with memory impairment, but not aged rats without memory impairment, show exaggerated neuron loss and atrophy of CBF neurons and decreased levels of cortical cholinergic markers (Biegon et al., 1986; Fischer et al., 1987; Koh and Loy, 1988; Koh et al., 1989; Markram and Segal, 1990). Recently, Cooper and coworkers (1994) demonstrated that retrograde transport of iodinated NGF from hippocampus declines in aged rats (no behavioral characterization). This study also revealed that CBF neurons in the aged rat exhibit diminished levels of the message for the high affinity Trk A receptor, but retain normal levels of the low affinity neurotrophin receptor (NTR) p75 mRNA. These observations suggest a relationship among age-related changes in retrograde transport of NGF, the expression of its signal-transducing high-affinity receptor, and morphological changes within CBF neurons.

2. NGF-RECEPTOR SUBTYPES

During the past several years a large body of evidence has accumulated demonstrating that NGF recognizes at least two classes of cell surface receptors (Meakin and Schooter, 1991), a fast-dissociating, low-affinity NGF receptor and a slow-dissociating, high-affinity NGF receptor (Fig. 1). The low-affinity receptor is a transmembrane glycoprotein of mol wt 75 kDa (*see* review in Bothwell, 1991). Since this receptor binds to virtually all members of the NGF family (Rodriguez-Tebar et al., 1990; Bothwell, 1991; Lindsay et al., 1994), it has been termed the p75 neurotrophin receptor ($p75^{NTR}$; Chao and Hempstead, 1995; Chao, 1995). In the classic model of NGF–$p75^{NTR}$ interactions, it was hypothesized that this complex is internalized and retrogradely transported to the CBF consumer neurons (Fig. 2), where the physiologic functions take place. Although recent molecular biology studies now indicate that the physiologic action of NGF requires NGF complexing to its high-affinity receptor (Fig. 2), several important functions have been suggested for the $p75^{NTR}$, including presentation or recruitment to concentrate NGF (Jing et al., 1992), signaling through a G-protein-mediated mechanism (Feinstein and Larhammer, 1990), and an association with molecules capable of signaling or providing substrates for the high-affinity receptor, Trk A (Parada et al., 1992).

The high-affinity Trk receptor is a 140-kDa transmembrane glycoprotein with a cytoplasmic protein kinase domain (Hempstead et al., 1991; for reviews, *see* Bothwell, 1991;

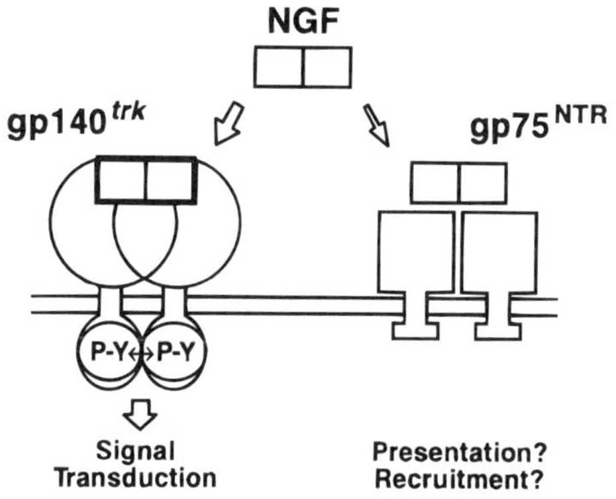

Fig. 1. Diagrammatic representation of the proposed molecular structure of the Trk A and p75^{NTR} receptors. NGF is represented as a dimeric molecule. P-Y indicates tyrosine-phosphorylated residues in the catalytic kinase domain (square boxes) of the gp^{140Trk} receptor. (Adapted with permission from Jing et al., 1992).

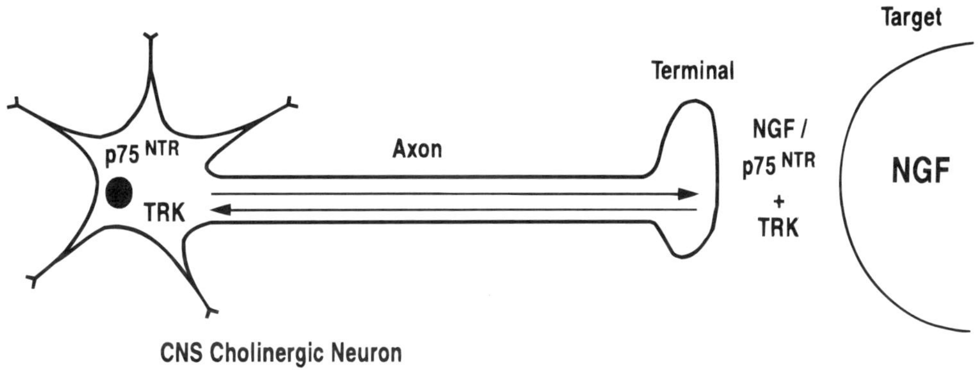

Fig. 2. Diagrammatic representation of the proposed interaction between a cholinergic basal forebrain (CBF) neuron and its NGF-producing target site. In the classic retrograde transport model, the p75^{NTR} is produced within a CBF neuron, anterogradely transported to its target site, binds with NGF, and is then retrogradely transported to cholinergic consumer neurons. It is now clear that this model must include the Trk A receptor, which is also produced by CBF neurons, and anterogradely transported to its target site, where it interacts with NGF. The complex is then transported retrogradely to these cholinergic neurons. Accumulating evidence indicates that the Trk A receptor participates in the signaling processes initiated by NGF.

Lindsay et al., 1994). The Trk receptor is essential for NGF signal transduction, which mediates its biological effects (Chao, 1992). It was at first controversial whether coexpression of both the p75^{NTR} and the Trk A receptor was necessary for NGF's biological effects to occur (*see* Jing et al., 1992; Parada et al., 1992). Mounting evidence now suggests that the presence of the Trk A receptor alone is sufficient to elicit cellular responses to NGF (Weskamp and Reichardt, 1991; Ibanez et al., 1992). Trk A, like most kinase growth-factor receptors, signals through receptor oligomerization (Heldin, 1995). Activation of the tyrosine kinase activity of Trk A via NGF binding leads to receptor trans- and autotyrosine phosphatidylation (PY; Fig. 1) and PY of second messengers, including phosphatidylinositol-3 kinase (PI-3; Soltoff et al., 1992). PI-3 kinase is involved in protein trafficking and endocytosis of ligand–receptor complexes (*see* Kaplan and Stephans, 1994). Since microinjection of NGF into cells does not induce NGF biological signals (Heumann et al., 1984), it is assumed that the binding to receptors located on the cell surface mediates these effects.

3. NGF IN HUMAN NEUROLOGIC DISEASE

It was originally hypothesized that degenerative events in human neurologic disorders may result from reductions in trophic support (Appel, 1981). This hypothesis was particularly favored in AD, in which reductions in cortical NGF were hypothesized to underlie CBF degeneration (Appel, 1981; Hefti and Weiner, 1983; Hefti et al., 1989). The initial observation that cortical NGF mRNA was unchanged in AD tempered enthusiasm for this hypothesis. The current explosion of information concerning the relative importance of Trk A in NGF signal transduction has reinvigorated interest in the role NGF may play in the degeneration of CBF neurons in AD. This Chapter will review the data concerning changes in NGF and its low- and high-affinity receptors in AD relative to normal aged individuals. The information gained from these experiments has led us to postulate that alterations in the production and utilization of the Trk A receptor is a major pathological event underlying CBF degeneration in AD.

3.1. NGF Levels in Normal Aged Human Cortex

In order to more fully understand disease-related changes in NGF, it is imperative to have a detailed understanding of its levels in the normal aged human brain. Together with our collegeaues at the University of Cincinnati and the University of Alabama, we evaluated the status of NGF levels within cortical production sites, using a specific bioassay for this neurotrophin (Crutcher et al., 1993; Scott et al., 1995). Based on control and AD hippocampal and cortical samples, we found that this assay was unaffected by postmortem delay. Tissue samples from the hippocampus also failed to reveal a gender difference in NGF levels in either aged control or AD patients. Aged human controls displayed high levels of NGF-like immunoreactivity within the hippocampal formation (1100 pg/gm tissue), amygdala (690 pg/mg), temporal cortex (450 pg/gm), frontal cortex (330 pg/gm), parietal cortex (330 pg/gm), occipital cortex (270 pg/gm), putamen (230 pg/gm), and cerebellum (60 pg/gm). Levels of NGF were also high in the caudate nucleus, medial thalamus, and choroid plexus. These findings are similar to the distribution of NGF-like immunoreactivity seen in the rat and monkey brain (*see* review by Scott and Crutcher, 1994). NGF immunoreactivity was also high (1100 pg/gm tissue) in the nucleus basalis of Meynert. Since there is little mRNA within the basalis, this probably reflects large amounts of NGF that have been retrogradely transported from its target sites.

3.2. NGF-Like Immunoreactivity in the Primate Hippocampus

In order to visualize the distribution of NGF within its cortical and hippocampal production sites, we immunostained tissue from adult monkeys and aged humans (Mufson et al., 1994) with a polyclonal antibody raised against NGF (Conner et al., 1992; Conner and Varon, 1992). Immunohistochemistry revealed a diffuse staining pattern within the neuropil of the hippocampus, with no evidence of immunoreactive cell bodies or fibers (Fig. 3). NGF immunoreactivity was most intense within the hilus (CA4) of the dentate

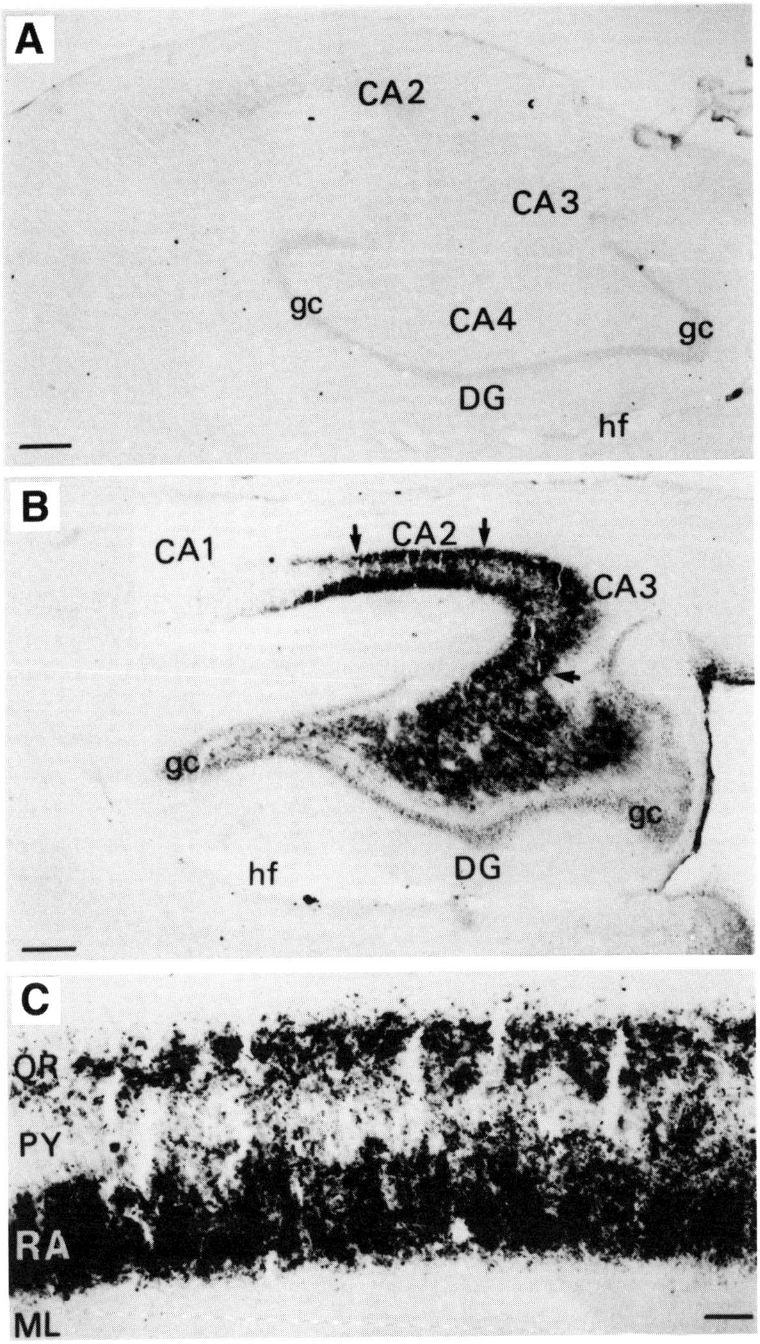

Fig. 3. Photomicrographs of NGF immunoreactivity within the monkey hippocampal formation. **(A)** Hippocampal formation immunoreacted with anti-NGF antibodies preadsorbed to purified β-NGF results in a complete absence of hippocampal staining. **(B)** NGF immunoreactivity within the neuropil of the hilar (CA4) and CA3–CA2 hippocampal subfields. **(C)** High-power photomicrograph of CA2 as demarcated by arrows in (B). Note the trilaminar appearance of the reaction product, which is most intense in stratum radiatum (RA), moderate in stratum oriens (OR), and light in the stratum pyramidal (PY). Scale bar in (A, B) = 200 μm; (C) = 40 μm. (Reproduced with permission from Mufson et al., 1994).

gyrus and extended to the CA3 and CA2 regions of the hippocampus (Fig. 3). Within CA2, the pattern of staining was trilaminar, with the heaviest labeling in stratum radiatum, intermediate labeling in stratum oriens, and lightest labeling in stratum pyramidal. At no level was NGF-like immunoreactivity seen within CA1 or adjacent subicular and entorhinal cortices. Although we did not observe NGF-immunoreactive cells, Conner and Varon (1992) demonstrated the presence of NGF-reactive neurons within the rat cortex, hippocampus, olfactory bulb, amygdala, striatum, thalamus, hypothalamus, and brainstem, as well as within noncholinergic neurons of the basal forebrain in rats pretreated with the axoplasmic transport blocker colchicine. The inability to block axoplasmic flow in human cases probably precluded our visualization of NGF immunoreactive somata.

3.3. *Retrogradely Transported NGF in the Aged Human Cholinergic Basal Forebrain*

We also used this NGF antibody (Conner and Varon, 1992) to immunohistochemically visualize retrogradely transported NGF within cholinergic neurons of the basal forebrain (Mufson et al., 1994, 1995). A continuum of NGF-immunoreactive neurons was seen within the subfields of the basal forebrain, including the medial septum (Ch1), vertical limb of the diagonal band (Ch2), and all subdivisions of the nucleus basalis (Ch4; Fig. 4A). The NGF reaction product most often appeared as a granular or punctate particulate, localized primarily within the cell body and within primary within dendritic processes (Fig. 4). The intensity of labeling ranged from a few granules to instances in which the staining was so intense that it appeared to fill the entire perikarya (Fig. 4A–C). We also observed a few scattered NGF-immunoreactive fibers with a beaded appearance coursing within basal forebrain subfields. Colocalization studies revealed that the NGF reaction product was exclusively found within basal forebrain neurons that were immunoreactive for $p75^{NTR}$ (Mufson et al., 1994).

3.4. *NGF Receptors in the Cholinergic Basal Forebrain*

Immunohistochemical and *in situ* hybridization studies in rats (Yan and Johnson, 1988; Batechelor et al., 1989; Steninger et al., 1993; Gibbs and Pfaff, 1994; Sobreveila et al., 1995; Holtzman et al., 1995), monkeys (Kordower et al., 1988), and humans (Hefti and Mash, 1989; Kordower et al., 1989; Mufson et al., 1989; Higgins and Mufson, 1989) have demonstrated neurons within all subfields of the CBF, including the septal/diagonal band complex (Ch1–Ch2) and nucleus basalis (Ch4), which contain both the $p75^{NTR}$ (Figs. 5–7) and the Trk receptors (Figs. 8 and 17). It is well established that these CBF neurons provide the major cholinergic innervation to the cortex and hippocampus (Mesulam et al., 1983). Codistribution and co-localization studies have shown a high degree of overlap between CBF neurons expressing the $p75^{NTR}$ and the specific cholinergic marker choline acetyltransferase (ChAT) (Kordower ct al., 1988, 1989, 1994b; Mufson et al., 1989, Sobreveila et al., 1995). This $p75^{NTR}$ antibody is an excellent marker for CBF neurons in the nonhuman primate (Kordower et al., 1988) and human (Mufson et al., 1989) brain.

In order to determine the extent to which Trk-immunoreactive neurons in the primate CBF colocalized with $p75^{NTR}$, we used a double chromagen immunohistochemical procedure. In these experiments, we stained tissue from adult and aged monkeys with an

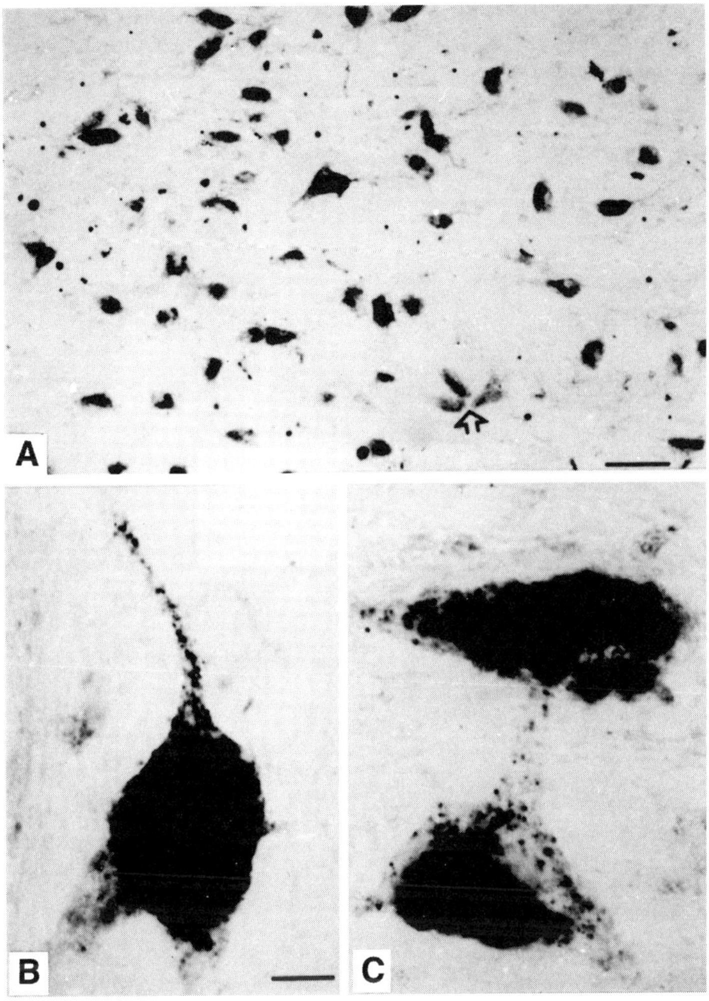

Fig. 4. NGF immunoreactivity within the anterolateral division of the human nucleus basalis (Ch4al). **(A)** Numerous NGF-positive neurons exhibiting differing levels of immunoreactivity. Open arrow indicates a cluster of lightly labeled neurons. **(B,C)** Higher magnification photomicrographs of nucleus basalis NGF-immunoreactive neurons. Scale bar in (A) = 100 µm; (B, C) = µm. (Reproduced with permission from Mufson et al., 1994).

antibody that reacts with all Trks, but primarily recognizes the Trk A receptor (Steininger et al., 1993; Kordower et al., 1994b; Chen et al., 1996). All monkeys displayed an extensive but incomplete co-localization between Trk- and $p75^{NTR}$-immunoreactivity within basal forebrain perikarya (Fig. 9). In addition to double-labeled neurons, many basal forebrain cells were observed that only expressed Trk-immunoreactivity (Fig. 9). Conversely, only a few basal forebrain neurons were seen that just expressed the $p75^{NTR}$ (Fig. 9). Colocalization studies revealed that between 68 and 73% of all basal forebrain neurons colocalized the Trk and $p75^{NTR}$. In contrast, between 23 and 28% of basal forebrain neurons expressed only the Trk receptor, but approx 4% of

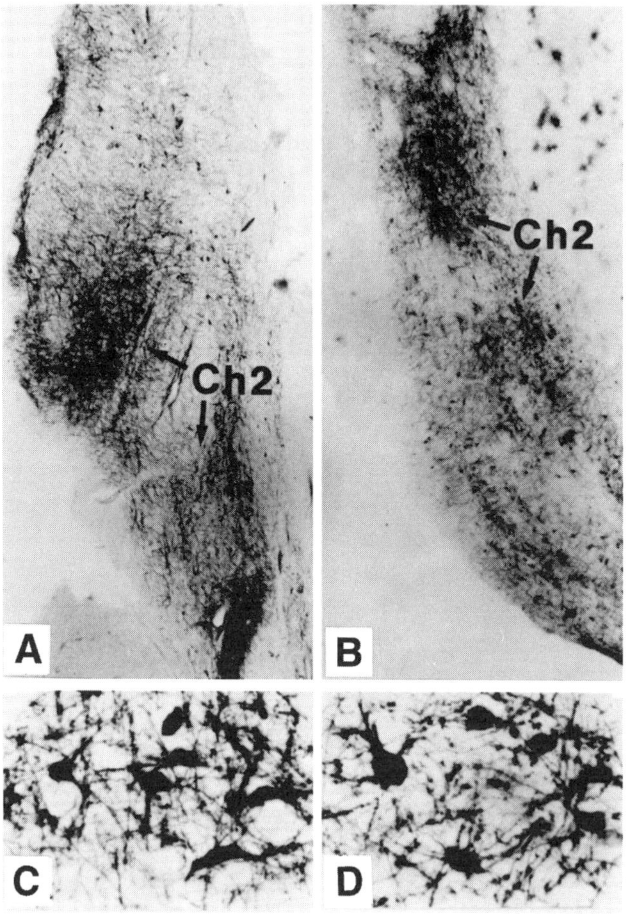

Fig. 5. Comparison of the distribution of p75^{NTR}-immunoreactive neurons within the septal–diagonal band (Ch2) of an aged control **(A)** and an AD **(B)** case. In contrast to Ch4 (*see* Fig. 11), the Ch2 region is relatively intact in AD. Higher magnifications p75^{NTR} Ch2-immunoreactive neurons and fibers of an aged control **(C)** and AD **(D)** case (×180).(Reproduced with permission from Mufson et al., 1989.)

the basal forebrain neurons only expressed the p75^{NTR}. Although there was some individual variability, a similar degree of Trk–p75^{NTR} colocalization was observed in all subdivisions of the basal forebrain. Separate quantitation of double-labeled neurons was performed for hippocampal-projecting septal–diagonal band (Ch1–Ch2) neurons and the cortical and amygdaloid-projecting neurons of the nucleus basalis (Ch4) subfields. Between 42 and 75% of the neurons within the septal–diagonal band complex colocalized the Trk and p75^{NTR}; 66–76% of neurons within the nucleus basalis subfield also displayed both Trk- and p75^{NTR}-immunoreactivity. This study also revealed that there was no age-related change in the degree of colocalization (Fig. 9). The reader is referred to recent articles by Gibbs and Pfaff (1994), Sobreveila et al. (1995), and Holtzman et al. (1995), using probes specific for the Trk A receptor for detailed descriptions of the expression of Trk A and its colocalization with p75^{NTR} within the rat CNS.

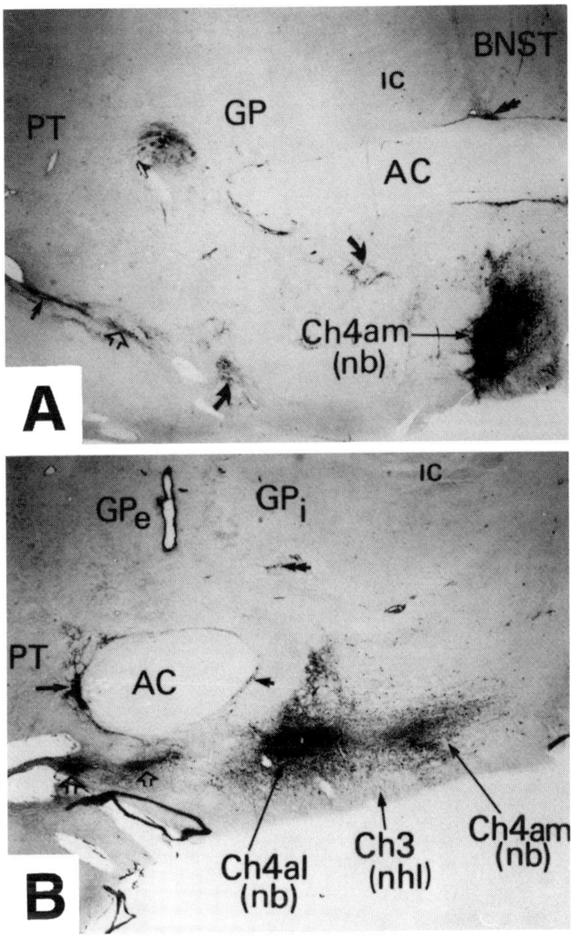

Fig. 6. p75^{NTR} immunohistochemistry in the brain of a normal-aged individual showing the location within the nucleus basalis of the positive neurons corresponding to Ch4am, Ch4al, and Ch3. Medial is to the left and dorsal toward the top. **(A)** p75^{NTR} staining in the rostral aspect of Ch4am. Curved arrows indicate p75^{NTR}-containing neurons embedded within fibers coursing through the olfactory tubercle. Single black arrow points to p75^{NTR}-positive fibers coursing within the external capsule, the open arrow to a cluster of p75^{NTR}-containing subputaminal neurons, and the double black arrow to a group of p75^{NTR}-immunoreactive neurons in the bed nucleus of the stria terminalis (×10). **(B)** p75^{NTR} immunoreactivity within the nucleus basalis corresponding to Ch4am, Ch4al, and Ch3. Open arrows indicate subputaminal p75^{NTR}-positive neurons, single arrow and arrowhead point to p75^{NTR}-immunoreactive neurons adjacent to the anterior commissure, and double arrow indicates interstitial p75^{NTR}-containing neurons within the internal medullary lamina (×10). (Reproduced with permission from Mufson et al., 1989.)

4. DEFECTS IN THE NGF SYSTEM IN ALZHEIMER'S DISEASE

4.1. Cytoskeletal Abnormalities Within p75^{NTR}-Containing CBF Neurons in Alzheimer's Disease

In AD, CBF neurons display extensive neurofibrillary degeneration. To determine the extent to which p75^{NTR}-containing CBF neurons exhibit cytoskeletal lesions, we

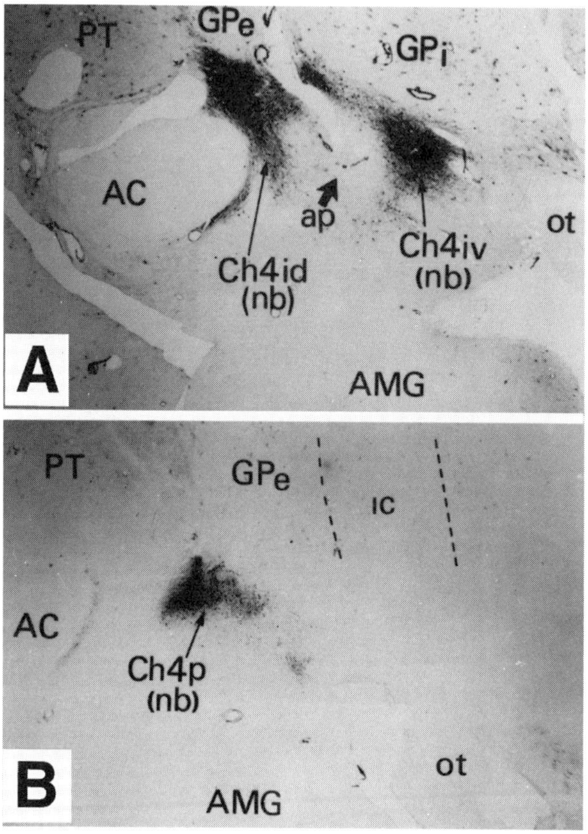

Fig. 7. NGF-receptor immunoreactivity in the brain of an aged normal individual showing the intermediate (Ch4i) and posterior subfields (Chp) of the nucleus basalis. **(A)** p75^{NTR} immunoreactivity within the intermediodorsal (Ch4id) and **(B)** intermedioventral (Ch4iv) subsectors of the nucleus basalis. Dashed lines indicate position of the internal capsule (×10). (Reproduced with permission from Mufson et al., 1989.)

stained basal forebrain sections from AD and aged controls using thioflavin-S or an antibody against paired helical filament (PHF). A few senile plaques were seen scattered in the AD basal forebrain. Of particular interest is the observation derived from our fluorescence microscopic analysis, which revealed a large number of p75^{NTR}-immunoreactive neurons that are invested with differing degrees of neurofibrillary material, suggestive of a continuum of degenerative events (Fig. 10C). Some neurons were thioflavin-S negative or only minimally fluorescent for this amyloid marker. Others were more heavily invested, suggesting a greater involvement in the disease process (Fig. 10D, E). In addition, many neurofibrillary tangle-bearing profiles observed in the basal forebrain were ghosts; that is, neurons appeared to be largely devoid of cytoplasm and organelles consisting of neurofibrillary remnants, suggestive of end-stage degeneration (Fig. 10D). The neurofibrillary tangles observed within the CBF neurons were globose in shape and fluoresced as a dense core mass (Fig. 10C). This is in contrast to the more wispy appearance of the neurofibrillary tangles observed within the cortex (Fig. 10A). This potentially represents differences in the pattern of amyloid aggregation between these two cell types.

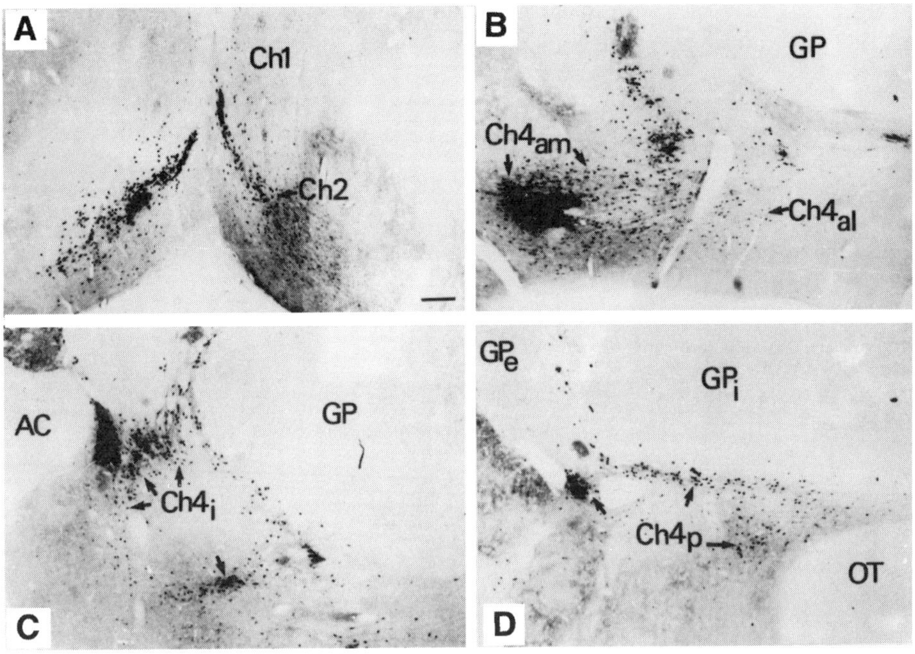

Fig. 8. Low-power photomicrographs illustrating the distribution of Trk-immunoreactive neurons in the subfields of the monkey basal forebrain **(A)** Ch1–Ch2 (septal–diagonal band) complex. **(B)** Anterior (Ch4am–al) nucleus basalis. **(C)** Intermediate (Ch4i) subdivision of nucleus basalis. **(D)** Posterior (Ch4p) subdivision of nucleus basalis. AC, anterior commissure; GP, globus pallidus; GPe, external division of globus pallidus; GPi, internal division of globus pallidus; OT, optic tract. Scale bars represent 1000 µm. (Reproduced with permission from Kordower et al., 1994c.)

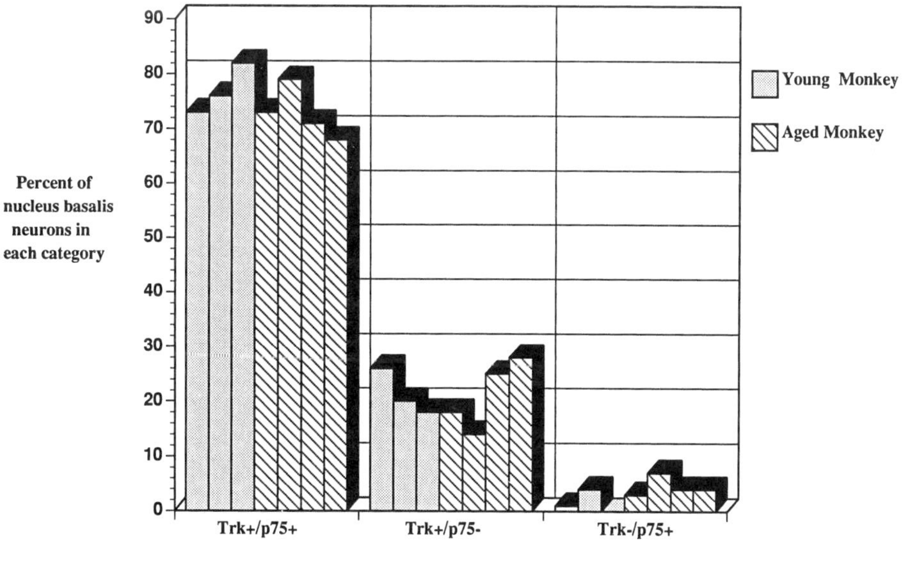

Fig. 9. Histogram showing the percent of nucleus basalis neurons colocalizing Trk and $p75^{NTR}$, or Trk and p75 alone in the young and aged monkey. Note the lack of an aged-related change in the number of neurons containing each marker (*see text* for details).

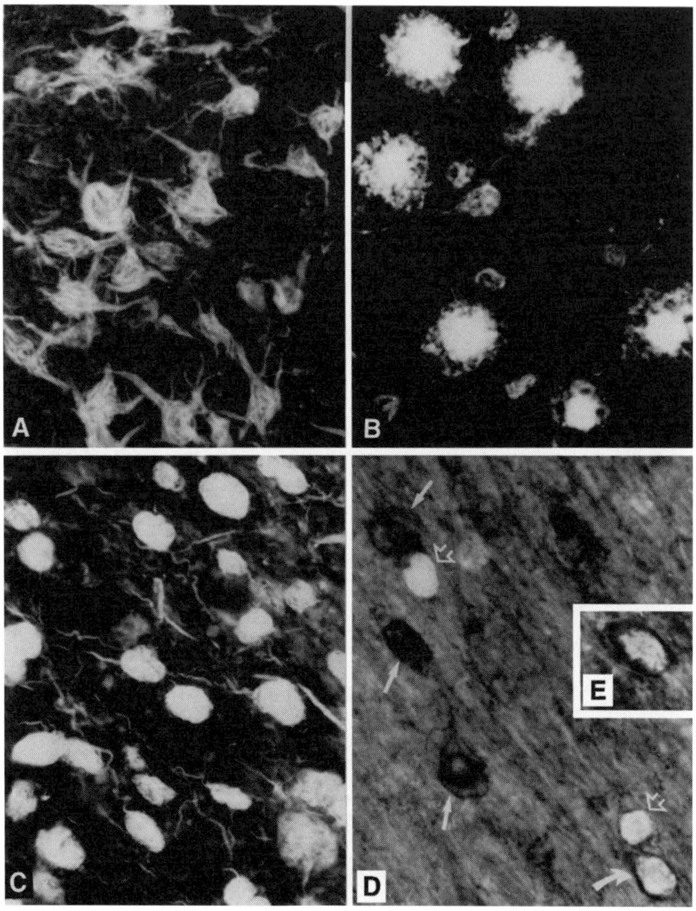

Fig. 10. Thioflavin-S stained sections showing neurofibrillary tangles (NFTs) and senile plaques (SP) in AD. **(A)** NFTs within layer II of the entorhinal cortex. **(B)** SPs within the superior temporal cortex. (**C**) NFTs within the nucleus basalis (Ch4am). **(D)** p75^{NTR} Ch4am-immunoreactive neurons counterstained with thioflavin-S. Solid white arrows indicate p75^{NTR}-containing neurons, open arrows indicate ghost tangles, and curved arrow denotes tangle-bearing p75^{NTR} neuron. **(E)** Another example of a tangle-bearing neuron containing p75^{NTR}. (A–E, ×380). (Reproduced with permission from Mufson et al., 1989.)

4.2. Reduction in p75^{NTR} Protein and mRNA-Containing CBF Neurons in Alzheimer's Disease

To determine the effect of AD on CBF neurons, we used immunohistochemcial and *in situ* hybridization protocols to visualize p75^{NTR} (Kordower et al., 1989; Mufson et al., 1989; Higgins and Mufson, 1989). Control cases averaged 190,000 p75^{NTR}-immunopositive neurons within the CBF. There was an apparent 1:1 relationship between the number of viable CBF neurons and the number of perikarya expressing the message for p75^{NTR}. There was a striking reduction in the number of CBF neurons expressing the protein or mRNA for p75^{NTR} within the subfields of the nucleus basalis (Figs. 5 and 11).

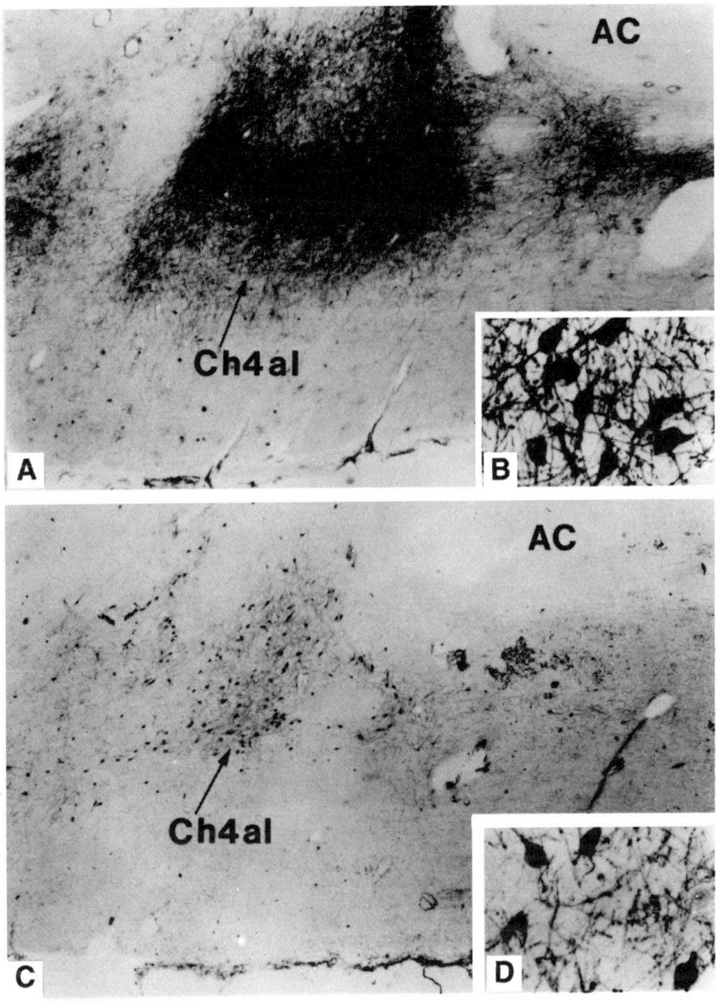

Fig. 11. Comparison of the loss of p75^{NTR} Ch4al immunoreactive neurons of the nucleus basalis in aged control patient **(A,B)** and AD **(C,D).** Note at the higher magnifications (insets), the dystrophic appearance of the Ch4al neurons and the decrease in p75^{NTR} immunoreactive neuropil staining. (A and C, ×19; B and D, ×171; AC, anterior commissure). (Reproduced with permission from Mufson et al., 1989).

Significant reductions in the number of p75^{NTR}-immunostained neurons were observed in the anteromedial (35.1%), anterolateral (76.4%), intermediate (62.1%), and posterior (76.5%) subdivisions of the nucleus basalis. Those remaining CBF neurons continue to express both the protein and mRNA for p75^{NTR} (Figs. 5, 11, and 12) and select AD cases exhibit an increase in p75^{NTR} mRNA expression (Fig. 17D; *see* Mufson et al., 1996). In contrast, the number of p75^{NTR}-immunoreactive neurons within the hippocampal projecting septal–diagonal band neurons were comparable in AD and aged controls (Fig. 5).

These observations indicate a regionally specific loss of p75^{NTR}-containing neurons within the basal forebrain in AD. In these cases, neurons projecting to the temporal

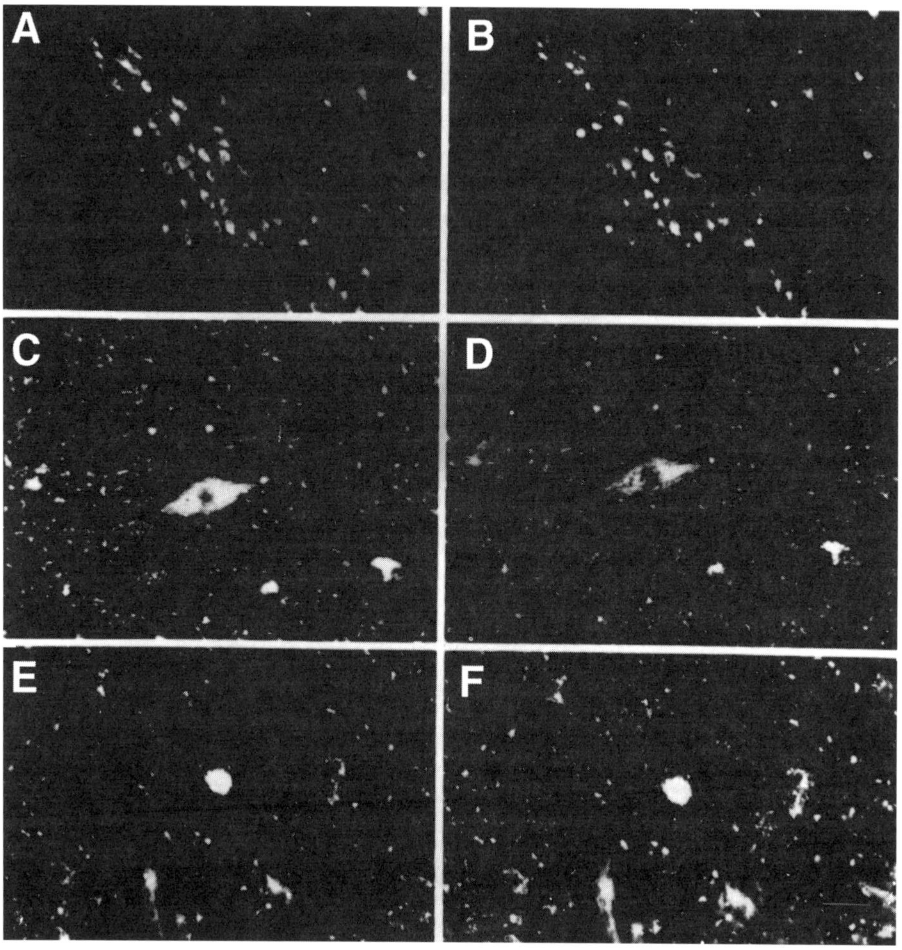

Fig. 12. Double-labeled profiles of basal forebrain neurons from an AD case stained for both ChAT (left panel) and $p75^{NTR}$ (right panel). **(A)** Nucleus basalis neurons stained for ChAT with Texas Red and **(B)** the virtually identical population stained for $p75^{NTR}$ with FITC. **(C,D)** are higher-power views demonstrating that healthy-appearing neurons within the nucleus basalis dual stained for ChAT (C) and $p75^{NTR}$ (D). Similarly, the neuron in **(E)** and **(F)** which is in a state of degeneration still co-localized ChAT (E) and $p75^{NTR}$ (F). Bar, 250 μm for (A) and (B) and 50 μm for (C) and (D). (Reproduced with permission from Kordower et al., 1989.)

cortex were most severely affected. In contrast, hippocampal-projecting basal forebrain neurons were affected to a lesser degree (*see also* Vogels et al., 1990). These findings suggest that efforts to rescue and restore telencephalic cholinergic neurotransmission with growth factors, like NGF, might best be focused on the hippocampal-projecting CBF neurons, which are atrophied, but present, in AD. On the other hand, neuron-replacement strategies might best be focused within the target areas of the anterolateral, intermediate, and posterior subfields of the nucleus basalis, since these cells are lost in the greatest numbers in AD.

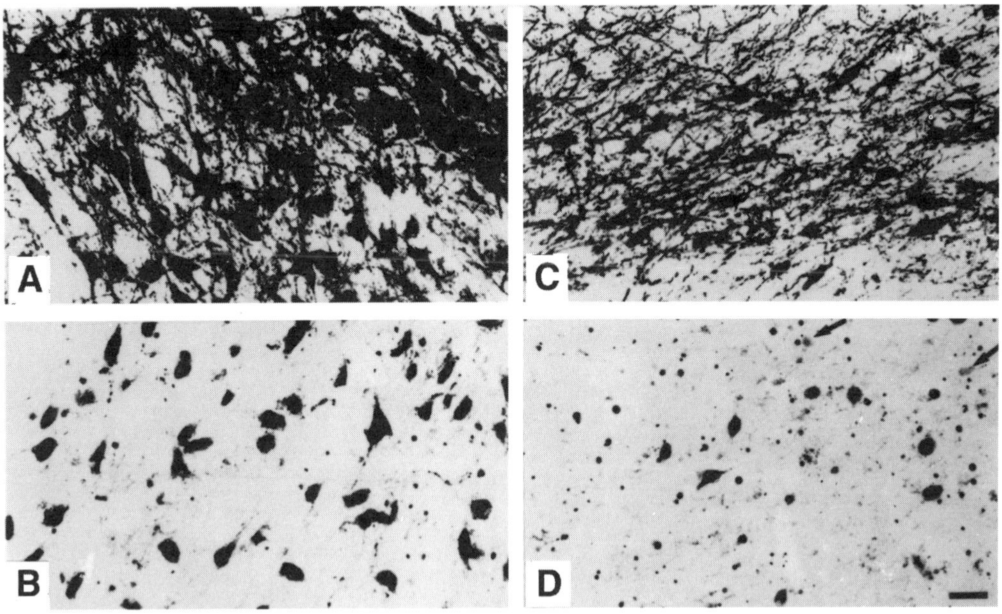

Fig. 13. Photomicrographs of sections immunostained for p75^{NTR} and NGF within the nucleus basalis from a normal aged control **(A,B)** and an AD **(C,D)** case. Note the dissociation in the expression of p75^{NTR} and NGF-immunoreactive neurons within the nucleus basalis in AD. (B) Paucity of intensely NGF immunoreactive neurons as compared to the extensive p75^{NTR} neuronal staining (A). The black arrows in (B) indicate examples of lightly NGF-immunostained neurons. Open arrows in (B) indicate NGF-immunostained neurons that exhibited an oval or ballooned appearance. Scale bars = 50 μm. (Reproduced with permission from Mufson et al., 1995.)

4.3. Defective Retrograde Transport of NGF in Alzheimer's Disease

4.3.1. NGF Within Cholinergic Basal Forebrain Neurons in Alzheimer's Disease

Currently, little is known about the ability of CBF-containing neurons to retrogradely transport NGF in AD. We combined NGF immunohistochemistry with quantitative optical densitometry to evaluate whether retrogradely transported NGF is altered within CBF consumer neurons in AD (Mufson et al., 1994, 1995). In aged control brains, there was virtually a complete overlap in the number and distribution of neurons immunoreactive for NGF and p75^{NTR} (Fig. 13A, B), illustrating that almost all nucleus basalis neurons retrogradely transport NGF. In contrast, there was a dissociation in the distribution of NGF- and p75^{NTR}-positive neurons within the nucleus basalis in AD. Although reduced in absolute number relative to age-matched controls, p75^{NTR}-immunoreactive perikarya were observed in the AD brain (Fig. 13C). However, many of these residual neurons were de-void of NGF staining in AD (Fig. 13D). Many CBF neurons appeared shrunken, with blunted dendrites, or displayed a ballooned appearance in AD (Fig. 13D). Of those neurons containing any detectable NGF immunoreactivity, optical density measurements revealed a significant 32% reduction in staining intensity in AD, compared to age-matched controls ($p < 0.001$; Fig. 14). In contrast, virtually all remaining CBF neurons were intensely p75^{NTR}-immunoreactive in AD (Fig. 13C), demonstrating the specificity of this finding.

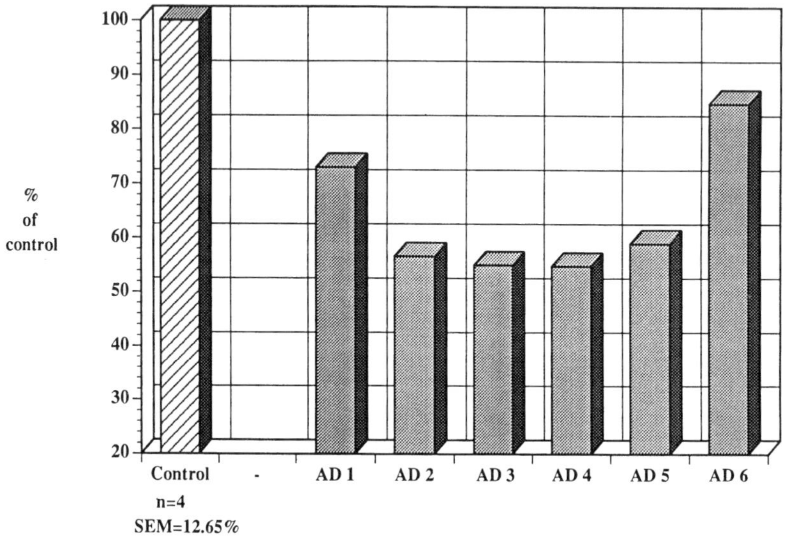

Fig. 14. Difference in optical density measurements of NGF-immunostained nucleus basalis neurons between normal aged controls and AD cases (*see text* for details). (Reproduced with permission from Mufson et al., 1995.)

In AD, CBF neurons displayed a significant reduction of NGF immunoreactivity, and many CBF neurons failed to exhibit NGF reactivity at all. These data suggest that CBF neurons in AD are receiving inadequate trophic support. Although the exact mechanism leading to a diminished NGF accumulation is not presently known, previous studies have indicated that this defect does not occur at the level of NGF synthesis (Higgins and Mufson, 1989; Ernfors et al., 1990).

4.3.2. NGF Cortical Levels Are Increased in Alzheimer's Disease

Based on the above findings of a reduction in retrogradely transported NGF within CBF neurons, we investigated whether NGF protein levels within the cortical production sites were altered in AD, using a sensitive two-site enzyme-linked immunosorbent assay (ELISA) (Crutcher et al., 1993; Scott et al., 1995). Evaluation of NGF levels revealed that protein levels were significantly increased in AD, compared to aged controls in the hippocampus, frontal cortex, superior temporal cortex, inferior parietal lobule, and occipital cortex (Fig. 15). These findings were similar to those previously reported by Crutcher et al. (1993), in which AD-related increases in frontal and occipital poles were observed. In contrast, tissue samples of the nucleus basalis of Meynert displayed a 50% AD-related reduction in NGF. This decline in NGF levels in the CBF is reminiscent of the reduction in NGF-like immunoreactivity we reported within neurons of the CBF in AD (*see* Fig. 14; Mufson et al., 1995). The accumulation of NGF seen within the AD cortex was not found in patients with Parkinson's disease (Fig. 16), suggesting that this response is disease-specific.

These findings indicate that there are increases in neocortical NGF levels, coupled with declines in the retrograde transport of this neurotrophin by the CBF consumer neurons in AD. The accumulation of NGF within the cortex indicates that NGF protein is abundantly available in AD brain.

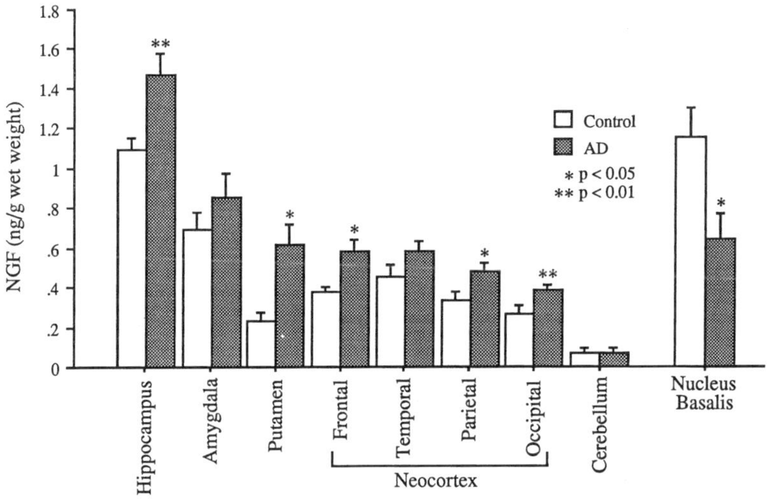

Fig. 15. NGF levels in AD across various regions of the human brain. Increases in NGF-like immunoreactivity activity relative to the control group are pronounced in neo- and limbic cortical regions, areas traditionally known to be affected in AD. Note the 50% decline in NGF within the nucleus basalis in AD. (Reproduced with permission from Scott et al., 1995.)

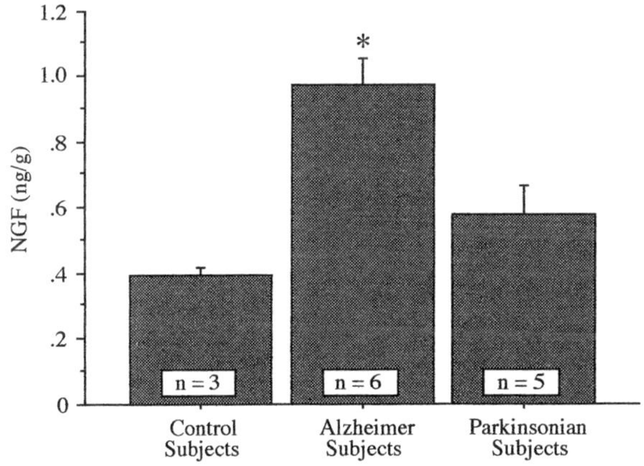

Fig. 16. NGF levels in the superior frontal gyrus (mean ± SEM). Increased levels were seen in this region in AD, but not Parkinson's disease, when compared with age-matched control samples. (Reproduced with permission from Scott et al., 1995.)

4.4. Downregulation of Trk A mRNA Within Cholinergic Basal Forebrain Neurons in Alzheimer's Disease

Increased NGF levels in AD cortex (Crutcher et al., 1993; Scott et al., 1995) suggest a defect in the utilization and/or axonal transport of NGF, despite normal levels of NGF synthesis. Since the high-affinity Trk receptor, which is synthesized in CBF neurons (Gibbs and Pfaff, 1994; Holtmzan et al., 1995; Shelton et al., 1995; Lesauteur et al., 1996; Mufson et al., 1996; Li, J.-M., Kordower, J. H., and Mufson, E. J. (1996) *Soc. Neurosci. Abstr.* **22,** 210), is essential for its signal transduction, a defect in this system

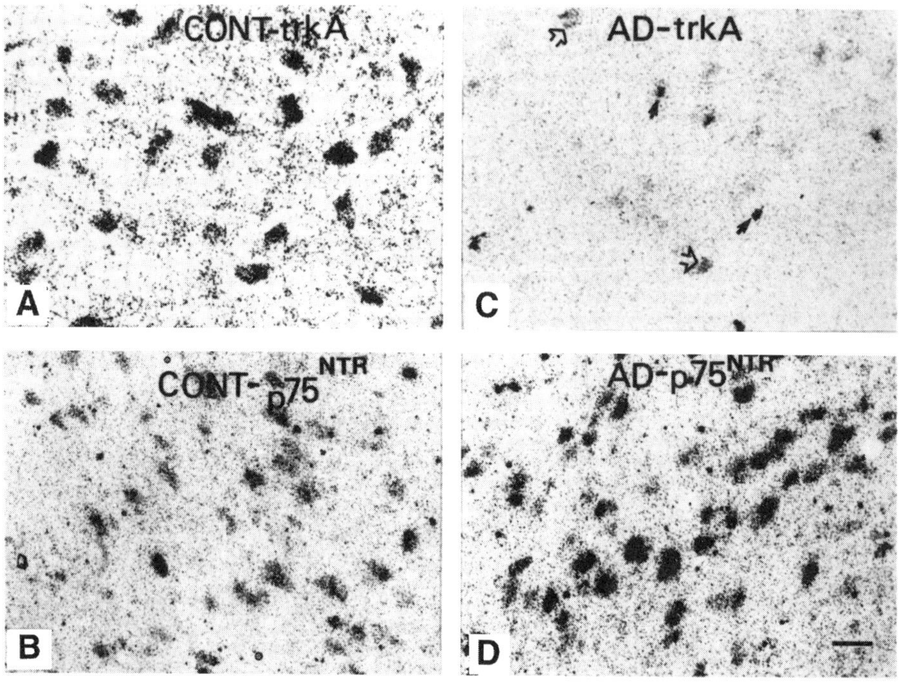

Fig. 17. Photomicrographs of Trk A and p75^{NTR} mRNA-positive neuronal hybridization within the nucleus basalis. **(A)** Trk A mRNA-hybridized neurons within the nucleus basalis of an aged control patient. **(C)** Reduction in Trk A mRNA expression in remaining neurons in AD patient. Black arrows indicate neurons exhibiting intense, and open arrows light, Trk A mRNA hybridization. **(B)** p75^{NTR} mRNA-hybridized neurons within the nucleus basalis of an aged control patient. **(D)** An example of more intense p75^{NTR} hybridization within nucleus basalis neurons in an AD patient. Note the approximate 1:1 relationship between Trk A (A) and p75^{NTR} (C) mRNA-positive cells in the aged controls. Scale bar in (D) = 70 μm for all panels. (Reproduced with permission from Mufson et al., 1996.)

would deprive NGF consumer neurons of needed trophic support, which then could result in atrophy and/or cell death. Therefore, we combined *in situ* hybridization, quantitative optical densitometry, and reverse transcription-polymerase chain reaction (RT-PCR) technology to evaluate whether Trk A mRNA expression is altered within CBF neurons in AD. Tissue hybridized for Trk A gene expression from aged control brains revealed numerous intensely labeled neurons scattered throughout the nucleus basalis (Fig. 17A). Adjacent sections immunostained for p75^{NTR} revealed an approx 1 : 1 relationship between the number of viable nucleus basalis cholinergic neurons and the number of neurons expressing detectable Trk A mRNA. In AD, there were many fewer Trk A mRNA-labeled neurons within the nucleus basalis (Fig. 17C), indicative of degeneration of this neuronal population in AD. Many remaining CBF neurons failed to express the message for Trk A. Of those AD neurons exhibiting detectable Trk A mRNA, optical density measurements revealed a significant (66%) reduction in hybridization intensity relative to aged controls. RT-PCR studies also revealed that Trk A mRNA levels were reduced in the nucleus basalis in AD relative to aged controls. This effect was specific for Trk A, since the gene expression for the housekeeping gene G3PDH was similar in AD and aged controls.

4.5. Is NGF Off Its Trk in Alzheimer's Disease?

The above findings, demonstrating decreases in retrogradely transported NGF and Trk A mRNA within CBF neurons, as well as an increase in cortical NGF protein levels, have led us to hypothesize that CBF degeneration may be caused, in part, by a defect in the high-affinity Trk A NGF receptor in AD (Fig. 19). Further support for this hypothesis is derived from the observations that both the protein and mRNA for $p75^{NTR}$ mRNA is stable or increased in AD (Fig. 17A, B; *see* Goedert et al., 1986; Kordower et al., 1989; Mufson et al., 1989; Ernfors et al., 1990). These findings indicate that at its production site, the $p75^{NTR}$ is not defective in AD. In addition to normal neuronal production of the $p75^{NTR}$ protein and mRNA, we have also found that residual basal forebrain axons continue to immunostain for the $p75^{NTR}$ in AD (Mufson et al., 1989). This suggests that the $p75^{NTR}$ may transport down the axon normally to its cortical target regions in the diseased brain, which indicates that general transport defects, per se, may not explain the reduction in NGF transport that we see.

At present, the role that the high-affinity Trk A receptor plays in the CBF hypofunction seen in AD is unknown. Studies have indicated that there is an age-related reduction in Trk A-receptor mRNA in aged rats (Cooper et al., 1994) similar to that seen in AD (Mufson et al., 1996). In aged rats that display atrophy of basal forebrain neurons, there is a significant (43%) decrease in Trk A, but not $p75^{NTR}$ mRNA within CBF neurons (Cooper et al., 1994). Furthermore, this animal investigation also found a reduction in the number of CBF neurons that retrogradely transported iodinated NGF, further supporting our hypothesis (Mufson et al., 1995) that reduced NGF transport may be secondary to diminished Trk A expression, resulting in degenerative processes within CBF neurons (Fig. 18). However, it remains to be determined whether there also is an alteration in the binding of Trk A with NGF at the level of its cortical production sites in AD (*see* Fig. 18).

5. NEUROPLASTICITY OF THE NGF-RECEPTOR SYSTEM IN ALZHEIMER'S DISEASE

The increased levels of NGF found within the cortex and the reduction of Trk A gene expression within CBF neurons in AD need to be considered, together with the findings that there are regenerative changes occurring in the AD brain related to the NGF–CBF system. For example, hippocampal fibers putatively derived from the NGF-consumer neurons of the septal–diagonal band subfield of the CBF reorganize in AD (Geddes et al., 1985), suggesting that neurotrophic factors mediate aspects of plasticity in aging and disease. Recently, we described two plasticity responses associated with the NGF system in AD dealing with a novel expression of $p75^{NTR}$ in the cortex and the hyperinnervation of galaninergic fibers on remaining CBF/$p75^{NTR}$ neurons in this disease.

5.1. De Novo Expression of $p75^{NTR}$-Containing Neurons in the Alzheimer's Disease Cortex

Using a monoclonal antibody against the $p75^{NTR}$, we observed a novel population of cortical neurons containing the $p75^{NTR}$ only in patients with AD (Fig. 19) and very old (>90 yr of age) normal adults (Mufson and Kordower, 1992; de Lacalle et al., 1994). These cells were preferentially distributed within the temporal association cortex, a region that displays severe degeneration in AD. A few such neurons were also located in the hippocampal and amygdaloid complexes. These neurons appeared healthy, with

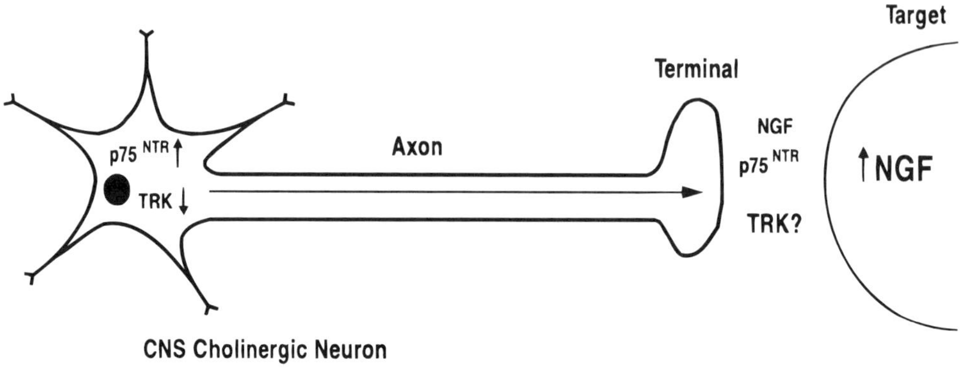

Fig. 18. Summary diagram of changes to the NGF system associated with the cholinergic basal forebrain (CBF) neurons and their target sites in Alzheimer's disease. Within CBF neurons, there is a decrease in the gene expression for the high-affinity Trk A receptor (downward arrow), but the low affinity remains stable or increases (upward arrow). Although there is an increase in NGF protein levels in cortical target sites (upward arrow), there is a decrease in retrogradely transported NGF by the CBF neurons. These findings have led us to hypothesize that CBF degeneration may be caused, in part, by a defect in the high-affinity Trk A receptor in AD. It remains to be determined whether there is a defect in the binding of Trk A with NGF, which is indicated by a **?** in this figure.

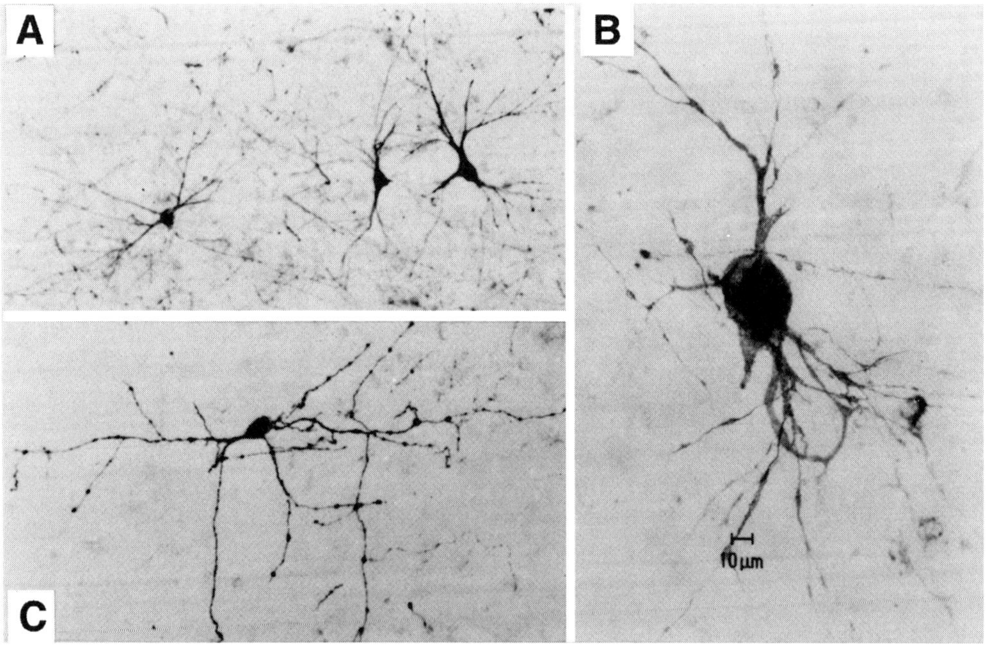

Fig. 19. Morphological profiles of mulitpolar p75^{NTR} immunoreactive neurons within the cortex in AD **(A–C)**. **(A)** Cluster of neurons within layer V of the inferior temporal cortex (×80). **(B)** Neurons within the basolateral amygdala (×100). **(C)** Stellate-like neuron with beaded dendritic processes in the inferior temporal cortex (×200). (Reproduced with permission from Mufson et al., 1992.)

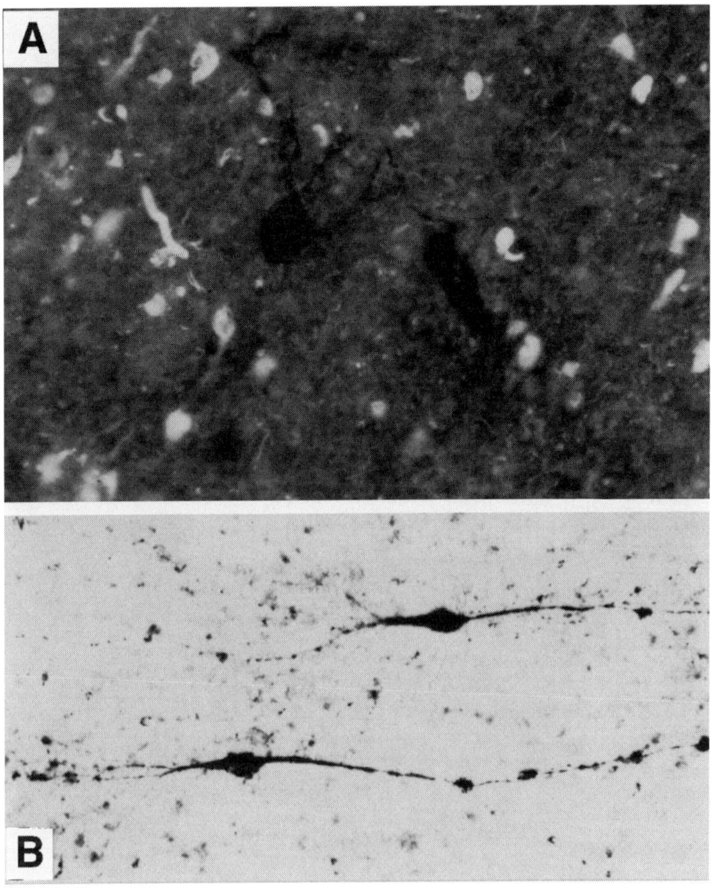

Fig. 20. (A) p75^{NTR}-immunoreacted section counterstained with thioflavin-S to demonstrate the lack of colocalization between the white fluorescent NFTs and the black immunoreactive cortical neuron (×350). **(B)** p75^{NTR}-immunoreactive neurons within the neocortex of a 21-wk-old human embryo (×350). (Reproduced with permission from Mufson et al., 1992.)

bipolar, fusiform, or multipolar morphologies, and did not contain neurofibrillary tangle-bearing material (Fig. 20). We have also observed p75^{NTR}-immunoreactive neurons in the developing human cortical subplate zone (Fig. 20; *see* Kordower and Mufson, 1992), many of which were morphologically similar to that seen in AD and advanced aging. This suggests that the expression of p75^{NTR}-containing cortical neurons in AD may be a re-expression of events that occur during embryonic development. On the other hand, it is well established that NGF and its receptor have a feed-forward relationship (Hefti et al., 1989). Infusion of NGF into the striatum (Gage et al., 1989) or lateral ventricle (Koliatsos et al., 1994) results in the expression of p75^{NTR}-immunoreactive neurons (but *see* Bartus et al., 1996). Thus, the increased synthesis of cortical NGF observed in AD (Crutcher et al., 1993; Scott et al., 1995) may provide trophic support for those cortical neurons' (re)-expressing p75^{NTR}-immunoreactivity in the diseased brain. It would be interesting to determine whether these neurons also express the high-affinity signal-transducing Trk A receptor and whether levels of NGF are elevated in the cortex of very old normal people in which there are p75^{NTR}-immunoreactive cortical neurons, compared to older individuals lacking these neurons.

5.2. Galanin Hyperinnervation of p75^{NTR} Cholinergic Basal Forebrain Neurons in Alzheimer's Disease

A consensus has been generated linking the neuropeptide galanin to CBF function in normal and pathologic states, including AD (*see* Crawley, 1996, for review). Galanin is a novel 29 (in humans, a 30)-amino acid peptide that was isolated from porcine small intestine (*see* Crawley, 1996, for review). Although it has been shown both immunohistochemically and biochemically to have a widespread distribution in the CNS (Melander et al., 1986; Kordower et al., 1992; Crawley, 1996, for review), its chemoanatomical distribution within the CBF exhibits a differential species-specific expression within the mammalian basal forebrain (*see* Kordower and Mufson, 1990; Benzing et al., 1993; Crawley, 1996, for review). In human brain, galanin does not colocalize with the cholinergic cells of the basal forebrain as it does in rats and monkeys (Melander, 1986; Kordower and Mufson, 1990; Benzing et al., 1993). Rather, it is localized within small galanin-immunoreactive interneurons, as well as within a widespread fiber network that courses through the CBF (Fig. 21A; Kordower et al., 1992; Benzing et al., 1993; Mufson et al., 1993). The finding that galanin inhibits acetylcholine (Fisone et al., 1987) and disrupts working memory in rats (*see* review in Crawley, 1996) has led to the suggestion that the this neuropeptide may play a key role in the regulation of cholinergic–p75^{NTR} basal forebrain function in AD. In AD, the hyperinnervation of remaining CBF–p75^{NTR} neurons by galaninergic fibers (Fig. 21B) suggests that this plasticity response may exacerbate cholinergic deficits and in turn enhance cognitive decline (Chan-Palay, 1988; Mufson et al., 1993) by altering ACh neurotransmission within remaining cholinergic neurons. This notion is supported by the finding that loss of cortical choline acetyltransferase is more severe than neuronal loss in the CBF in AD (Perry, 1980). Since NGF induces expression of both ChAT (Heumen et al., 1984) and galanin (Kaplan et al., 1991) in cultured pheochromocytoma (PC12) cells, it is possible that galanin may regulate the cellular expression of ChAT, but also components of the NGF system in vivo.

The above findings indicate that NGF-related systems exhibit regenerative events, which may be a response to the extensive neurodegeneration seen in the AD brain. The observation of a *de novo* expression of p75^{NTR} within cortical neurons is among the first evidence of neurotrophin-related trophism in AD. The galanin hyperinnervation response to CBF degeneration suggests that this peptide may antagonize possible trophic effects of NGF by altering NGF-receptor-mediated events associated with cholinergic function. Therefore, understanding whether galanin effects NGF or its receptors within the CBF will provide greater insight into the pathologic process(es) affecting cholinergic cell dysfunction in AD. Taken together, these findings suggest that such regenerative events may need to be considered when devising therapeutic treatments related to the use of NGF for the treatment of AD.

6. STRATEGIES FOR TREATING CBF DEGENERATION IN ALZHEIMER'S DISEASE

For many years NGF has been suggested as a primary or adjunctive therapy for the treatment of the cholinergic deficit in AD. The first AD patient to receive intraventricular

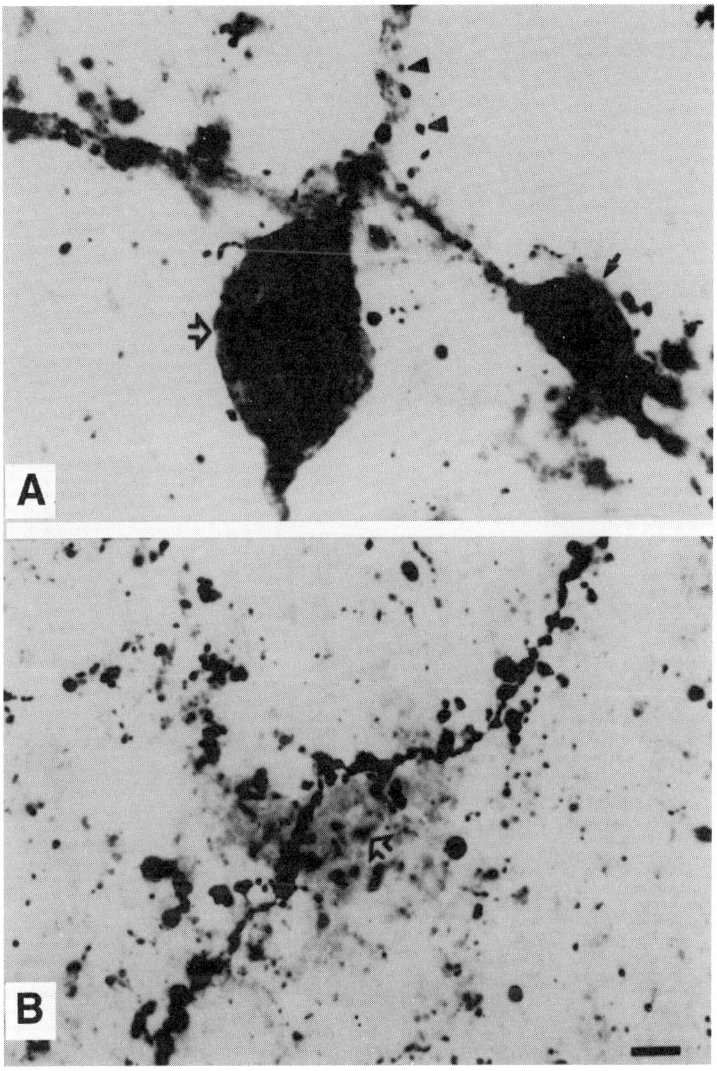

Fig. 21. Photomicrographs showing the relationship between galaninergic profiles and $p75^{NTR}$-containing neurons in human aged and AD basal forebrain. **(A)** Septal–diagonal band complex $p75^{NTR}$-containing cholinergic neuron (open arrow) decorated by fine galanin-immunoreactive fibers (arrowheads) in an aged control case. Note a galanin-immunoreactive neuronal process emanating from a galanin-containing neuron (small arrow), which reaches the dendrite of the $p75^{NTR}$-positive neuron. **(B)** Hypertrophied galanin-immunoreactive fibers surrounding a $p75^{NTR}$-positive neuron (open arrow) within septal–diagonal band complex in AD. Bar in B = 15 μm.

infusions of mouse NGF was reported to improve modestly on a task of episodic memory, although the patient remained unchanged on other tasks (Olson, 1993). This patient also experienced severe NGF-related side effects. Since human NGF (hNGF) does not readily cross the blood–brain barrier and long-term hNGF is likely to be required for treatment of AD patients, methods need to be devised for the chronic

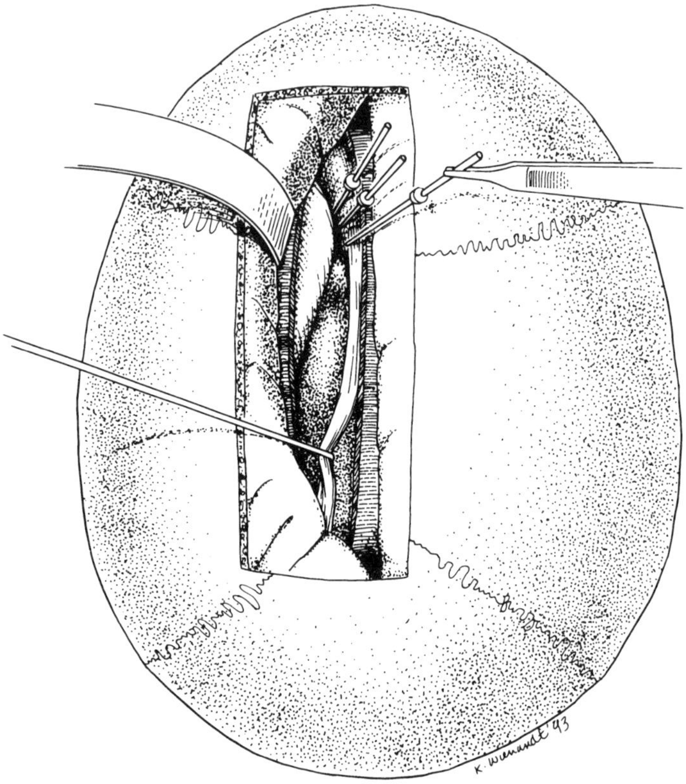

Fig. 22. Drawing of the dorsal aspect of primate brain illustrates transection of the fornix and placement of polymer capsules containing baby hamster kidney (BHK) cells. After craniotomy, the cerebral hemisphere is retracted and the fornix is unilaterally transected using a microdissector. The polymer capsules are placed within the lateral ventricle located between the caudate **(left)** and septal area **(right)** with microforceps. The upper portion of the capsules consists of a silastic tether that is clipped to permit closure of the dura. (Reproduced with permission from Emerich et al., 1994.)

delivery of this neurotrophin. Recently, we have been evaluating two novel approaches in animal models of CBF degeneration: grafting encapsulated cells that have been genetically modified to secrete NGF, and peripheral administration of NGF conjugated to an antibody directed against the transferrin receptor.

6.1. Grafting Encapsulated Cells Secreting Human NGF

Grafting of encapsulated cells secreting human NGF was employed in a series of experiments to rescue axotomized basal forebrain neurons in young (Emerich et al., 1994) and aged (Kordower et al., 1994c) monkeys. Figure 22 is a schematic drawing showing the surgical approach used to intraventricularly transplant polymer-encapsulated baby hamster kidney (BHK) fibroblasts that were genetically modified to secrete hNGF or identical grafts that lacked the NGF construct in axotomized monkeys. The findings derived from the aged monkey study will be reviewed, since these observations are extremely relevant to CBF degeneration seen in AD. Evaluation of these monkeys for AD-like pathology revealed numerous amyloid-containing immunoreactive plaques within the temporal and

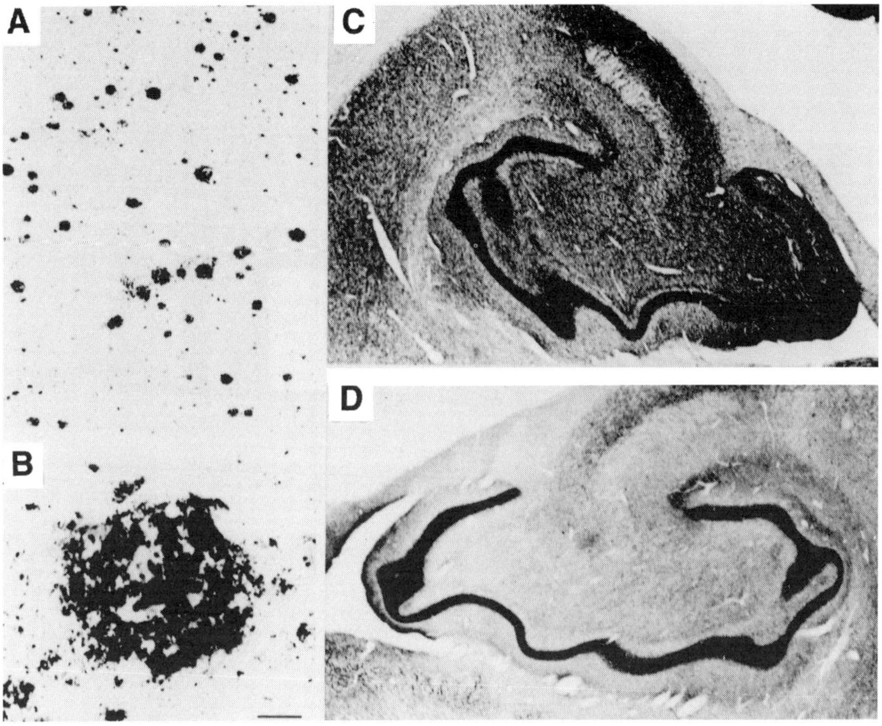

Fig. 23. **(A)** Low- and high- **(B)** power photomicrographs through the temporal neocortex of a 26-year-old monkey, demonstrating the presence and morphology of numerous β-amyloid-immunoreactive senile plaques. **(C)** The pattern of AChE staining within the hippocampus of a 25-yr-old monkey contralateral to the fornix transection. **(D)** The same section as illustrated in (B) illustrates a comprehensive loss of AChE-containing fibers ipsilateral to the fornix transection. Scale bar in (B) denotes the following magnifications: (A) = 200 μm; (B) = 50 μm; (C,D) = 500 μm. (Reproduced with permission from Kordower et al., 1994c.)

parietal cortex, amygdala, and hippocampus (Fig. 23A,B). NGF treatment had no effect on the number or distribution of amyloid plaques in comparison to control animals. Histologic evaluation of the lesion site revealed a complete unilateral transection of the fornix and a loss of AChE-containing fibers within the ipsilateral hippocampus (Fig. 23C,D). Monkeys receiving BHK-control grafts displayed a significant reduction (53%) of $p75^{NTR}$-containing neurons within the medial septum ipsilateral to the transplant (Fig. 24A). Many remaining neurons within the septum appeared atrophic relative to $p75^{NTR}$-containing neurons on the contralateral side. In contrast, the ability of grafted NGF-secreting cell to rescue CBF-containing neurons was striking. The lesion-induced degeneration of septal neurons was significantly attenuated ($p < 0.001$) in monkeys receiving grafts of polymer-encapsulated BHK-hNGF cells (Fig. 24A). These animals exhibited only a modest 7–23% reduction in ChAT and $p75^{NTR}$-immunoreactive neurons within the medial septum ipsilateral to the lesion, compared to the contralateral side (Fig. 24B,C). In addition to maintaining the viability and/or expression of cholinergic markers in septal neurons, BHK-hNGF grafts induced a robust sprouting of fibers expressing the cholinergic markers $p75^{NTR}$, ChAT, and AChE within the ipsilateral septal

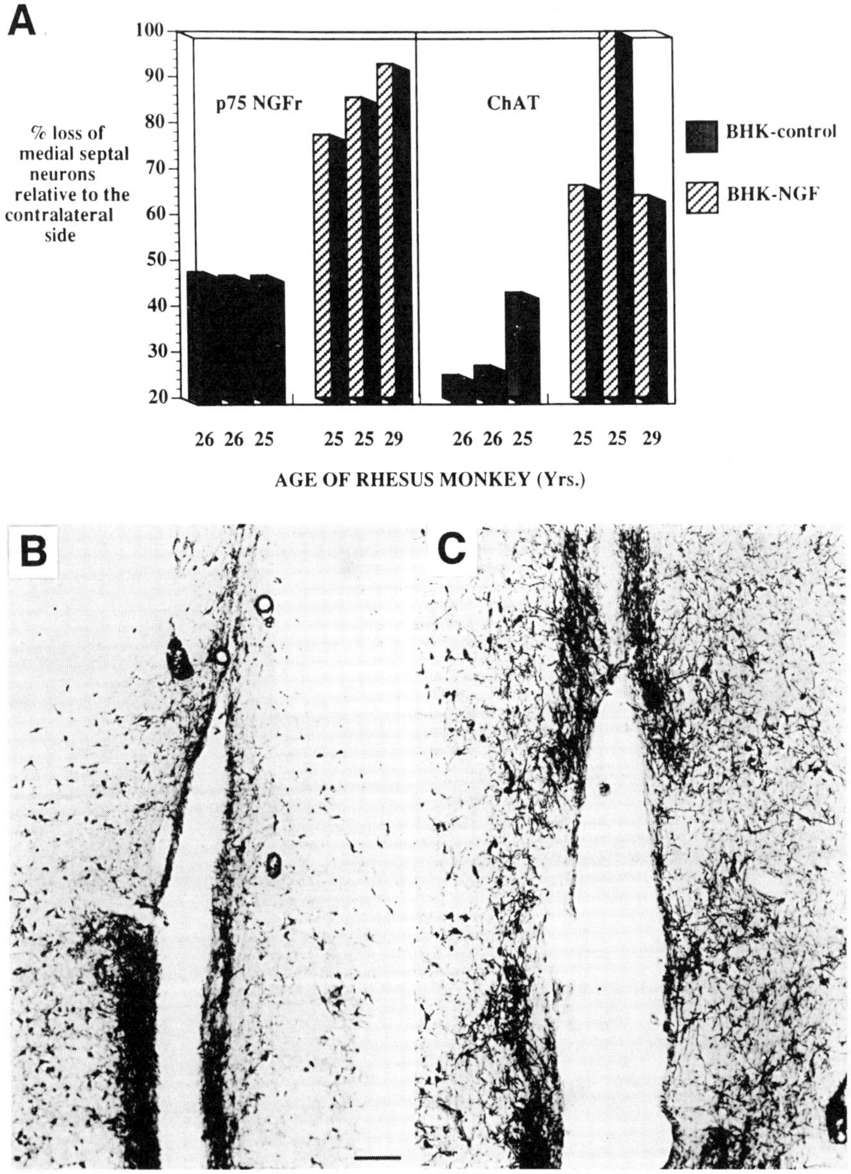

Fig. 24. (A) Quantification of the loss of p75[NTR] and ChAT-immunostained neurons in BHK-control and BHK-hNGF-grafted aged monkeys. **(B)** p75NTR-stained section through the medial septum of a monkey receiving a BHK-control graft. Note the extensive loss of neurons ipsilateral (right) to the lesion relative to the intact (left) side. **(C)** Monkeys receiving BHK-hNGF grafts displayed a symmetrical pattern of p75NTR staining within the medial septum with minimal neuronal loss. Scale bar in (B) represents 100 mm for (B) and (C). (Reproduced with permission from Kordower et al., 1994c.)

region (Fig. 25). In contrast, none of the BHK-control grafted monkeys exhibited a cholinergic sprouting response (Fig. 25B). These monkeys appeared lethargic relative to control grafted monkeys, and this behavioral change lasted for the duration of the experiment.

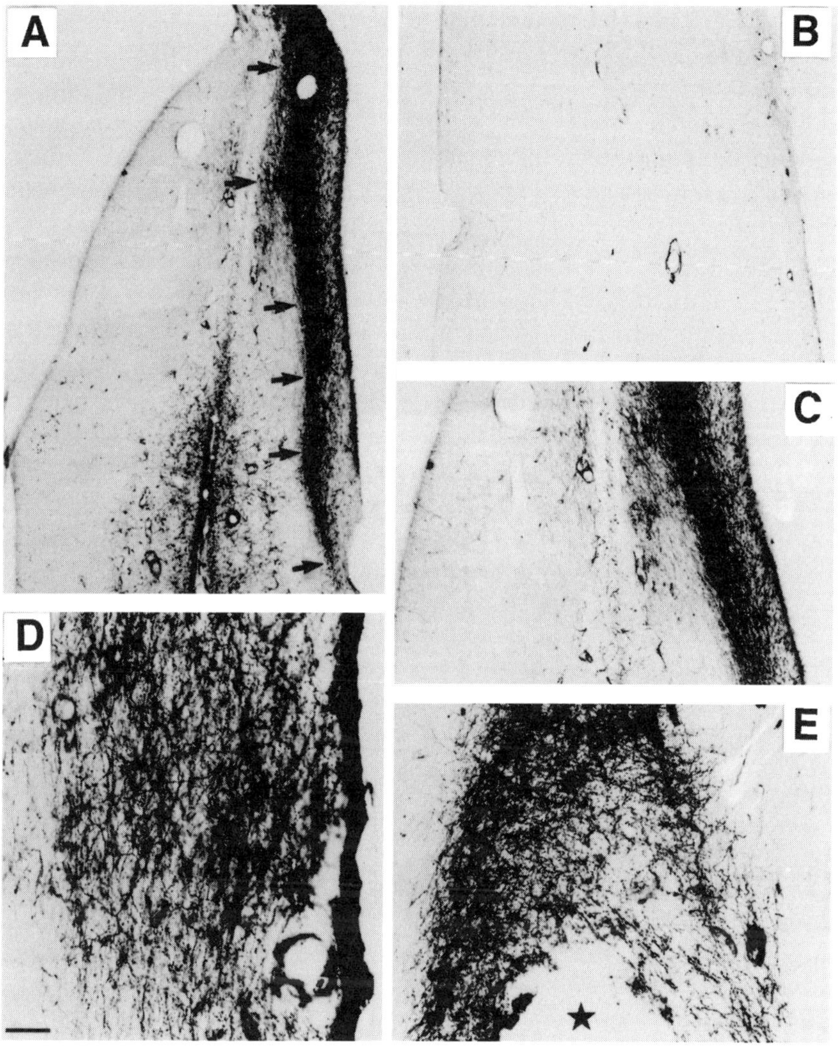

Fig. 25. (A) Dense unilateral p75^{NTR} immuoreactive sprouting response (arrows) within the septum proximal to a BHK-hNGF graft in a 29-yr-old monkey. **(B)** No such staining pattern is observed within the septum of a 25-yr-old monkey receiving a BHK-control graft. **(C,D)** The morphology of the p75^{NTR}-immunoreactive sprouting response is illustrated, which can be resolved into individual fibers that ramify against the ependyma of the ventricle. **(E)** In one animal, the capsule penetrated into the underlying parenchyma (star). A dense network of p75^{NTR}-immunoreactive fibers was also seen in this location adjacent to the implant. Scale bar in (D) denotes the following magnifications: (A) = 500 µm; (B, C) = 250 µm; (D, E) = 50 µm. (Reproduced with permission from Kordower et al., 1994c.)

It is important to determine whether the implanted BHK-hNGF-secreting cells continued to produced NGF over the time-course of the experiment. Therefore, just prior to sacrifice of the monkeys, the grafted capsules were retrieved. Numerous BHK cells were seen within the capsules, which now produced hNGF at a rate of 9.6 ng/24 h per capsule as compared to 44.65 ng of NGF/24 h per capsule prior to implantation. This lower level of NGF production was shown to be biologically relevant, bacause media obtained from the capsules after grafting for 1 mo in aged monkeys induced a robust differentiation of PC12A cells in vitro, in a manner similar to that seen with exposure to 50 ng of hNGF.

These findings demonstrate the feasibility of grafting encapsulated cells that have been genetically modified to secrete hNGF in the aged primate brain. This encapsulation procedure permits the use of xenografts without the need for potentially toxic treatments using agents that suppress immune function. The pores in the polymer capsule are sufficiently large to allow hNGF and nutrients' bidirectional passage, but are sufficiently small to prevent immune cells from entering and destroying the grafted BKH cells. Overall, these findings demonstrate that hNGF supports the viability of degenerating, aged primate CBF neurons (Kordower et al., 1994c), confirming earlier findings in young adult monkeys (Emerich et al., 1994). Recently, grafted BHK cells modified to secrete NGF were shown to continue to produce NGF for at least 1 yr after implantation in the rat (Winn et al., 1996). This further supports the concept that grafts of genetically modified cells sequestered in a polymer capsule may be a viable method of administering hNGF or other molecules to the human brain for the treatment of the cholinergic deficit seen in dementing illnesses, such as AD.

6.2. Peripheral Administration of NGF Conjugated to an Antitransferrin Receptor Antibody

Although trophic factors, such as NGF, may be powerful tools for modulating the function and viability of specific neuronal populations, their clinical utility may be limited by their inability to cross the blood–brain barrier. As a result, administration of neurotrophic factors to specific targets in the CNS requires invasive neurosurgical procedures, such as intraventricular injections (Williams et al., 1986; Kromer, 1987; Hefti et al., 1989) or the grafting of neurotrophin-secreting cells into CNS sites (Strömberg et al., 1990; Schumacher et al., 1991; Kawaja et al., 1992; Emerich et al., 1994; Kordower et al., 1994c), as described above. Recently, a novel noninvasive efficient means of facilitating the transport of macromolecules, like NGF, across the blood–brain barrier has been developed (Friden et al., 1991, 1993; Kordower et al., 1993, 1994a; Granholm et al., 1993, 1994). This delivery system takes advantage of the endogenous mechanism responsible for the active transport of iron into the brain by binding to the transferrin receptor (TfR), which is expressed on brain capillary endothelial cells at a high density. Transferrin, a plasma glygoprotein, binds iron in the blood and functions as the major iron-transport protein. Iron is delivered to cells through the interaction of the iron–transferrin complex with specific receptors located on the cell surface (Granholm et al., 1994), and is internalized through receptor-mediated endocytosis. When conjugated to an antibody directed against TfR, proteins normally excluded from the brain can cross the blood–brain barrier following peripheral administration (Friden et al., 1991, 1993; Kordower et al., 1994a; Backman et al., 1995). It has been shown that NGF can be covalently linked to an antibody against the transferrin receptor (termed OX-26), while retaining its biological activity (Friden et al., 1993). This OX-26-NGF conjugate can be delivered across the blood–brain barrier (Fig. 26) and exert physiological effects previously observed with direct administration of NGF into brain (Friden et al., 1993; Granholm et al., 1994).

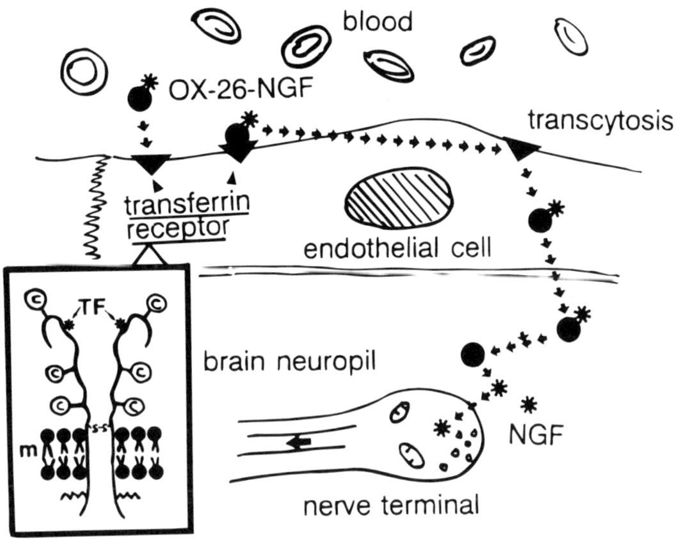

Fig. 26. Schematic illustration of the hypothesized binding and transfer of the OX-26-NGF conjugate on brain capillary endothelial cells. The conjugate is thought to bind to the tranferrin receptor on the surface of the endothelial cell, and then to be transferred by transcytosis into the brain parenchyma. Inset: Demonstrates the transferrin receptor location in the cell membrane (m) and the two sites for transferrin (TF) binding on the extracellular portion of the receptor. (Reproduced with permission from Granholm et al., 1994.)

6.3. *Effects of OX-26-NGF on Cholinergic Systems*

The ability of the intravenously administered NGF conjugate to support the viability of central cholinergic neurons has been demonstrated previously in several model systems dealing with cholinergic innervation in the CNS. Initial studies demonstrated that IV injections of the NGF conjugate increased the viability and phenotypic expression of CBF neurons following transplantation into the anterior chamber of the eye (Friden et al., 1993; Granholm et al., 1994; Backman et al., 1995). Within the CNS, we recently demonstrated that IV NGF-conjugate treatment specifically prevented the degeneration of cholinergic striatal neurons in a rodent model of Huntington's disease (Kordower et al., 1994c). This study demonstrates that 2 wk of NGF-conjugate administration prevents the loss of ChAT-immunoreactive, but not somatostatin-immunoreactive or diaphorase-containing, neurons, which normally occurs following intrastriatal injections of quinolinic acid.

6.4. *OX-26-NGF Rescues Cholinergic Basal Forebrain Neuron Degeneration in an Animal Model of Alzheimer's Disease*

Degeneration of CBF cortical projection system is a hallmark of patients with AD. In an animal model that mimics this pattern of degeneration, we tested whether IV administration of an OX-26-NGF conjugate could reverse the atrophy of CBF neurons following lesion of their cortical target sites by epidural application of *N*-methyl-D-aspartic acid (NMDA) (Charles et al., 1996). Seventy-five days postlesion, cholinergic nucleus basalis neurons were atrophic ipsilateral to the cortical insult relative to the contralateral side in control rats receiving IV injections of vehicle or a nonconjugated

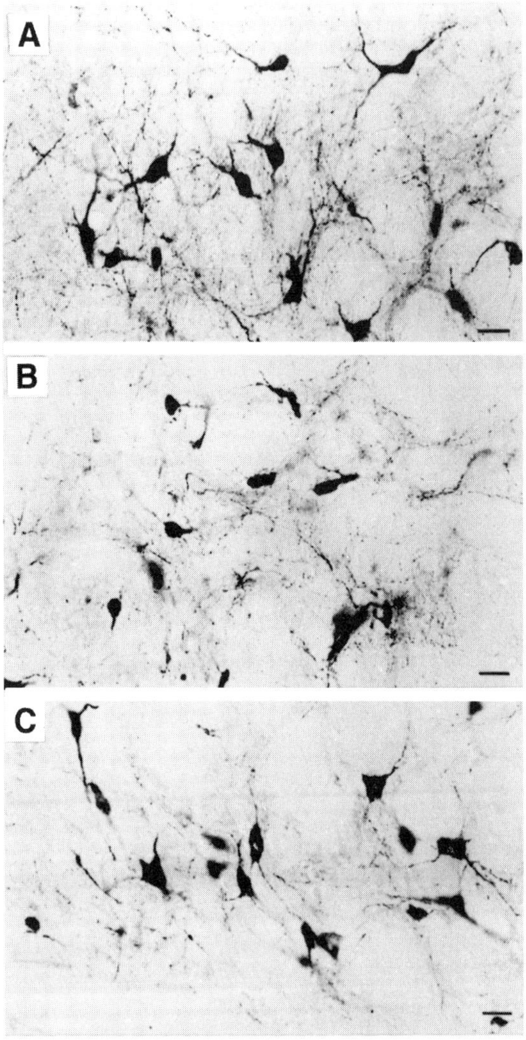

Fig. 27. Photomicrographs of Trk A-immunoreactive neurons within nucleus basalis. **(A)** Note the large, healthy-appearing magnocellular perikarya seen within the intact nucleus basalis. **(B)** These neurons are shrunken in vehicle-treated animals ipsilateral to the lesion. **(C)** In contrast, Trk A-immunoreactive nucleus basal neurons are of normal size when treated with the OX-26–NGF conjugate. Scale bar represents 40 μm. (Reproduced with permission from Charles et al., 1996.)

mixture of OX-26 and NGF. In contrast, IV injections of the OX-26-NGF conjugate restored the size of nucleus basalis perikarya to within normal limits relative to the non-lesioned contralateral side conjugate, as determined by both $p75^{NTR}$ and Trk A immunohistochemistry (Fig. 27). The magnitude of this effect is similar to what has been reported in other studies (Sofroniew et al., 1983, 1987; Figueiredo et al., 1995). Immunohistochemical studies using rat serum albumen antisera indicated that the blood–brain barrier was closed at the time of treatment, indicating that this trophic effect did not result from NGF crossing through a compromised blood–brain barrier at the site of the lesion.

These data demonstrate that systemic administration of a neurotrophic factor–antibody conjugate, intended to circumvent the blood–brain barrier, can provide trophic influences on atrophic neurons within the nucleus basalis. The lesion model used here does not induce cell death, but rather shrinks the size of basal forebrain perikarya modeling an early pathological process seen in AD. The findings presented support the emerging concept that NGF covalently conjugated to a protein vector delivery system, the OX-26 rat antibody against the transferrin receptor, can facilitate the transfer of impermeable therapeutic compounds, such as NGF, across the blood–brain barrier in models of AD-like CBF degeneration.

7. CONCLUDING COMMENTS

Although it has been hypothesized that alterations in the expression of NGF may underlie the pathophysiology of the CBF neurodegeneration seen in AD (Appel, 1981; Hefti et al., 1989), the data presented in this chapter suggest a more complex scenario. Investigations in our laboratories, using an antibody that recognizes endogenous NGF (Conner et al., 1992; Mufson et al., 1994), revealed a reduction in transported NGF-like immunoreactivity within remaining CBF consumer neurons in AD (Mufson et al., 1995). In contrast, NGF levels in cholinerigc basal forebrain cortical target regions are reported to be unchanged (Goedert et al., 1986) or increased (Crutcher et al., 1993; Scott et al., 1995) in AD. Similarly, levels of the $p75^{NTR}$ has been described as either stable or elevated in AD (Higgins and Mufson, 1989; Ernfors et al., 1990). This suggested that there is impaired internalization and/or retrograde transport of target-derived NGF associated with a defect in the complexing of this trophin with its high-affinity signal-transducing Trk A receptor in AD. In support of this hypothesis, we found a significant reduction in Trk A mRNA expression in remaining CBF neurons in AD (Mufson et al., 1996). Taken together, these findings suggest that NGF accumulates in the cortex, caused in part by a defect in the production, utilization, and/or binding of NGF with its Trk A signal-transducing receptor in AD. Therefore, reductions in Trk A mRNA expression within CBF neurons in AD may signal the crucial neuropathological change that underlies CBF degeneration. It has been shown that cortical damage induces an upregulation of Trk A within the basal forebrain (Figueiredo et al., 1995) and that activation of the Trk A alone can prevent the degeneration of axotomized CBF neurons in rats (Lucidi-Phillipi et al., 1996). These observations suggest that signaling via the Trk A receptor is a pivotal trophic event underlying the viability of CBF neurons. Therefore, a need may exist to develop compounds that do not necessarily regulate endogenous NGF production, but rather its high-affinity Trk A receptor. Strategies aimed at developing small molecules that mimic the effects of the Trk A receptor, combined with infusions of NGF or grafts of NGF-secreting cells, may forestall the CBF degeneration seen in AD.

ACKNOWLEDGMENTS

We wish to thank our collaborators who participated in some of the studies described in this manuscript. We would also like to thank S. Jaffar, A. Munarriz, S. Quijano, and N. Doogan for histologic, photographic, and secretarial assistance, respectively. This work was supported by AG10161, AG09466, AG10668, Alzheimer's Disease Association, Washington Square Foundation, and the Illinois Department of Public Health.

REFERENCES

Appel, S. H. (1981) A unifying hypothesis for the cause of amyotrophic lateral sclerosis, Parkinsonism, and Alzheimer's disease. *Ann. Neurol.* **10,** 499–505.

Backman, C., Biddle, P. T., Ebendal, T., Friden, P. M., Gerhardt, G. A., Henry, M. A., et al. (1995) Effects of transferrin receptor antibody-NGF conjugate on young and aged septal transplants in oculo. *Exp. Neurol.* **132,** 1–15.

Barbacid, M. (1994) The Trk family of neurotrophin receptors. *J. Neurobiol.* **25,** 1386–1403.

Bartus, R. T., Dean, R. L., Beer, B., and Lippa, A. S. (1982) The cholinergic hypothesis of geriatric memory dysfunction. *Science* **217,** 408–417.

Bartus, R. T., Dean, R. L., Abelleira, S., Charles, V., and Kordower, J. H. (1996) Dissociation of p75 receptors and nerve growth factor neurotrophic effects: lack of p75 immunoreactivity in striatum following physical trauma, excitotoxicity and NGF administration. *Restor. Neurol.* **10,** 49–59.

Batchelor, P. E., Armstrong, D. M., Blaker, S. N., and Gage, F. H. (1989) Nerve growth factor receptor and choline acetyltransferase colocalization in neurons within the rat forebrain: response to fimbria-fornix transection. *J. Comp. Neurol.* **284,** 187–204.

Benzing, W. C., Kordower, J. H., and Mufson, E. J. (1993) Galanin immunoreactivity within the primate basal forebrain: evoluntionary change between monkeys and apes. *J. Comp. Neurol.* **283,** 189–194.

Biegon, A., Greenberger, V., and Segal, M. (1986) Quantitative histochemistry of brain acetylcholine and learning rate in the aged rat. *Neurobiol. Aging* **7,** 215–217.

Bierer, L. M., Haroutunian, V., Gabreil, S., Knott, P. J., Carlin, L. S., Purohit, D. P., et al. (1995) Neurochemical correlates of dementia in Alzheimer's disease: relative importance of cholinergic deficits. *J. Neurochem.* **64,** 749–760.

Bothwell, M. (1991) Keeping track of neurotrophin receptors. *Cell* **65,** 915–918.

Chan-Palay, V. (1988) Galanin hyperinnervates surviving neurons of the human basal nucleus of meynert in dementias of alzheimer's and parkinson's disease: a hypothesis for the role of galanin in accentuating cholinergic dysfunction in dementia. *J. Comp. Neurol.* **273,** 543–557.

Chao, M. V. (1992) Neurotrophin receptors: a window into neuronal differentiation. *Neuron.* **9,** 583–593.

Chao, M. V. (1995) The p75 neurotrophin receptor. *J. Neurobiol.* **25,** 1375–1385.

Chao, M. V. and Hempstead, B. L. (1995) p75 and Trk: a two-receptor system. *Trends Neurosci.* **18,** 321–326.

Charles, V., Mufson, E. J., Friden, P. M., Bartus, R. T., and Kordower, J. H. (1997) Atrophy of cholinergic basal forebrain neurons following excitotoxic cortical lesions is reversed by intravenous administration of an NGF conjugate. *Brain Res.* **728,** 193–203.

Chen, E.-Y., Mufson, E. J., and Kordower, J. H. (1996) Trk and p75 NGF receptor systems in the developing human brain. *J. Comp. Neurol.* **369,** 591–618.

Collier, T. J. and Coleman, P. (1991) Divergence of biological and chronological aging: evidence from rodent studies. *Neurobio. Aging* **12,** 685–693.

Conner, J. M., Muir, D., Varon, S., Hagg, T., and Manthorpe, M. (1992) The localization of nerve growth factor-like immunoreactivity in the adult rat basal forebrain and hippocampal formation. *J. Comp. Neurol.* **319,** 454–462.

Conner, J. M. and Varon, S. (1992) Distribution of nerve growth factor-like immunoreactive neurons in the adult rat brain following colchicine treatment. *J. Comp. Neurol.* **326,** 347–362.

Cooper, J. D., Lindholm, D., and Sofroniew, M. V. (1994) Reduced transport of [^{125}I] nerve growth factor by cholinergic neurons and down-regulated Trk A expression in the medial septum of aged rats. *Neuroscience* **62,** 625–639.

Crawley, J. N. (1996) Galanin-acetylcholine interactions: relevance to memory and Alzheimer's disease. *Life Sci.* **58,** 2185–2199.

Crutcher, K. A., Scott, S. A., Liang, S., Everson, W. V., and Weingartner, J. (1993) Detection of NGF-like activity in the human brain tissue: increased levels in Alzheimer's disease. *J. Neurosci.* **13,** 2540–2550.
de Lacalle, S., Lim, C., Sobreviela, T., Mufson, E. J., Hersh, L. B., and Saper, C. B. (1994) Cholinergic innervation in the human hippocampal formation including the entorhinal cortex. *J. Comp. Neurol.* **345,** 321–344.
Emerich, D. F., Winn, S. R., Hammang, J. P., Baetge, E. E., and Kordower, J. H. (1994) Grafts of polymer-encapsulated cells genetically modified to secrete nerve growth factor rescue basal forebrain neurons in nonhuman primates. *J. Comp. Neurol.* **349,** 148–165.
Ernfors, P., Lindefors, N., Chan-Palay, V., and Persson, H. (1990) Cholinergic neurons of the nucleus basalis express elevated levels of nerve growth factor receptor mRNA in senile dementia of the Alzheimer's type. *Dementia* **1,** 138–145.
Feinstein, D. and Larhammer, D. (1990) Identification of a conserved protein motif in a group of growth factor receptors. *FEBS Lett.* **272,** 7–11.
Figueiredo, B. C., Skup, M., Bedard, A. M., Tetzlaff, W., and Cuello, A. C. (1995) Differential expression of p140-Trk, p75NGFr, and growth-associated phosphoprotein-43 genes in nucleus basalis magnocellularis, thalamus, and adjacent cortex following neocortical infarction and nerve growth factor treatment. *Neuroscience* **68,** 29–45.
Fischer, W., Wictorin, K., Bjorklund, A., Williams, L. R., Varon, S., and Gage, F. H. (1987) Amelioration of cholinergic neuron atrophy and spatial memory impairment in aged rats by nerve growth factor. *Nature* **329,** 65–68.
Fisone, G., Wu, C. F., Consolo, S., Nordstrom, O., Brynne, N., Bartfai, T., Melander, T., and Hokfelt, T. (1987) Galanin inhibits acetylcholine relaease in the ventral hippocampus of the rat: histochemical, autoradiographic, in vivo, and in vitro studies. *Proc. Natl. Acad. Sci. USA* **84,** 7339–7343.
Friden, P. M., Walus, L. R., Watson, P., Doctrow, S. R., Kozarich, J. W., Backman, C. (1993) Blood–brain barrier penetration and in vivo activity of an NGF conjugate. *Science* **259,** 373–377.
Friden, P. M., Walus, L. R., Musso, G. F., Taylor, M. A., Malfroy, B., and Starzyk, R. M. (1991) Anti-transferrin receptor antibody and antibody-drug conjugates cross the blood–brain barrier. *Proc. Natl. Acad. Sci. USA* **88,** 4771–4775.
Gage, F. H., Batchelor, P., Chen, K. S., Chin, D., Higgins, G. A., Koh, S., et al. (1989) NGF receptor reexpression and NGF-mediated cholinergic neuronal hypertrophy in the damaged adult neostriatum. *Neuron.* **2,** 1177–1184.
Geddes, J. W., Monaghan, D. T., Cotman, C. W., Lott, I. T., Kim, R. C., and Chui, H.-C. (1985) Plasticity of hippocampal circuitry in Alzheimer's disease. *Science* **230,** 1179–1181.
Gibbs, R. B. and Pfaff, D. W. (1994) In situ hybridization detection of Trk A mRNA in brain: distribution, colocalization with p75NGFR and up-regulation by nerve growth factor. *J. Comp. Neurol.* **341,** 324–339.
Gnahn, H., Hefti, F., Heumann, R., Schwabb, M. E., and Thoenen, H. (1983) NGF-mediated increase of choline acetyltranferase (ChAT) in the neonatal forebrain: evidence for a physiological role of NGF in the brain? *Dev. Brain Res.* **9,** 45–52.
Goedert, M., Fine, A., Hunt, S. P., and Ulrich, A. (1986) NGF RNA in peripheral and rat central tissues and in the human central nervous system: lesion effects in the rat brain and levels in Alzheimer's disease. *Mol. Brain Res.* **1,** 85–92.
Granholm, A.-C., Backman, C., Bloom, F., Ebendal, T., Gerhardt, G. A., Hoffer, B., et al. (1993) NGF and anti-transferrin receptor antibody conjugate: short- and long-term effects on survival of cholinergic neurons in intraocular septal transplants. *J. Pharmacol. Exp. Ther.* **268,** 448–459.
Granholm, A.-C., Biddle, P. T., Backman, C., Ebenberg, T., Gerhardt, G., Hoffer, B., et al. (1994) Peripheral administration of nerve growth factor to an anti-transferrin receptor antibody cholinergic neuron survival in intraocular forebrain transplants, in *Providing*

Pharmacological Access to the Brain: Alternative Approaches (Flanagan, T. R., Emerich, D. F., and Winn, S. R., eds.). Academic, New York, pp. 71–92.

Hefti, F. and Weiner, W. J. (1983) Is Alzheimer's disease caused by lack of nerve growth factor? *Ann. Neurol.* **13,** 109–110.

Hefti, F., Hartikka, J., Eckenstein, F., Gnahn, H., Heumann, R., and Schwab, M. (1985) Nerve growth factor (NFG) increases choline acetyltransferase but not survival or fiber outgrowth of cultured fetal septal cholinergic neurons. *Neuroscience* **14,** 55–68.

Hefti, F., Hartikka, J., Salviaterra, P., Weiner, W. J., and Mash, D. C. (1986) Localization of nerve growth factor receptors on cholinergic neurons of the human basal forebrain. *Neurosci. Lett.* **69,** 275–281.

Hefti, F. and Mash, D. C. (1989) Localization of nerve growth factor receptors in the normal human brain and in Alzheimer's disease. *Neurobiol. Aging* **10,** 75–87.

Hefti, F., Hartikka, J., and Knusel, B. (1989) Function of neurotrophic factors in the adult and aging brain and their possible use in the treatment of neurodegenerative disease. *Neurobiol. Aging* **10,** 515–533.

Heldin, C.-H. (1995) Dimerization of cell surface receptors in signal transduction. *Cell* **80,** 213–223.

Hempstead, B. L., Martin-Zanca, D., Kaplan, D. R., Parada, L. R., and Chao, M. V. (1991) High-affinity NGF binding requires coexpression of the Trk proto-oncogene and the low-affinity NGF receptor. *Nature* **350,** 678–683.

Heumann, R., Schwab, M., Merkl, R., and Thoenen, H. (1984) Nerve growth factor-mediated induction of choline acetyl transferase in PC12 cells: evaluation of the site of action of nerve growth factor and the involvement of lysosomal degradation prodcutrs of nerve growth factor. *J. Neurosci.* **4,** 3030–3050.

Higgins, G. A. and Mufson, E. J. (1989) NGF receptor gene expression is decreased in the nucleus basalis in Alzheimer's disease. *Exp. Neurol.* **106,** 222–236.

Holtzman, D. M., Kilbridge, J., Li, Y., Cunningham Jr., E. T., Lenn, N. J., Clary, D. O., Reichardt, L. F., and Mobley, W. C. (1995) Trk A expression in the CNS: evidence for the existence of several novel NGF-responsive CNS neurons. *J. Neurosci.* **15,** 1567–1676.

Ibanez, C. F., Ebendal, T., Barbany, G., Murray-Rust, J., Blundell, T. L., and Persson, H. (1992) Similarities and differences in the way neurotrophins interact with the Trk receptors in neuronal and nonneuronal cells. *Cell* **69,** 329–341.

Ip, N., Tapley, P., Glass, D. J., Greene, L. A., and Yancopoulos, G. D. (1993) Similarities and differences in the way neurotrophins interact with the Trk receptors in neuronal and non-neuronal cells. *Neuron* **10,** 137–149.

Ip, N., Ibanez, C., Nye, S., McClain, J., Jones, P. F., Gies, D. R., et al. (1992) Mammalian neurotrophin-4: structure, chromosomal localization, tissue distribution, and receptor specificity. *Proc. Natl. Acad. Sci. USA* **89,** 3060–3064.

Jing, S., Tapley, P., and Barbacid, M. (1992) Nerve growth factor mediates signal transduction through Trk homodimer receptors. *Neuron* **9,** 1067–1079.

Kaplan, D. R., Hempstead, B. L., Martin-Zanca, D., Chao, M. V., and Parada, L. F. (1991) The Trk proto-ongogene product: signal transducting receptor for nerve growth factor. *Science* **252,** 554–558.

Kaplan, D. R. and Stephens, R. M. (1994) Neurotrophin signal transduction by the Trk receptor. *J. Neurobiol.* **25,** 1404–1417.

Kawaja, M. D., Rosenberg, M. B., Yoshida, K., and Gage, F. H. (1992) Somatic gene transfer of nerve growth factor promotes the survival of axotomized septal neurons and the regeneration of the axons in adult rats. *J. Neurosci.* **12,** 2849–2864.

Koh, S., Chang, P., Collier, T. J., and Loy, R. (1989) Loss of NGF receptor immunoreactivity in basal forebrain neurons of aged rats: correlation with spatial memory impairment. *Brain Res.* **498,** 397–404.

Koh, S. and Loy, R. (1988) Age-related loss of nerve growth factor sensitivity in rat basal forebrain neurons. *Brain Res.* **440,** 396–401.

Koliatsos, V. E., Price, D. L., Gouras, G. K., Cayouette, M. H., Burton, L. E., and Winslow, J. W. (1994) Highly selective effects of nerve growth factor, brain-derived neurotrophic factor, and neurotrophin-3 on intact and injured basal forebrain magnocellular neurons. *J. Comp. Neurol.* **343,** 247–262.

Kordower, J. H., Bartus, R. T., Bothwell, M., Schatteman, G., and Gash, D. M. (1988) Nerve growth factor receptor immunoreactivity in the nonhuman primate (Cebus apella): distribution, morphology, and colocalization with cholinergic enzymes. *J. Comp. Neurol.* **277,** 465–486.

Kordower, J. H., Gash, D. M., Bothwell, M., Hersh, L. B., and Mufson, E. J. (1989) Nerve growth factor receptor and choline acetyltransferase remain colocalized in the nucleus basalis (Ch4) of Alzheimer's patients. *Neurobiol. Aging* **10,** 287–294.

Kordower, J. H., Le, H. K., and Mufson, E. J. (1992) Galanin-immunoreactivity in the primate central nervous system. *J. Comp. Neurol.* **319,** 479–500.

Kordower, J. H. and Mufson, E. J. (1990) Galanin-like immunoreactivity within the primate basal forebrain: differential staining patterns between humans and monkeys. *J. Comp. Neurol.* **294,** 281–292.

Kordower, J. H. and Mufson, E. J. (1992) NGF receptor immunoreactive neurons within the developing human cerebral cortex and hippocampal complex. *J. Comp. Neurol.* **323,** 25–41.

Kordower, J. H., Mufson, E. J., Granholm, A. C., Hoffer, B., and Friden, P. M. (1993) Delivery of trophic factors to the primate brain. *Exp. Neurol.* **124,** 21–30.

Kordower, J. H., Charles, V., Bayer, R., Bartus, R. T., Putney, S., Walus, L., and Friden, P. M. (1994a) Intravenous administration of an NGF conjugate prevents neuronal degeneration in a model of Huntington's disease. *Proc. Natl. Acad. Sci. USA* **91,** 9077–9080.

Kordower, J. H., Chen, E.-Y., Sladek Jr., J. R., and Mufson, E. J. (1994b) Trk-immunoreactivity in the monkey central nervous system I: forebrain. *J. Comp. Neurol.* **349,** 20–35.

Kordower, J. H., Winn, S. R., Liu, Y.-T., Mufson, E. J., Sladek, J. R., Hammang, J. P., Baetge, E. E., and Emerich, D. F. (1994c) The aged monkey basal forebrain: rescue and sprouting of degenerating cholinergic neurons following grafts of polymer-encapsulated human NGF secreting cells. *Proc. Natl. Acad. Sci. USA* **91,** 10,898–10,902.

Koster, G. R., Winkler, C., Raulf, F., Lottspeich, F., Schartl, M., and Thoenen, H. (1994) Neurotrophin-6 is a new member of the nerve growth family. *Nature* **372,** 266–269.

Kromer, L. F. (1987) Nerve growth factor treatment after brain injury prevents neuronal death. *Science* **235,** 214–216.

Lesauteur, L., Malirtchouk, S., Lejeune, H., Quirion, R., and Saragovi, H. U. (1996) Potent human p40-Trk agonists derived from an anti-receptor monoclonal antibody. *J. Neurosci.* **16,** 1308–1316.

Levi-Montalcini, R. and Angeletti, P. U. (1968) Nerve growth factor. *Physiol. Rev.* **8,** 534–569.

Li, J.-M, Kordower, J. H., and Mufson, E. J. (1996) Trk A, but not $p75^{NTR}$, gene expression is reduced within basal forebrain neurons in Alzheimer's disease. *Soc. Neurosci. Abstracts* **22,** 210.

Lindsay, R. M., Wiegand, S. J., Altar, C. A., and Distephano, P. S. (1994) Neurotrophic factors: from molecule to man. *Trends Neurosci.* **17,** 182–190.

Lucidi-Phillipi, C. A., Clary, D. O., Reichardt, L. F., and Gage, F. (1996) Trk A activation is sufficient to rescure axotomized cholinergic neurons. *Neuron.* **16,** 653–663.

Maisonpierre, P. C., Bellusco, L., Friedman, B., Alderson, R. F. (1990) NT-3, BDNF in the developing rat nervous system: parallel as well as reciprocal patterns of expression. *Neuron* **4,** 501–509.

Martinez, H. J., Dryfuss, C. F., Jonakait, G. M., and Black, I. B. (1985) Nerve growth factor promotes cholinergic development in brain striatal cultures. *Proc. Natl. Acad. Sci. USA* **82,** 7777–7781.

Markram, H. and Segal, M. (1990) Regional changes in NGF receptor immunohistochemical labeling in the septum of the aged rat. *Neurobiol. Aging* **11,** 481–484.

Meakin, S. O., and Shooter, E. M. (1991) Molecular investigation of the high-affinity nerve growth factor receptor. *Neuron* **6,** 153–163.

Melander, T., Hokfelt, T., and Rokaeus, A. (1986) Distribution of galanin-like immunoreactivity in the rat central nervous system. *J. Comp. Neurol.* **248,** 475–517.

Mesulam, M. M., Mufson, E. J., Levey, A. I., and Wainer, B. H. (1983) Cholinergic innervation of cortex by the basal forebrain: cytochemistry and cortical connections of the septal area, diagonal band nuclei, nucleus basalis (substantia innominata), and hypothalamus in the Rhesus monkey. *J. Comp. Neurol.* **214,** 170–197.

Mobley, W. C., Tennekoon, J. L., Buchannan, K., and Johnston, M. V. (1985) Choline acetyltransferase activity in the striatum of neonatal rats is increased by nerve growth factor. *Science* **229,** 284–287.

Mufson, E. J., Bothwell, M., and Kordower, J. H. (1989) Loss of nerve growth factor receptor-containing neurons in Alzheimer's disease: a quantitative analysis across subregions of the basal forebrain. *Exp. Neurol.* **105,** 221–232.

Mufson, E. J., Cochran, E., Benzing, W. C., and Kordower, J. H. (1993) Galaninergic innervation of the cholinergic vertical limb of the diagonal band (Ch2) and bed nucleus of the striia terminalis in aging, Alzheimer's disease and Down's syndrome. *Dementia* **4,** 237–250.

Mufson, E. J., Connor, J. M., and Kordower, J. H. (1995) NGF and Alzheimer's disease: defective retrograde transport to the nucleus basalis. *Neuroreport* **6,** 1063–1066.

Mufson, E. J., Connor, J., Varon, S., and Kordower, J. H. (1994) Nerve growth factor immunoreactivity in the basal forebrain and hippocampus in primates and Alzheimer's disease. *J. Comp. Neurol.* **341,** 507–519.

Mufson, E. J. and Kordower, J. H. (1992) Cortical neurons express the receptor for NGF in advanced age and Alzheimer's disease. *Proc. Natl. Acad. Sci. USA* **89,** 569–573.

Mufson, E. J., Li, J.-M., Sobreviela, T., and Kordower, J. H. (1996) Decreased trk A gene expression within basal forebrain neurons in Alzheimer's disease. *Neuro Report* **8,** 25–29.

Olson, L. (1993) NGF and the treatment of Alzheimer's disease. *Exp. Neurol.* **124,** 5–15.

Parada, L. F., Tsoulfas, P., Tessarollo, L., Blair, J., Reid, S. W., and Soppet, D. (1992) The Trk family of tyrosine kinases-receptors for NGF- related neurotrophins. *Cold Spring Harbor Symp. Quant. Biol.* **57,** 43–51.

Perry, E. K. (1980) The cholinergic in old age and Alzheimer's disease. *Age Ageing* **9,** 1–8.

Richardson, P. M., Verge Issa, V. M. K., and Riopelle, R. J. (1986) Distribution of neuronal receptor for nerve growth factor in the rat. *J. Neurosci.* **6,** 2312–2321.

Rodriguez-Tebar, A., Dechant, G., and Barde, Y.-A. (1990) Binding of brain-derived neurotrophic factor to the nerve growth factor receptor. *Neuron* **4,** 487–492.

Schumacher, J. M., Short, M. P., Hyman, B. T., Breakefield, X. O., and Isacson, O. (1991) Intracerebral implantation of nerve growth factor-producing fibroblasts protects striatum against neurotoxic levels of excitatory amino acids. *Neuroscience* **45,** 561–570.

Schwaab, M. E., Otten, U., Agid, Y., and Thoenen, H. (1979) Nerve growth factor in the rat CNS. Absence of specific retrograde transport and tyrosine hydroxylase induction in the locus coeruleus and substantia nigra. *Brain Res.* **168,** 473–483.

Scott, S. A. and Crutcher, K. A. (1994) Nerve grwoth factor and Alzheimer's disease. *Rev. Neurosci.* **5,** 179–211.

Scott, S. A., Mufson, E. J., Weingartner, J. A., Skau, K. A., and Crutcher, K. A. (1995) Nerve growth factor in Alzheimer's disease: increased levels throughout the brain coupled with declines in nucleus basalis. *J. Neurosci.* **15,** 6213–6221.

Scott, S. A., Weingartner, J. A., Liang, S., and Crutcher, K. A. (1994) Increased NGF-like activity in young but not aged rat hippocampus after septal lesions. *Neurobiol. Aging* **15,** 337–346.

Seiler, M. and Schwaab, M. E. (1984) Specific retrograde transport of nerve growth factor from neocortex to nucleus basalis in rat. *Brain Res.* **300,** 33–39.

Shelton, D. L., Sutherland, J., Gripp, J., Camerato, T., Armanini, M., Philips, H., et al. (1995) Human Trks-molecular cloning, tissue distribution and expression of extracellular domain immunoadhesions. *J. Neurosci.* **15,** 477–491.

Sobreviela, T., Clary, D. O., Reichardt, L. F., Brandabur, M. M., Kordower, J. H., and Mufson, E. J. (1995) Trk A immunoreactive profiles in the central nervous system: colocalization with neurons containing p75 nerve growth factor receptor, choline acetyltransferase and serotonin. *J. Comp. Neurol.* **350,** 587–611.

Sofroniew, M. V., Pearson, R. C. A., Eckenstein, F., Cuello, A. C., and Powell, T. P. S. (1983) Retrograde changes in cholinergic neurons in the basal forebrain of the rat following cortical damage. *Brain Res.* **289,** 370–374.

Sofroniew, M. V., Pearson, R. C. A., and Powell, T. P. S. (1987) The cholinergic nuclei of the basal forebrain of the rat: normal structure, development and experimentally induced degeneration. *Brain Res.* **371,** 310–331.

Soltoff, S. P., Rabin, S. L., Cnatley, L. C., and Kaplan, D. R. (1992) Nerve growth factor promotes the activation of phosphatidylinosito-3 kinase and its association with the Trk tyrosine-kinase. *J. Biol. Chem.* **267,** 17,472–17,477.

Springer, J. E., Koh, S. H., Tayrien, M. W., and Loy, R. (1987) Basal forebrain magnocellular neurons stain for nerve growth factor receptor: correlation with cholinergic cell bodies and effects of axotomy. *J. Neurosci. Res.* **17,** 111–118.

Steininger, T. L., Wainer, B. H., Klein, R., Barbicid, M., and Palfrey, H. C. (1993) High affinity nerve growth factor receptor (Trk) immunoreactivity is localized in cholinergic neurons of the basal forebrain and striatum in the adult rat brain. *Brain Res.* **612,** 330–335.

Strömberg, I., Wetmore, C. J., Ebendal, T., Ernfors, P., Persson, H., and Olson, L. (1990) Rescue of basal forebrain cholinergic neurons after implantation of genetically modified cells producing recombinant NGF. *J. Neurosci. Res.* **25,** 405–411.

Thoenen, H., and Barde, D. Y. A. (1980) Physiology of nerve growth factor. *Physiol. Rev.* **60,** 1284–1335.

Vogels, O. J. M., Broere, C. A. J., Ter Laak, H. J., Ten Donkelaar, H. J., Nieuwenhuys, R., and Schulte, B. P. M. (1990) Cell loss and shrinkage in the nucleus basalis Meynert complex in Alzheimer's disease. *Neurobiol. Aging* **11,** 3–13.

Weskamp, G. and Reichardt, L. F. (1991) Evidence that biological activity of NGF is mediated through a novel subclass of high affinity receptors. *Neuron* **6,** 649–663.

Whitehouse, P. J., Price, D. L., Struble, R. G., Clark, A. W., Coyle, J. T., and DeLong, M. R. (1983) Alzheimer's disease and senile dementia: loss of neurons in the basal forebrain. *Science* **215,** 1237–1239.

Whitemore, S. R. and Seiger, A. (1987) The expression, localization and functional significance of B-nerve growth factor in the central system. *Brain Res. Rev.* **12,** 439–464.

Will, L. R. and Hefti, F. (1985) Behavioral and neurochemical effects of chronic intraventricular injections of nerve growth factor in adult rats with fimbrial lesions. *Behav. Brain Res.* **17,** 17–24.

Williams, L. R., Varon, S., Peterson, G. M., Wictorian, K., Fisher, W., Björklund, A., and Gage, F. H. (1986) Continuous infusion of nerve growth factor prevents basal forebrain neuronal cell death after fimbria-fornix transection. *Proc. Natl. Acad. Sci. USA* **83,** 9231–9236.

Winn, S. R., Linder, M. D., Lee, A., Haggett, G., Francis, J. M., and Emerich, D. F. (1996) Polymer-encapsulated genetically modified cells continue to secrete human nerve growth factor for over one year in rat ventricles: behavioral and anatomical consequences. *Exper. Neurol.* **140,** 126–138.

Yan, Q. and Johnson Jr., E. M. (1988) An immunohistochemical study of the nerve growth factor receptor in developing rats. *J. Neurosci. 8,* 3481–3498.

3

Molecular Mechanisms of Neuroprotection from Neuronal Death by Trophic Factor Deprivation

Thomas L. Deckwerth

1. INTRODUCTION

Trophic factors and other neurotrophic agents control neuronal survival during the different stages of nervous system development, with some neuronal populations losing >50% of all born neurons as a result of diverse developmental mechanisms (reviewed by Oppenheim, 1991). These mechanisms include the control of survival during precursor proliferation and differentiation through afferent innervation, and after target innervation by limited amounts of target-derived trophic factors.

A limited number of neuronal types has been favored for the study of the molecular mechanisms of death by trophic-factor deprivation. Cultures of peripheral neurons, such as dorsal root ganglion (DRG) and trigeminal sensory, sympathetic, and parasympathetic ciliary neurons (reviewed by Oppenheim, 1991; Johnson and Deckwerth, 1993), and some central nervous system (CNS) neurons, such as cerebellar granule cells (e.g., D'Mello et al., 1993), are easily maintained because of simple and well-defined trophic factor requirements, and are induced to die on withdrawal of trophic support. These cultured neurons can be genetically manipulated by microinjection of expression vectors (e.g., Garcia et al., 1992). In vivo, the well-defined anatomical location and morphological appearance of peripheral nervous system (PNS) neurons (e.g., Snider, 1994), such as facial and spinal motor neurons (e.g., Sendtner et al., 1996), and second-order acoustic neurons (Rubel et al., 1990) facilitates their quantitative analysis. Their target-derived trophic support can be discontinued by axotomy (Oppenheim, 1991), or by using neurons from mice deficient in trophic factors, their receptors (reviewed by Snider, 1994), or the target tissue itself (e.g., Herrup and Busser, 1995). Most recently, cell-death pathways are being analyzed by applying these experimental systems to mice with mutations in cell-death pathway components (e.g., Martinou et al., 1994; Deckwerth et al., 1996). Finally, cell lines, such as PC12 cells, which differentiate after nerve growth factor (NGF) exposure into sympathetic-like neurons, can be genetically altered by selection for spontaneous mutants or transfection with expression vectors (e.g., Loeb and Greene, 1993), and may be induced to die by serum deprivation when undifferentiated, or by NGF deprivation after differentiation (e.g., Rukenstein et al., 1991; Troy et al., 1996).

From: Neuroprotective Signal Transduction *edited by M. P. Mattson*
Humana Press Inc., Totowa, NJ

Several distinct features are common to neuronal death by trophic factor deprivation. Removal of trophic support initiates a sequence of degenerative events that starts with changes in signal transduction, reductions in the metabolic rates of glucose uptake and macromolecular synthesis, and the induction of a stress response. An almost invariant feature is that death depends on ongoing macromolecular synthesis and is modulated by Bcl-2 family members. Considerable neuronal atrophy occurs prior to the onset of cell death. The terminal destruction of the cell is initiated by activation of members of the ICE family of cysteine proteases, which irreversibly cleave cytoplasmic and nuclear substrates, and is accompanied by the disintegration of the neurites. In most neurons, apoptotic features, such as DNA fragmentation and chromatin condensation, accompany this terminal phase and lead to death. The whole degenerative process, as well as genetic and pharmacological means of intervention, are described in detail in the following chapters.

2. TROPHIC FACTOR RECEPTORS INVOLVED IN NEURONAL SURVIVAL AND DEATH

Initially trophic factors promote survival and suppress neuronal degeneration by binding to and activating receptors on the neuronal surface. In contrast, the dissociation of the ligand inactivates the receptors, stops survival promotion, and induces degenerative events that ultimately trigger death. In most neurons, the receptors for neurotrophins, GDNF, ciliary neurotrophic factor (CNTF), and leukemia inhibitory factor (LIF) relay the presence or absence of trophic factors to the signal transduction cascades in control of survival.

2.1. Signaling of Neuronal Survival Through Neurotrophin Trk Receptors

The three Trk receptors, Trk A, Trk B, and Trk C, bind specifically to the neurotrophins NGF, brain-derived neurotrophic factor (BDNF) and NT-4, and NT-3, respectively (reviewed by Bothwell, 1995). They are required and sufficient for most of the survival responses of neurons to neurotrophins. The massive neuronal death seen in three Trk-deficient mouse strains during development dramatically illustrates the necessity of Trk receptors to neuronal health (reviewed by Snider, 1994). Likewise, naive PC12 cells deficient in Trk A lose all responses to NGF, including the response following serum deprivation that usually promotes survival; on the other hand, transfection of Trk A into these cells restores the ability of NGF to promote survival (Loeb and Greene, 1993).

Binding of the NGF homodimer to the Trk A receptor crosslinks two Trk A proteins and activates the cytoplasmic tyrosine kinase domains to transphosphorylate each other on tyrosine residues. Trk A dimerization is indispensible for survival in response to NGF, since monovalent anti-Trk A Fab fragments, which block NGF binding, prevent tyrosine autophosphorylation and survival of sympathetic neurons in the presence of NGF (Clary et al., 1994). Conversely, divalent anti-Trk A IgG crosslinks Trk A monomers, induces tyrosine phosphorylation, and mimicks NGF by supporting the survival of sympathetic neurons. Constitutive tyrosine kinase activation and Trk A autophosphorylation appear to be critical for survival, since inhibition of the Trk A tyrosine kinase by the alkaloid K252a disables autophosphorylation and prevents NGF from protecting serum-deprived PC12 cells (Rukenstein et al., 1991). Conversely, stimulation of the Trk A tyrosine kinase with the ganglioside GM1 promotes autophosphorylation and supports the survival of serum-deprived PC12 cells in the absence of NGF (Ferrari et al., 1995a). Finally, Trk A remains constitutively autophosphorylated in the presence of NGF in sympathetic neurons (Franklin et al., 1995; Nobes et al., 1996) and becomes completely dephosphorylated within 2 h after NGF removal, with neuronal death following 24–48 h later.

2.2. *Multiple Roles of the p75 Receptor in Neuronal Survival and Death*

In contrast to the Trk receptors, the p75 receptor binds all neurotrophins equally well and its effects on neuronal survival are more subtle (reviewed by Bothwell, 1995). The best-established roles for p75 are its ability to potentiate Trk receptor activation when the ligand is present in limiting amounts and to enhance the selectivity of Trk receptors for their preferred ligands. Survival of cultured p75-deficient sensory and sympathetic neurons or of wild-type DRG sensory neurons exposed to p75-antisense oligonucleotides is slightly reduced during the period of target-dependent neuronal death, when the concentration of NGF is limiting (Lee et al., 1994; Barrett and Bartlett, 1994). p75 potentiates neuronal survival by increasing Trk A autophosphorylation, an effect that does not require the presence of the cytoplasmic domain of p75 (Hantzopoulos et al., 1994). These results leave open whether p75 acts by concentrating neurotrophins at the cell surface, by promoting neurotrophin-binding to Trk, or by facilitating Trk-dimerization. In addition to potentiating Trk autophosphorylation in response to their preferred neurotrophin, p75 diminishes the ability of Trk receptors to respond to their nonpreferred ligand. p75-deficient SCG neurons show increased survival at limiting NT-3 concentrations (Lee et al., 1994), and PC12 cells with reduced p75 levels respond to NT-3 with increased neurite outgrowth (Benedetti et al., 1993). The enhanced selectivity of Trk receptors for their preferred ligands may serve to improve the quality of the pairwise matches between a single target and multiple innervating neuronal populations responsive to different neurotrophins.

A less well-established role of p75 is its ability to actively promote neuronal death in the absence of neurotrophins. The p75 receptor belongs to the Fas/TNFR family of cytokine receptors, which share limited sequence homology within the "death domain," a cytoplasmic sequence required for the ligand-dependent cytotoxicity of Fas and TNFR1 (reviewed by Boldin et al., 1995; Nagata and Golstein, 1995). Two observations suggest that p75 may act in the opposite way, that is, unligated p75 may exert a constitutive death-promoting activity, which is suppressed on binding NGF. First, during the period when neurons are losing their survival dependence for target-derived trophic factors, the survival of NGF-deprived DRG neurons improves when neuronal p75 expression is reduced by antisense oligonucleotides (Barrett and Bartlett, 1994). Second, expression of p75 in a Trk A- and p75-deficient neural cell line enhances serum deprivation-induced death in the absence of NGF; exposure to NGF not only reduces the kinetics of serum deprivation-induced neuronal death in p75-expressing cells, but even enhances their survival relative to p75-deficient cells (Rabizadeh et al., 1993). This result is consistent with the promotion of neuronal death by unligated p75, and in addition raises the possibility that binding of NGF to p75 may actually produce a survival-promoting signal as well.

Besides the latter observation, only circumstantial evidence indicates that p75 alone might be able to promote neuronal survival after binding of neurotrophins. Binding of NGF and NT-3 to p75 induces the transient hydrolysis of sphingomyelin in p75-expressing T9 glioblastoma cells, a response that is mediated by the cytoplasmic domain of p75 (Dobrowsky et al., 1994). Trk A blocks signaling by p75, since simultaneous activation of Trk A autophosphorylation by NGF inhibits p75-mediated sphingomyelin hydrolysis in these cells (Dobrowsky et al., 1995). Although no evidence yet shows that these signaling pathways function in neuronal cells, the ability of NGF to induce sphingomyelin hydrolysis warrants further attention, since ceramide, the product of sphingomyelin hydrolysis, can be neuroprotective and does delay the death of sympathetic neurons after NGF deprivation (Ito and Horigome, 1995; Goodman and Mattson, 1996). It remains to be seen whether activation of p75 by NGF can be shown to promote neuronal survival directly by inducing ceramide production. Such a pathway could be relevant for neuronal survival in experimental and clinical situations in which p75-expressing neurons that lack expression of a particular Trk receptor are exposed to pharmacological doses of the neurotrophin preferred by that receptor.

2.3. Signaling of Neuronal Survival and Death by the Tripartite CNTF Receptor

The tripartite CNTF receptor mediates the cellular responses to CNTF, LIF, and probably to at least one unknown ligand, and may induce both survival and death-promoting pathways in neurons. The cytokines LIF and CNTF activate the two signaling subunits LIFRb and gp130, two transmembrane proteins that dimerize on ligand binding (reviewed by Stahl and Yancopoulos, 1994). Although LIF binds both receptor subunits directly, only a complex between CNTF and the glycosyl phosphatidylinositol (GPI)-anchored extracellular membrane protein CNTFRα may activate LIRFβ and gp130. Mice deficient in CNTFRα or LIFRβ show widespread motor neuron loss at birth, suggesting that these receptor subunits are required for maintaining the viability of motor neurons during development (DeChiara et al., 1995; Li et al., 1995). The physiological ligand that activates the two receptor subunits must be different from CNTF and LIF, since mice deficient in either one or both factors do not show any motor neuron dysfunction at birth (Sendtner et al., 1996). LIFRβ and gp130 utilize the Jak–STAT pathway for signaling, but no studies have yet identified individual Jaks and STATs required for neuronal survival.

In sympathetic neurons, CNTF and LIF may act as cytotoxic or neurotrophic factors, depending on the developmental stage. CNTF and LIF are poor neurotrophic factors for freshly dissociated neonatal rat SCG neurons (Burnham et al., 1994), and they induce apoptosis, with pharmacological characteristics similar to NGF deprivation in NGF-maintained neurons (Kessler et al., 1993; Burnham et al., 1994). During the first postnatal week, the neurotrophic potency of CNTF and LIF increases, and their cytotoxicity decreases, so that most NGF-deprived SCG neurons of P6 rats survive with LIF or CNTF for 3 d, and hardly any death of NGF-maintained neurons occurs in the presence of these factors any longer (Kotzbauer et al., 1994). The molecular basis for the toxicity of CNTF and LIF to neonatal cultures is unknown, except for the demonstration that the treatment of sympathetic neurons with phosphatidyl inositol phospholipase C (PI-PLC) diminishes CNTF toxicity, presumably by releasing CNTFRα from the neuronal surface (Kessler et al., 1993). The similarities between death by NGF deprivation and by CNTF/LIF exposure raise the possibility that signaling from the CNTF receptor complex directly interferes with the survival-promoting signaling from Trk A receptors, so that the cell effectively experiences NGF deprivation. Changes in receptor subunit transcription do not occur during this switch from cytotoxicity to neurotrophism, raising the possibility that developmental changes in downstream signaling pathways may account for the switch (Kotzbauer et al., 1994). Whether the switch has a physiological role is unclear; in contrast to the motor nuclei, the SCG of neonatal CNTFRα-deficient mice appear normal by histology and volumetric analysis, indicating that signaling through this receptor subunit alone is not critical for the survival of sympathetic neurons during development (DeChiara et al., 1995).

2.4. GDNF, Neurturin, and the GDNF Receptor

The GDNF receptor consists of two different subunits and mediates the survival response of neurons to GDNF (Durbec et al., 1996; Jing et al., 1996; Treanor et al., 1996; Trupp et al., 1996) and, most likely, to other related trophic factors. The receptor tyrosine kinase Ret is the signaling subunit capable of dimerization and autophosphorylation. It does not seem to bind GDNF directly, however, but is activated by GDNF only in a complex with GDNFRα, an extracellular protein tethered to the outer lipid bilayer by a GPI anchor. GDNF requires GDNFRα for neuronal survival, since treatment of sensory, sympathetic, dopaminergic, and motor neurons with PI-PLC, which releases GPI-anchored proteins such as GDNFRα, diminishes the neurotrophic potency of GDNF (Treanor et al., 1996). The most definitive data to date for a role of Ret in neuronal survival remain the observations that Ret-deficient mice lose all sympathetic precursors in the SCG anlage between E10.5 and E12.5 (Durbec et al., 1996), and that GDNF induces Ret autophosphorylation in cultures of embryonic spinal cord motor neurons, whose survival is supported by GDNF (Jing et al., 1996).

Recently, the GDNF homolog neurturin (NTN) has been purified and cloned, based on its ability to support the survival of NGF-deprived rat sympathetic neurons (Kotzbauer et al., 1996). In addition, NTN also promotes the survival of embryonic BDNF-deprived nodose and NGF-deprived DRG neurons in culture. Although its receptor has not been identified, exposure of NGF-deprived sympathetic neurons to NTN activates the MAP kinases ERK1 and ERK2, as do GDNF and NGF, suggesting that these factors share similar signaling pathways.

3. SIGNAL TRANSDUCTION REGULATION OF NEURONAL SURVIVAL WITH AND WITHOUT TROPHIC FACTORS

Signal transduction by trophic factor receptors has been reviewed recently in great detail (e.g., Segal and Greenberg, 1996). This chapter focuses solely on those aspects of signal transduction that are critical for neuronal survival and death. These signaling cascades can be specific for different receptors and neuronal types, which may prevent any crosstalk among incompatible components. This has been nicely demonstrated by Allsopp et al. (1993a), who showed that NGF does not promote the survival of CNTF-deprived ciliary neurons expressing Trk A and p75; ectopic expression of Trk A allowed BDNF-deprived trigeminal mesencephalic neurons to survive with NGF. Furthermore, although sympathetic neurons continue to express p75 and Trk A receptors beyond the period of developmental neuronal death, and remain able to respond to NGF withdrawal with soma atrophy and reduction in neuritic arborization, NGF withdrawal fails to induce neuronal death (Koike and Tanaka, 1991), suggesting that signal transduction events specific for triggering neuronal death by trophic factor deprivation can be developmentally regulated.

3.1. Ras and Neurofibromin-1

Increasing evidence suggests that activated Ras, a plasma membrane-bound guanine nucleotide-binding protein, constitutes an important mediator of the survival response of peripheral neurons to neurotrophins, CNTF, LIF, cAMP, and elevated cytosolic calcium. On activation by NGF, the Trk A tyrosine kinase phosphorylates the adapter protein Shc, which forms a trimeric complex with the adapter protein Grb2 and the guanine nucleotide exchange factor (GEF) SOS. SOS activates Ras by catalyzing the exchange of GDP for GTP bound to Ras. Treatment of DRG neurons and PC12 cells with NGF or sympathetic neurons with NGF, LIF, or CNTF increases the cellular levels of activated (GTP-bound) Ras protein (Ng and Shooter, 1993; Schwarzschild et al., 1994; Carter et al., 1995). Conversely, levels of activated Ras decline by 75% within 3 h after removal of NGF from differentiated PC12 cells (Mesner et al., 1995). Constitutively active oncogenic Ras protein, when introduced into freshly dissociated chicken DRG, nodose, or ciliary neurons by trituration, or when overexpressed in PC12 cells, enhances neuronal survival in the absence of NGF, BDNF, CNTF, or serum, respectively (Borasio et al., 1989; Rukenstein et al., 1991). Likewise, rat sympathetic neurons are supported by oncogenic Ras (Nobes et al., 1996). Conversely, neutralization of endogenous Ras by trituration with anti-Ras antibodies blocks the survival of chicken DRG and nodose neurons in the presence of NGF or BDNF, respectively (Borasio et al., 1993), and diminishes the survival of rat sympathetic neurons maintained by NGF, CNTF, or LIF (Nobes and Tolkovsky, 1995).

Although these observations support a role for Ras in the mediation of neuronal survival by trophic factors, a number of reports failed to observe this effect of Ras under their experimental conditions. For example, overexpression of dominant negative Ras did not diminish the protection of serum-deprived PC12 cells by NGF (Yao and Cooper, 1995), and introduction of oncogenic Ras or of an anti-Ras antibody into chicken sympathetic neurons failed to replicate the observations in sensory neurons (Borasio et al., 1993). These observations remain unexplained, but all other data suggest that Ras constitutes a major point of convergence of the pathways by which neurotrophins, CNTF, and LIF exert their survival-promoting effects on peripheral neurons.

In addition to being activated by oncogenic mutations or by the GEF SOS, levels of GTP-bound Ras may be increased by inactivating GTPase-activating proteins (GAP). Mice deficient for the gene of the GAP neurofibromin-1 (NF-1) display hyperplasia of the sympathetic ganglia, and neurons from their DRG, nodose, and trigeminal ganglia are largely independent of NGF and BDNF for survival at E12.5–13.5 (Vogel et al., 1995). NF-1-deficient sympathetic neuroblasts can be maintained in vitro in the absence of NGF for weeks and never become dependent on NGF for survival. These results suggest that adequate GAP activity by NF-1 is a prerequisite for the ability of trophic factors to control neuronal survival by regulating the levels of activated Ras.

3.2. MAP Kinase (ERK) Pathway

A well-studied signaling cascade initiated by various growth factors leads from activated Ras via the serine–threonine kinase Raf and the dual-specificity MAPK/ERK kinases (MEK) MKK1 and MKK2 to the serine–threonine MAP kinases ERK1 and ERK2. The MAP kinases are close to the end of this signaling cascade, since they can translocate to the nucleus and phosphorylate transcription factors involved in the regulation of immediate early and delayed response genes. MAP kinase activity is stimulated by NGF in PC12 cells and sympathetic neurons, and is maintained elevated as long as the cells are exposed to NGF (Virdee and Tolkovsky, 1995; Xia et al., 1995; Yao and Cooper; 1995; Creedon et al., 1996). Removal of NGF from sympathetic neurons reduces MAP kinase activity to <25% within 2 h. Whether the survival response of PC12 cells to NGF requires MAP kinase activation may depend on the differentiation state in PC12 cells. Transient overexpression of a constitutively active mutant of the MEK MKK1, which would be expected to activate MAP kinase, retards death of differentiated PC12 cells after NGF withdrawal (Xia et al., 1995); expression of dominant negative Ras in naive PC12 cells prevented MAP kinase activation, but did not affect NGF protection of serum-deprived PC12 cells (Yao and Cooper, 1995). In contrast to these potentially conflicting results in PC12 cells, all available data argue against a role for MAP kinase activation in sympathetic neuronal survival. Treatment of NGF-maintained sympathetic neurons with the MEK inhibitor PD 098059 inhibits MAP kinase activity, but does not diminish survival (Creedon et al., 1996). Furthermore, the correlation between MAP kinase activity and sympathetic neuronal survival is poor for agents other than NGF. Although maintaining neuronal survival, LIF and CNTF stimulate MAP kinase activity only transiently, and exposure to the cAMP analog CPTcAMP does not stimulate MAP kinase at all (Virdee and Tolkovsky, 1995; Creedon et al., 1996).

3.3. *PI-3 Kinase Pathway*

Stimulation of Trk A-expressing cells activates phosphatidylinositol-3 kinase (PI-3K) (reviewed by Rodriguez-Viciana et al., 1996). Its 85 kDa regulatory subunit interacts with phosphotyrosine residues on Trk A and its 110 kDa catalytic subunit may bind to Ras. Whether PI-3K acts upstream or downstream of Ras is currently controversial. PI-3K has been proposed to be involved in NGF-mediated survival, since the PI-3K inhibitors wortmannin and LY 294002 kill serum-deprived PC12 cells maintained alive by NGF (Yao and Cooper, 1995). In addition, transfection of PC12 cells with the PDGF receptor stimulates PI-3K activity and rescues PC12 cells from serum deprivation on PDGF treatment, but a mutant receptor unable to bind PI-3K is inactive. It remains to be seen whether these results extend to trophic factor-dependent neurons.

3.4. *Stress-Activated Jun Kinase/p38 MAP Kinase Pathways*

Exposure of cells to stress, such as UV irradiation, oxidants, or glucose deprivation, may activate the Jun kinase pathway (reviewed by Woodgett et al., 1996), whose terminal kinase cascade consists of the Raf homolog MEKK1, the MEK homolog MKK4, and the MAP kinase homolog Jun kinase (JNK, SAPK). Parallel to MKK4 and JNK, there exists another stress-activated pathway consisting of the MEK homolog MKK3 and p38 MAP kinase. JNK and p38 activate the AP-1 transcription factor components c-Jun and ATF-2 by phosphorylation of serines and threonines located in the N-terminal transactivation domains. Activated c-Jun enhances its own transcription greatly by an amplifying feed-forward loop, which may increase levels of AP-1 and enhance the transcription of AP-1-responsive genes.

Like the stressors mentioned above, withdrawal of trophic support activates the Jun kinase and p38 pathways in trophic factor-dependent neurons. NGF-deprivation induces c-Jun transcription in neonatal sympathetic neurons and differentiated PC12 cells, consistent with the initiation of a stress response (Estus et al., 1994; Mesner et al., 1995). A similar increase is seen in cerebellar granule neurons after combined potassium and serum deprivation (Miller and Johnson, 1996), and in apoptotic cells during naturally occurring cell death (Ferrer et al., 1996). Microinjection of neutralizing anti-c-Jun antibodies or expression of a dominant negative c-Jun mutant retards NGF-deprivation-induced death of sympathetic neurons, indicating that c-Jun protein needs to be synthesized and form transcriptionally active AP-1 complexes in order to promote neuronal death (Estus et al., 1994; Ham et al., 1995). Consistent with a proapoptotic role for c-Jun, overexpression of native c-Jun alone induces apoptosis in NGF-maintained sympathetic neurons. This requirement for transcription and translation of c-Jun explains, at least in part, why inhibitors of macromolecular synthesis protect NGF-deprived sympathetic neurons.

Some of the events leading to increased c-Jun transcription have been examined in differentiated PC12 cells deprived of NGF (Xia et al., 1995; Park et al., 1996). JNK and p38 MAP kinase activity increase greatly after NGF deprivation; when NGF-deprived PC12 cells are treated with agents, such as forskolin, insulin, or bFGF, that prevent their death, the increase in JNK and p38 activity is also prevented. Likewise, in sympathetic neurons, NGF deprivation increases nuclear levels of c-Jun protein phosphorylated in its transactivation domain (Deckwerth, unpublished). Taken together, these data demonstrate that trophic factor deprivation induces a stress response that activates the transcription and leads to the synthesis of active c-Jun protein in dependent neurons. Further studies are required to link this signaling segment to upstream and downstream events, such as inactivated trophic factor receptors and the activation of other genes required for the induction of apoptosis, respectively.

4. REGULATION OF NEURONAL SURVIVAL BY TROPHIC AGENTS OTHER THAN TROPHIC FACTORS

4.1. Control of Neuronal Survival by Cytoplasmic Calcium Ions

Modest elevation of cytoplasmic calcium ion concentrations allows many types of PNS and CNS neurons to survive in the absence of trophic factors, including sympathetic and cerebellar granule neurons (reviewed by Franklin and Johnson, 1992). According to the calcium set-point hypothesis, at a narrow range of slightly elevated cytoplasmic calcium concentrations, neurons are independent of trophic factors for survival. Elevation of cytoplasmic calcium may be achieved by a number of different pharmacological manipulations, e.g., by depolarization of the plasma membrane with elevated extracellular potassium ion concentrations, which causes calcium influx through L-type calcium channels, or by activating NMDA receptors. Inhibition of the calcium-sequestering calcium ATPase of the endoplasmatic reticulum by thapsigargin raises cytoplasmic calcium and protects trophic factor-deprived sympathetic and potassium-deprived cerebellar granule neurons (Lampe et al., 1995; Levick et al., 1995). The physiological basis for the ability of membrane-depolarizing stimuli to elevate cytoplasmic calcium levels and promote survival is not clear for most neurons, but for at least some may be related to the ability of afferent innervation to control numbers of target neurons by means of electrical excitation (reviewed by Oppenheim, 1991).

Several pieces of evidence suggest that the molecular mechanisms of death following reduction of elevated calcium levels and deprivation of trophic factor are similar. Combined potassium and serum deprivation of cerebellar granule neurons, maintained with elevated potassium, reduces levels of cytoplasmic calcium; it also induces neuronal death, which is apoptotic, depends on ongoing macromolecular synthesis, and can be inhibited by trophic agents, such as insulin-like growth factor (IGF-1) and CPTcAMP (D'Mello et al., 1993). Furthermore, the changes in energy and macromolecular metabolism and the induction of the stress response are reminiscent of those in NGF-deprived sympathetic neurons (Miller and Johnson, 1996; *see* Section 5.). The similarity of the responses suggests that the signaling pathways activated by NGF and elevated cytoplasmic calcium may converge rather quickly on a common survival-promoting signal. This is confirmed by the observation that both NGF and increased cytoplasmic calcium act at the same time after onset of NGF withdrawal to halt degeneration of sympathetic neurons (Franklin et al., 1995). This common signal is located downstream of the activation of the Trk A receptor; it is also identical to or located upstream of Ras, since potassium depolarization fails to induce Trk A autophosphorylation, but does activate Ras and the MAP kinase pathway (Rosen et al., 1994; Franklin et al., 1995). Four different signaling cascades have been described for PC12 cells and cortical neurons that may link elevated cytoplasmic calcium to Ras activation (Finkbeiner and Greenberg, 1996). None of these pathways, however, have yet identified the calcium sensor responsible for activation of these signaling cascades.

4.2. *Promotion of Survival by Elevated Levels of cAMP*

Agents that elevate levels of cAMP may substitute for trophic factors and protect many PNS and CNS neurons from apoptosis. cAMP is a second messenger generated by adenylate cyclase in response to the activation of G-protein-coupled receptors and activates cAMP-dependent protein kinase (PKA). For example, activation of PKA by elevated cAMP levels protects sympathetic neurons from death by NGF deprivation (Rydel and Greene, 1988), PC12 cells from serum deprivation (Rukenstein et al., 1991), and cerebellar granule neurons from potassium–serum deprivation (D'Mello et al., 1993).

Comparison of the time-courses of rescue of NGF-deprived sympathetic neurons with NGF and cAMP indicates that both NGF and the elevation of cAMP act at similar times after onset of NGF withdrawal to halt neuronal degeneration (Edwards et al., 1991; Deckwerth and Johnson, 1993). This suggests that the signaling pathways of NGF and cAMP may converge on a common survival-promoting signal. Similar to cytoplasmic calcium, this signal may be identical to Ras, since introduction of an anti-Ras antibody into sympathetic neurons by trituration interferes with the survival-promoting effect of cAMP (Nobes and Tolkovsky, 1995).

Although the simplest interpretation of these data suggests that cAMP may act independent of other trophic support in sympathetic neurons, it is possible that, rather than generating a survival-promoting signal on its own, cAMP actually acts as a gate to amplify a weak constitutive neuroprotective signal above a critical survival-ensuring threshold (Iyengar, 1996). Gating by cAMP has been demonstrated for cultures of retinal ganglion cells, whose marginal survival by a cocktail of trophic factors is greatly enhanced by cAMP-elevating agents (Meyer-Franke et al., 1995).

4.3. *Signaling by Reactive Oxygen Species and Survival Promotion by Superoxide Dismutase*

In addition to being under the control of classic second messenger systems, such as calcium ions and cAMP, neuronal death by trophic factor deprivation is also modulated by agents that change the cellular redox equilibrium. The redox equilibrium is controlled by enzymes that detoxify reactive oxygen species (ROS), such as Cu, Zn-superoxide dismutase (SOD), catalase, and glutathione peroxidase, and by small antioxidant molecules, such as glutathione, ascorbate, and vitamin A (reviewed by Halliwell and Gutteridge, 1989).

Three observations suggest that ROS may possess a signaling role in the induction of neuronal apoptosis by trophic factor deprivation. First, NGF increases the expression and activity of several antioxidant enzymes and increases glutathione levels in PC12 cells, indicating that NGF may strengthen neuronal antioxidant defense systems and thus prevent ROS accumulation (Pan and Perez-Polo, 1993). Second, NGF deprivation induces a brief increase in ROS in sympathetic neurons, as measured with a redox-sensitive fluorescein derivative (Greenlund et al., 1995a). This increase in ROS peaks after 3 h and diminishes thereafter. Third, glutathione levels increase by 50% after 12 h of NGF deprivation, reminiscent of the common cellular response to mild oxidative stress, which can also be observed after exposing sympathetic neurons to the pro-oxidant tert-butyl hydroxyanisole (Deckwerth and Johnson, 1995).

The transient ROS signal appears to be important for the progression of the degenerative process, because its modulation by pro- and antioxidant treatments affects the rate of induction of apoptosis. Microinjection of SOD protein or expression of SOD from a microinjected expression vector or SOD-expressing adenovirus-retarded sympathetic neuronal death after NGF deprivation, but only if SOD protein was injected before the transient increase in ROS levels (Greenlund et al., 1995a; Jordan et al., 1995). Conversely, expression of antisense SOD mRNA accelerated neuronal death by NGF deprivation. Taken together, these data suggest that the levels of endogeneous SOD protein may regulate the ROS signal and thereby control the induction of apoptosis. Whether the ROS signal is involved in the induction of the stress-activated Jun kinase pathway and the decline in metabolic rates (*see* Section 5.) is not known.

4.4. *Survival Promotion by N-Acetyl* L*-Cysteine and Other Sulfhydryl Reagents*

Extensive surveys of different antioxidants revealed that millimolar concentrations of the glutathione-elevating agent *N*-acetyl L-cysteine (NAC) retard the death of NGF-deprived sensory and sympathetic neurons, as well as of PC12 cells (Mayer and Noble, 1994; Deckwerth and Johnson, 1995; Ferrari et al., 1995b). Although NAC may be deacetylated to cysteine and converted to glutathione, the five- to 10-fold increase of glutathione levels induced by NAC in sympathetic neurons and PC12 cells is caused by a different mechanism, since the D-isomer of NAC produces the same effect (Deckwerth and Johnson, 1995; Ferrari et al., 1995b). Furthermore, the promotion of survival of NGF-deprived sympathetic neurons and PC12 cells is not mediated through the increase in glutathione, since both isomers of NAC promote neuronal survival in the presence of the γ-glutamyl synthetase inhibitor buthionine sulfoximine (Deckwerth and Johnson, 1995; Yan et al., 1995). Only the reduced sulfhydryl group is required for the neuroprotective effect, since sulfhydryl compounds other than NAC are neuroprotective as well, and oxidation of the sulfhydryl group to a disulfide or substitution with a hydroxyl group abolishes neuroprotection (Deckwerth and Johnson, 1995). Protection of serum-deprived PC12 by NAC is prevented by inhibitors of RNA synthesis, suggesting that NAC maintains or induces the transcription of survival-promoting genes (Yan et al., 1995). The neuroprotective effect of NAC seems to be unrelated to its antioxidant properties, however, since most antioxidants other than sulfhydryl reagents fail to prevent death by trophic factor deprivation (Ferrari et al., 1995b).

4.5. Similarities Between the Control of the Cell Cycle and Neuronal Apoptosis

Several recent studies have explored the hypothesis that the control of G1/S-phase transition of dividing cells and apoptosis in differentiated postmitotic neurons is regulated in a similar way (Heintz, 1993). Among the most readily testable predictions of this hypothesis are that agents that promote the induction of S-phase should activate neuronal death, and that cell cycle inhibitors should be neuroprotective. For example, the induction of c-Jun by NGF deprivation and the prevention of neuronal death by the inactivation of c-Jun with neutralizing antibodies is parallel to the c-Jun induction during G1/S transition of mitogen-treated dividing cells and the G1 arrest induced by anti-c-Jun antibodies, respectively (Kovary and Bravo, 1991; Estus et al., 1994). Expression of dominant-negative Ras inhibits both PC12 cell proliferation and the induction of apoptosis in serum-deprived naive or NGF-deprived differentiated PC12 cells (Ferrari and Greene, 1994). This result suggests that Ras has a dual role in promoting both the induction of apoptosis and mediating the survival response of trophic factors (*see* Section 3.1.). The similarities between the apoptotic and mitotic control mechanisms seem to be confined to the G1 phase, since cyclin-dependent kinase (CDK)-inhibitors and G1/S transition blockers, but not inhibitors of S, M, and G2 phases, inhibit death of PC12 cells and sympathetic neurons after NGF deprivation (Farinelli and Greene, 1996; Park et al., 1996).

The strongest evidence for a role of cell cycle regulators in the control of neuronal apoptosis is provided by the observation that NGF deprivation increases the steady-state mRNA levels of the G1-cyclin D1 in sympathetic neurons about 10-fold after 15-h (Freeman et al., 1994). A similar induction of cyclin D1 is seen in cerebellar granule cells dying after their synaptic target, the Purkinje cell, has died in lurcher or staggerer mutant mice (Herrup and Busser, 1995). An induction of cyclin D1 has also been observed in differentiated neuroblastoma cells dying after serum withdrawal (Kranenburg et al., 1996). Kranenburg and colleagues also showed that overexpression of cyclin D1 could induce apoptosis on its own, that serum withdrawal led to the activation of the cyclin D1-dependent kinase CDK4, and that the transient expression of the CDK-inhibitor p16 prevented death by serum deprivation. These results suggest that cyclin D1 may have a pivotal role in mediating neuronal death by trophic factor deprivation. Studies using cyclin D1-deficient animals (Sicinski et al., 1995) should provide a rigorous assessment of the importance of cyclin D1 in the different models of neuronal death by trophic factor deprivation.

4.6. Dependence of Neuronal Degeneration on Ongoing Macromolecular Synthesis

Although inhibitors of protein and RNA synthesis are potent inhibitors of neuronal death by trophic factor deprivation in vivo and in vitro (reviewed by Johnson and Deckwerth, 1993) and of cerebellar granule neuron death by potassium–serum deprivation (D'Mello et al., 1993), the molecular mechanism of neuroprotection has been controversial. Two new hypotheses and sets of results warrant a reassessment, since the last extensive discussion of potential mechanisms (Johnson and Deckwerth, 1993).

First, Ratan et al. (1994) demonstrated that inhibition of protein synthesis by cycloheximide supports glutathione levels in cortical neurons, which protected these neurons from oxidative stress caused by the inhibition of cystine uptake. They suggested that the protection of trophic factor-deprived neurons by inhibitors of protein synthesis might likewise be mediated by an elevation of glutathione levels. Several reasons make this hypothesis appear unlikely. First, in their model, cystine deprivation depletes glutathione, and cycloheximide acts by maintaining glutathione levels but sparing cysteine for glutathione synthesis. Second, death by trophic factor deprivation is not caused by glutathione depletion, but both NGF deprivation of sympathetic neurons alone (Section 4.3.) and in combination with cycloheximide actually increase glutathione levels modestly (Deckwerth and Johnson, 1995; Deckwerth, unpublished). Neuronal survival, however, is only maintained when cycloheximide is present. Finally, concentrations of NAC, which elevate glutathione to a similar level as cycloheximide concentrations that confer complete protection, protect only a small fraction of all neurons from death by NGF deprivation.

A second hypothesis, proposed by Ferrari and Greene (1994), states that, during their differentiation and acquisition of a postmitotic phenotype, neurons may downregulate cell cycle components that are also required for the induction of apoptosis. This suggestion is supported by the requirement for increased levels of functional c-Jun protein and the increase in cyclin D1 levels in the induction of death by trophic factor deprivation (*see* Sections 3.4. and 4.5.), observations that provide the first direct evidence that ongoing macromolecular synthesis may be required in order to permit the synthesis of proteins that are indispensible for the successful induction of apoptosis.

5. METABOLIC CHANGES ASSOCIATED WITH WITHDRAWAL OF TROPHIC SUPPORT

In addition to promoting survival, trophic factors also exert anabolic effects that alter energy and macromolecular metabolism. Increases in total cellular protein or morphological features, such as increased neuronal soma size and neurite outgrowth, can be attributed to the anabolic effects of trophic factors. Changes in metabolic rates and soma size have been examined in NGF-deprived sympathetic neurons, potassium–serum-deprived cerebellar granule neurons, and deafferented second-order acoustic neurons of the nucleus magnocellularis in chicken. Survival of magnocellular neurons is maintained by a trophic agent continuously released from active presynaptic nerve terminals; a candidate for this agent is glutamate, which may promote survival by binding to metabotropic glutamate receptors that activate PKA (Lachica et al., 1995). In all three neuronal types, removal of trophic support leads to precipitous declines in the rates of glycolysis, protein, and RNA synthesis, as measured by the uptake of 2-deoxyglucose or the incorporation of amino acids or uridine into protein or RNA, respectively (Rubel et al., 1990; Deckwerth and Johnson, 1993; Miller and Johnson, 1996). In sympathetic and cerebellar granule neurons, these declines in metabolic rates occur concurrent with the ROS signal and with early genetic events, such as the induction of c-Jun (*see* Section 3.4.), and precede the onset of measurable neuronal atrophy and the time at which sympathetic neurons lose the ability to respond to the readdition of NGF by surviving (Deckwerth and Johnson, 1993). At the same time-point, cerebellar neurons fail to remain viable after readdition of potassium and serum (Miller and Johnson, 1996), and increased mitochondrial proliferation spontaneously starts in some of the magnocellular neurons, an event that is critical for halting the degenerative process in those neurons (Hyde and Durham, 1994).

The reductions in metabolic rates alone are not sufficient to cause neuronal death. Bax-deficient sympathetic neurons survive deprivation of NGF for weeks, displaying remarkable atrophy that renewed exposure to NGF can reverse (Deckwerth et al., 1996; *see* Section 6.). The rate of protein synthesis of NGF-deprived sympathetic neurons maintained with the ICE family protease inhibitor Boc-Asp(OMe)-fmk is only 13% of that of NGF-maintained neurons (Deshmukh et al., 1996), yet the neurons remain alive for days, demonstrating that massive declines in these metabolic parameters are compatible with neuronal survival.

6. MODULATION OF NEURONAL SURVIVAL BY PROTEINS OF THE BCL-2 FAMILY

The Bcl-2 family of proteins consists of apoptosis-promoting (e.g., Bax, Bak, Bcl-xS) and survival-maintaining (e.g., Bcl-2 Bcl-xL) members, which function, at least in part, through competing homo- and heterodimer formation (reviewed by Korsmeyer, 1995). Multiple members are expressed in a given cell type, and the relative expression levels determine the cellular response to a death-inducing stimulus. The role of Bcl-2 family members in the control of neuronal death after trophic factor withdrawal is well established. For example, overexpression of the neuroprotective members Bcl-2 and Bcl-xL prevents the death of sympathetic neurons after NGF deprivation (Garcia et al., 1992; Frankowski et al., 1995; Greenlund et al., 1995a), as well as death of neonatal motor neurons during development or after axotomy (Dubois-Dauphin et al., 1994; Martinou et al., 1994). Conversely, elimination of Bcl-2 by targeted deletion accelerates the kinetics of sympathetic neuronal death after NGF withdrawal (Greenlund et al., 1995b) and Bcl-x-deficient embryonic telencephalic neurons display an enhanced susceptibility to serum deprivation (Roth et al., 1996). Not all trophicfactor-dependent death is responsive to Bcl-2, however, since Bcl-2 overexpression is ineffective in protecting CNTF-dependent ciliary neurons or CNTF-maintained trigeminal sensory neurons from CNTF deprivation (Allsopp et al., 1993b). Among the proapoptotic members, Bcl-xS reduces the neuroprotection afforded by Bcl-2 (Martinou et al., 1995), and overexpression of Bak accelerates NGF deprivation-induced death (Farrow et al., 1995). Most strikingly, diminished expression of Bax by gene targeting or antisense oligonucleotide exposure slows the kinetics of neuronal death to such an extent that, even in the complete absence of Bax, sympathetic and facial motor neurons survive completely without NGF or after axotomy, respectively (Deckwerth et al., 1996; Gillardon et al., 1996). Although these salvaged Bax-deficient neurons atrophy considerably, the soma atrophy of sympathetic neurons is reversible on re-exposure to NGF, indicating that degeneration in these neurons is arrested at a point at which they remain responsive to trophic factors. Moreover, Bax-deficient mice possess more facial motor and sympathetic neurons than wild-type animals, indicating that naturally occurring death during development is diminished. The number of SCG neurons is already increased in neonatal animals, suggesting that Bax may control the survival of sympathetic neurons or neuroblasts prior to the period of target-dependent neuronal death. Taken together, these results demonstrate that Bcl-2 family members control the activation of most neuronal deaths by trophic factor deprivation, and that, at least in sympathetic and motor neurons, Bax is of singular importance for triggering apoptosis after trophic factor withdrawal.

Bcl-2 regulates neuronal survival in the absence of trophic factors, but it does not appear to be required for the mediation of neuronal survival by trophic factors. Unlike the neurotrophin or Trk-knockouts, Bcl-2 deficient mice have normal numbers of sympathetic SCG, sensory DRG, and facial motoneurons after the period of target-dependent neuronal death (Michaelidis et al., 1996). Furthermore, BDNF or CNTF rescue the same number of neonatal axotomized facial motor neurons in Bcl-2-deficient and wild-type mice. Whether Bcl-xL is required for neuronal survival by means of trophic factors is not known; however, the massive death of differentiating neurons in the developing CNS of Bcl-x-deficient mice would be consistent with such a role (Roth et al., 1996).

The molecular mechanism of action by Bcl-2 family members is not well understood. Most of the functional studies have been conducted with Bcl-2 in nonneuronal cells and in paradigms other than trophic factor deprivation. A review of these data is beyond the scope of this chapter. Any neuronal hypothesis needs to accommodate the observation that neurons saved from trophic factor deprivation by overexpression of Bcl-2 or inactivation of Bax remain functional: e.g., Bcl-2-overexpressing motor neurons retain their electrophysiological properties (Alberi et al., 1996), and Bax-deficient neurons extend neurites in the absence of NGF and respond to NGF with enhanced neurite outgrowth and soma hypertrophy (Deckwerth et al., 1996).

7. INDUCTION OF NEURONAL APOPTOSIS BY CASPASES (ICE FAMILY PROTEASES)

The cysteine proteases of the caspase family (interleukin 1β-converting enzyme [ICE] family) cleave peptide bonds that follow aspartate residues (reviewed by Henkart, 1996). They appear to be constitutively present in most cells as inactive precursors and are activated by proteolysis. Two major subfamilies (ICE and CPP32 subfamilies) are distinguished by their primary sequence homology and by the tertiary structure of their active site, which together explain differences in their substrate specificity. The aspartate-based substrate specificity and the catalytic mechanism have led to the design of peptide inhibitors specific for caspases, some of which are capable of targeting one of the subfamilies preferentially. Naturally occurring inhibitors include the viral serpin CrmA, with a preference for ICE, and the more promiscuous baculovirus protein p35.

Good evidence exists that caspases are required for neuronal death by trophic factor deprivation. Expression of CrmA and p35 in sensory DRG and sympathetic neurons, respectively, retards neuronal death after NGF withdrawal in vitro (Gagliardini et al.,

1994; Martinou et al., 1995). Extracellular application of ICE inhibitors based on the tetrapeptide sequence YVAD inhibit death of chicken motor neurons after trophic factor withdrawal in vitro and during the period of target-dependent neuronal death in vivo (Milligan et al., 1995). The same compound diminishes caspase activity in cerebellar granule neurons and prevents their death after potassium–serum deprivation (Schulz et al., 1996). Likewise, NGF deprivation-induced death of sympathetic neurons is prevented by the caspase inhibitor Boc-Asp(OMe)-fmk (BAF) (Deshmukh et al., 1996). Introduction of the active site hexapeptide IQACRG into the cytoplasm of PC12 cells and sympathetic neurons by an antennapedia vector peptide has been reported to suppress death after serum and NGF withdrawal, possibly by binding to potential cleavage sites on substrates or on inactive cysteine protease precursors (Troy et al., 1996). These studies demonstrate that caspases are required for trophic factor deprivation-induced neuronal death; more specific inhibitors or genetic approaches are required to identify any particular protease as critical for triggering neuronal apoptosis.

In contrast to other peptide-based inhibitors, BAF lacks any preference for either ICE or CPP32, and thus appears especially suitable for the study of the functional consequences of caspase inhibition, but the identity of the critical protease(s) remains obscure (Deshmukh et al., 1996). BAF maintains sympathetic neurons in the absence of NGF for days without any evidence of nuclear apoptotic changes. The cell bodies of NGF-deprived sympathetic neurons protected from death by BAF atrophy slightly, but remain viable and capable of responding to readdition of NGF with neurite outgrowth and soma hypertrophy. BAF neither affects the decrease in neuronal protein synthesis induced by NGF deprivation, nor prevents the transient increase in neuronal glutathione levels (Deckwerth, unpublished) nor the increase in neuronal c-Jun mRNA levels. This indicates that caspases have no role in the early signal transduction, metabolic decreases, or changes in gene expression triggered by trophic factor withdrawal. c-Fos mRNA and protein are greatly increased in most BAF-saved neurons; in dying NGF-deprived neurons, the increase in c-Fos mRNA is restricted to a fraction of all apoptotic nuclei in NGF-deprived neurons (Estus et al., 1994). This suggests that c-Fos accumulation actually precedes chromatin condensation, that c-Fos mRNA is rapidly degraded on the manifestation of nuclear apoptosis, and that caspases act shortly after c-Fos induction to induce nuclear apoptosis. NGF acts at the same time as BAF to arrest the degenerative process induced by NGF deprivation, raising the possibility that NGF, like BAF, may be able to directly suppress caspase activity.

To date, the molecular order of the action of Bcl-2 family members and caspases has been determined for several different apoptotic stimuli, except for serum trophic-factor deprivation. The unanimous result from these studies is that Bcl-2 family members prevent the activation of cysteine proteases, if they arc protective at all (Chinnaiyan et al., 1996). Given the evolutionary conservation of the pathway involving Bcl-2 and ICE family proteases, at least down to nematodes, any deviation from this order for trophic factor deprivation-induced neuronal death seems unlikely. To further advance the understanding of the regulation of neuronal death, it will be critical to determine what functional differences exist between sympathetic neurons saved by Bax deficiency and BAF, and how Bcl-2 family members control caspases activation.

8. APOPTOTIC DESTRUCTION OF THE NEURONAL SOMA

Unlike other well studied models of cell death, few studies have examined the biochemical changes associated with the terminal destruction of neuronal soma and neurites on trophic factor withdrawal. The available data suggest that a large number of the underlying mechanisms should be similar. For example, the nuclei of NGF-deprived sympathetic or potassium-deprived cerebellar granule neurons demonstrate the chromatin condensation, nuclear fragmentation, and oligonucleosomal DNA fragmentation characteristic of apoptotic cell death (Edwards et al., 1991; Deckwerth and Johnson, 1993; D'Mello et al., 1993). Furthermore, the participation of ICE family proteases in the induction of neuronal apoptosis suggests that substrates identified in nonneuronal cells may likewise be cleaved in neurons. Thus, many of the molecular changes responsible for the apoptotic phenotype can be expected to be similar in neurons and nonneuronal cells.

In addition to the extensive similarities between neuronal and nonneuronal apoptosis, the specialized morphology and function of differentiated neurons suggest differences as well. One unique neuronal feature is indicated by the nature of the degeneration of neuronal axons after NGF withdrawal from sympathetic neurons. On NGF deprivation, axon disintegration is under its own genetic control and procedes autonomously from the apoptotic death of the neuronal soma, as if the neurites were disconnected from their soma by axotomy (Deckwerth and Johnson, 1994). Sympathetic neuronal axons from mice carrying the dominant slow Wallerian degeneration (Wlds) allele resist NGF deprivation-induced destruction and, unlike the neuronal somas, remain viable by the criteria that they possess an intact plasma membrane, maintain metabolic functions characteristic for a living cell, and require an active energy metabolism for their continued viability. The identity of the Wlds allele and its molecular mechanism of action remain unknown.

9. CONCLUSION

The paradigm of trophic factor withdrawal stands out among the many death-inducing insults to neurons, in that it requires the active participation of the neuronal metabolism in the their demise, reflected by the metabolic changes, the induction of a stress response, and the requirement for ongoing macromolecular synthesis. Because of the participation of many signaling pathways and a large part of neuronal metabolism, neuronal death by trophic factor deprivation is an ideal system to define critical life-sustaining processes and to test neuroprotective compounds that act upstream of the induction of apoptosis. Despite these early degenerative changes, neurons retain many of their functional properties, including the ability to respond to trophic factors with survival, until Bcl-2 and caspases become critical for the induction of apoptosis. This suggests that, in a neurodegenerative disease or after an injury in which Bcl-2 and caspases are required for effecting neuronal death, inhibitors of these two control mechanisms should be able to stabilize degenerating neurons prior to the onset of irreversible damage. Therefore, with elimination of the diverse proximal defects responsible for triggering neuronal degeneration, these inhibitors promise to salvage damaged neurons by providing the chance for trophic factors and upstream-acting neuroprotective compounds to initiate full functional recovery.

ACKNOWLEDGMENTS

I would like to thank E. M. Johnson for his generous support and J. Russell and E. M. Johnson for their constructive criticism of the manuscript.

REFERENCES

Alberi, S., Raggenbas, M., De Bilbao, F., and Dubois-Dauphin, M. (1996) Axotomized neonatal motoneurons overexpressing the Bcl2 proto-oncogene retain functional electrophysiological properties. *Proc. Natl. Acad. Sci. USA* **93,** 3978–3983.

Allsopp, T. E., Robinson, M., Wyatt, S., and Davies, A. M. (1993a) Ectopic Trk A expression mediates a NGF survival response in NGF-independent sensory neurons but not in parasympathetic neurons. *J. Cell Biol.* **123,** 1555–1566.

Allsopp, T. E., Wyatt, S., Paterson, H. F., and Davies, A. M. (1993b) The proto-oncogene bcl-2 can selectively rescue neurotrophic factor-dependent neurons from apoptosis. *Cell* **73,** 295–307.

Barrett, G. L. and Bartlett, P. F. (1994) The p75 nerve growth factor receptor mediates survival or death depending on the stage of sensory neuron development. *Proc. Natl. Acad. Sci. USA* **91,** 6501–6505.

Benedetti, M., Levi, A., and Chao, M. V. (1993) Differential expression of nerve growth factor receptors leads to altered binding affinity and neurotrophin responsiveness. *Proc. Natl. Acad. Sci. USA* **90,** 7859–7863.

Boldin, M. P., Varfolomeev, E. E., Pancer, Z., Mett, I. L., Camonis, J. H., and Wallach, D. (1995) A novel protein that interacts with the death domain of Fas/APO1 contains a sequence motif related to the death domain. *J. Biol. Chem.* **270,** 7795–7798.

Borasio, G. D., John, J., Wittinghofer, A., Barde, Y. A., Sendtner, M., and Heumann, R. (1989) Ras p21 protein promotes survival and fiber outgrowth of cultured embryonic neurons. *Neuron* **2,** 1087–1096.

Borasio, G. D., Markus, A., Wittinghofer, A., Barde, Y. A., and Heumann, R. (1993) Involvement of ras p21 in neurotrophin-induced response of sensory, but not sympathetic neurons. *J. Cell Biol.* **121,** 665–672.

Bothwell, M. (1995) Functional interactions of neurotrophins and neurotrophin receptors. *Annu. Rev. Neurosci.* **18,** 223–253.

Burnham, P., Louis, J. C., Magal, E., and Varon, S. (1994) Effects of ciliary neurotrophic factor on the survival and response to nerve growth factor of cultured rat sympathetic neurons. *Dev. Biol.* **161,** 96–106.

Carter, B. D., Zirrgiebel, U., and Barde, Y. A. (1995) Differential regulation of p21ras activation in neurons by nerve growth factor and brain-derived neurotrophic factor. *J. Biol. Chem.* **270,** 21,751–21,757.

Chinnaiyan, A. M., Orth, K., O'Rourke, K., Duan, H., Poirier, G. G., and Dixit, V. M. (1996) Molecular ordering of the cell death pathway. *Bcl-2* and Bcl-xL function upstream of the CED-3-like apoptotic proteases. *J. Biol. Chem.* **271,** 4573–4576.

Clary, D. O., Weskamp, G., Austin, L. R., and Reichardt, L. F. (1994) Trk A cross-linking mimics neuronal responses to nerve growth factor. *Mol. Biol. Cell* **5,** 549–563.

Creedon, D. J., Johnson, E. M., Jr., and Lawrence, J. C. (1996) Mitogen-activated protein kinase-independent pathways mediate the effects of nerve growth factor and cAMP on neuronal survival. *J. Biol. Chem.* **271,** 20,713–20,718.

DeChiara, T. M., Vejsada, R., Poueymirou, W. T., Acheson, A., Suri, C., Conover, J. C., et al. (1995) Mice lacking the CNTF receptor, unlike mice lacking CNTF, exhibit profound motor neuron deficits at birth. *Cell* **83,** 313–322.

Deckwerth, T. L. and Johnson, E. M., Jr. (1993) Temporal analysis of events associated with programmed cell death (apoptosis) of sympathetic neurons deprived of nerve growth factor. *J. Cell Biol.* **123,** 1207–1222.

Deckwerth, T. L. and Johnson, E. M., Jr. (1994) Neurites can remain viable after destruction of the neuronal soma by programmed cell death (apoptosis). *Dev. Biol.* **165,** 63–72.
Deckwerth, T. L. and Johnson, E. M., Jr. (1995) Changes in glutathione metabolism suggest a critical role for the redox equilibrium in neuronal death after trophic factor deprivation. *J. Cell. Biochem.* **21B(Suppl),** 92.
Deckwerth, T. L., Elliott, J. L., Knudson, C. M., Johnson, E. M., Jr., Snider, W. D., and Korsmeyer, S. K. (1996) Bax is required for neuronal death after trophic factor deprivation and during development. *Neuron* **17,** in press.
Deshmukh, M., Vasilakos, J., Deckwerth, T. L., Lampe, P. A., Shivers, B. D., and Johnson, E. M., Jr. (1996) Genetic and metabolic status of NGF-deprived sympathetic neurons saved by an inhibitor of ICE-family proteases. *J. Cell Biol.* **135,** 1341–1354.
D'Mello, S. R., Galli, C., Ciotti, T., and Calissano, P. (1993) Induction of apoptosis in cerebellar granule neurons by low potassium: Inhibition of death by insulin-like growth factor I and cAMP. *Proc. Natl. Acad. Sci. USA* **90,** 10,989–10,993.
Dobrowsky, R. T., Werner, M. H., Castellino, A. M., Chao, M. V., and Hannun, Y. A. (1994) Activation of the sphingomyelin cycle through the low-affinity neurotrophin receptor. *Science* **265,** 1596–1599.
Dobrowsky, R. T., Jenkins, G. M., and Hannun, Y. A. (1995) Neurotrophins induce sphingomyelin hydrolysis. Modulation by co-expression of p75NTR with Trk receptors. *J. Biol. Chem.* **270,** 22,135–22,142.
Dubois-Dauphin, M., Frankowski, H., Tsujimoto, Y., Huarte, J., and Martinou, J. C. (1994) Neonatal motoneurons overexpressing the bcl-2 protooncogene in transgenic mice are protected from axotomy-induced cell death. *Proc. Natl. Acad. Sci. USA* **91,** 3309–3313.
Durbec, P. L., Larsson-Blomberg, L. B., Schuchardt, A., Costantini, F., and Pachnis, V. (1996) Common origin and developmental dependence on c-ret of subsets of enteric and sympathetic neuroblasts. *Development* **122,** 349–358.
Edwards, S. N., Buckmaster, A. E., and Tolkovsky, A. M. (1991) The death programme in cultured sympathetic neurones can be suppressed at the posttranslational level by nerve growth factor, cyclic AMP, and depolarization. *J. Neurochem.* **57,** 2140–2143.
Estus, S., Zaks, W. J., Freeman, R. S., Gruda, M., Bravo, R., and Johnson, E. M., Jr. (1994) Altered gene expression in neurons during programmed cell death: identification of c-jun as necessary for neuronal apoptosis. *J. Cell Biol.* **127,** 1717–1727.
Farinelli, S. E. and Greene, L. A. (1996) Cell cycle blockers mimosine, ciclopirox, and deferoxamine prevent the death of PC12 cells and postmitotic sympathetic neurons after removal of trophic support. *J. Neurosci.* **16,** 1150–1162.
Farrow, S. N., White, J. H., Martinou, I., Raven, T., Pun, K. T., Grinham, C. J., Martinou, J. C., and Brown, R. (1995) Cloning of a bcl-2 homologue by interaction with adenovirus E1B 19K. *Nature* **374,** 731–733.
Ferrari, G. and Greene, L. A. (1994) Proliferative inhibition by dominant-negative Ras rescues naive and neuronally differentiated PC12 cells from apoptotic death. *EMBO J.* **13,** 5922–5928.
Ferrari, G., Anderson, B. L., Stephens, R. M., Kaplan, D. R., and Greene, L. A. (1995a) Prevention of apoptotic neuronal death by GM1 ganglioside. Involvement of Trk neurotrophin receptors. *J. Biol. Chem.* **270,** 3074–3080.
Ferrari, G., Yan, C. Y., and Greene, L. A. (1995b) N-acetylcysteine (D- and L-stereoisomers) prevents apoptotic death of neuronal cells. *J. Neurosci.* **15,** 2857–2866.
Ferrer, I., Olive, M., Ribera, J., and Planas, A. M. (1996) Naturally occurring (programmed) and radiation-induced apoptosis are associated with selective c-Jun expression in the developing rat brain. *Eur. J. Neurosci.* **8,** 1286–1298.
Finkbeiner, S. and Greenberg, M. E. (1996) Ca^{2+}-dependent routes to ras: mechanisms for neuronal survival, differentiation, and plasticity? *Neuron* **16,** 233–236.
Franklin, J. L. and Johnson, E. M., Jr. (1992) Suppression of programmed neuronal death by sustained elevation of cytoplasmic calcium. *Trends Neurosci.* **15,** 501–508.

Franklin, J. L., Sanz-Rodriguez, C., Juhasz, A., Deckwerth, T. L., and Johnson, E. M., Jr. (1995) Chronic depolarization prevents programmed death of sympathetic neurons in vitro but does not support growth: Requirement for Ca^{2+} influx but not Trk activation. *J. Neurosci.* **15,** 643–664.

Frankowski, H., Missotten, M., Fernandez, P. A., Martinou, I., Michel, P., Sadoul, R., and Martinou, J. C. (1995) Function and expression of the Bcl-x gene in the developing and adult nervous system. *Neuroreport* **6,** 1917–1921.

Freeman, R. S., Estus, S., and Johnson, E. M., Jr. (1994) Analysis of cell cycle-related gene expression in postmitotic neurons: selective induction of Cyclin D1 during programmed cell death. *Neuron* **12,** 343–355.

Gagliardini, V., Fernandez, P. A., Lee, R. K., Drexler, H. C., Rotello, R. J., Fishman, M. C., and Yuan, J. (1994) Prevention of vertebrate neuronal death by the crmA gene. *Science* **263,** 826–828.

Garcia, I., Martinou, I., Tsujimoto, Y., and Martinou, J. C. (1992) Prevention of programmed cell death of sympathetic neurons by the *bcl*-2 proto-oncogene. *Science* **258,** 302–304.

Gillardon, F., Zimmermann, M., Uhlmann, E., Krajewski, S., Reed, J. C., and Klimaschewski, L. (1996) Antisense oligodeoxynucleotides to Bax mRNA promote survival of rat sympathetic neurons in culture. *J. Neurosci. Res.* **43,** 726–734.

Goodman, Y. and Mattson, M. P. (1996) Ceremide protects hippocampal neurons against excitotoxic and oxidative insulots, and amyloid b-peptide toxicity. *J. Neurochem.* **66,** 869–872.

Greenlund, L. J. S., Deckwerth, T. L., and Johnson, E. M., Jr. (1995a) Superoxide dismutase delays neuronal apoptosis: a role for reactive oxygen species in programmed neuronal death. *Neuron* **14,** 303–315.

Greenlund, L. J. S., Korsmeyer, S. J., and Johnson, E. M., Jr. (1995b) Role of BCL-2 in the survival and function of developing and mature sympathetic neurons. *Neuron* **15,** 649–661.

Halliwell, B. and Gutteridge, J. M. (1989) *Free Radicals in Biology and Medicine.* Clarendon, Oxford.

Ham, J., Babij, C., Whitfield, J., Pfarr, C. M., Lallemand, D., Yaniv, M., and Rubin, L. L. (1995) A c-Jun dominant negative mutant protects sympathetic neurons against programmed cell death. *Neuron* **14,** 927–939.

Hantzopoulos, P. A., Suri, C., Glass, D. J., Goldfarb, M. P., and Yancopoulos, G. D. (1994) The low affinity NGF receptor, p75, can collaborate with each of the Trks to potentiate functional responses to the neurotrophins. *Neuron* **13,** 187–201.

Heintz, N. (1993) Cell death and the cell cycle: a relationship between transformation and neurodegeneration? *Trends Biochem. Sci.* **18,** 157–159.

Henkart, P. A. (1996) ICE family proteases: mediators of all apoptotic cell death? *Immunity* **4,** 195–201.

Herrup, K. and Busser, J. C. (1995) The induction of multiple cell cycle events precedes target-related neuronal death. *Development* **121,** 2385–2395.

Hyde, G. E. and Durham, D. (1994) Increased deafferentation-induced cell death in chick brainstem auditory neurons following blockade of mitochondrial protein synthesis with chloramphenicol. *J. Neurosci.* **14,** 291–300.

Ito, A. and Horigome, K. (1995) Ceramide prevents neuronal programmed cell death induced by nerve growth factor deprivation. *J. Neurochem.* **65,** 463–466.

Iyengar, R. (1996) Gating by cyclic AMP: expanded role for an old signaling pathway. *Science* **271,** 461–463.

Jing, S. Q., Wen, D. Z., Yu, Y. B., Holst, P. L., Luo, Y., Fang, M., et al. (1996) GDNF-induced activation of the Ret protein tyrosine kinase is mediated by GDNFR-a, a novel receptor for GDNF. *Cell* **85,** 1113–1124.

Johnson, E. M., Jr., and Deckwerth, T. L. (1993) Molecular mechanisms of developmental neuronal death. *Annu. Rev. Neurosci.* **16,** 31–46.

Jordan, J., Ghadge, G. D., Prehn, J. H., Toth, P. T., Roos, R. P., and Miller, R. J. (1995) Expression of human copper/zinc-superoxide dismutase inhibits the death of rat sympathetic neurons caused by withdrawal of nerve growth factor. *Mol. Pharmacol.* **47,** 1095–1100.

Kessler, J. A., Ludlam, W. H., Freidin, M. M., Hall, D. H., Michaelson, M. D., Spray, D. C., Dougherty, M., and Batter, D. K. (1993) Cytokine-induced programmed death of cultured sympathetic neurons. *Neuron* **11,** 1123–1132.

Koike, T. and Tanaka, S. (1991) Evidence that nerve growth factor dependence of sympathetic neurons for survival in vitro may be determined by levels of cytoplasmic free Ca^{2+}. *Proc. Natl. Acad. Sci. USA* **88,** 3892–3896.

Korsmeyer, S. J. (1995) Regulators of cell death. *Trends Genet.* **11,** 101–105.

Kotzbauer, P. T., Lampe, P. A., Estus, S., Milbrandt, J., and Johnson, E. M., Jr. (1994) Postnatal development of survival responsiveness in rat sympathetic neurons to leukemia inhibitory factor and ciliary neurotrophic factor. *Neuron* **12,** 763–773.

Kotzbauer, P. T., Lampe, P. A., Heuckeroth, R. O., Golden, J. P., Creedon, D. J., Johnson, E. M., Jr., and Milbrandt, J. (1996) Identification of neurturin, a GDNF-related factor which promotes the survival of multiple neuronal populations. *Nature* **384,** 467–470.

Kovary, K. and Bravo, R. (1991) The jun and fos protein families are both required for cell cycle progression in fibroblasts. *Mol. Cell Biol.* **11,** 4466–4472.

Kranenburg, O., van der Eb, A. J., and Zantema, A. (1996) Cyclin D1 is an essential mediator of apoptotic neuronal cell death. *EMBO J.* **15,** 46–54.

Lachica, E. A., Rübsamen, R., Zirpel, L., and Rubel, E. W. (1995) Glutamatergic inhibition of voltage-operated calcium channels in the avian cochlear nucleus. *J. Neurosci.* **15,** 1724–1734.

Lampe, P. A., Cornbrooks, E. B., Juhasz, A., Johnson, E. M., Jr., and Franklin, J. L. (1995) Suppression of programmed neuronal death by a thapsigargin-induced Ca^{2+} influx. *J. Neurobiol.* **26,** 205–212.

Lee, K. F., Davies, A. M., and Jaenisch, R. (1994) p75-deficient embryonic dorsal root sensory and neonatal sympathetic neurons display a decreased sensitivity to NGF. *Development* **120,** 1027–1033.

Levick, V., Coffey, H., and D'Mello, S. R. (1995) Opposing effects of thapsigargin on the survival of developing cerebellar granule neurons in culture. *Brain Res.* **676,** 325–335.

Li, M., Sendtner, M., and Smith, A. (1995) Essential function of LIF receptor in motor neurons. *Nature* **378,** 724–727.

Loeb, D. M. and Greene, L. A. (1993) Transfection with Trk restores "slow" NGF binding, efficient NGF uptake, and multiple NGF responses to NGF-nonresponsive PC12 cell mutants. *J. Neurosci.* **13,** 2919–2929.

Martinou, J. C., Dubois-Dauphin, M., Staple, J. K., Rodriguez, I., Frankowski, H., Missotten, M., et al. (1994) Overexpression of BCL-2 in transgenic mice protects neurons from naturally occurring cell death and experimental ischemia. *Neuron* **13,** 1017–1030.

Martinou, I., Fernandez, P. A., Missotten, M., White, E., Allet, B., Sadoul, R., and Martinou, J. C. (1995) Viral proteins E1B19K and p35 protect sympathetic neurons from cell death induced by NGF deprivation. *J. Cell Biol.* **128,** 201–208.

Mayer, M. and Noble, M. (1994) *N*-acetyl-L-cysteine is a pluripotent protector against cell death and enhancer of trophic factor-mediated cell survival in vitro. *Proc. Natl. Acad. Sci. USA* **91,** 7496–7500.

Mesner, P. W., Epting, C. L., Hegarty, J. L., and Green, S. H. (1995) A timetable of events during programmed cell death induced by trophic factor withdrawal from neuronal PC12 cells. *J. Neurosci.* **15,** 7357–7366.

Meyer-Franke, A., Kaplan, M. R., Pfrieger, F. W., and Barres, B. A. (1995) Characterization of the signaling interactions that promote the survival and growth of developing retinal ganglion cells in culture. *Neuron* **15,** 805–819.

Michaelidis, T. M., Sendtner, M., Cooper, J. D., Airaksinen, M. S., Holtmann, B., Meyer, M., and Thoenen, H. (1996) Inactivation of bcl-2 results in progressive degeneration of motoneurons, sympathetic and sensory neurons during early postnatal development. *Neuron* **17,** 75–89.

Miller, T. and Johnson, E. M., Jr. (1996) Metabolic and genetic analyses of apoptosis in potassium/serum-deprived rat cerebellar granule cells. *J. Neurosci.* **16,** 7487–7495.

Milligan, C. E., Prevette, D., Yaginuma, H., Homma, S., Cardwell, C., Fritz, L. C., et al. (1995) Peptide inhibitors of the ICE protease family arrest programmed cell death of motoneurons in vivo and in vitro. *Neuron* **15,** 385–393.

Nagata, S. and Golstein, P. (1995) The Fas death factor. *Science* **267,** 1449–1456.

Ng, N. F. and Shooter, E. M. (1993) Activation of p21ras by nerve growth factor in embryonic sensory neurons and PC12 cells. *J. Biol. Chem.* **268,** 25,329–25,333.

Nobes, C. D. and Tolkovsky, A. M. (1995) Neutralizing anti-p21ras Fabs suppress rat sympathetic neuron survival induced by NGF, LIF, CNTF and cAMP. *Eur. J. Neurosci.* **7,** 344–350.

Nobes, C. D., Reppas, J. B., Markus, A., and Tolkovsky, A. M. (1996) Active p21 ras is sufficient for rescue of NGF-dependent rat sympathetic neurons. *Neuroscience* **70,** 1067–1079.

Oppenheim, R. W. (1991) Cell death during development of the nervous system. *Annu. Rev. Neurosci.* **14,** 453–501.

Pan, Z. and Perez-Polo, R. (1993) Role of nerve growth factor in oxidant homeostasis: glutathione metabolism. *J. Neurochem.* **61,** 1713–1721.

Park, D. S., Farinelli, S. E., and Greene, L. A. (1996) Inhibitors of cyclin-dependent kinases promote survival of post-mitotic neuronally differentiated PC12 cells and sympathetic neurons. *J. Biol. Chem.* **271,** 8161–8169.

Rabizadeh, S., Oh, J., Zhong, L. T., Yang, J., Bitler, C. M., Butcher, L. L., and Bredesen, D. E. (1993) Induction of apoptosis by the low-affinity NGF receptor. *Science* **261,** 345–348.

Ratan, R. R., Murphy, T. H., and Baraban, J. M. (1994) Macromolecular synthesis inhibitors prevent oxidative stress-induced apoptosis in embryonic cortical neurons by shunting cysteine from protein synthesis to glutathione. *J. Neurosci.* **14,** 4385–4392.

Rodriguez-Viciana, P., Marte, B. M., Warne, P. H., and Downward, J. (1996) Phosphatidylinositol 3′ kinase: one of the effectors of Ras. *Philos. Trans. R. Soc. Lond. B. Biol. Sci.* **351,** 225–232.

Rosen, L. B., Ginty, D. D., Weber, M. J., and Greenberg, M. E. (1994) Membrane depolarization and calcium influx stimulate MEK and MAP kinase via activation of Ras. *Neuron* **12,** 1207–1221.

Roth, K. A., Motoyama, N., and Loh, D. Y. (1996) Apoptosis of Bcl-x-deficient telencephalic cells in vitro. *J. Neurosci.* **16,** 1753–1758.

Rubel, E. W., Hyson, R. L., and Durham, D. (1990) Afferent regulation of neurons in the brain stem auditory system. *J. Neurobiol.* **21,** 169–196.

Rukenstein, A., Rydel, R. E., and Greene, L. A. (1991) Multiple agents rescue PC12 cells from serum-free cell death by translation- and transcription-independent mechanisms. *J. Neurosci.* **11,** 2552–2563.

Rydel, R. E. and Greene, L. A. (1988) cAMP analogs promote survival and neurite outgrowth in cultures of rat sympathetic and sensory neurons independently of nerve growth factor. *Proc. Natl. Acad. Sci. USA* **85,** 1257–1261.

Schulz, J. B., Weller, M., and Klockgether, T. (1996) Potassium deprivation-induced apoptosis of cerebellar granule neurons: a sequential requirement for new mRNA and protein synthesis, ICE-like protease activity, and reactive oxygen species. *J. Neurosci.* **16,** 4696–4706.

Schwarzschild, M. A., Dauer, W. T., Lewis, S. E., Hamill, L. K., Fink, J. S., and Hyman, S. E. (1994) Leukemia inhibitory factor and ciliary neurotrophic factor increase activated Ras in a neuroblastoma cell line and in sympathetic neuron cultures. *J. Neurochem.* **63,** 1246–1254.

Segal, R. A. and Greenberg, M. E. (1996) Intracellular signalling pathways activated by neurotrophic factors. *Annu. Rev. Neurosci.* **19,** 463–489.

Sendtner, M., Götz, R., Holtmann, B., Escary, J. L., Masu, Y., Carroll, P., et al. (1996) Cryptic physiological trophic support of motoneurons by LIF revealed by double gene targeting of CNTF and LIF. *Current Biol.* **6,** 686–694.

Sicinski, P., Donaher, J. L., Parker, S. B., Li, T., Fazeli, A., Gardner, H., et al. (1995) Cyclin D1 provides a link between development and oncogenesis in the retina and breast. *Cell* **82,** 621–630.

Snider, W. D. (1994) Functions of the neurotrophins during nervous system development: what the knockouts are teaching us. *Cell* **77,** 627–638.

Stahl, N. and Yancopoulos, G. D. (1994) The tripartite CNTF receptor complex: activation and signaling involves components shared with other cytokines. *J. Neurobiol.* **25,** 1454–1466.

Treanor, J. J. S., Goodman, L., de Sauvage, F., Stone, D. M., Poulsen, K. T., Beck, C. D., et al. (1996) Characterization of a multicomponent receptor for GDNF. *Nature* **382,** 80–83.

Troy, C. M., Stefanis, L., Prochiantz, A., Greene, L. A., and Shelanski, M. L. (1996) The contrasting roles of ICE family proteases and interleukin-1b in apoptosis induced by trophic factor withdrawal and by copper/zinc superoxide dismutase down-regulation. *Proc. Natl. Acad. Sci. USA* **93,** 5635–5640.

Trupp, M., Arenas, E., Fainzilber, M., Nilsson, A.-S., Sieber B.-A., Grigoriou, M., et al. (1996) Functional receptor for GDNF encoded by the c-ret proto-oncogene. *Nature* **81,** 785–789.

Virdee, K. and Tolkovsky, A. M. (1995) Activation of p44 and p42 MAP kinases is not essential for the survival of rat sympathetic neurons. *Eur. J. Neurosci.* **7,** 2159–2169.

Vogel, K. S., Brannan, C. I., Jenkins, N. A., Copeland, N. G., and Parada, L. F. (1995) Loss of neurofibromin results in neurotrophin-independent survival of embryonic sensory and sympathetic neurons. *Cell* **82,** 733–742.

Woodgett, J. R., Kyriakis, J. M., Avruch, J., Zon, L. I., Zanke, B., and Templeton, D. J. (1996) Reconstitution of novel signalling cascades responding to cellular stress. *Philos. Trans. R. Soc. London Ser. B: Biol. Sci.* **351,** 135–141.

Xia, Z., Dickens, M., Raingeaud, J., Davis, R. J., and Greenberg, M. E. (1995) Opposing effects of ERK and JNK-p38 MAP kinases on apoptosis. *Science* **270,** 1326–1331.

Yan, C. Y., Ferrari, G., and Greene, L. A. (1995) *N*-acetylcysteine promoted survival of PC12 cells is glutathione-independent but transcription-dependent. *J. Biol. Chem.* **270,** 26,827–26,832.

Yao, R. and Cooper, G. M. (1995) Requirement for phosphatidylinositol-3 kinase in the prevention of apoptosis by nerve growth factor. *Science* **267,** 2003–2006.

4
Gene Induction and Neuronal Apoptosis

Steve Estus

1. INTRODUCTION

To begin to identify neuroprotective mechanisms, we have focused on elucidating genetic changes leading to neuronal death, with a particular emphasis on neuronal apoptosis. In this chapter, we shall discuss the contribution of apoptosis to neuronal loss during development and during conditions of human pathology, an intracellular signaling cascade implicated as necessary for neuronal death in at least one model system, i.e., nerve growth factor (NGF)-deprived sympathetic neurons, events that likely lead to the activation of these genes, and a possible mechanism whereby these activations of the genes may lead to neuronal death. During this discussion, we will highlight points at which neuroprotective signaling pathways may interfere with, and thereby block, cell death.

2. PROGRAMMED CELL DEATH (PCD) AND APOPTOSIS CHARACTERIZE MATURATION OF THE NERVOUS SYSTEM

Maturation of the nervous system is characterized by extensive apoptotic cell death. Shortly after the period of neuroblast proliferation, many neurons, commonly about 50%, die during the time functional connections with their target tissues are being made (reviewed in Oppenheim, 1991). The purpose for this process appears to be to match the number of innervating neurons to the size of their targets. Target- or locally derived trophic factors typically control neuronal survival during, and subsequent to, this period. In the most widely accepted scenario, trophic factors are secreted in minute amounts by their target, bind to receptors on the innervating nerve terminals, and are internalized and retrogradely transported to the neuronal cell body. Presumably during the period of naturally occurring neuronal death, cells that fail to obtain sufficient trophic factor die, but those that obtain sufficient factor survive.

2.1. *Neurons Undergoing PCD Manifest Hallmarks of Apoptosis*

Neurons undergoing this naturally occurring, or programmed, cell death (PCD) typically manifest the hallmarks of apoptosis, a type of cell death defined largely by certain morphologic criteria (Wyllie et al., 1980; Clarke, 1990). The principal criteria that have been used to define cell death as apoptotic are the condensation of nuclear chromatin

From: Neuroprotective Signal Transduction *edited by M. P. Mattson*
Humana Press Inc., Totowa, NJ

and the fragmentation of nuclear DNA into oligonucleosomal-sized fragments (multiples of approx 180 bp). This type of cell death is found in many cell types during development, and, in the adult animal, particularly in nonneuronal cells. Apoptosis may be triggered in response to either physiological stimuli, e.g., insufficient trophic factor stimulation, or nonphysiological stimuli (*see* Section 2.2.). After the initiating stimulus, cells transiently show blebbing of the plasma membrane. Subsequent to the nuclear changes described above, cytoplasmic organelles start to degenerate, and both nucleus and cytoplasm become compartmentalized into membrane-bound apoptotic bodies that, in intact tissues, are engulfed by phagocytic cells. These morphologic criteria constitute the definition of apoptosis.

2.2. *Studies Suggest Apoptosis Is a Cellular Process Within PCD*

That PCD includes an apoptotic-like process is supported by both in vivo and in vitro data. The most-studied model of trophic factor deprivation-induced PCD is based on the physiological role of the prototypical neurotrophic factor, NGF. Depriving immature sympathetic (Levi-Montalcini and Booker, 1960) or certain sensory neurons (Johnson et al., 1980) of NGF results in massive cell death. In the case of sympathetic neurons, NGF deprivation results in massive cell death, even in the adult animal (Gorin and Johnson, 1980). That this PCD occurs by an apoptotic process, a form of programmed cell death, is supported by several studies. Early electron micrographs of sympathetic neurons after in vivo injection of antibodies to NGF (anti-NGF) appear apoptotic, although they were not called this at the time (Levi-Montalcini et al., 1969). Since this work, apoptosis itself has been defined, and compelling focused evidence that neuronal PCD involves an apoptotic-like process has emerged from in vitro models. Most prevalent are studies in cultured sympathetic neurons undergoing NGF deprivation, showing that neuronal cell death is accompanied by membrane blebbing, chromatin condensation, and DNA fragmentation into oligonucleosomes (Edwards et al., 1991; Deckwerth and Johnson, 1993). Thus, the death of neurons during normal development appears to be an example of apoptosis.

That apoptosis is a cellular process subsumed within PCD is suggested by studies examining the dependence of each process on macromolecular synthesis. That neuronal PCD is an active genetic process is suggested by observations that death after trophic factor deprivation, either in vitro or in vivo, typically requires ongoing RNA and protein synthesis (Martin et al., 1988; Oppenheim et al., 1990; Scott and Davies, 1990). However, the final event in PCD, apoptosis, can be triggered in apparently all cell types by the kinase inhibitor staurosporine, even in the presence of trophic factor support or protein synthesis inhibitors (Jacobson et al., 1994). Furthermore, apoptosis can be induced in certain cell types in the absence of transcription (Jacobson et al., 1994). Considered together, these data suggest two alternative possibilities: Apoptosis induced by agents in the absence of transcription involves a fundamentally different process than apoptosis during PCD; or, PCD is a cellular process wherein initial events requiring RNA and protein synthesis serve to trigger apoptosis, which, once initiated, proceeds without additional macromolecular synthesis. Considering the morphologic similarities in death in each paradigm, the latter hypothesis appears more feasible. Hence, we currently hypothesize that PCD is a cellular process that culminates in apoptosis.

3. PCD AND APOPTOSIS CONTRIBUTE TO NEURONAL LOSS IN MULTIPLE PATHOLOGIC SITUATIONS

Neuronal apoptosis has been suggested to underlie neuronal loss in neurodegenerative diseases, stroke, and epilepsy, but strong evidence in support of this possibility is only now emerging. Several recent reports examining the incidence of neurons manifesting fragmented chromatin have suggested that apoptosis contributes to cellular loss in Alzheimer's disease (AD) (Su et al., 1994; Lassmann et al., 1995), Huntington disease (Portera-Calliau et al., 1995), and kainic acid-induced epileptic seizure models (Pollard et al., 1994). Although these results may be controversial (Migheli et al., 1994), they represent an indication that apoptosis likely contributes to human diseases, such as AD. Apoptosis and PCD have also been implicated as contributing to neuronal death in both focal and global models of ischemia, because this death is accompanied by DNA fragmentation and is reduced by protein synthesis inhibitors (Goto et al., 1990; Shigeno, 1990, Linnik et al., 1993; MacManus et al., 1993, 1994).

3.1. Role in Clinical Manifestions of Neurodegenerative Diseases

Neuronal loss likely contributes to the clinical manifestations of neurodegenerative diseases, such as AD. Characteristic neuronal loss is a hallmark of AD (reviewed in Price et al., 1992). For example, within the hippocampus and medial temporal lobe of AD patients, the entorhinal cortex and CA1 and CA2 pyramidal neurons exhibit neurofibrillary tangles (NFT) and are reduced in number (Ball, 1977; Tomlinson et al., 1981; Hyman et al., 1984). These changes are significant because these areas are important in the establishment of memory (reviewed in Zola-Morgan and Squire, 1993). Other populations are also susceptible to loss, including certain cholinergic, noradrenergic, serotoninergic, and glutamatergic neurons (reviewed in Price et al., 1992). AD is marked by neuronal degeneration and loss that likely contributes to the clinical manifestations of AD.

3.2. Multiple Stimuli Induce Apoptosis

Multiple stimuli can induce neurons to undergo apoptosis. Although we have focused on trophic factor deprivation in our research, other neuronal apoptosis-inducing stimuli include a variety of oxidative stressors (Bonfoco et al., 1995), neurotoxins, such as amyloid β (Forloni et al., 1993b; Loo et al., 1993; Watt et al., 1994) and the Creutzfeldt-Jacob prion protein (Forloni et al., 1993a), and agents that alter kinase activity, such as staurosporine. Moreover, we recently demonstrated that sympathetic neurons, at least under certain conditions, can be induced to undergo PCD by agents normally considered growth factors or cytokines and that changes can occur in neurons that determine whether these factors initiate or prevent death. Both ciliary neurotrophic factor (CNTF) and leukemia inhibitory factor (LIF) can induce sympathetic neurons to undergo PCD in the presence of NGF, or, alternatively, inhibit the PCD induced by NGF deprivation, depending on the maturity of the cultured neurons and/or their length of time in culture (Kotzbauer et al., 1994). Kessler and colleagues (Kessler et al., 1993) also observed that CNTF and LIF can induce PCD. The changes that occur in the cell to cause this switch are unknown, and whether this is relevant to events that occur normally or pathologically is not known. Clearly, neuronal PCD can be triggered by agents ranging from neurotoxins to growth factors.

4. GENE EXPRESSION APPEARS CRITICAL TO NEURONAL PCD

The basic hypothesis underlying our work is that cell death may be viewed as a normal process of cellular differentiation; as with differentiation into many other phenotypes, death requires the altered expression of specific genes in a coordinated fashion. During normal development, this genetic process is physiologically appropriate and typically triggered by trophic-factor deprivation. During pathologic situations, other stimuli activate this process at a point that may or may not coincide with that of trophic-factor deprivation. For example, since apoptosis induced by staurosporine appears independent of macromolecular synthesis (Jacobson et al., 1994), this agent likely initiates apoptosis at a position downstream from that initiated by trophic factor deprivation.

4.1. Four Lines of Evidence Suggest that Neuronal PCD Results from Altered Gene Transcription

4.1.1. NGF Deprivation Induces a Genetic Cascade

First, NGF deprivation induces a genetic cascade (Estus et al., 1994; Freeman et al., 1994; Ham et al., 1995). These cascades, which include a delayed induction of immediate early genes (IEGs), begins with c-*jun* and the cell cycle regulator cyclin D1, prior to overt changes in neuronal viability, followed by c-*fos, fosB, jun B*, and *ngfi-a*, coincident with initial declines in neuronal viability, and conclude with transin and collagenase, concurrent with dramatic declines in neuronal viability.

4.1.2. In NGF Deprivation Model, c-jun Is Essential to Apoptosis

Second, one of the genes that is induced in these cascades, c-*jun*, is necessary for apoptosis, at least in the NGF-deprivation model. As described previously (Estus et al., 1994), we tested the role of members of the Jun and Fos families in neuronal apoptosis induced by NGF deprivation by microinjecting neutralizing antibodies into neurons. Antibodies directed against the Jun family (c-Jun, Jun B, and Jun D) or the Fos family (c-Fos, Fos B, Fra-1, and Fra-2) protected neurons from apoptosis. Moreover, an antibody preparation specific for c-Jun also blocked death. Hence, these results provide the first support for the hypothesis that mammalian neuronal apoptosis depends on the activation and function of specific gene products, i.e., c-Jun. These results have recently been confirmed by others working with a dominant negative form of c-Jun (Ham et al., 1995).

4.1.3. Activation of c-jun Is Common in Mammalian Cell Death

Third, activation of c-*jun* is a common element in other instances of mammalian cell death. For example, c-*jun* has been implicated in the PCD of certain lymphoid cell lines (Colotta et al., 1992). Moreover, Gluckman and his colleagues examined IEG expression in a central nervous system (CNS) model of neuronal PCD involving ischemic injury (Dragunow et al., 1993, 1994) and found that c-Jun expression was induced for long periods in neurons likely to dic, but c-Fos and Fos B expression was more temporally restricted; this pattern is grossly similar to that observed in the sympathetic neuron model (Estus et al., 1994). At least a portion of this neuronal death appears apoptotic because the characteristic nucleosomal ladder and DNA fragmentation *in situ* is detected (Linnik et al., 1993; MacManus et al., 1993, 1994) and the size of the infarcted area is reduced by cycloheximide treatment (Linnik et al., 1993).

Also, Cotman and coworkers have found increased immunoreactivity of the Jun family in NFT-bearing neurons in AD tissue (Anderson et al., 1994); although their antibody did not distinguish among Jun family members (c-Jun, Jun B, and Jun D), this result suggests that c-Jun is induced in neurons that are typically vulnerable to death in AD. Indeed, a preliminary report suggests that Jun is induced in neurons manifesting DNA fragmentation in AD brain (Anderson et al., 1995). Hence, genes that we have associated with Aβ-induced neuronal apoptosis are also associated with neurons susceptible to loss in AD.

4.1.4. *Genetic Control of PCD Is Suggested by Studies of Analogous Systems in "Lower" Species*

Fourth, several genes have been identified as necessary for PCD in experimental systems involving lower species, suggesting, by analogy, that mammalian PCD may also be under genetic control. Two genes have been identified as necessary for PCD in *Caenorhabditis elegans*, namely *ced-3* and *ced-4*, but another gene, *ced-9*, appears to block PCD (Ellis et al., 1991). Although *ced-9* shows homology to a mammalian gene, *bcl-2*, in both sequence and protective function (Hockenbery, 1994), *ced-3* shows homology to the ICE family of proteases in sequence and toxic function (Bredesen, 1995). Two genes have also been associated with cell death in *Drosophila*, i.e., *reaper* and *hid* (White et al., 1994; Steller, 1995). Although the functions of the corresponding gene products remain unclear, their necessity for cell death suggests the existence of a genetic pathway, probably more complex, leading to PCD in mammals.

In summation, c-Jun is induced during, and appears necessary for, a cascade of molecular events that constitute PCD and that culminate in apoptosis. However, the role of c-Jun in neurons is not restricted to death pathways, in that c-Jun is also induced in other paradigms that do not involve neuronal death, e.g., c-Jun is induced during the long-term survival of axotomized retinal neurons (Robinson, 1994). Considering these data together, since c-Jun is necessary for neuronal death, and yet is not normally sufficient, then situations can arise in neurons whereby the neuron is made vulnerable to the actions of c-Jun activation; in these situations, c-Jun activation becomes toxic.

4.2. *c-Jun Characteristics Suggest Possible Toxic Mechanism*

An understanding of a possible toxic mechanism of c-Jun activation may be obtained by considering several primary characteristics of c-Jun (Angel and Karin, 1991; Morgan and Curran, 1991; Cochran, 1993).

4.2.1. *Induction Typically Occurs in an Immediate Early Manner*

First, the induction of *c-jun* typically occurs in an immediate early manner, i.e., its induction is independent of new protein synthesis and occurs within minutes of the initial stimulus. Enhanced transcription of *c-jun* typically results from activated c-Jun acting in a feed-forward fashion (Angel et al., 1988); cytoplasmic c-Jun is activated by altered phosphorylation or redox status, translocates to the nucleus, binds as a dimer to consensus motif in the *c-jun* promoter, and increases the transcription of *c-jun* mRNA (Abate et al., 1990; Woodgett et al., 1993).

4.2.2. *Sole Functional Attribute of c-jun: to Modulate Gene Transcription*

The second primary characteristic of c-Jun, and its sole known functional attribute, is its ability to modulate gene transcription. The active transcription factor, loosely termed activator protein-1 (AP-1), is formed by dimerization of c-Jun with itself or with other members of the Jun family (JunB and Jun D) or Fos family (c-Fos, FosB, Fra-1, and Fra-2). All are members of the "leucine-zipper" superfamily of transcription factors (Angel and Karin, 1991; Morgan and Curran, 1991; Cochran, 1993). The active dimer modulates gene transcription by binding to a consensus motif (5′-TGA G/C TCA-3′) within the target gene promoter. However, the ability of c-Jun to activate this promotor element depends, in part, on its dimerization partner, e.g., although c-Fos/c-Jun heterodimers are strong activators of transcription of collagenase, Fra-1/c-Jun heterodimers are much less effective (Suzuki et al., 1991). Hence, the activity of c-Jun is modulated, in part, by the levels of other transcription factors within the neuron.

4.2.3. c-jun Is Necessary for Cell-Cycle Re-Entry

Third, c-Jun is necessary for cell-cycle re-entry and, indeed, is a proto-oncogene, since constitutive expression can lead to cell transformation. When quiescent fibroblasts are serum-stimulated to re-enter the cell cycle, c-Jun is induced and is necessary for successful re-entry, as shown by antibody microinjection studies (Kovary and Bravo, 1991, 1992). The constitutive overexpression of c-Jun, or of its viral oncogenic homolog, v-Jun, can lead to cellular transformation (Angel and Karin, 1991; Morgan and Curran, 1991; Cochran, 1993).

5. MECHANISMS WHEREBY ACTIVATION OF THESE GENES MAY LEAD TO NEURONAL DEATH

5.1. Oxidative Stress May Tranduce Apoptotic Stimulus

Oxidative stress may transduce the apoptotic stimulus between NGF deprivation and an apoptotic genetic cascade. It has been observed that NGF deprivation causes cellular oxidative stress approx 3 h after trophic factor removal, as discerned by an oxidation-sensitive fluorochrome (Greenlund et al., 1995). This stress is critical to cell death, because death is blocked by agents that ameliorate oxidative stress. For example, antioxidants, such as *N*-acetyl cysteine, protect trophic-factor deprived neurons (Deckwerth T. L. and Johnson, E. M. (1994) *Soc. Neurosci. Abstr.* **20,** 682; Yan et al., 1996). Moreover, superoxide dismutase, an enzyme that degrades superoxide, inhibits death induced by NGF-deprivation (Greenlund et al., 1995). Significantly, this oxidative stress occurs prior to c-*jun* induction. This indicates that cellular antioxidant mechanisms, and their regulation, are critical to neuronal apoptosis.

The second set of observations suggesting that oxidative stress transduces the signals from NGF deprivation to gene induction is that oxidative stress activates pre-existing c-Jun by two mechanisms. First, oxidative stress, such as UV irradiation or hydrogen peroxide treatment, initiates a cascade of kinase activation that culminates in the phosphorylation of Jun kinase (JNK) (Derijard et al., 1994; Kyriakis et al., 1994; Pombo et al., 1994; Sanchez et al., 1994), which then activates pre-existing cytoplasmic c-Jun by phosphorylating c-Jun on ser-63 and ser-73. Second, Ref-1 activates c-Jun during oxidative stress via a critical cysteine in c-Jun (Abate et al., 1990). Ref-1 is activated by both increased *ref-1* transcription and by posttranslational activation of pre-existing Ref-1 (Ausserer et al., 1994). Hence, two pathways have been identified whereby oxidative stress leads to c-Jun induction. The role of *ref-1* during apoptosis remains unclear, but the activation of the JNK pathway has been implicated as necessary for neuronal apoptosis, in that JNK antagonists inhibit PC12 cell apoptosis after serum deprivation

(Xia et al., 1995). We have found that JNK is activated in NGF-deprived sympathetic neurons in a fashion essentially identical to that reported for the PC12 cells (C. van Rooyen and S. Estus, unpublished observations), suggesting by analogy that JNK may also be critical for the death of primary neurons. The phosphatase(s) that inactivate c-Jun, or the JNK pathway, have not been identified, but clearly represent a potential means to antagonize this apoptotic pathway.

5.2. *Aberrant Cell Cycle Regulation and Apoptosis*

Our observation that c-*jun* and related genes are induced during neuronal apoptosis is consistent with the possibility that neurons are induced to undergo apoptosis because of conflicting growth regulatory signals, i.e., these gene products stimulate the postmitotic neurons to attempt re-entry into the cell cycle. Indeed, apoptosis and the cell cycle have been recognized to share morphologic elements for some time (reviewed in Ucker, 1991); the data presented here strengthen the molecular link between these two processes. For example, c-*jun, junB*, c-*fos*, and *fos B* are induced during neuronal apoptosis and also on cell-cycle re-entry in fibroblasts; at least c-Jun appears necessary for both NGF-deprivation-induced apoptosis and cell cycle re-entry (Kovary and Bravo, 1991). Another shared molecular marker is cyclin D1 (Freeman et al., 1994), which we identified as induced during the PCD of sympathetic neurons, and which is also induced during the G_1 phase of the cell cycle; cyclin D1 is necessary for cell-cycle progression (Baldin et al., 1993). Although the point(s) where apoptosis and the cell cycle diverge requires clarification, Fra-1 and Fra-2, which can act to inhibit c-Jun and c-Fos (Suzuki et al., 1991), are induced subsequent to c-*jun* and c-*fos* during cell cycle re-entry (Cohen and Curran, 1988), but not during neuronal apoptosis; during cell cycle re-entry, c-Jun initially forms heterodimers with c-Fos and FosB, which are gradually displaced by Fra-1 and Fra-2 (Kovary and Bravo, 1992). Hence, the lack of *fra-1* induction during apoptosis may represent the loss of a critical regulatory factor.

To summarize, elements often associated with cell-cycle re-entry and progression are induced in neurons undergoing apoptosis. For postmitotic neurons, these elements may well represent a conflicting growth regulatory signal. By analogy with other systems wherein such signals induce cells to undergo apoptosis, e.g., c-*myc* overexpression in fibroblasts made quiescent by serum deprivation (Askew et al., 1991; Evan et al., 1992), large T antigen expression in postmitotic cells (Al-Ubaidi et al., 1992; Feddersen et al., 1992; Hammang et al., 1993), or p53 expression in transformed cells (Shaw et al., 1992; Ryan et al., 1993; Yonish-Rouach et al., 1993), these conflicting signals may initiate neuronal apoptosis.

6. SUMMARY

To summarize this chapter: Neuronal PCD is an important process that contributes to both physiologically appropriate and inappropriate neuronal death; trophic factor deprivation triggers PCD in neurons in vivo or maintained in vitro; the induction of specific genes, including c-*jun*, accompanies this neuronal apoptosis; oxidative stress probably plays a role in transducing NGF deprivation to intracellular genetic changes; c-Jun is required for neuronal death, at least in neurons undergoing NGF-deprivation; and the actions of c-Jun may involve inducing the postmitotic neurons to attempt cell cycle re-entry. This sequence of events is outlined in Fig. 1.

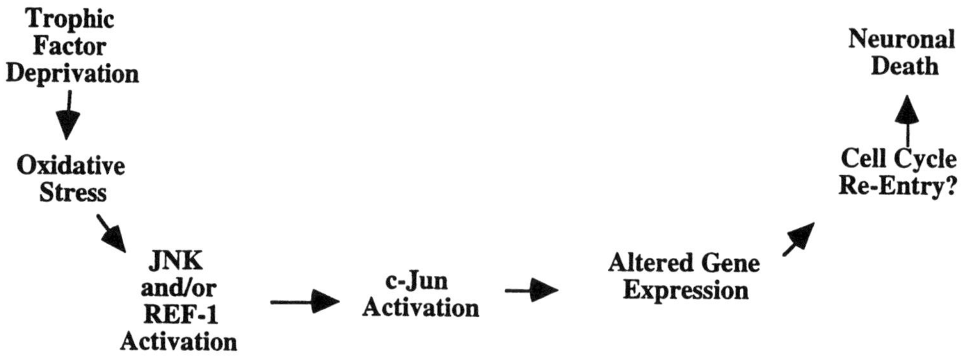

Fig. 1. Summary of our current model of intracellular events culminating in neuronal apoptosis induced by trophic factor deprivation. Agents that induce oxidative stress directly may activate the apoptosis pathway by acting downstream of trophic factor deprivation per se.

REFERENCES

Abate, C., Patel, L., Rauscher, F. J. 3., and Curran, T. (1990) Redox regulation of fos and jun DNA-binding activity in vitro. *Science* **249,** 1157–1161.

Al-Ubaidi, M. R., Hollyfield, J. G., Overbeek, P. A., and Baehr, W. (1992) Photoreceptor degeneration induced by the expression of simian virus 40 large tumor antigen in the retina of transgenic mice. *Proc. Natl. Acad. Sci. USA* **89,** 1194–1198.

Anderson, A. J., Cummings, B. J., and Cotman, C. W. (1994) Increased immunoreactivity for Jun- and Fos-related proteins in Alzheimer's disease: association with pathology. *Exp. Neurol.* **125,** 286–295.

Anderson, A. J., Su, J. H., and Cotman, C. W. (1995) DNA damage in Alzheimer's disease: colocalization with c-Jun immunoreactivity, and effects of brain area and postmortem delay. *Soc. Neurosci. Abstracts* **21,** 1009.

Angel, P., Hatori, K., Smeal, T., and Karin, M. (1988) The c-jun proto-oncogene is positively regulated by its product, Jun/AP-1. *Cell* **55,** 875–885.

Angel, P. and Karin, M. (1991) The role of Jun, Fos and the Ap-1 complex in cell proliferation and transformation. *Biochim. Biophys. Acta* **1072,** 129–157.

Askew, D. S., Ashmun, R. A., Simmons, B. C., and Cleveland, J. L. (1991) Constitutive c-myc expression in an IL-3-dependent myeloid cell line suppresses cell cycle arrest and accelerates apoptosis. *Oncogene* **6,** 1915–1922.

Ausserer, W. A., Bourrat-Floeck, B., Green, C. J., Laderoute, K. R., and Sutherland, R. M. (1994) Regulation of c-jun expression during hypoxic and low-glucose stress. *Mol. Cell. Biol.* **14,** 5032–5042.

Baldin, V., Lukas, J., Marcote, M. J., Pagano, M., and Draetta, G. (1993) Cyclin D1 is a nuclear protein required for cell cycle progression in G1. *Genes Dev.* **7,** 812–821.

Ball, M. J. (1977) Neuronal loss, neurofibrillary tangles and granulovacuolar degeneration in the hippocampus with ageing and dementia. *Acta Neuropathol.* **37,** 111–118.

Bonfoco, E., Krainc, D., Ankarcrona, M., Nicotera, P., and Lipton, S. A. (1995) Apoptosis and necrosis: two distinct events induced, respectively, by mild and intense insults with N-methyl-D-aspartate or nitric oxide/superoxide in cortical cell cultures. *Proc. Natl. Acad. Sci. USA* **92,** 7162–7166.

Bredesen, D. E. (1995) Neural apoptosis. *Ann. Neurol.* **38,** 839–851.

Clarke, P. G. (1990) Developmental cell death: morphological diversity and multiple mechanisms. *Anat. Embryol.* (Berlin), **181,** 195–213.

Cochran, B. H. (1993) Regulation of immediate early gene expression. *NIDA Res. Monogr. Ser.* **125,** 3–24.

Cohen, D. R. and Curran, T. (1988) fra-1: a serum-inducible, cellular immediate-early gene that encodes a fos-related antigen. *Mol. Cell. Biol.* **8,** 2063–2069.

Colotta, F., Polentarutti, N., Sironi, M., and Mantovani, A. (1992) Expression and involvement of c-fos and c-jun protooncogenes in programmed cell death induced by growth factor deprivation in lymphoid cell lines. *J. Biol. Chem.* **267,** 18,278–18,283.

Deckwerth, T. L. and Johnson, E. M., Jr. (1993) Temporal analysis of events associated with programmed cell death (apoptosis) of sympathetic neurons deprived of nerve growth factor (NGF). *J. Cell Biol.* **123,** 1207–1222.

Derijard, B., Hibi, M., Wu, I., Barrett, T., Su, B., Deng, T., Karin, M., and Davis, R. J. (1994) JNK1: a protein kinase stimulated by UV light and Ha-Ras that binds and phosphorylates the c-Jun activation domain. *Cell* **76,** 1025–1037.

Dragunow, M., Beilharz, E., Sirimanne, E., Lawlor, P., Williams, C., Bravo, R., and Gluckman, P. (1994) Immediate-early gene protein expression in neurons undergoing delayed detah, but not necrosis, following hypoxic-ischaemic injury to the rat brain. *Mol. Brain Res. 25,* 19–33.

Dragunow, M., Young, D., Hughes, P., MacGibbon, G., Lawlor, P., Singleton, K., Sirimanne, E., Beilharz, E., and Gluckman, P. (1993) Is c-Jun involved in nerve cell death following status epilepticus and hypoxic-ischaemic brain injury? *Mol. Brain Res.* **18,** 347–352.

Edwards, S. N., Buckmaster, A. E., and Tolkovsky, A. M. (1991) The death programme in cultured sympathetic neurones can be suppressed at the posttranslational level by nerve growth factor, cyclic AMP, and depolarization. *J. Neurochem.* **57,** 2140–2143.

Ellis, R. E., Yuan, J. Y., and Horvitz, H. R. (1991) Mechanisms and functions of cell death. *Annu. Rev. Cell Biol.* **7,** 663–698.

Estus, S., Zaks, W., Freeman, R., Gruda, M., Bravo, R., and Johnson, E. (1994) Altered gene expression in neurons during programmed cell death; identification of c-jun as necessary for neuronal apoptosis. *J. Cell Biol.* **127,** 1717–1727.

Evan, G. I., Wyllie, A. H., Gilbert, C. S., Littlewood, T. D., Land, H., Brooks, M., Waters, C. M., Penn, L. Z., and Hancock, D. C. (1992) Induction of apoptosis in fibroblasts by c-myc protein. *Cell* **69,** 119–128.

Feddersen, R. M., Ehlenfeldt, R., Yunis, W. S., Clark, H. B., and Orr, H. T. (1992) Disrupted cerebellar cortical development and progressive degeneration of Purkinje cells in SV40 T antigen transgenic mice. *Neuron* **9,** 955–966.

Forloni, G., Angeretti, N., Chiesa, R., Monzani, E., Salmona, M., Bugiani, O., and Tagliavini, F. (1993a) Neurotoxicity of a prion protein fragmant. *Nature* **362,** 543–546.

Forloni, G., Chiesa, R., Smiroldo, S., Verga, L., Salmona, M., Tagliavini, F., and Angeretti, N. (1993b) Apoptosis mediated neurotoxicity induced by chronic application of β amyloid fragment 25–35. *Neuroreport* **4,** 523–526.

Freeman, R. S., Estus, S., and Johnson, E. M., Jr. (1994) Analysis of cell cycle-related gene expression in postmitotic neurons: selective induction of cyclin D1 during programmed cell death. *Neuron* **12,** 343–355.

Gorin, P. D. and Johnson, E. M., Jr. (1980) Effects of long-term nerve growth factor deprivation on the nervous system of the adult rat: an experimental approach. *Brain Res.* **198,** 27–42.

Goto, K., Ishige, A., Sekiguchi, K., Iizuka, S., Sugimoto, A., and Yuzurihara, M. (1990) Effects of cycloheximide on delayed neuronal death in rat hippocampus. *Brain Res.* **534,** 299–302.

Greenlund, L. J. S., Deckwerth, T. L., and Johnson, E. M. (1995) Superoxide dismutase delays neuronal apoptosis: a role for reactive oxygen species in programmed neuronal death of protects sympathetic neurons from NGF deprivation induced apoptosis. *Neuron* **14,** 303–315.

Ham, J., Babij, C., Whitfield, J., Pfarr, C. M., Lallemand, D., Yaniv, M., and Rubin, L. L. (1995) A c-Jun dominant negative mutant protects sympathetic neurons against programmed cell death. **Neuron 14,** 927–939.

Hammang, J. P., Behringer, R. R., Baetge, E. E., Palmiter, R. D., Brinster, R. L., and Messing, A. (1993) Oncogene expression in retinal horizontal cells of transgenic mice results in a cascade of neurodegeneration. *Neuron* **10,** 1197–1209.

Hockenbery, D. M. (1994) bcl-2 in cancer, development and apoptosis. *J. Cell. Sci.* **18 (Suppl),** 51–55.
Hyman, B., T., Van Hoesen, G. W., Damasio, A. R., and Barnes, C. L. (1984) Alzheimer's disease: cell-specific pathology isolates the hippocampal formation. *Science* **225,** 1168–1170.
Hyman, B. T., Van, H. G., Kromer, L. J., and Damasio, A. R. (1986) Perforant pathway changes and the memory impairment of Alzheimer's disease. *Ann. Neurol.* **20,** 472–81.
Jacobson, M. D., Burne, J. F., and Raff, M. C. (1994) Programmed cell death and Bcl-2 protection in the absence of a nucleus. *EMBO J.* **13,** 1899–1910.
Johnson, E. M., Jr., Gorin, P. D., Brandeis, L. D., and Pearson, J. (1980) Dorsal root ganglion neurons are destroyed by exposure in utero to maternal antibody to nerve growth factor. *Science* **210,** 916–918.
Kessler, J. A., Ludlam, W. H., Freidin, M. M., Hall, D. H., Michaelson, M. D., and Spray, D. C., Dougherty, M., and Batter, D. K. (1993) Cytokine induced programmed death of cultured sympathetic neurons. *Neuron* **11,** 1123–1132.
Kotzbauer, P. T., Lampe, P. A., Estus, S., Milbrandt, J., and Johnson, E. M., Jr. (1994) Postnatal development of survival responsiveness in rat sympathetic neurons to leukemia inhibitory factor and ciliary neurotrophic factor. *Neuron* **12,** 763–773.
Kovary, K. and Bravo, R. (1991) The Jun and Fos protein families are both required for cell cycle progression in fibroblasts. *Mol. Cell. Biol.* **11,** 4466–4472.
Kovary, K. and Bravo, R. (1992) Existence of different Fos/Jun complexes during the G0-to-G1 transition and during exponential growth in mouse fibroblasts: differential role of Fos proteins. *Mol. Cell. Biol.* **12,** 5015–5023.
Kyriakis, J. M., Banerjee, P., Nikolakaki, E., Dai, T., Rubie, E. A., Ahmad, M. F., Avruch, J., and Woodgett, J. R. (1994) The stress-activated protein kinase subfamily of c-Jun kinases. *Nature* **369,** 1546–1560.
Lassmann, H., Bancher, C., Breitschopf, H., Wegiel, J., Bobinski, M., Jellinger, K., and Wisniewski, H. M. (1995) Cell death in Alzheimer's disease evaluated by DNA fragmentation in situ. *Acta Neuropathol.* (Berlin) **89,** 35–41.
Levi-Montalcini, R. and Booker, B. (1960) Destruction of the sympathetic ganglia in mammals by an anti-serum to the nerve-growth promoting factor. *Proc. Natl. Acad. Sci. USA* **46,** 384–391.
Levi-Montalcini, R., Caramia, F., and Angeletti, P. U. (1969) Alterations in the fine structure of nucleoli in sympathetic neurons following NGF-antiserum treatment. *Brain Res.* **12,** 54–73.
Linnik, M. D., Zobrist, R. H., and Hatfield, M. D. (1993) Evidence supporting a role for programmed cell death in focal cerebral ischemia in rats. *Stroke* **24,** 2002–2009.
Loo, D. T., Copani, A., Pike, C. J., Whittemore, E. R., Walencewicz, A. J., and Cotman, C. W. (1993) Apoptosis is induced by β-amyloid in cultured central nervous system neurons. *Proc. Natl. Acad. Sci. USA* **90,** 7951–7955.
MacManus, J. P., Buchan, A. M., Hill, I. E., Rasquinha, I., and Preston, E. (1993) Global ischemia can cause DNA fragmentation indicative of apoptosis in rat brain. *Neurosci. Lett.* **164,** 89–92.
MacManus, J. P., Hill, I. E., Huang, Z. G., Rasquinha, I., Xue, D., and Buchan, A. M. (1994) DNA damage consistent with apoptosis in transient ischaemic neocortex. *Neuro report* **5,** 493–496.
Martin, D. P., Schmidt, R. E., DiStefano, P. S., Lowry, O. H., Carter, J. G., and Johnson, E. M., Jr. (1988) Inhibitors of protein synthesis and RNA synthesis prevent neuronal death caused by nerve growth factor deprivation. *J. Cell Biol.* **106,** 829–844.
Migheli, A., Cavalla, P., Marino, S., Schiffer, D. (1994) A study of apoptosis in normal and pathologic nervous tissue after in situ end-labeling of DNA strand breaks. *J. Neuropathol. Exp. Neurol.* **53,** 606–616.
Morgan, J. I. and Curran, T. (1991) Stimulus-transcription coupling in the nervous system: involvement of the inducible proto-oncolgenes fos and jun. *Annu. Rev. Neurosci.* **14,** 421–451.

Oppenheim, R. W. (1991) Cell death during development of the nervous system. *Annu. Rev. Neurosci.* **14,** 453–501.

Oppenheim, R. W., Prevette, D., Tytell, M., and Homma, S. (1990) Naturally occurring and induced neuronal death in the chick embryo in vivo requires protein and RNA synthesis: evidence for the role of cell death genes. *Dev. Biol.* **138,** 104–113.

Pollard, H., Cantagrel, S., Charriaut-Marlangue, C., Moreau, J., and Ben Ari, Y. (1994) Apoptosis associated DNA fragmentation in epileptic brain damage. *Neuro report* **5,** 1053–1055.

Pombo, C. M., Bonventre, J. V., Avruch, J., Woodgett, J. R., Kyriakis, J. M., and Force, T. (1994) The stress-activated protein kinases are major c-Jun amino-terminal kinases activated by ischemia and reperfusion. *J. Biol. Chem.* **269,** 26, 546–26, 551.

Portera-Calliau, C., Hedreen, J. C., Price, D. L., and Koliatsos, V. E. (1995) Evidence for apoptotic cell death in Huntington disease and excitotoxic animal models. *J. Neurosci.* **15,** 3775–3787.

Price, D. L., Martin, L. J., Clatterbuck, R. E., Koliatsos, V. E., Sisodia, S. S., Walker, L. C., and Cork, L. C. (1992) Neuronal degeneration in human diseases and animal models. *J. Neurobiol.* **23,** 1277–1294.

Robinson, G. A. (1994) Immediate early gene expression in axotomized and regenerating retinal ganglion cells of the adult rat. *Brain Res.* **24,** 43–54.

Ryan, J. J., Danish, R., Gottlieb, C. A., and Clarke, M. F. (1993) Cell cycle analysis of p53-induced cell death in murine erythroleukemia cells. *Mol. Cell. Biol.* **13,** 711–719.

Sanchez, I., Hughes, R. T., Mayer, B. J., Yee, K., Woodgett, J. R., Avruch, J., Kyriakis, J. M., and Zon, L. I. (1994) Role of SAPK/ERK kinase-1 in the stress-activated pathway regulating transcription factor c-Jun. *Nature* **372,** 794–798.

Scott, S. A. and Davies, A. M. (1990) Inhibition of protein synthesis prevents cell death in sensory and parasympathetic neurons deprived of neurotrophic factor in vitro. *J. Neurobiol.* **21,** 630–638.

Shaw, P., Bovey, R., Tardy, S., Sahli, R., Sordat, B., and Costa, J. (1992) Induction of apoptosis by wild-type p53 in a human colon tumor-derived cell line. *Proc. Natl. Acad. Sci. USA* **89,** 4495–4499.

Shigeno, T. (1990) Reduction of delayed neuronal death by inhibition of protein synthesis. *Neurosci. Lett.* **120,** 117–119.

Steller, H. (1995) Mechanisms and genes of cellular suicide. **Science 267,** 1445–1449.

Su, J. H., Anderson, A. J., Cummings, B. J., and Cotman, C. W. (1994) Immunohistochemical evidence for apoptosis in Alzheimer's disease. *Neuroreport* **5,** 2529–2533.

Suzuki, T., Okuno, H., Yoshida, T., Endo, T., Nishina, H., and Iba, H. (1991) Difference in transcriptional regulatory function between c-fos and Fra-2. *Nucleic Acids Res.* **19,** 5537–5542.

Tomlinson, B. E., Irving, D., and Blessed, G. (1981) Cell loss in the locus coerulus in senile dementia of Alzheimer type. *J. Neurol. Sci.* **49,** 4419–4428.

Ucker, D. S. (1991) Death by suicide: one way to go in mammalian cellular development? *New Biol.* **3,** 103–109.

Watt, J. A., Pike, C. J., Walencewicz-Wasserman, A. J., and Cotman, C. W. (1994) Ultrastructural analysis of beta-amyloid-induced apoptosis in cultured hippocampal neurons. *Brain Res.* **661,** 147–156.

White, K., Grether, M. E., Abrams, J. M., Young, L., Farrell, K., and Steller, H. (1994) Genetic control of programmed cell death in Drosophila. *Science* **264,** 677–683.

Woodgett, J. R., Pulverer, B. J., Nikolakaki, E., Plyte, S., Hughes, K., Franklin, C. C., and Kraft, A. S. (1993) Regulation of Jun/AP-1 oncoproteins by protein phosphorylation, in *Advances in Second Messenger Research* (Brown, B. L. and Dobson, P. R. M., eds.) Raven, NY, pp. 261–269.

Wyllie, A. H., Kerr, J. F. R., and Currie, A. R. (1980) Cell death: the significance of apoptosis. *Int. Rev. Cytol.* **68,** 251–306.

Xia, Z., Dickens, M., Raingeaud, J., Davis, R. J., and Greenberg, M. E. (1995) Opposing effects of ERK and JNK-p38 MAP kinases on apoptosis. *Science* **270,** 1326–1331.

Yan, C. Y. I., Ferrari, G., and Greene, L. A. (1996) N-acetylcysteine-promoted survival of PC12 cells is glutathione-independent but transcription-dependent. *J. Biol. Chem.* **270,** 26,827–26,832.

Yonish-Rouach, E., Grunwald, D., Wilder, S., Kimchi, A., May, E., Lawrence, J. J., May, P., and Oren, M. (1993) p53-mediated cell death: relationship to cellcycle control. *Mol. Cell. Biol.* **13,** 1415–1423.

Zola-Morgan, S., and Squire, L. R. (1993) Neuroanatomy of memory. *Anu. Rev. Neurosci.* **16,** 547–563.

5
Clinical Potential of Compounds That Stimulate Nerve Growth Factor Production

Atsumi Nitta, Shoei Furukawa, and Toshitaka Nabeshima

1. INTRODUCTION

Neuronal atrophy or death occurs in aging and neurological diseases. Recent research has identified some neurotrophins, such as nerve growth factor (NGF), brain-derived neurotrophic factor (BDNF), neurotrophin-3 (NT-3), NT-4/5, and NT-6, and their involvement in various processes in neuronal growth, differentiation, and maintenance has been clarified (Leibrock et al., 1989; Hohn et al., 1990; Maisonpierre et al., 1990; Hallbook et al., 1991; Gotz et al., 1994). These neurotrophins are closely linked to injured cells and are capable of initiating a cascade of events in neurons and glia designed to prevent further damage. In many types of injury, both in the central nervous system (CNS) and in the peripheral nervous system (PNS), the induction of neurotrophins has been observed. The amounts of NGF increase after sciatic nerve lesion or transection (Lindholm et al., 1987), fimbria fornix transection (Gasser et al., 1986; Weskamp et al., 1986), electrolytic lesion of the septohippocampal pathway (Hefti, 1986; Oderfeld-Nowak et al., 1992), needle injection (Ballarin et al., 1991), denervation of the nigral dopaminernergic neuronal system (Nitta et al., 1992), limbic seizures (Gall and Issackson, 1989), transection of the optic nerve (Lu et al., 1991), excitotoxic destruction of hippocampal neurons (Bakhit et al., 1991), bilateral decortication (Lorez et al., 1988), and the evocation of aggressive behavior in adult male mice (Spillantini et al., 1989).

Although methods of determination for the proteins of other neurotrophins have not yet been developed, changes in mRNA of other neurotrophins in some cases of neuronal injury have been demonstrated. NT-3 mRNA increases after treatment with excitotoxins and decreases following ischemia (Lindvall, 1992; Rocamora et al., 1993). BDNF mRNA increases after the insult of trauma and the infusion of kainate (Ballarin et al., 1991).

The reports above suggest that these neurotrophins may play important roles in neuroprotection from neuronal injury. It has been demonstrated that neurotrophins can prevent damage from a variety of neuronal insults: NGF prevents the neuronal damage induced by glucose deprivation (Cheng and Mattson, 1991), ischemia (Shigeno et al., 1991), axotomy (Hefti, 1986; Williams et al., 1986), excitotoxins (Frim et al., 1993; Ogawa et al., 1993), hydrogen peroxide (Jackson et al., 1990), iron (Zhang et al., 1993),

From: Neuroprotective Signal Transduction *edited by M. P. Mattson*
Humana Press Inc., Totowa, NJ

cyanide (Mattson and Cheng, 1994), and diabetes neuropathy (Fernyhough et al., 1995). It has also been reported that BDNF prevents neuronal damage induced by 6-hydroxydopamine (Spina et al., 1992) and axotomy (Yan et al., 1992). NT-3 prevents the death of adult central noradrenergic neurons from the lesion induced by 6-hydroxydopamine (Arenas and Persson, 1994). These results indicate that some neurotrophins have a potent protective activity on neuronal degeneration, in both the CNS and PNS.

Neurotrophins are macromolecules that cannot pass through the blood–brain barrier, suggesting a limited availability of these proteins for therapeutic use in diseases characterized by neuronal degeneration in the CNS. Since neurotrophins are easily metabolized by peptidases when they are administered peripherally, the neuroprotective effects of neurotrophins, even in the PNS, cannot be expected. The insertion of a neurotrophin-delivering catheter into the area of neuronal dysfunction would strengthen the effects, but ethically, direct and continuous infusion into the CNS or PNS is not feasible in terms of the quality of life of the patients. We have speculated that compounds with a low mol wt, which elicit stimulation of neurotrophin synthesis, could be much more useful and practical for therapeutic purposes.

To investigate the physiological functions of the compounds that induce neurotrophin synthesis, the neurotrophin mRNA and protein levels following administration of the stimulators should be established. Some aspects of the regulatory mechanisms of the proteins involved have already been clarified. At present, the mRNA of any protein can be detected and measured, but not all protein levels can be measured. NGF is such a well-known neurotrophic factor that many studies on it have been carried out worldwide. Enzyme immunoassay (EIA) methods have been completely established (Furukawa et al., 1983), and aspects of the regulatory mechanism of NGF in vivo and in vitro, as well as its physiological roles in the CNS and PNS, have been clarified. However, the proteins of other neurotrophins cannot be detected at present, and physiological studies for them have not yet been developed. In this chapter, we focus on NGF and describe our novel approach to the development of therapeutic agents that induce NGF synthesis and ameliorate neuronal dysfunction in some dementia models in rats.

2. STIMULATORS OF NGF SYNTHESIS OBSERVED IN VITRO AND IN VIVO

Many previous studies have demonstrated that certain types of cells, such as peripheral fibroblasts (Pantazis et al., 1977; Lindholm et al., 1988; Schwartz, 1988), NIH3T3 cells (Green, 1991), Schwann cells (Ohi et al., 1990), primary hippocampal neurons (Lindholm et al., 1992), and astroglial cells (Furukawa et al., 1986b), synthesize NGF in culture systems. Since the effects of many compounds on NGF synthesis have been studied in only one of several unrelated systems (Table 1), it is usually not possible to compare their effects directly. As shown in Table 1, NGF is induced by dexamethazone in primary cultured neurons (Lindholm et al., 1992), but not in astrocytes. The stimulatory effects of IL-1β on NGF synthesis is quite different among cell types. This cytokine, which was first reported to induce NGF in nonneuronal cells of the sciatic nerve (Lindholm et al., 1987), was later shown to induce NGF mRNA in primary fibroblasts (Lindholm et al., 1992), astrocytes (Spranger et al., 1990), and mesangial cells (Steiner et al., 1991), but not in Schwann cells (Matsuoka et al., 1991) or neurons (Vige et al., 1992). In mixed-culture systems, cortical neurons inhibited the IL-1β-elicited secreted

Table 1
Stimulators for NGF Synthesis/Secretion in Various Cultured Cells

NGF stimulators in the primary astroglial cells of rat or mouse	
Dopamine	(Furukawa et al., 1987)
Epinephrine	(Furukawa et al., 1987)
Isoproterenol	(Furukawa et al., 1987)
Derivatives of 1,4-benzoquinone	(Takeuchi et al., 1990)
Phorbol 12-myristate 13-acetate	(Neveu et al., 1992)
Hericenones C, D, and E	(Kawagishi et al., 1991)
Interleukin 1-β (IL1-β)	(Lindholm et al., 1987)
IL-4	(Awatsuji et al., 1993)
IL-5	(Awatsuji et al., 1993)
Acidic fibroblast growth factor (aFGF)	(Ono et al., 1991)
Basic fibroblast growth factor (bFGF)	(Ono et al., 1991; Yoshida and Gage, 1991)
Tumor necrosis factor-α (TNF-α)	(Ono et al., 1991; Yoshida and Gage, 1991)
Epidermal growth factor (EGF)	(Yoshida and Gage, 1991)
Transforming growth factor-β1 (TGF-β1)	(Lindholm et al., 1990)
NGF stimulators in the primary neuronal cells of rat or mouse	
Isoproterenol	(Schwartz and Mishler, 1990)
Dexamethasone	(Lindholm et al., 1992)
Aldosterone	(Lindholm et al., 1992)
Prostaglandin E2	(Friedman et al., 1990)
Lipopolysaccharide	(Friedman et al., 1990)
NGF stimulators in the fibroblast L-M cells of mouse	
Dopamine	(Furukawa et al., 1987)
Epinephirine	(Furukawa et al., 1987)
DOPA	(Furukawa et al., 1986a)
Norepinephirine	(Furukawa et al., 1986a)
4-Methylcatechol and related structure	(Furukawa et al., 1986a, 1987)
Fellutamide A	(Yamaguchi et al., 1993)
Pyrroloquinoline quinone	(Yamaguchi et al., 1996)
Kansuinin A	(Yamaguchi et al., 1996)
Ingenol triacetate	(Yamaguchi et al., 1996)
NGF stimulators in the fibroblast L929 cells of mouse	
1,25-Dihydroxyvitamin D3	(Wion et al., 1991)
Retinoic acid	(Wion et al., 1987)
Phorbol 12-myristate 13-acetate	(D'Mello et al., 1990)

NGF normally observed in pure cultures of cortical astrocytes (Vige et al., 1992). These results may be inconsistent with the in vivo findings that NGF induction by IL-1β in the hippocampus and basal forebrain was shown in astrocytes, but constitutive NGF synthesis remained high in neurons (Oderfeld-Nowak et al., 1992). These data suggest that a given NGF-inducing compound may evoke differential effects on NGF synthesis in selected tissues, depending on the distribution of cell types in that region.

An in vivo study has shown that excitotoxin-induced destruction of the dorsal hippocampus neurons increased NGF-like immunoreactivity with the activation of astrocytes (Bakhit et al., 1991). In neurological disorders, such as Alzheimer's disease (AD), there is a dramatic increase in the expression of glial fibrillary acidic protein (GFAP) in cortical astrocytes, with an obvious laminar pattern, indicating that astrocytes in AD patients' brains are activated (Duffy et al., 1980; Delacourte, 1990). Based on these results, we hypothesize that stimulators of NGF synthesis in astrocytes may be effective in neuronal disease. As noted above, it is difficult to select the culture system and conditions for the screening of stimulators for NGF synthesis in vitro. Therefore, we have focused on the compounds that increase the NGF contents in the media of quiescent astroglial cells as stimulators of NGF synthesis. Cultured astrocytes have been shown to produce NGF (Furukawa et al., 1986b).

3. IDEBENONE AND PROPENTOFYLLINE, STIMULATORS FOR NGF SYNTHESIS/SECRETION

3.1. *Effects in Cultured Primary Astroglial Cells*

Several compounds have been reported to enhance NGF synthesis/secretion in cultured mouse astroglial cells (Table 1). Catechols, such as dopamine, epinephrine, and isoproterenol, increase NGF contents in media cultured with astroglial cells (Table 1; Furukawa et al., 1986a, 1987; Schwartz and Mishler, 1990). Growth factors, such as acidic fibroblast growth factor (Fukumoto et al., 1991; Ono et al., 1991; Yoshida and Gage, 1991), bFGF (Yoshida and Gage, 1991), tumor necrosis factor-α (TNF-α) (Steiner et al., 1991), epidermal growth factor (Yoshida and Gage, 1991), and hericenones C, D, and E (Kawagishi et al., 1991), also stimulate NGF synthesis/secretion in cultured astroglial cells. Low-mol-wt compounds are more useful for therapeutic purposes than are growth factors or cytokines, since they can maintain adequate stability in the bloodstream to produce long-lasting effects. Compounds that possess a simple structure for chemical modifications facilitate the design of efficacious drug-delivery systems and enhance the potency of the drug; they also have a low mol wt and pass through the blood–brain barrier (Furukawa and Furukawa, 1990). Several compounds of low mol wt upregulate NGF synthesis/secretion of cultured astroglial cells. A xanthine derivative, propentofylline, stimulates NGF synthesis/secretion in quiescent astroglial cells in a dose-dependent manner (Shinoda et al., 1990). Another group of compounds with similar activities are 1,4-benzoquinone and its derivatives (Takeuchi et al., 1990; Nitta et al., 1993b). Similarly, 4-alkylcatechols positively influence NGF synthesis in cultured quiescent astroglial cells (Furukawa et al., 1990). Furthermore, 4-methylcatechol (4-MC), which is composed of a catechol ring with a methyl group as a side-chain at position 4, also stimulates NGF synthesis/secretion in cultured astroglial cells. These low-mol-wt chemical compounds are very important candidates for therapeutic drugs for neurological diseases (chemical structures of each compound are shown in Fig. 1).

3.2. *Effects in Forebrain-Lesioned and Aged Rats*

3.2.1. *The Pharmacological Effects on NGF Contents*

The repeated administration of idebenone (10 and 20 mg/kg, po) for 21 d increased the NGF level in basal forebrain-lesioned (Nitta et al., 1994) and aged rat brains (Nitta et al., 1993a), although it did not affect the NGF level in young rat brains (Nitta et al.,

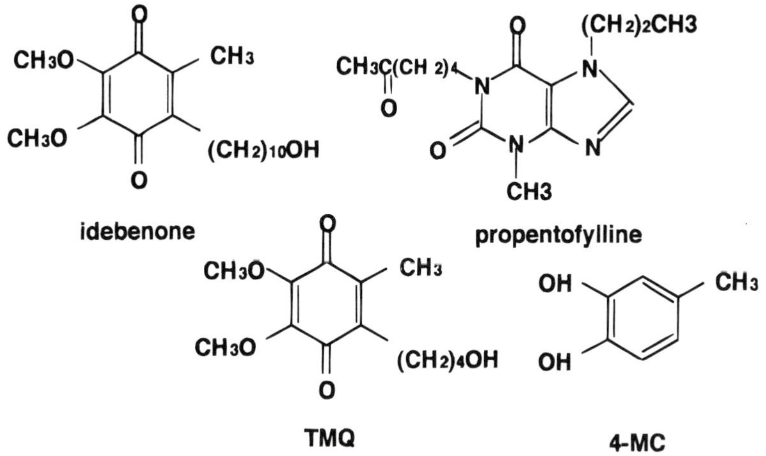

idebenone; 6-(10-hydroxydecyl)-2,3-dimethoxy-5-methyl-1,4-benzoquinone
propentofylline; 3,7-dihydro-3-methyl-1-(5-oxo-hexyl)-7-propyl-1H-purine-2,6-dione
TMQ; 6-(4-hydroxybutyl)-2,3,5-trimethyl -1,4-benzoquinone
4-MC; 4-methylcatechol

Fig. 1. Chemical structures of stimulators for NGF synthesis.

1993a) (Tables 2 and 3). It is interesting that idebenone was effective only in the damaged (basal forebrain-lesioned or aged) rat brain. Propentofylline also increased the NGF level in aged rat brains (Table 2; Nabeshima et al., 1993). Excitotoxin-induced destruction of dorsal hippocampal neurons increased NGF-like immunoreactivity, which was accompanied by the activation of astrocytes (Bakhit et al., 1991). That astroglial cells are activated by aging is shown by the finding that glial fibrillary acidic protein (GFAP) and its mRNA increase with age in rodents (Goss et al., 1991; O'Callaghan and Miller, 1991). It is thus conceivable that the effects of idebenone and propentofylline on NGF level in the brain may be related to the degree of astrocyte activation. However, thus far, there is no direct evidence for NGF production by astrocytes in vivo. Further studies should therefore be undertaken, using *in situ* hybridization techniques, to clarify the cellular localization of NGF synthesis. In AD patients' brains, there is a dramatic increase in the expression of GFAP in cortical astrocytes, with an obvious laminar pattern, indicating that astrocytes in AD patients are activated (Duffy et al., 1980; Delacourte, 1990). We therefore believe that stimulators of NGF synthesis in astrocytes may be effective in treating patients with neurodegenerative disorders such as AD.

3.2.2. *Effects on Learning and Memory*

As described above, idebenone and propentofylline increased the NGF levels in the brains of aged and basal forebrain-lesioned rats. To investigate whether the increased NGF levels could ameliorate cholinergic neuronal dysfunction or impairment of learning and memory, we carried out behavioral and biochemical experiments (Fuji et al., 1993; Nitta et al., 1994).

In a water-maze task, the latencies to the platform in the sham-operated group in the first training session did not differ from those in the basal forebrain-lesioned groups.

Table 2
Effects of Idebenone and Propentofylline on Brain NGF Content in Aged Rats[a]

Treatment	N	NGF content (pg/g wet wt) Olfactory bulb	Frontal cortex	Parietal cortex	Septum	Hippocampus
Young rats + Vehicle	4	1773.3 ± 337.0	3525.0 ± 562.9	2958.3 ± 454.8	507.7 ± 122.9	5450.0 ± 689.3
Aged rats + Vehicle	4	1947.0 ± 488.4	560.5 ± 263.0[c]	1980.0 ± 280.1[b]	426.0 ± 39.1	2957.5 ± 677.7[b]
+ Idebenone (10 mg/kg)	4	1471.0 ± 166.1	1994.4 ± 219.9[d]	3048.0 ± 375.9[d]	381.0 ± 81.1	4172.0 ± 292.2
+ Idebenone (20 mg/kg)	4	1420.0 ± 264.9	1556.0 ± 147.4[d]	2681.0 ± 149.5	468.0 ± 74.7	3990.0 ± 746.6
+ Propentofylline (10 mg/kg)	4	1186.7 ± 80.1	1850.0 ± 101.0	3286.7 ± 413.3[d]	486.0 ± 68.9	2723.3 ± 799.9
+ Propentofylline (25 mg/kg)	4	1311.2 ± 271.2	2184.0 ± 224.0[d]	2564.0 ± 384.7	1080.8 ± 574.2	2860.0 ± 441.1

[a] Values represent means ± SE. NGF was measured by enzyme immunoassay after a 21-successive-day administration of idebenone (10 and 20 mg/kg) or propentofylline (10 and 25 mg/kg).
[b] $p < 0.05$,
[c] $p < 0.005$ vs vehicle-treated young rats,
[d] $p < 0.05$ vs vehicle-treated aged rats.

Table 3
Effects of Idebenone on Brain NGF Content in Young Rats[a]

Treatment	N	NGF content (pg/g wet wt) Olfactory bulb	Frontal cortex	Parietal cortex	Septum	Hippocampus
Young rats + Vehicle	4	1052.0 ± 144.8	4227.5 ± 1087.7	3880.0 ± 698.6	315.8 ± 14.0	5253.3 ± 141.4
Aged rats + Vehicle	4	1452.3 ± 121.5	3946.7 ± 897.0	3083.1 ± 1072.7	286.5 ± 96.1	4933.3 ± 450.4

[a] Values represent means ± SE. NGF was measured by enzyme immunoassay after a 21-successive-day administration of idebenone (20 mg/kg).

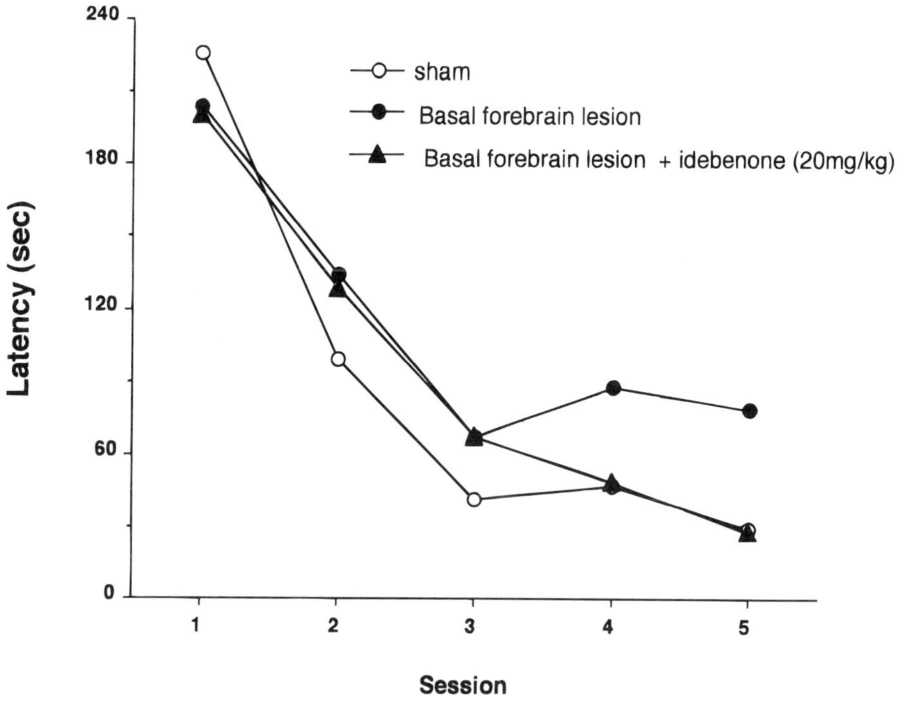

Fig. 2. Effects of idebenone on the performance of rats in a water maze task. Each value represents the mean of the latency to escape onto the hidden platform. The tasks were carried out on d 21–25 after the first basal forebrain lesion was produced.

However, repeated training slowly shortened the latencies in the basal forebrain-lesioned groups; those in the sham-operated group were rapidly shortened. The administration of idebenone attenuated the basal forebrain-lesion-induced impairment of learning (Fig. 2; Nitta et al., 1994). In the spatial probe trial, carried out after the final training to examine whether the rats had learned the position of the platform, the basal forebrain-lesioned groups, compared with the sham group, showed a significant decrease in the time spent in the quadrant in which the platform had been located (platform-quadrant) during trial. Propentofylline reversed this decreased time to the level of that in the sham group (Fuji et al., 1993).

The rats in the sham-operated group had a very long step-through latency during the retention test for the passive avoidance task. In the vehicle-treated basal forebrain-lesioned rats, the step-through latency was shorter than that in the vehicle-treated sham-operated rats. Basal forebrain-lesioned rats treated with idebenone and propentofylline showed a significantly longer step-through latency than that shown in the vehicle-treated group (Fuji et al., 1993; Nitta et al., 1994).

Successive oral administration of stimulators of NGF synthesis/secretion recovers the behavioral deficits in learning and memory in basal forebrain-lesioned rats (Nitta et al., 1993b; 1994). Choline acetyltransferase activities, one of the marker enzymes for the cholinergic neuronal system, was reduced by basal forebrain lesion in several brain areas of rats. The repeated administration of idebenone or propentofylline prevented the cholinergic dysfunction. The improved ability to learn and remember following administration of idebenone and propentofylline, as described above, was concomitant with cholinergic recovery.

We found that 6-(4-hydroxybutyl)-2,3,5-trimethyl-1,4-benzoquinone (TMQ; chemical structure shown in Fig. 1) stimulates NGF synthesis/secretion in mouse astroglial cells. The compound also had improving effects on memory in the water-maze and passive-avoidance tasks, and prevented cholinergic dysfunction in basal forebrain-lesioned rats (Nitta et al., 1993b).

3.2.3. *Prevention of Dysfunction Induced by Infusion of Anti-NGF Antibody*

As described above, the peripheral administration of idebenone and propentofylline increases brain NGF content in vivo and reverses the reduced cholineacetyltransferase activity and defects in learning ability induced by basal forebrain lesion. However, direct evidence that these compounds exerted their effects on cognitive dysfunction via the stimulation of NGF synthesis has not been obtained, since these compounds also have other effects (Mrsulja et al., 1983; Stefanovich and Nagata, 1983; Nagaoka et al., 1984; Nagata et al., 1984; Yamazaki et al., 1984; Banati et al., 1993).

To address the multieffect issue, we have developed a new dementia model in rats, induced by depleting endogenous NGF by immunological means; in this model, specific deficits are shown in cholinergic function (Nitta et al., 1993c). With this dementia model, we can investigate whether the above stimulators exerted their effects on cognitive dysfunction via the stimulation of NGF synthesis (Nitta et al., 1996).

We developed our dementia model by infusing anti-NGF antibody into the septum of rats; behavioral deficits and a reduction in cholinergic activity were induced (Nitta et al., 1993c). Propentofylline ameliorated the behavioral deficits in the water-maze and passive-avoidance tasks (Nitta et al., 1996).

In the water-maze task, the latencies in the control group on the first training trial were not different from those of the anti-NGF-antibody infusion group. Repeated training slowly shortened the latencies in the anti-NGF antibody infusion group, but rapidly shortened the latencies in the control group. The administration of propentofylline attenuated the anti-NGF antibody infusion-induced impairment of learning. In the fourth and eighth trials, the latencies of propentofylline (10 mg/kg)-treated rats were shorter than those of the vehicle-treated rats (Fig. 3, Nitta et al., 1996).

In the passive-avoidance task, the control rats had a very long step-through latency during the retention test. The anti-NGF antibody-infused rats had a shorter step-through latency than did the controls. The propentofylline (10 mg/kg)-treated rats showed a significantly longer step-through latency than did the anti-NGF antibody-infused rats (Nitta et al., 1996).

Furthermore, propentofylline prevented the severe damage observed in the cholinergic neuronal system in this model. Choline acetyltransferase activity in the hippocampus was reduced to 76% of the control level by the infusion of anti-NGF antibody. However, treatment with propentofylline (10 and 25 mg/kg) prevented the reduction in choline acetyltransferase activity in the hippocampus (Table 4; Nitta et al., 1996). Cholinesterase activity in the hippocampus of rats infused with anti-NGF antibody was markedly reduced, to 33% of the control level. Again, treatment with propentofylline (25 mg/kg) also prevented the reduction in cholinesterase activity in these rats (Table 4; Nitta et al., 1996).

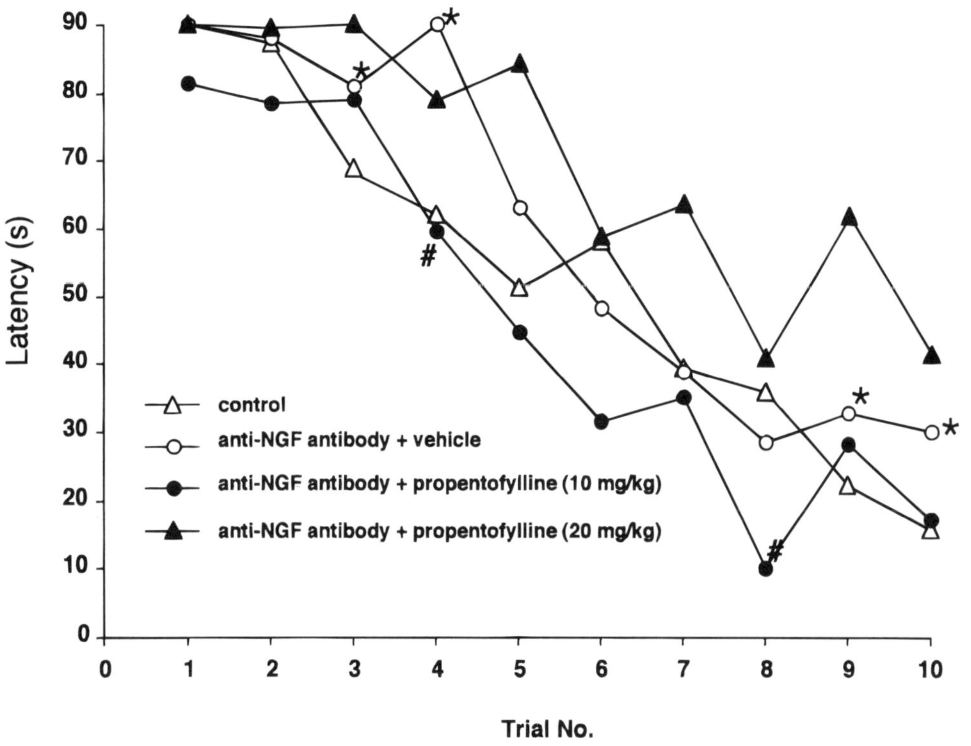

Fig. 3. Effects of propentofylline on the performance in the water maze task of rats that received a continuous infusion of anti-NGF antibody into the septum. $*p < 0.05$ vs control group and $\#p < 0.05$ vs anti-NGF antibody-infused group.

Idebenone had ameliorating effects similar to those of propentofylline on the impairment of learning and memory; it also induced a cholinergic reduction in the anti-NGF antibody-infused model (Table 4; Nitta et al., unpublished results). These results suggest that the ameliorating effects of propentofylline and idebenone on the impairment of learning and memory and on the deficits of cholinergic function in this dementia model may be related to their stimulatory effects on NGF synthesis.

4. EFFECTS OF 4-METHYLCATECHOL, AN NGF STIMULATOR, ON PNS IN VIVO

Furukawa et al. (1990) have demonstrated that 4-MC (chemical structure shown in Fig. 1), a derivative of catecholamine that probably lacks adrenergic action, stimulates the synthesis/secretion of NGF in cultured astroglial cells. The IP administration of 4-MC induced an increase in NGF protein and mRNA in the adult rat heart and submaxillary gland (Kaechi et al., 1992, 1995). The increase in NGF protein was successively translocated from the sciatic nerve to sensory or sympathetic ganglia. Furthermore, Hanaoka et al. (1992) have demonstrated the effects of the administration of 4-MC in the streptozotocin (STZ)-induced diabetic rat, which is the most commonly used animal model of human diabetic neuropathy: a decrease of NGF and a reduction of motor nerve conduction velocity (MNCV). Treatment with 4-MC in the diabetic rats for 4 wk, starting from the STZ injection, elevated the NGF content and prevented the

Table 4
Effects of Idebenone and Propentofylline on the Reduced Activity of Choline Acetyltransferase and Cholinesterase in the Hippocampus of Rats Receiving a Continuous Infusion of Anti-NGF Antibody[a]

Treatment		*N*	Choline acetyltransferase activity (nmol/g/min)	Cholinesterase activity (μmol/ngr/h)
	+ Vehicle	5	3702.2 ± 152.7	1702.1 ± 573
Anti-NGF antibody	+ Vehicle	5	2806.2^{b} ± 294.1	559.3^{b} ± 74
	+ Idebenone (10 mg/kg)	5	4227.6^{c} ± 773.8	979.5 ± 265
	+ Idebenone (20 mg/kg)	5	4203.9^{c} ± 207.9	1016.6^{c} ± 315
	+ Propentofylline (10 mg/kg)	5	4153.2^{c} ± 252.7	917.3 ± 299
	+ Propentofylline (25 mg/kg)	5	4541.1^{c} ± 442.7	1383^{c} ± 303

[a] Values represent means ± SE. Choline acetyltransferase and cholinesterase were measured after a 21-successive-day administration of idebenone (10 and 20 mg/kg) or propentofylline (10 and 25 mg/kg).

[b] $p < 0.05$ vs control rats.

[c] $p < 0.05$ vs vehicle-treated anti-NGF antibody-infused rats.

reduction in MNCV, but no effect on high blood glucose levels was seen (Hanaoka et al., 1992). These findings suggest that decreased NGF levels in the sciatic nerve of the experimental diabetic rat may be involved in the development of the diabetic neuropathic process, and that 4-MC, which can elevate endogenous NGF levels in vivo, may compensate for the inhibitory effect of STZ on the NGF synthesis in progressive diabetic neuropathy.

5. EXPECTATIONS FOR OTHER NEUROTROPHINS IN THE CLINICAL FIELD

At present there is no firm evidence to support the hypothesis that any neurodegenerative disease results from a decrement in the level of a neurotrophin, a neurotrophin receptor, or the signal transduction pathway of a neurotrophin. BDNF may play more important roles than does NGF, since the expression of BDNF mRNA is 50 times as strong as that of NGF mRNA in rat brain (Hofer et al., 1990). A recent study has shown that there is a loss of BDNF mRNA expression in the hippocampus of AD patients (Phillips et al., 1991). On the one hand, this decrease in BDNF mRNA might account for the loss of cholinergic neurons in AD because of a reduction in target-derived BDNF; on the other hand, it may account for the loss of neurons within the hippocampal formation and related areas of temporal cortex afferent to the hippocampus. Therefore, BDNF must be a key protein for the establishment of therapeutic methods for neurological disease using neurotrophic factors. In fact, some neurologists suggest that BDNF is also effective in Parkinson's and amyotrophic lateral sclerosis disease (Barinaga, 1994). Furthermore, one study has suggested that the coadministration of BDNF and ciliary neurotrophic factor is effective in motor neurons in an animal model for amyotrophic lateral sclerosis (Mitsumoto et al., 1994). Yet, thus far, few studies on the relationship between BDNF and neurological diseases have been performed, since methods of measurement of BDNF have not been established. Recently, however, an EIA system for BDNF has been developed (Nawa et al., 1995), so further studies on the relationship between BDNF and neurological disorders may be expected in the near future.

6. CONCLUSIONS

Effective therapeutic methods have not been established for many neurological disorders, including AD, Parkinson's disease, neuropathy induced by diabetes, and so on. The impairments in neuronal disease patients may be ameliorated with the use of neurotrophins. Our novel approach, peripheral administration of stimulators for the synthesis/secretion of neurotrophins, could be useful for clinical trials of these neurological diseases.

ACKNOWLEDGMENTS

This work was supported in part by an SRF Grant for Biomedical Research, by a Grant-in-Aid for Science Research from the Ministry of Education, Science, and Culture of Japan (No. 05454148, 07557009, and 09780727), by a grant for Gerontological Science Research from the Ministry of Health and Walfare of Japan (No. 91A-2406 and 94A-2405), and by a grant from the Uehara Memorial Foundation.

REFERENCES

Arenas, E. and Persson, H. (1994) Neurotrophin-3 prevents the death of adult central non-adrenergic neurons in vivo. *Nature* **367,** 368–371.

Awatsuji, H., Furukawa, Y., Hirota, M., Murakami, Y., Nii, S., Furukawa, S., and Hayashi, K. (1993) Interleukin-4 and -5 as modulators of nerve growth factor synthesis/secretion in astrocytes. *J. Neurosci. Res.* **34,** 539–545.

Bakhit, C., Armanini, M., Bennett, G. L., Wong, W. L. T. Hansen, S. E., and Taylor, R. (1991) Increase in glia-derived NGF following destruction of hippocampal neurons. *Brain Res.* **560,** 76–83.

Ballarin, M., Ernfors, P., Lindefors, N., and Persson, H. (1991) Hippocampal damage and kainic acid injection induce a rapid increase in mRNA for BDNF and NGF in the rat brain. *Exp. Neurol.* **114,** 35–43.

Banati, R. B., Gehrmann, J., Schubert, P., and Kreutzberg, G. W. (1993) Cytotoxicity of microglia. *Glia* **7,** 111–118.

Barinaga, M. (1994) Neurotrophic factors enter the clinic. *Science* **263,** 772–774.

Cheng, B. and Mattson, M. P. (1991) NGF and bFGF protect rat hippocampal and human cortical neurons against hypoglycemic damage by stabilizing calcium homeostasis. *Neuron* **7,** 1031–1041.

Delacourte, A. (1990) General and dramatic glial reaction in Alzheimer brain. *Neurology* **40,** 33–37.

D'Mello, S. R. and Heinrich, G. (1990) Induction of NGF by 12-*O*-teradecanoyl phorbol-13-acetate. *J. Neurochem.* **55,** 718–721.

Duffy, P. E., Rapport, M., and Graf, L. (1980) Glial fibrillary acidic protein and Alzheimer type senile dementia. *Neurology* **30,** 778–782.

Fernyhough, P., Diemel, L. T., Hardy, J., Brewster, W. J., Mohiuddin, L., and Tomlinson, D. R. (1995) Human recombinant nerve growth factor replaces deficient neurotrophic support in the diabetic rat. *Eur. J. Neurosci.* **7,** 1107–1110.

Friedman, W. J., Larkfors, L., Ayer-Leliever, C. Ebendal, T., Olson. L., and Persson, H. (1990) Regulation of β-NGF expression by inflammatory mediators in hippocampal cultures. *J. Neurosci. Res.* **27,** 374–382.

Frim, D. M., Short, M. P., Rosenberg, W. S., Simpson, J., Breakfield, X. O., and Isacson, O. (1993) Local protective effects of nerve growth factor-secreting fibroblast against excitotoxic lesions in the rat striatum. *J. Neurosurg.* **78,** 267–273.

Fuji, K., Hiramatsu, M., Kameyama, T., and Nabeshima, T. (1993) Effects of repeated administration of propentofylline on memory impairment produced by basal forebrain lesion in rats. *Eur. J. Pharmacol.* **236,** 411–417.

Fukumoto, H., Kakihana, M., and Suno, M. (1991) Recombinant human bFGF induces secretion of NGF in cultured rat astroglial cells. *Neurosci. Lett.* **122,** 221–224.

Furukawa, S., Kamo, I., Furukawa, Y., Akazawa, S., Satoyoshi, E., Itoh, K., and Hayashi, K. (1983) A highly sensitive enzyme immunoassay for mouse β nerve growth factor. *J. Neurochem.* **40,** 734–744.

Furukawa, Y., Furukawa, S., Ikeda, F., Satoyoshi, E., and Hayashi, K. (1986a) Aliphatic side chain of catecholamine potentiates the stimulatory effect of the catechol part on the synthesis of NGF. *FEBS Lett.* **208,** 258–262.

Furukawa, S., Furukawa, Y., Satoyoshi, E., and Hayashi, K. (1986b) Synthesis and secretion of nerve growth factor by mouse astroglial cells in culture. *Biochem. Biophys. Res. Commum.* **136,** 57–63.

Furukawa, S., Furukawa, Y., Satoyoshi, E., and Hayashi, K. (1987) Regulation of NGF synthesis/secretion by catecholamine in cultured astroglial cells. *Biochem. Biophys. Res. Commun.* **147,** 1048–1054.

Furukawa, S. and Furukawa, Y. (1990) Nerve growth factor synthesis and its regulatory mechanisms: an approach to therapeutic induction of nerve growth factor synthesis. *Cerebrovascular Brain Metab. Rev.* **2,** 328–344.

Furukawa, Y., Fukazawa, N., Miyama, Y., Hayashi, K., and Furukawa, S. (1990) Stimulatory effect of 4-alkylcatechols and their diacetylated derivatives on the synthesis of nerve growth factor. *Biochem Pharmacol.* **40,** 2337–2342.

Gall, C. M. and Issackson, P. J. (1989) Limbic seizures increase neuronal production of mRNA for NGF. *Science* **245,** 758–761.

Gasser, U. E., Weskamp, G., Otten, U., and Dravid, A. R. (1986) Time course of the elevation of NGF content in the hippocampus and septum following lesions of the septohippocampal pathway in rats. *Brain Res.* **376,** 351–356.

Goss, J. R., Finch, C. E., and Morgen, D. G. (1991) Age-related changes in glial fibrillary acidic protein mRNA in the mouse brain. *Neurobiol. Aging* **12,** 165–170.

Gotz, R., Koster, R., Winkler, C., Raulf, F., Lottspeich, F., Schartl, M., and Thoenen, H. (1994) Neurotrophin-6 is a new member of nerve growth factor family. *Nature* **372,** 266–269.

Green, J. E. (1991) Trans activation of NGF in transgenic mice containing the human T-cell lymphotrophic virsus type I tax gene. *Mol. Cell. Biol.* **11,** 4635–4641.

Hallbook, F., Ibanez, C. F., and Persson, H. (1991) Evaluationary studies of the nerve growth factor family reveal a novel member abundantly expressed in Xenopus ovary. *Neuron* **6,** 845–858.

Hanaoka, Y., Ohi, T., Furukawa, S., Furukawa, Y., Hayashi, K., and Matsukura, S. (1992) Effects of 4-methylcatechol on sciatic nerve growth factor level and motor nerve conduction velocity in experimental diabetic neuropathic process. *Exp. Neurol.* **115,** 292–296.

Hefti, F. (1986) Nerve growth factor promotes survival of septal cholinergic neurons after fimbrial transections. *J. Neurosci.* **11,** 2155–2162.

Hofer, M., Pagliusi, R., Hohn, A., Leibrock, J., and Barde, Y.-A. (1990) Regional distribution of brain-derived neurotrophic factor mRNA in the adult mouse brain. *EMBO J.* **9,** 2459–2464.

Hohn, A., Leibrock, J., Bailey, K., and Barde, Y.-A. (1990) Identification and characterization of a novel menber of the nerve growth factor/brain-derived neurotrophic factor family. *Nature* **344,** 339–341.

Jackson, G. R., Apffel, L., Werrbach-Perez, K., and Perez-Polo, J. R. (1990) Role of nerve growth factor in oxidant-antioxidant balance and neuronal injury. I. Stimulation of hydrogen peroxide resistance. *J. Neurosci. Res.* **25,** 360–368.

Kaechi, K., Furukawa, Y., Ikegami, R., Nakamura, N., Omae, F., Hashimoto, Y., Hayashi, K., and Furukawa, S. (1992) Pharmacological induction of physiologically active nerve growth factor in rat peripheral nervous system. *J. Pharmacol. Exp. Ther.* **264,** 321–326.

Kaechi, K., Ikegami, R., Nakamura, N., Nakajima, M., Furukawa, Y., and Furukawa, S. (1995) 4-methylcatechol, an inducer of nerve growth factor synthesis, enhances peripheral nerve regeneration across nerve gaps. *Exp. Neurol.* **272,** 1300–1304.

Kawagishi, H., Ando, M., Sakamoto, H., Yoshida, F., Ojima, F., Ishiguro, Y., Ukai, N., and Furukawa, S. (1991) Hericenones C, D and E, stimulators of nerve growth factor (NGF)-synthesis, from the mushroom *Hericium erinaceum. Tetrahedran Lett.* **32,** 4561–4564.

Leibrock, J., Lottspeich, F., Hohn, A., Hofer, M., Hengerer, B., Masiakowski, P., Thoenen, H., and Barde, Y. A. (1989) Molecular cloning and expression of brain-derived neurotrophic factor. *Nature* **341,** 149–151.

Lindholm, D., Castren, E., Hengerer, B., Zafra, F., Berninger, B., and Thoenen, H. (1992) Differential regulation of NGF synthesis in neurons and astrocytes by glucocorticoid hormons. *Eur. J. Neurosci.* **4,** 404–410.

Lindholm, D., Hengerer, B., Zafra, F., and Thoenen, H. (1990) Transforming growth factor-β1 stimulates expression of NGF in the rat CNS. *Neuroreports* **1,** 9–12.

Lindholm, D., Heumann, R., Hengerer, B., and Thoenen, H. (1988) Interleukin 1 increases stability and transcription of mRNA encoding nerve growth factor in cultures rat fibroblasts. *J. Biol. Chem.* **263,** 16,348–16,351.

Lindholm, D., Heumann, R., Meyer, M., and Thoenen, H. (1987) Interleukin-1 regulates synthesis of NGF in non-neuronal cells of rat sciatic nerve. *Nature* **350,** 658–659.

Lindvall, O. (1992) Differential regulation of mRNAs for nerve growth factor, brain-derived neurotrophic factor, and neurotrophin 3 in the adult rat brain following cerebral ischemia and hypoglycemic coma. *Proc. Natl. Acad. Sci. USA* **89,** 648–652.

Lorez, H. P., Von Franenber, M., Weskamp, G., and Otten, U. (1988) Effect of bilateral decortication on NGF content in basal nucleus and neostriatum of adult rat brain. *Brain Res.* **454,** 355–360.

Lu, B., Yokoyama, M., Dreyfus, C. F., and Black, I. B. (1991) NGF gene expression in actively growing brain glia. *J. Neurosci.* **11,** 318–326.

Maisonpierre, P. C., Belluscio, L., Squinto, S., Ip, N. Y., Furth, M. E., Lindsay, R. M., and Yancopoulos, G. D. (1990) Neurotrophin-3: a neurotrophic factor related to NGF and BDNF. *Science* **247,** 1446–1451.

Matsuoka, I., Meyer, M., and Thoenen, H. (1991) Cell-type-specific regulation of NGF synthesis in non-neuronal cells: comparison of Schwann cells with other cell types. *J. Neurosci.* **10,** 3165–3177.

Mattson, M. P. and Cheng, B. (1994) Growth factors protect neurons against excitotoxic/ischemia insults by stabilizing calcium homeostasis. *Stroke* **24,** 1136–1140.

Mitsumoto, H., Ikeda, K., Klinkosz, B., Cedarbaum, J. M., Wong, V., and Lindsay, R. M. (1994) Arrest of motor neuron disease in wobbler mice cotreated with CNTF and BDNF. *Science* **265,** 1107–1110.

Mrsulja, B. B., Micic, D. V., and Djuricic, B. M. (1983) Gerbil stroke model: an approach to the study of therapeutic aspects of postischemic brian edema. *Adv. Biosci.* **43,** 45–62.

Nabeshima, T., Nitta, A., and Hasegawa, T. (1993) Impairment of learning and memory and the accessory sympton in aged rats as senile dementia model (3): oral administration of propentofylline produces recovery of reduced NGF content in the brain of aged rats. *Jpn. J. Psychopharmacol.* **13,** 89–95.

Nagaoka, A., Shino, A., Kakihana, M., and Iwatsuka, H. (1984) Inhibitory effect of idebenone (CV-2619), a novel compound, on vascular lesion in hypertensive rats. *Jpn. J. Pharmacol.* **36,** 291–299.

Nagata, A., Suno, M., Shibota, M., and Kakihana, M. (1984) Effects of idebenone (CV-2619) on neurological deficits, local cerebral blood flow, and energy metabolism in rats with experimental cerebral ischemia. *Folia Pharmacol. Jpn.* **84,** 303–309.

Nawa, H., Carnahan, J., and Gall, C. (1995) BDNF protein measured by a novel enzyme immunoassay in normal brain and after seizure: partial disagreement with mRNA levels. *Eur. J. Neurosci.* **7,** 1527–1535.

Neveu, I., Jehan, F., Houlgatte, R., Wion, D., and Brachet, P. (1992) Activation of nerve growth factor synthesis in primary glial cells by phorbol 12-myristate 13-acetate: role of protein kinase C. *Brain Res.* **570,** 316–322.

Nitta, A., Furukawa, Y., Hayashi, K., Hiramatsu, M., Kameyama, T., Hasegawa, T., and Nabeshima, T. (1992) Denervation of dopaminergic neurons with 6-hydroxydopamine increases NGF content in rat brain. *Neurosci. Lett.* **144,** 152–156.

Nitta, A., Hasegawa, T., and Nabeshima, T. (1993a) Oral administration of idebenone, a stimulator of NGF synthesis, recovers reduced NGF content in aged rat brain. *Neurosci. Lett.* **163,** 219–222.

Nitta, A., Murase, K., Furukawa, Y., Hayashi, K., Hasegawa, T., and Nabeshima, T. (1993b) Effects of oral administration of a stimulator for nerve growth factor synthesis in basal forebrain-lesioned rats. *Eur. J. Pharmacol.* **250,** 23–30.

Nitta, A., Murase, K., Furukawa, Y., Hayashi, K., Hasegawa, T., and Nabeshima, T. (1993c) Memory impairment and neural dysfunction after continuous infusion of anti-nerve growth factor antibody into the septum in adult rats. *Neuroscience* **57,** 495–499.

Nitta, A., Murakami, Y., Furukawa, Y., Kawatsura, W., Hayashi, K., Yamada, K., Hasegawa, T., and Nabeshima, T. (1994) Oral administration of idebenone induces nerve growth factor in the brain and improves learning and memory in basal forebrain-lesioned rats. *Naunyn-Schmiedebergs Arch. Pharmacol.* **349,** 401–407.

Nitta, A., Ogihara, Y., Onishi, J., Hasegawa, T., Furukawa, S., Nabeshima, T. (1996) Propentofylline prevents neuronal dysfunction induced by infusion of anti-nerve growth factor antibody into the septum. *Eur. J. Pharmacol.* **307,** 1–6.

O'Callaghan, J. P. and Miller, D. B. (1991) The concentration of glial fibrillary acidic protein increases with age in the mouse and rat brain. *Neurobiol. Aging* **12,** 171–174.

Oderfeld-Nowak, B., Bacia, A., Gradkowska, M., Fusco, M., Vantini, G., Leon, A., and Aloe, L. (1992) In vivo activated brain astrocytes may produce and secrete nerve growth factor-like molecules. *Neurochem. Int.* **21,** 455–461.

Ogawa, S., Nabeshima, T., Kameyama, T., and Hayashi, K. (1993) Effects of nerve growth factor (NGF) in rats with basal forebrain lesions. *Jpn. J. Pharmacol.* **61,** 141–144.

Ohi, T., Furukawa, S., Hayashi, K., and Matsukura, S. (1990) Ganglioside stimulation of NGF synthesis in cultured rat Schwann cells. *Biochem. Int.* **20,** 739–746.

Ono, T., Saito, H., Kishimoto, T., Okumoto, T., and Miyamoto, K. (1991) Stimulation of biosynthesis of NGF by aFGF in cultured mouse astrocytes. *Neurosci. Lett.* **126,** 18–20.

Pantazis, N. J., Blanchard, M. H., Arnason, B. G. W., and Young, M. (1977) Molecular properties of the nerve growth factor secreted by L cells. *Proc. Natl. Acad. Sci. USA* **74,** 1492–1496.

Phillips, H. S., Hains, J. M., Armanini, M., Laramee, G. R., Johnsos, S. A., and Winslow, J. W. (1991) BDNF mRNA is decreased in the hippocampus of individuals with Alzheimer's disease. *Neuron* **7,** 695–702.

Rocamora, N., Massieu, L., Boddeke, H. W., Mengod, G., and Palacios, J. M. (1993) Neuronal death and neurotrophin gene expression: long-lasting stimulation of neurotrophin-3 messenger RNA in the degenerating CA1 and CA4 pyramidal cell layers. *Neuroscience* **53,** 905–908.

Schwartz, J. P. and Mishler, K. (1990) β-adrenergic receptor regulation, through cyclic AMP, of NGF expression in rat cortical and cerebellar astrocytes. *Cell. Mol. Neurobiol.* **10,** 447–457.

Shigeno, T., Mima, T., Takakura, K., Graham, D. I. Kato, G., Hashimoto, Y., and Furukawa, S. (1991) Amelioration of delayed neuronal death in the hippocampus by nerve growth factor. *J. Neurosci.* **11,** 2914–2919.

Shinoda, I., Furukawa, Y., and Furukawa, S. (1990) Stimulation of NGF synthesis/secretion by propentophylline in cultured mouse astroglial cells. *Biochem. Pharmacol.* **39,** 1813–1816.

Spillantini, M. G., Aloe, L., Alleva, E., De Simone, R., Goederet, M., and Levi-Montalcini, R. (1989) NGF mRNA and protein increase in hypothalamus in mouse model of aggression. *Proc. Natl. Acad. Sci. USA* **86,** 8555–8559.

Spina, M. B., Squinto, S. P., Miller, J., Lindsay, R. M., and Hymanm, C. (1992) Brain-derived neurotrophic factor protects dopamine neurons against 6-hydroxydopamine and N-methyl-4-phenylpyridinium ion toxicity: involvement of glutathione system. *J. Neurochem.* **59,** 99–106.

Spranger, M., Lindholm, D., Bandtlow, C., Heumann, R., Graham, H., Naher-Noe, M., and Thoenen, H. (1990) Regulation of nerve growth factor (NGF) synthesis in the rat central nervous system: comparison between the effects of interleukin-1 and various growth factors in astrocyte cultures and in vivo. *Eur. J. Neurosci.* **2,** 69–76.

Stefanovich, V. and Nagata, K. (1983) Effects of propentofylline and cinepazide on cerebral energy metabolism in nitrogen breathing rats. *IRCS Med. Sci.* **11,** 972.

Steiner, P., Pfeilschifter, J., Boeckh, C., Radeke, H., and Otten, U. (1991) Interleukin-1β and tumor necrosis factor-alpha synergistically stimulate NGF synthesis in rat mesangial cells. *Am. J. Physiol.* **261,** F792–F798.

Takeuchi, R., Murase, K., Furukawa, Y., Furukawa, S., and Hayashi, K. (1990) Stimulation of nerve growth factor syntesis/secretion by 1,4-benzoquinone and its derivatives in cultured mouse astroglia. *FEBS Lett.* **261,** 63–66.

Vige, X., Tang, B., and Wise, B. C. (1992) Cortical neurons inhibit basal and interleukin-1-stimulated astroglial cell secretion of nerve growth factor. *Brain Res.* **591,** 345–350.

Weskamp, G., Gasser, V. E., Dravid, A. R., and Otten, U. (1986) Fimbria fornix lesion increases NGF content in adult rat septum and hippocampus. *Neurosci. Lett.* **70,** 121–126.

Williams, L. R., Varon, S., Peterson, G. M., Wictorin, K., Fischer, W., Bjorklund, A., and Gage, F. H. (1986) Continuous infusion of nerve growth factor prevents basal forebrain neuronal death after fimbria fornix transection. *Proc. Natl. Acad. Sci. USA* **83,** 9231–9235.

Wion, D., Houglatte, R., Barbot, N., Barrand, P., Dicou, E., and Brachet, P. (1987) Retionic acid increases the expression of NGF gene in mouse L cells. *Biochem. Biophys. Res. Commun.* **149,** 510–514.

Wion, D., Macgrogan, D., Neveu, I., Jehan, F., Houglatte, R., and Brachet, P. (1991) 1,25-dihydroxyvitamin D3 is a potent inducer of NGF synthesis. *J. Neurosci. Res.* **28,** 110–114.

Yamaguchi, K., Tsuji, T., Uemura, D., and Kondo, K. (1996) Cyclooxygenase induction is essential for NGF synthesis enhancement by NGF inducers in L-M cells. *Biosci. Biotechnol. Biochem.* **60,** 92–94.

Yamaguchi, K., Tsuji, T., Waruri, S., Yazawa, K., Kondo, K., Shigemori, H., and Kobayashi, J. (1993) Stimulation of NGF synthesis and secretion by Fellutamide A in vitro. *Biosci. Biotechnol. Biochem.* **57,** 195–199.

Yamazaki, N., Take, Y., Nagaoka, A., and Nagawa, Y. (1984) Beneficial effect of idebenone (CV-2619) on cerebral ischemia-induced amnesia in rats. *Jpn. J. Pharmacol.* **36,** 349–356.

Yan, Q., Elliott, J., and Snider, W. D. (1992) Brain-derived neurotrophic factor rescues spinal motor neurons from axotomy-induced cell death. *Nature* **360,** 753–755.

Yoshida, K. and Gage, F. H. (1991) Fibroblast growth factors stimulate NGF synthesis and secretion by astrocytes. *Brain Res.* **538,** 118–126.

Zhang, Y., Tatsuno, T., Carney, J. M., and Mattson, M. P. (1993) Basic FGF, NGF, and IGF-II protect hippocampal and cortical neurons against iron-induced degeneration. *J. Cereb. Blood Flow Metab.* **13,** 378–388.

6

Preclinical Trials of Basic Fibroblast Growth Factor (bFGF) in Animal Models of Stroke

Ilknur Ay and Seth P. Finklestein

1. INTRODUCTION

The availability of recombinant growth factors has made possible the testing of these polypeptides as treatments in experimental models of stroke and head injury (Sabel et al., 1992; Berlove and Finklestein, 1993; Finklestein, 1995, 1996; Chen and Finklestein, 1996). Neurotrophic growth factors have several properties that make them attractive in this regard. First, they promote neuronal survival and, conversely, they protect against neuronal death. Second, they promote neuronal outgrowth and synapse formation. These properties of neurotrophic growth factors are evident in neuronal tissue culture and in the developing brain. The fact that these same principles may apply in the mature damaged brain provides a rationale for the use of these agents in experimental models of stroke and brain injury.

Indeed, within the past few years a growing literature has supported the concept that exogenously administered neurotrophic growth factors can limit the extent of cerebral damage and/or enhance functional recovery in animal models of stroke and brain injury. Factors used in these studies include members of the neurotrophin, fibroblast growth factor (FGF), transforming growth factor-β (TGF-β), and insulin-like growth factor (IGF) superfamilies (Chen and Finklestein, 1996). This chapter will focus on studies using basic fibroblast growth factor (bFGF), perhaps the best-studied trophic factor in animal models of stroke and brain injury. Indeed, these preclinical studies have recently led to human clinical safety trials of bFGF in acute stroke, the first trials to examine polypeptide growth factor treatment in this context.

Basic FGF is an 18-kD-a polypeptide found in relatively high concentrations in the mammalian brain (Baird, 1994). Brain bFGF mRNA and protein levels increase markedly during early postnatal development (Caday et al., 1990; Powell et al., 1991). In the mature brain, bFGF is largely cell-associated and found within a few select neuronal populations and in astroglial cells scattered throughout the brain. Following mechanical brain injury or ischemia, levels of bFGF mRNA and protein are markedly upregulated in tissue surrounding focal brain wounds or infarcts (Finklestein et al., 1988; Speliotes et al., 1996). This upregulation reaches its maximum level within a few days and is localized primarily to reactive astroglial cells. The role of endogenous bFGF expression

From: Neuroprotective Signal Transduction *edited by M. P. Mattson*
Humana Press Inc., Totowa, NJ

following brain injury or stroke remains unclear. Because of its multipotential trophic effects on several cell types (neurons, as well as glia and endothelial cells), increased bFGF expression may play a role in neuronal sprouting, glial proliferation, and angiogenesis (vascular proliferation), all of which occur following brain injury or stroke (Speliotes et al., 1996). On the other hand, the bFGF gene lacks a classical leader sequence, so it is unclear how and under what circumstances bFGF is released from cells in which it is made.

In vitro, bFGF supports the survival and outgrowth of neurons from several regions of the embryonic rat brain, including cortex and hippocampus (Morrison et al., 1986; Walicke, 1988). As a corollary to these survival-promoting effects, a number of investigators have also reported that bFGF, in nanomolar concentrations, protects cultured neurons from succumbing to a number of exogenous toxins and insults. At least four groups have reported that bFGF protects neurons against excitatory amino acid toxicity (Mattson et al., 1989; Fernandez et al., 1992; Freese et al., 1992; Finklestein et al., 1993; Louis et al., 1993). Other investigators have reported that bFGF protects against hypoglycemia, anoxia, nitric oxide, free radicals, phorbol esters, and excess intracellular calcium (Mattson, 1990; Cheng and Mattson, 1991; Boniece and Wagner, 1993; Louis et al., 1993; Maiese et al., 1993; Zhang et al., 1993; Maiese and Boccone 1995). The precise molecular mechanisms of neuroprotection by bFGF have not been determined, but appear to depend on new gene expression and protein synthesis (Mattson et al., 1989, 1995). In cultured embryonic hippocampal neurons, bFGF downregulates the expression of an NMDA receptor protein, upregulates the expression of the GluR1 subunit of the AMPA receptor, increases superoxide dismutase (SOD) and glutathione reductase activities, and stabilizes mitochondrial function (Cheng et al., 1995; Mattson et al., 1993a,b, 1995). All of these effects may contribute to neuroprotection (Fig. 1).

2. PRECLINICAL STUDIES OF bFGF AS A CYTOPROTECTIVE AGENT IN ACUTE STROKE

In addition to its cytoprotective effects, bFGF is also a potent vasodilator. In rats and rabbits, iv administration of bFGF as a bolus of 300–1000 ng produces hypotension (Cuevas et al., 1991). Topical administration of bFGF to the pial surface (5–200 ng/mL) produces dilation of the cerebral arterioles, an effect that is mediated through nitric oxide (NO) (Regli et al., 1994; Rosenblatt et al., 1994)

For these two reasons—namely, its cytoprotective and vasoactive properties—we hypothesized that the exogenous administration of bFGF might limit the degree of cerebral damage following cerebral ischemia. In initial studies that established proof of principle, bFGF was administered by continuous icv infusion to mature rats for 3 d before focal infarcts were made by permanent unilateral occlusion of the middle cerebral artery (MCA) and temporary bilateral carotid artery occlusion (Chen model). Under these circumstances, a 25% reduction in infarct volume was observed in bFGF, compared to vehicle-treated animals (Koketsu et al., 1994). In subsequent studies, bFGF was administered systemically (intravenously) at times after the onset of focal ischemia. In a model of permanent MCA occlusion by intra-arterial suture, iv administration of bFGF (45 μg/kg/h for 3 h), beginning at 30 min after the onset of ischemia, produced a 40–50% reduction in infarct volume, and a corresponding improvement in neurological outcome (Fisher et al., 1995). Subsequent studies have shown that the effective dose

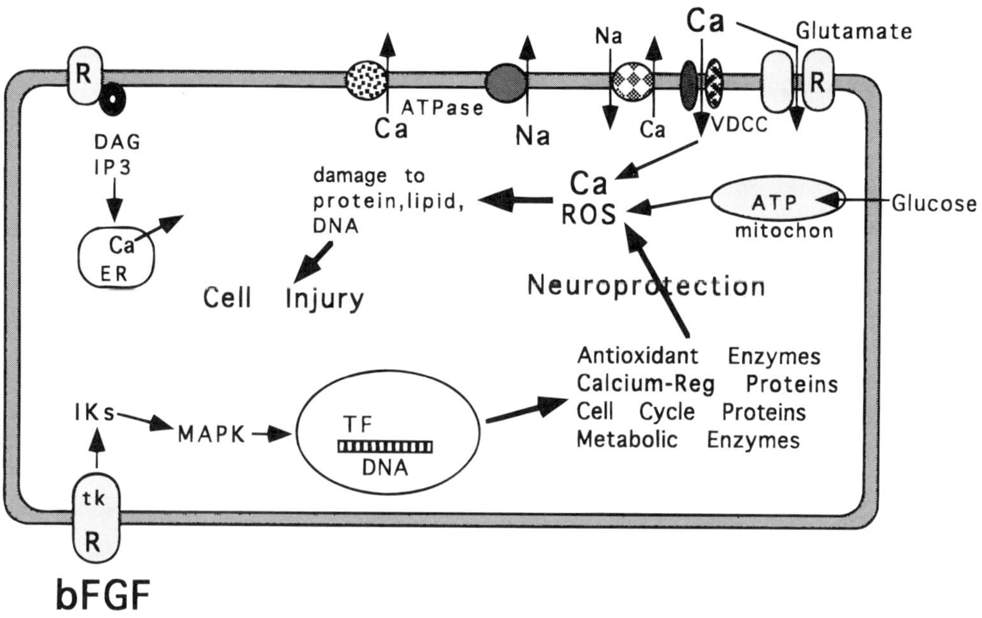

Fig. 1. Signal transduction pathway of bFGF involved in protecting neurons against ischemic injury. Cerebral ischemia results in energy depletion caused by reduced glucose and oxygen availability, with resulting failure of plasma membrane ion-motive ATPases that remove calcium (Ca) and sodium (Na). This leads to membrane depolarization, calcium influx through glutamate receptor channels and voltage-dependent calcium channels, generation of reactive oxygen species (ROS), and membrane depolarization. Ischemia may also promote inositol phospholipid hydrolysis resulting in liberation of diacylglycerol (DAG) and inositol triphosphate (IP3), and calcium release from endoplasmic reticulum (ER). ROS and calcium damage cellular proteins, lipids, and nucleic acids, ultimately resulting in cell injury and death. Cell surface receptors for bFGF possess intrinsic tyrosine kinase activity, and, once activated by bFGF, initiate a cascade of phosphorylation reactions involving intermediate kinases (IK), mitogen-activated protein kinases (MAPK), and transcription factors (TF). Activation of this pathway results in expression of neuroprotective gene products, such as antioxidant enzymes and calcium-regulating proteins, which suppress accumulation of ROS and calcium.

range is 5–50 µg/kg/h for 3 h, and that infarct-reducing effects are seen when the infusion is begun up to 3 h after the onset of permanent ischemia (Ren and Finklestein, 1997). In a model of temporary (2 h) MCA occlusion followed by reperfusion in rats, the iv administration of bFGF (45 µg/kg/h for 3 h), starting at the time of reperfusion, also resulted in a 40–50% reduction in infarct volume and an improvement in neurological outcome (Jiang et al., 1996).

Other investigators (Tanaka et al., 1995) have reported a similar reduction in infarct volume following intracarotid bFGF administration in a model of permanent MCA occlusion in rats. Reduction in infarct volume by iv bFGF administration following MCA occlusion in cats has also been reported (Bethel et al., 1996). Intravenous bFGF also reduces the volume of cerebral contusions following traumatic brain injury in rats (Dietrich et al., 1996).

The mechanism of infarct reduction by bFGF remains unclear. Possibilities include effects on cerebral blood flow (CBF) and/or direct protection of vulnerable brain cells. In rats, intra-arterial or iv bFGF does increase CBF in the ischemia penumbra at doses that reduce infarct volume (Tanaka et al., 1995; Chen and Finklestein, unpublished data). In cats, bFGF does not appear to increase rCBF at infarct-reducing doses (Bethel et al., 1996). Perhaps the most definitive evidence suggesting that changes in rCBF may not be an essential mechanism of infarct reduction by bFGF comes from studies using genetically altered knockout mice deficient in the gene for endothelial nitric oxide synthase (eNOS). Since the vasoactive effects of bFGF in rodents are mediated through endothelial NO production, mice deficient in eNOS would not be expected to show vasodilation in response to bFGF. Indeed, following MCA occlusion, eNOS-deficient mice do not show increased rCBF in the ischemic penumbra, as do wild-type controls (Huang et al., 1996). Nonetheless, the degree of infarct reduction is equivalent in eNOS knockout and wild-type animals (Huang et al., 1996). These data suggest that, in wild-type animals, increased rCBF is an epiphenomenon and not an essential mechanism of infarct reduction by bFGF.

Intravenously administered bFGF appears to cross the damaged blood–brain barrier after ischemia to directly enter ischemic rat brain tissue (Fisher et al., 1995). Although only a small percentage (0.01%) of the total administered dose appears to enter the brain, concentrations as high as 1–10 ng/mL may be reached in brain following administration of infarct-reducing doses (Fisher et al., 1995). These concentrations are equivalent to those necessary for the trophic effects of bFGF in vitro (Walicke, 1988, Mattson et al., 1989; Finklestein et al., 1993). It is conceivable that at such concentrations bFGF directly triggers receptor-mediated signal transduction pathways on vulnerable cells, resulting in protective gene expression. Such genes may include free radical scavenging enzymes and Ca^{2+} binding proteins, among others (Fig. 1). Clearly, further work is required to elucidate mechanisms of infarct reduction by bFGF.

The above preclinical studies, as well as extensive toxicity testing in animals, has led to the recent initiation of a human clinical safety trial of iv bFGF in acute stroke. This is a double-blind, placebo-controlled dose escalation study in patients with thromboembolic occlusion of a major cerebral artery recruited within 12 h of onset. Five bFGF dose groups will be studied, 1, 3, 9, and 25 μg/kg/h for 3 h, and a final group of 5 μg/kg/h for 24 h. Extensive safety data will be obtained, as well as preliminary efficacy data, acquired using standard stroke rating scales and techniques of diffusion/perfusion MR imaging (Sorensen et al., 1996).

3. PRECLINICAL STUDIES OF bFGF IN ENHANCING FUNCTIONAL RECOVERY FROM STROKE

Recent studies have also addressed the potential utility of bFGF as an agent to enhance functional recovery long after stroke has occurred. In these studies, bFGF is given at late times after infarction that are beyond the therapeutic window during which infarct volume can be reduced.

Although damaged mammalian brain tissue does not regenerate after stroke, functional recovery of neurological deficits often occurs. The substrate of such recovery may involve, in part, new neuronal sprouting and synapse formation in undamaged parts of brain. For example, considerable evidence both in animals and humans suggests that the contralateral undamaged hemisphere may play an important role in recovery following unilateral cerebral injury (Weiller et al., 1992; Jones and Schallert, 1994; Cramer et al., 1996). For example, Schallert and colleagues have shown increased dendritic growth in the contralateral homotypic cortex following ablation lesions of the sensorimotor cortex in rats (Jones and Schallert, 1994). This dendritic sprouting parallels functional recovery of the affected forelimb. Moreover, splinting of the normal forelimb inhibits dendritic sprouting as well as recovery of the impaired limb, suggesting that use of neuronal circuits on the intact side may be important for recovery (Jones and Schallert, 1994).

Basic FGF might enhance such recovery in at least two ways: First, bFGF might protect neurons distant from the site of injury projecting to the injury site. Such neurons might die back if deprived of their trophic support from their targets. Indeed, this principle has been shown by Yamada et al. (1991), who have shown that neuronal degeneration in thalamic nuclei following cortical infarction in rats can be prevented by intracisternal administration of bFGF. As a second mechanism, bFGF might directly promote neuronal sprouting and new synapse formation in brain. For example, Otto and Unsicker (1990) have shown increased sprouting of remaining dopaminergic fibers following MPTP lesions of the nigrostriatal system in animals receiving intracerebral bFGF.

For these reasons, we hypothesized that exogenous administration of bFGF might promote functional recovery following stroke. In recent studies (Kawamata et al., 1996), mature rats received bFGF by repeated biweekly intracisternal injections (1 μg/injection), beginning at 1 d and lasting for 4 wk following unilateral cerebral infarction caused by MCA occlusion. These infarcts involve the dorsolateral cerebral cortex (including the sensorimotor cortex) and underlying striatum. Basic FGF was administered intracisternally in order to gain ready access to the intact hemisphere, which may play an important role in recovery. Following stroke surgery, all animals had marked deficits in placing behavior of the contralateral forelimb and hindlimb. In vehicle-treated animals, a slow recovery of placing behavior of the contralateral limbs was found over the first month following stroke. In bFGF-treated animals, however, a marked enhancement of contralateral forelimb and hindlimb placing was found (Kawamata et al., 1996).

In these initial studies, the intracisternal bFGF dose administered (8 μg total) was associated with weight loss in some animals (Kawamata et al., 1996). However, at half of this dose, the effects on behavioral recovery were preserved, and no weight loss was observed (Kawamata et al., 1997). As expected in these studies, there was no difference in infarct volume between bFGF- and vehicle-treated animals, since the first bFGF dose was given at 24 h after infarction, which is beyond the therapeutic window during which bFGF can reduce infarct size (Yamada et al., 1991; Tenjin et al., 1995; Kawamata et al., 1996). These studies support the hypothesis that bFGF enhances functional recovery following stroke. The mechanism of recovery enhancement remains to be elucidated, but may depend on protection against distant neuronal death, as well as enhancement of neuronal sprouting and synapse formation.

In summary, recent preclinical studies have examined the potential efficacy of bFGF treatment in animal models of stroke. These have included both studies of reduction in infarct volume following iv bFGF treatment in acute stroke, as well as studies of behavioral recovery following intracisternal bFGF treatment in chronic stroke. The first strategy has led to ongoing human clinical trials of iv bFGF as a cytoprotective therapy in acute stroke. Because of their neurotrophic and cytoprotective effects, growth factors may represent a new strategy for the treatment of stroke.

REFERENCES

Baird, A. (1994) Fibroblast growth factors: activities and significance of non-neurotrophin neurotrophic growth factors. *Curr. Opinion Neurobiol.* **4,** 78–86.

Berlove, D. J. and Finklestein, S. P. (1993) Growth factors and brain injury, in *Growth Factors, Peptides and Receptors* (Moody, T. W., ed.), Plenum, New York, pp. 137–143.

Bethel, A., Kirsch, J. R., Koehler, R. C., Finklestein, S. P., and Traystman, R. J. (1997) Intravenous basic fibroblast growth factor decreases brain injury resulting from focal ischemia in cats. *Stroke* **28,** 609–615.

Boniece, I. R., and Wagner, J. A. (1993) Growth factors protect PC12 cells against ischemia by a mechanism that is independent of PKA, PKC, and protein synthesis. *J. Neurosci.* **13,** 4220–4228.

Caday, C. G., Klagsbrun, M., Fanning, P. J., Mirzabegian, A., and Finklestein, S. P. (1990). Fibroblast growth factor (FGF) levels in the developing rat brain. *Dev. Brain Res.* **52,** 241–246.

Chen, K., and Finklestein, S. P. (1996) Neurotrophic factors, in *Ischemic Stroke: From Basic Mechanisms to New Drug Development* (Hus, C. Y., ed.), Karger, Basel, in press.

Cheng, B. and Mattson, M. P. (1991) NGF and bFGF protect rat hippocampal and human cortical neurons against hypoglycemic damage by stabilizing calcium homeostasis. *Neuron* **7,** 1–20.

Cheng, B., Furukawa, K., O'Keefe, J. A., Goodman, Y., Kihiko, M., Fabian, T., and Mattson, M. P. (1995) Basic fibroblast growth factor selectively increases AMPA-receptor subunit GluR1 protein level and differentially modulates Ca^2 responses to AMPA and NMDA in hippocampal neurons. *J. Neurochem.* **65,** 2525–2536.

Cramer, S. C., Kaplan, J., Nelles, G., Benson, R., Finklestein, S. P., and Rosen, B. R. (1996) Patterns of brain activation using functional MRI in subjects recovered from stroke (abstr.). *NeuroImage* **3,** S478.

Cuevas, P., Carceller, F., Ortega, S., Zazo, M., Nieto, I., and Gimenez-Gallego, G. (1991) Hypotensive activity of fibroblast growth factor. *Science* **254,** 1208–1210.

Dietrich, W. D., Alonso, O., Busto, R., and Finklestein, S. P. (1996) Post-treatment with intravenous basic fibroblast growth factor reduces histopathological damage following fluid-percussion brain injury in rats. *J. Neurotrauma* **6,** 309–315.

Fernandez, M. T., Torreblanca, A., Gascon, S., and Novelli, A. (1992) Differentiation independent protection of basic fibroblast growth factor from NMDA and non-NMDA receptor mediated neurotoxicity in cerebellar neurons. *J. Cell Biol.* **115,** 419a.

Finklestein, S. P. (1995) Growth factors in stroke, in *Brain Ischemia: Basic Concepts and Clinical Relevance* (Caplan, L. R., ed.), Springer-Verlag, London, pp. 37–41.

Finklestein, S. P. (1996) The potential use of neurotrophic growth factors in the treatment of cerebral ischemia, in *Cellular and Molecular Mechanisms of Ischemic Brain Damage* (Siesjo, B. K. and Wieloch, T., eds.), Lippincott-Raven, Philadelphia, pp. 413–418.

Finklestein, S. P., Apostolides, P. J., Caday, C. G., Prosser, J., Philips, M. F., and Klagsbrun, M. (1988) Increased basic fibroblast growth factor (bFGF) immunoreactivity at the site of focal brain wounds. *Brain Res.* **460,** 253–259.

Finklestein, S. P., Kemmou, A., Caday, C. G., and Berlove, D. J. (1993) Basic fibroblast growth factor protects cerebrocortical neurons against excitatory amino acid toxicity *in vitro. Stroke* **24,** I141–I143.

Fisher, M., Meadows, M.-E., Do, T., Weise, J., Trubetskoy, V., Charette, M., and Finklestein, S. P. (1995) Delayed treatment with intravenous basic fibroblast growth factor reduces infarct size following permanent focal cerebral ischemia in rats. *J. Cereb. Blood Flow Metab.* **15,** 953–959.

Freese, A., Finklestein, S. P., and DiFiglia, M. (1992) Basic fibroblast growth factor protects striatal neurons *in vitro* from NMDA-receptor mediated excitotoxicity. *Brain Res.* **575,** 351–355.

Huang, Z., Chen, K., Huang, P. L., Finklestein, S. P., and Moskowitz, M. A. (1996) Basic fibroblast growth factor ameliorates focal ischemic injury in type III NOS deficient mice by rCBF-independent mechanisms. *Am. J. Physiol.* **272,** H1401–H1405.

Jiang, N., Finklestein, S. P., Do, T., Caday, C. G., Charette, M., and Chopp, M. (1996) Delayed intravenous administration of basic fibroblast growth factor (bFGF) reduces infarct volume in a model of focal cerebral ischemia/reperfusion in the rat. *J. Neurol. Sci.* **139,** 173–179.

Jones, T. A., and Schallert, T. (1994) Use-dependent growth of pyramidal neurons after neocortical damage. *J. Neurosci.* **14,** 2140–2152.

Kawamata, T., Alexis, N. E., Dietrich, W. D., and Finklestein, S. P. (1996) Intracisternal basic fibroblast growth factor (bFGF) enhances behavioral recovery following focal cerebral infarction in the rat. *J. Cereb. Blood Flow Metab.* **16,** 542–547.

Kawamata, T., Schallert, T. J., Dietrich, W. D., Benowitz, L. I., Gotts, J. E., Cocke, R. R., and Finklestein, S. P. (1997) Intracisternal basic fibroblast growth factor (bFGF) enhances functional recovery and upregulates the expression of a molecular marker of neuronal sprouting following focal cerebral infarction (abstr.). *Stroke* **28,** 250.

Koketsu, N., Berlove, D. J., Moskowitz, M. A., Kowall, N. W., Caday, C. G., and Finklestein, S. P. (1994) Pretreatment with intraventricular basic fibroblast growth factor (bFGF) decreases infarct size following focal cerebral ischemia in rats. *Ann. Neurol.* **35,** 451–457.

Louis, J. C., Magal, E., Gerdes, W., and Seifert, W. (1993) Survival-promoting and protein kinase C-regulating roles of basic FGF for hippocampal neurons exposed to phorbol ester, glutamate and ischaemia-like conditions. *Eur. J. Neurosci.* **5,** 1610–1621.

Maiese, K., and Boccone, L. (1995) Neuroprotection by peptide growth factors against anoxia and nitric oxide toxicity requires modulation of protein kinase C. *J. Cereb. Blood Flow Metab.* **15,** 440–449.

Maiese, K., Boniece, I., DeMeo, D., and Wagner, J. A. (1993) Peptide growth factors protect against ischemia in culture by preventing nitric oxide toxicity. *J. Neurosci.* **13,** 3034–3040.

Mattson, M. P. (1990) Excitatory amino acids, growth factors, and calcium: a teeter-totter model for neural plasticity and degeneration, in *Excitatory Amino Acids and Neural Plasticity* (Ben-Ari, T., ed.), Plenum, New York, pp. 211–220.

Mattson, M. P., Murrain, M., Guthrie, P. B., and Kater, S. B. (1989) Fibroblast growth factor and glutamate: Opposing roles in generation and degeneration of hippocampal neuroarchitecture. *J. Neurosci.* **9,** 3728–3740.

Mattson, M. P., Kumar, K. N., Wang, H., Cheng, B., and Michaelis, E. K. (1993a) Basic FGF regulates the expression of a functional 71 kDa NMDA receptor protein that mediates calcium influx and neurotoxicity in hippocampal neurons. *J. Neurosci.* **13,** 4575–4588.

Mattson, M. P., Zhang, Y., and Bose, S. (1993b) Growth factors prevent mitochondrial dysfunction, loss of calcium homeostasis and cell injury, but not ATP depletion in hippocampal neurons deprived of glucose. *Exp. Neurol.* **121,** 1–13.

Mattson, M. P., Lovell, M. A., Furukawa, K., and Markesbery, W. R. (1995) Neurotrophic factors attenuate glutamate-induced accumulation of peroxides, elevation of intracellular Ca^{2+} concentration, and neurotoxicity and increase antioxidant enzyme activities in hippocampal neurons. *J. Neurochem.* **65,** 1740–1751.

Morrison, R. S., Sharma, A., DeVellis, J., and Bradshaw, R. A. (1986) Basic fibroblast growth factor supports the survival of cerebral cortical neurons in primary culture. *Proc. Natl. Acad. Sci. USA* **83,** 7537–7541.

Otto, D., and Unsicker, K. (1990) Basic FGF reverses chemical and morphological deficits in the nigrostriatal system of MPTP-treated *J. Neurosci.* **10,** 1912–1921.

Powell, P. P., Finklestein, S. P., Dionne, C. A., Jaye, E. M., and Klagsbrun, M. (1991) Temporal differential and regional expression mRNA for basic fibroblast growth factor in the developing and adult rat brain. *Mol. Brain Res.* **11,** 71–77.

Regli, L., Anderson, R. E., and Meyer, F. B. (1994) Basic fibroblast growth factor increases cortical blood flow in vivo. *Brain Res.* **665,** 155–157.

Ren, J. and Finklestein, S. P. (1997) Time dependence of infarct reduction by intravenous basic fibroblast growth factor (bFGF) following focal cerebral ischemia in rats. *Eur. J. Pharmacol.* **327,** 11–16.

Rosenblatt, S., Irikura, K., Caday, C. G., Finklestein, S. P., and Moskowitz, M. A. (1994) Basic fibroblast growth factor (bFGF) dilates rat pial *J. Cereb. Blood Flow Metab.* **14,** 70–74.

Sabel, B. A., Vantini, G., and Finklestein, S. P. (1992) The role of neurotrophic factors in the treatment of neurological disorders, in *Neurological Disorders: Novel Experimental and Therapeutic Strategies* (Vecsei, L., Freese, A., Schwartz, K. J., and Beal, M. F., eds.), Ellis Horwood, New York, pp. 113–180.

Sorensen, A. G., Buonanno, F. S., Schwamm, L., Lev, M. H., Huang-Hellinger, F. R., Reese, T. R., et al. (1996) Accuracy of diffusion and perfusion MR imaging in diagnosis of stroke. *Radiology* **199,** 391–401.

Speliotes, E. K., Caday, C. G., Do, T., Weise, J., Kowall, N. W., and Finklestein, S. P. (1996) Increased expression of basic fibroblast growth factor (bFGF) following focal cerebral infarction in the rat. *Mol. Brain Res.* **39,** 31–42.

Tanaka, R., Miyasaka, Y., Yada, K., Ohwada, T., and Kameya, T. (1995) Basic fibroblast growth factor increases regional cerebral blood flow and reduces infarct size after experimental ischemia in a rat model. *Stroke* **26,** 2154–2159.

Tenjin, H., Anderson, R. E., and Meyer, F. B. (1995) Treatment with basic fibroblastic growth factor following focal cerebral ischemia does not prevent neuronal injury. *J. Neurol. Sci.* **128,** 66–70.

Walicke, P. A. (1988) Basic and acidic fibroblast growth factors have trophic effects on neurons from multiple CNS regions. *J. Neurosci.* **8,** 618–627.

Weiller, C., Chollet, F., Friston, K. J., Wise, R. J., and Frackowiak, R. S. (1992) Functional reorganization of the brain in recovery from striatocapsular infarction in man. *Ann. Neurol.* **31,** 463–472.

Yamada, K., Kinoshita, A., Kohmura, E., Sakaguchi, T., Taguchi, J., Kataoka, K., and Hayakawa, T. (1991) Basic fibroblast growth factor prevents thalamic degeneration after cortical infarction. *J. Cereb. Blood Flow Metab.* **11,** 472–478.

Zhang, Y., Tatsuno, T., Carney, J. M., and Mattson, M. P. (1993) Basic FGF, NGF, and IGFs protect hippocampal and cortical neurons against iron-induced degeneration. *J. Cereb. Blood Flow Metab.* **13,** 378–388.

7

Transforming Growth Factor-β Signaling and Neuroprotection

Relevance to Ischemic Brain Injury, Alzheimer's, and Parkinson's Disease

Kerstin Krieglstein and Josef Krieglstein

1. TGF-β IN GENERAL

Transforming growth factors-beta (TGF-β) are multifunctional cytokines with widespread distribution. They are known to modulate important events, such as growth and differentiation (Roberts and Sporn, 1990). TGF-β was discovered in the early 1980s (Roberts et al., 1981). The initial purification of the 25-kDa protein was achieved by using its ability to induce normal rat kidney fibroblasts to grow and form colonies of cells in soft agar in the presence of epidermal growth factor (EGF; Roberts et al., 1981, 1983). This transforming capacity provided the original name for the protein: transforming growth factor. However, the name is somewhat misleading, since it is known by now that TGF-β affects many functions in nearly all cells (for review, *see* Roberts and Sporn, 1990; Kingsley, 1994; Krieglstein et al., 1995a). The initially purified material actually consisted of two components named TGF-α and TGF-β (Anzano et al., 1983). Elucidating the amino acid sequence revealed that these two proteins are nonhomologous: TGF-α belongs to the EGF family (Todaro et al., 1980; Derynck et al., 1984); TGF-β, as TGF-β1, became the prototype of a constantly expanding superfamily of its own. The cloning of TGF-β1 and the elucidation of its precursor structure (Assoian et al., 1983; Derynck et al., 1985) has led to the identification of three other forms of TGF-β: TGF-β2 (deMartin et al., 1987; Madisen et al., 1988), -β3 (Derynck et al., 1988), and -β5 (Xenopus; Kondaiah et al., 1990); TGF-β4 seems to be the chick homolog of TGF-β1 (Jakowlew et al., 1988).

1.1. Structure and Superfamily

The isolated TGF-β protein is a disulfide-linked homodimeric protein with a molecular mass of 25-kDa (Assoian et al., 1983). Cloning of the TGF-β isoforms revealed some common structural features. TGF-β is synthesized as a large precursor molecule with an amino-terminal signal sequence and a prodomain, containing the carboxy-terminal-located mature part of the protein that is generated by proteolytic cleavage. There is little sequence similarity found within precursor molecules, but the processed,

From: Neuroprotective Signal Transduction *edited by M. P. Mattson*
Humana Press Inc., Totowa, NJ

C-terminal-located mature part is highly conserved within isoforms and species (*see* Roberts and Sporn, 1990; Burt and Paton, 1992). Other important criteria are seven distinctly spaced cystine residues located in the mature part of the protein.

These structural determinants are the basis for classification as a TGF-β superfamily member. Figure 1 shows a list of more than 40 members. Some of them can be grouped into distinct subfamilies with highly related sequences, i.e. TGF-β, activin, and BMP families (Kingsley, 1994; Massague et al., 1994; Griffith et al., 1996). Recent elucidation of the three-dimensional structure of TGF-β2 (Doapin et al., 1992; Schlunegger and Grütter, 1992) and BMP7/OP1 (Griffith et al., 1996), using X-ray crystallography, and TGF-β1 in solution, using multinuclear magnetic resonance spectroscopy (Hinck et al., 1996), could convincingly show that all three share a common polypeptide fold, although there is a limited sequence identity (TGF-β2/BMP7 36%; Griffith et al., 1996).

Within one monomere fold, six of the seven cystine residues form three disulfide bridges, resulting in the characteristic cystine knot motif. Emanating from the knot, four strands of antiparallel β-sheets form two fingerlike projections on one side and an α-helix on the opposite side. However, the structure of the cystine knot family is found not only in members of the TGF-β superfamily, but also in nerve growth factor (NGF) and platelet-derived growth factor, forming a structural superfamily of growth factors (McDonald and Hendrickson, 1993). Although all three types of growth factors are dimeric, the modes of dimerization are distinct and involve different monomer interfaces (McDonald and Hendrickson, 1993). Furthermore, the orientation of the subunits are arranged differently, i.e., in the case of TGF-βs, they are oriented in antiparallel arrangements, in contrast to the parallel ones of NGF (Murzin and Chothia, 1992; Swindells, 1992).

1.2. Receptors

Almost every cell in the body contains binding sites for TGF-β. Crosslinking experiments using ^{125}I-TGF-β1 identified several membrane-bound binding molecules. These proteins were termed by their increasing mol wt TGF-β receptor type I (TβR-I, 65–70 kDa), type II (TβR-II, 85–110 kDa) and type III (200 kDa). Somewhat later, additional TGF-β-binding cell surface molecules have been identified (type IV–IX receptors; Massague, 1992). Cloning of the receptors identifies TβR-I and TβR-II as transmembrane serine–threonine kinase signaling molecules; TβR-III is a nonsignaling transmembrane proteoglycan, called betaglycan (Lopez-Casillas et al., 1991; Wang et al., 1991). In order to transmit a signal, formation of a heteromeric complex of type I–type II receptors is required (Lin et al., 1992; Wrana et al., 1992, 1994; Franzen et al., 1993; Massague et al., 1994; Miyazono et al., 1994). TGF-β first binds to TβR-II, followed by the recruitment of TβR-I into the complex. TβR-I by itself has no affinity to TGF-β (Lin et al., 1992). TβR-II activates TβR-I by phosphorylation, enabling propagation of the signal (Wrana et al., 1994; Wieser et al., 1995). The family of serine–threonine kinases serving as type I or type II signaling receptors for members of the TGF-β superfamily has now grown to almost 20 members (Massague et al., 1994; ten Dijke et al., 1996). Until recently, it was believed that signaling via a heteromeric serine–threonine kinase complex is one unique feature of the TGF-β superfamily members, in contrast to almost all other growth factors and cytokines utilizing tyrosine kinase receptors. However, identification of the receptors for glial cell line-derived neurotrophic factor (GDNF), a very distant member of the TGF-β superfamily, as the orphan receptor c-Ret

TGF ß family	TGF-ß2 TGF-ß3 TGF-ß1 TGF-ß5 *xenopus* TGF-ß4 *chicken*
activin family	Activin A Activin B Activin C Activin D
DVR Groupe	nodal fugacin *xenopus* Vg1 GDF-1 GDF-3/Vgr2 SCREW UNIVIN BMP-10
	BMP-9 dorsalin
	BMP-3/Osteogenin BMP-3b GDF-10 ADMP *xenopus*
	BMP-9
	GDF-5/CDMP-1 CDMP-2 GDF-6/BMP-13 GDF-7/BMP-12
	BMP-2 BMP-4 dpp
	BMP-5 BMP-6/Vgr1 BMP-7/OP1 BMP-8/OP2 OP3 60A
distant	MIS GDF-9 Inhibin-α lefty
	GDNF Neurturin
ancestral	prothoracicotropic hormone (PTTH)

Fig. 1. An overview of the current status of the members of the TGF-β superfamily. Superfamily members are grouped according to Kingsley (1994) and Griffith and coworkers (1996) Sequences are referred to Celeste and Wozney, 1994; Wozney and Celeste, 1993; Arora et al., 1994; Chang et al., 1994; Kingsley, 1994; Özkaynak and Oppermann, 1994; Stenzel et al., 1994; Storm et al., 1994; Ecochard et al., 1995; Hötten et al., 1994, 1995; Moos et al. 1995; Noguti et al., 1995; Oda et al., 1995; Griffith et al., 1996; Ikada et al., 1996; Kotzbauer et al., 1996; Meno et al., 1996; Takao et al., 1996.

(Durbec et al., 1996; Trupp et al., 1996) and GDNFR-α, a glycosylphosphatidylinositol (GPI)-anchored protein (Jing et al., 1996; Treanor et al., 1996), introduces tyrosine kinase signaling mechanisms. This finding is surprising and prompts two questions: whether other TGF-β superfamily members signal via tyrosine kinase receptors as well; or whether GDNF, which is only very distantly related to any other TGF-β superfamily member, belongs to another cystine knot growth factor family (Massague, 1996).

The signal transduction pathway of the TGF-β-mediated growth inhibitory effect on epithelial cells, downstream of the serine–threonine receptors, is now being analyzed. Components identified so far are cytoplasmatic proteins associated to the ligand-free (inactive) TβR-I, which, like the immunophillin FKBP12, are thought to block, or like the α-subunit of p21ras farnesyltransferase released on phosphorylation of TβR-I, activate propagation of the signal (Wang et al., 1996a,b). p21ras (Mulder and Morris, 1992), TAK1, a member of the MAPKK family (Yamaguchi et al., 1995), and its specific activator (TAB1; Shibuya et al., 1996), represent further downstream targets. However, the majority of reports have focused on defining components that are regulated by TGF-β within the nucleus. For example, TGF-β decreases c-*myc*, $p34^{cdc2}$, *cdk4*, and B-*myb* expression (Mulder et al., 1988, 1990; Mulder and Brattain, 1988; Pietenpol et al., 1990a,b; Ewen et al., 1993; Elben et al., 1994; Satterwhite et al., 1994), decreases the phosphorylation of the retinoblastoma protein and $p34^{cdc2}$ (Laiho et al., 1990, 1991; Howe et al., 1991), increases the expression of immediate early genes, members of the Jun-family, c-*fos*, *hic-5*, and *tsc-22* (Pertovaara et al., 1989; Kim et al., 1990; Shibanuma et al., 1992, 1994; Hsu et al., 1994; Ohba et al., 1994; Subramaniam et al., 1995), stimulates phosphorylation of cAMP-responsive element binding protein (Kramer et al., 1991), prevents *cdk2* activation (Koff et al., 1993), and regulates several G1 cyclins and cell cycle-associated cyclin-*cdk* inhibitors (Mulder and Morris, 1992; Geng and Weinsberg, 1993; Polyak et al., 1994; Slingerland et al., 1994).

1.3. Biological Activities

TGF-βs are found in most organs and tissues of vertebrates as well as of invertebrates; for example, in muscle, endothelium, blood platelets, and immune cells, and in mesenchymal cells, including bone and connective tissues (Roberts and Sporn, 1990). TGF-βs are multifunctional growth factors and can affect diverse functions in a single cell, such as proliferation, production of cell surface and extracellular matrix molecules, and programmed cell death. The nature of its actions on a particular target cell is crucially dependent on contextual conditions, i.e., cell type and its state of differentiation, and on other growth factors present (Nathan and Sporn, 1991). Furthermore, the prodomain of TGF-β can be secreted as a biological inactive (latent) complex (Pircher et al., 1986; Miyazono et al., 1988; Wakefield et al., 1988) from which the active part can be generated by cleavage or low pH (Lyons et al., 1990). TGF-β is also found in a complex with α2-macroglobulin in serum (O'Connor-McCourt and Wakefield, 1987), which is thought to represent a scavenger mechanism for excess TGF-β (O'Connor-McCourt and Wakefield, 1987; Huang et al., 1988). These mechanisms of activation and clearance represent important means for regulating the availability of extracellular, biologically active TGF-β.

In cells of mesenchymal origin, like fibroblasts or osteoblasts, TGF-β stimulates mitosis. However, this effect is most likely indirect and involves upregulation of the receptors for EGF (Assoian et al., 1984) or induction of platelet-derived growth factor (Leof et al., 1986). In epithelial cells, TGF-β inhibits mitosis, probably by interfering directly with the cell cycle and blocking the transition from late G1 to S-phase (Alexandrow and Moses, 1995). TGF-β prevents the hyperphosphorylation of the retinoblastoma-gene product, which is necessary for the propagation from G1 to S-phase, via interaction with cyclins and cyclin-dependent kinases (Laiho et al., 1990; Howe et al., 1991; Koff et al., 1993; Alexandrow and Moses, 1995).

A prominent function of TGF-β on specialized tissues is that of being a major regulator of extracellular matrix accumulation and composition. TGF-β has been shown to activate gene transcription, and increases synthesis and secretion of extracellular matrix proteins and integrins; on the other hand, it decreases synthesis of proteases that degrade extracellular matrix, and increases synthesis of protease inhibitors that block the activity of these enzymes. Among the matrix molecules that have been shown to be regulated by TGF-β are type I, III, IV, and V collagen, fibronectin, thrombospondin, osteopontin, tenascin, elastin, and chondroitin–dermatan sulfate proteoglycans (for review, *see* Roberts and Sporn, 1990; Centrella et al., 1994).

2. TGF-β ISOFORMS IN THE ADULT CNS

2.1. Distribution

The first immunohistochemical analyses carried out using antibodies against TGF-β1 initially suggested that TGF-β-immunoreactivity (IR) was lacking in the central nervous system (CNS), except in the meninges and choroid plexus. With the identification of the TGF-β isoforms -β2 and -β3 and the generation of isoform-specific antibodies (Flanders et al., 1988, 1990, 1991), it became apparent that TGF-β2 and -β3 are widely distributed within the developing and adult peripheral and CNS (Flanders et al., 1991; Unsicker et al., 1991).

In the adult nervous system, TGF-β2 and -β3 IR are widespread and occur in many neuron populations, as well as in astrocytes, Schwann cells, and the neuroendocrine cells of the adrenal medulla (Unsicker et al., 1991). TGF-β2 and -β3 mRNAs are found in all areas of the CNS, including cerebral cortex, hippocampus, striatum, brainstem, cerebellum, and spinal cord. In the CNS, TGF-β2 and -β3 IR entirely overlap and are generally found in large multipolar neurons. Highest densities of IR neuronal perikarya were found in spinal cord and hindbrain motoneurons, neurons within the substantia nigra, red nucleus, habenular nuclei, hippocampal pyramidal neurons, and dentate gyrus granule cells, hypothalamus, amygdaloid complex, large cortical neurons within layers II, III, and V, retinal ganglion cells, and a subpopulation of the cerebellar Purkinje cells (Unsicker et al., 1991). Most dorsal root ganglionic neurons, their satellite cells, and Schwann cells in peripheral nerves are also labeled. In sciatic nerve, TGF-β3 mRNA was highly expressed, but TGF-β2 mRNA was undetectable (Unsicker et al., 1991; Rufer et al., 1994).

2.2. Functions

Using highly enriched or purified neurons, i.e., chicken dorsal root ganglionic or ciliary ganglionic neurons, a survival-promoting effect of TGF-β could not be demonstrated (Flanders et al., 1991; Unsicker et al., 1996a). However, when more complex culture conditions exist, TGF-β seems to affect neuronal survival. TGF-β1 is able to promote survival of spinal cord motoneurons when cultured on a feeder layer of cortical astrocytes (Martinou et al., 1990). Similarly, neonatal sensory neurons grown with non-neuronal cells in the presence of serum are supported by TGF-β1 and -β2 (Chalazonitis et al., 1992). This effect seems to be mediated by NGF, since neutralizing antibodies to NGF abolished the neurotrophic effect of TGF-β. In purified embryonic chick sensory neurons cultivated under serum-free conditions, NGF induces the synthesis of TGF-β3 mRNA (Krieglstein and Unsicker, 1996). When neutralizing antibodies against TGF-β were added, the number of neurons surviving in the presence of NGF was greatly reduced, suggesting a paracrine interaction of NGF and TGF-β (Krieglstein and Unsicker, 1996). Another mechanism to effect neuronal survival is the regulation of the target-derived neurotrophic factors. Buchman and coworkers (1994) investigated the regulation of expression of NGF and neurotrophin-3 (NT-3) in the target areas of NGF- and NT-3-dependent neurons, showing that, depending on age, TGF-β induced NGF and decreased NT-3 mRNA. These findings raise the possibility that TGF-βs play a role in regulating the pattern of NGF and NT-3 expression in certain neuronal target fields.

Classical neurotrophic factors, such as the neurotrophins, are retrogradely transported along the axon in order to exert their actions in the cell body (Bothwell, 1995). However, TGF-βs do not seem to be axonally transported in neurons (Blottner et al., 1996).

3. TGF-βS IN THE LESIONED CNS

The central role of TGF-βs as potent neuromodulatory molecules, acting as both immunosuppressive and proinflammatory agents, is firmly established (for review, *see* Wahl, 1992; Kiefer et al., 1995). TGF-β also occupies a central position in organizing the nervous system responses to trauma, degeneration, and toxic impairment (Finch et al., 1993; Logan and Berry, 1993; Mattson et al., 1996).

3.1. Expression

Expression of TGF-βs is altered in neural lesions and neurodegenerative diseases. In general, it can be said that TGF-β1 mRNA is highly elevated over an extended period of time, whereas TGF-β2 and -β3 mRNAs are hardly affected. Following a localized surgical cerebral injury, induction of TGF-β1 mRNA occurs in microglia/macrophages and astrocytes around the wound margins, but also in endothelial cells of meninges, hippocampal fissure, and choroid plexus (Lindholm et al., 1992; Logan et al., 1992). Two to three days after the lesion, mRNA and protein peak and then decrease between 7 and 14 d, and become localized within the glial scar at that time. Elevation of TGF-β1 mRNA has also been reported to be caused by deafferentating lesions of hippocampus and striatum (Morgan et al., 1993; Pasinetti et al., 1993). Similarly, excitotoxins induce TGF-β1 mRNA in the hippocampus (Morgan et al., 1993). In several other lesion paradigms, TGF-β1 mRNA is also increased, for example, following transection of the facial nerve in the facial nucleus (Nichols et al., 1991), or in axotomized auditory and dorsal root ganglionic neurons (Lefebvre et al., 1992; Delree et al., 1993; Rogister et al., 1993). Peripheral nerve lesions, crush, and transection are also accompanied by a pronounced induction of TGF-β1 mRNA (Scherer et al., 1993; Rufer et al., 1994). Normalization of expression has been proposed to be induced by axonal contacts (Scherer et al., 1993).

There is ample evidence that TGF-βs are also involved in human neurodegenerative diseases or inflammatory lesions. In Alzheimer's disease and adult Down's syndrome, subsets of senile plaques (van der Wal et al., 1993), and in multiple sclerosis, endothelial cells (Sriram et al., 1991) are IR for TGF-β1. Finally, upregulation of TGF-β1 is also important in other inflammatory lesions, such as those caused by AIDS and experimental autoimmune encephalitis (Wahl et al., 1991; Da Cunha et al., 1993; Kiefer et al., 1993a).

3.2. *Microglial Response*

As mentioned earlier, TGF-β1 is expressed in activated microglia (for review, *see* Morgan et al., 1993). TGF-β has multiple effects on the monocyte/macrophage lineage. These include induction of chemotaxis (Yao et al., 1990) and IL-1 production (Wahl et al., 1987), inhibition of interferon-γ-induced macrophage activation, and induction of class II MHC antigens (Czarniecki et al., 1988; Tsunawaki et al., 1988; Ding et al., 1990). TGF-β may also regulate microglia functions by suppressing acid phosphatase activity, superoxide anion formation, and cytokine (IL-1, IL-6, TNF-α) production by activated microglia (Suzumura et al., 1993).

3.3. *Regulation of Astrocyte Cell Proliferation and Differentiation*

Astroglial cells in culture express all three TGF-β isoforms (Lindholm et al., 1992; Morganti-Kossman et al., 1992; Flanders et al., 1993). TGF-β itself, but also EGF, FGF, and IL-1α, are involved in the upregulation of TGF-β expression (Eccleston et al., 1989; Da Cunha and Vitkovic, 1992; Lindholm et al., 1992; Da Cunha et al., 1993; Flanders et al., 1993). Effects of TGF-β on cytoskeletal and cell surface–extracellular matrix composition and on astrocyte proliferation suggest important functions in the induction of an activated/reactive astrocytic phenotype (Labourdette et al., 1990; Saad et al., 1991; Baghdassarvian et al., 1993; Flanders et al., 1993; Hunter et al., 1993; Pasinetti et al., 1993; Laping et al., 1994). In vivo, TGF-β infusion induces the mRNA of glial fibrillary acidic protein (GFAP) and vimentin, two astroglial-specific intermediate filaments (Laping et al., 1994). In vitro, TGF-β increases plasminogen activator activity and, together with FGF-2, collagen I mRNA (Flanders et al., 1993). Induction of fibronectin expression by TGF-β can only be observed in vivo (Flanders et al., 1993; Pasinetti et al., 1993). Furthermore, TGF-β1 and -β2 increase L1 and decrease N-CAM expression (Saad et al., 1991). TGF-β has also been shown to stimulate NGF expression in astrocytes, which is consistent with its role as an activator of astroglial functions (Lindholm et al., 1990). Along this line, TGF-β1 is also chemotactic for cultured astrocytes, which may explain its capacity to recruit astrocytes into a glial scar (Morganti-Kossman et al., 1992; Logan et al., 1994).

4. NEUROPROTECTIVE POTENCY OF TGF-βS

4.1. *Cerebral Ischemia*

4.1.1. *Pathogenetic Processes*

Brain damage caused by ischemia cannot be attributed to one single factor; rather, several pathogenetic mechanisms are involved. Loss of cellular calcium homeostasis, enhanced production of free radicals, and accumulation of lipid mediators are the main processes that come into play, as well as an ensuing inflammatory response, particularly during the recirculation period of the tissue (Peruche and Krieglstein, 1993; Siesjö et al., 1995; Ferger and Krieglstein, 1996). The initial event during ischemia is energy failure, and the next is neuronal membrane depolarization. The decrease in the energy state and the increase in the intracellular Ca^{2+} concentration ($[Ca^{2+}]_i$) act synergistically to degrade macromolecules and subsequently the structure of the neuronal membrane. However, unless oxygen is resupplied, degradation of cell structures is restricted. Reoxygenation triggers a second series of reactions that lead to very toxic compounds, such as oxygen radicals and nitric oxide (NO) (Siesjö et al., 1995). Nitric oxide synthase (NOS) is activated by Ca^{2+}-dependent calmodulin, leading to a burst of production of NO. Further steps in this cascade could lead to the formation of reactive oxygen species, such as the hydroxyl radical ($OH^{\cdot}$), superoxide anion ($O_2^{\cdot-}$) and hydrogen peroxide (H_2O_2). The combination of $NO^{\cdot}$ and $O_2^{\cdot-}$ yields peroxynitrite. Thus, ischemia with reoxygenation can cause critical cellular damage that may rapidly lead to neuronal death. The overall protein synthesis is depressed over a long period of recirculation, but new mRNAs are expressed and translated into proteins. Thus, the expression of RNA transcripts for neurotrophins suggests an endogenous mechanism of protection (for review, *see* Krieglstein, 1996).

Neurons are the cell type most sensitive against ischemia of brain tissue, but microglia and astrocytes are also affected. The most characteristic feature of microglial cells is their rapid activation in response to brain injury, one of the earliest and most sensitive signs of ongoing pathology in the brain (Kiefer, 1994). Activated glial cells can remove deleterious debris, but also promote tissue repair by secreting growth factors, and thus facilitate the return to tissue homeostasis (Kreutzberg, 1996). Microglia activation was demonstrated 20 min after ischemia, far preceding proliferation, which occurs within 48 h after ischemia (Gehrmann et al., 1995). The microglial reaction after ischemia is closely correlated with the development of delayed neuronal death and resembles a double-edged sword. On the one hand, microglial-derived growth factors may protect the neurons against damage caused by ischemia. On the other hand, microglia cytotoxicity may contribute to, and even promote, neuronal damage (Gehrmann et al., 1995).

In the rat model of global cerebral ischemia, it has been shown that TGF-β1 is expressed by activated microglia (Lehrmann et al., 1995) and simultaneously reduces microglial activation. Thus, microglia might control their own activity state by expressing TGF-β1 (Kiefer et al., 1993b). The spatial distribution pattern of TGF-β1 correlates well with the distribution pattern of activated microglia. A massive expression of TGF-β1 mRNA occurs at sites of neuronal degeneration in the hippocampal CA1 subfield (Armstead et al., 1993; Lehrmann et al., 1995). TGF-β1 inhibits proliferation of astrocytes and suppresses the mitotic effects of FGF and EGF on astrocytes, but it enhances its own synthesis, providing evidence for the possible existence of an autocrine loop in the brain (Lindholm et al., 1990).

TGF-β1 is the lesion form of the TGF-β family and acts as a wound-healing factor by enhancing angiogenesis and extracellular matrix formation (Armstead et al., 1993). Its synthesis by activated microglia represents an endogenous CNS response, which aims at self-limiting tissue damage after an ischemic insult and may promote repair mechanisms. Thus, evidence has accumulated that TGF-β1 plays a significant role in the pathogenesis of brain tissue after injury (Fig. 2). The expression fits the category of delayed response genes (Wang et al., 1995). Wießner et al. (1993) demonstrated that TGF-β1 mRNA increased 3 d after transient global ischemia and became most pronounced after 7 d. TGF-β1 expression 7 d after ischemia was demonstrated in the regions where neuronal degeneration and subsequent microglial and astrocytic activation

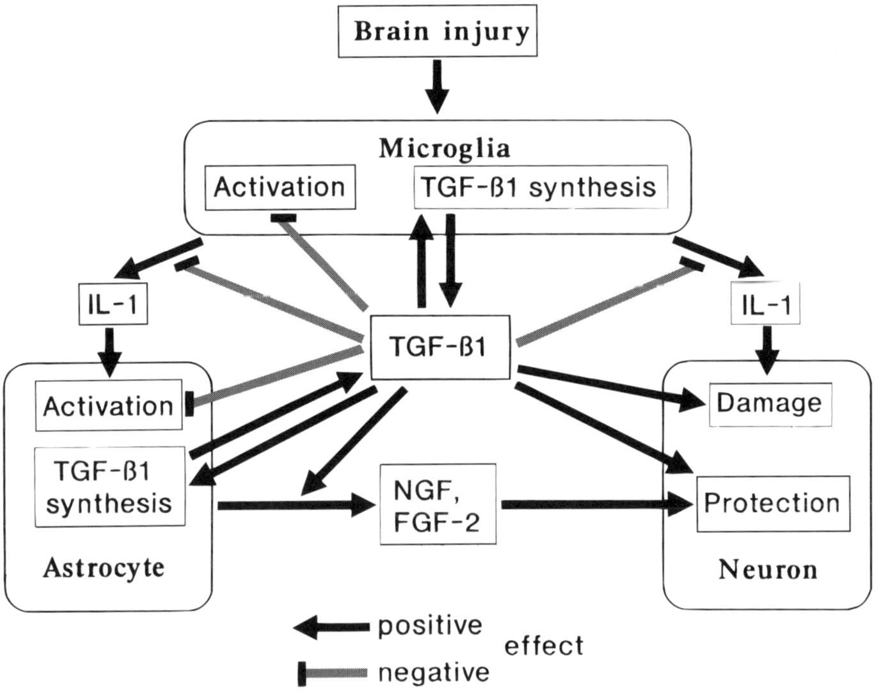

Fig. 2. The critical role of TGF-β1 for glial function and neuroprotection after brain injury. A simplified presentation of the crosstalk between microglia, astrocytes, and neurons.

occurred. The late induction of TGF-β1 mRNA after ischemia points to an involvement in the persistent glial response rather than in the initial glial activation (Lehrmann et al., 1995). Astrocytes are activated by microglia-derived cytokines, but TGF-β1 inhibits astrocytic proliferation and has a downregulating effect on microglia, as evidenced by its inhibition of microglial cytotoxicity. A possible function of TGF-β1 in the postischemic brain may be related to the regulation of glia cell activation, in particular, that of astrocytes, after being released from activated microglia (Fig. 2). The late induction of TGF-β1 mRNA after ischemia (3–5 d after) could indicate that TGF-β1 immediately down-regulates astrocytosis and astrocytic activation (as assessed by GFAP immunostaining), peaks 3 d after ischemia, and then persists at a relatively constant level over the next 5 wk (Wießner et al., 1993; Lehrmann et al., 1995).

TGF-β1 seems to play a similar role in brain tissue after ischemic stroke in humans. Krupinski et al. (1996) examined brain samples from patients who suffered from ischemic stroke for the expression of TGF-β1 mRNA. The ischemic penumbra contained the highest level of TGF-β1 mRNA.

Even if the neuroprotective potency of TGF-β1 after ischemia holds true, delayed neuronal death occurs despite the massive expression of TGF-β1 in the vicinity of injured neurons. One might speculate that the other toxic compounds overcome the neuroprotective effect of TGF-β1, or that it is expressed too late to rescue the dying neurons. Therefore, administration of TGF-β1 early after cerebral ischemia could be an appropriate therapeutic strategy. It could suppress cytotoxic glial reactions and stimulate its own synthesis and that of other growth factors, such as NGF and FGF-2. In addition,

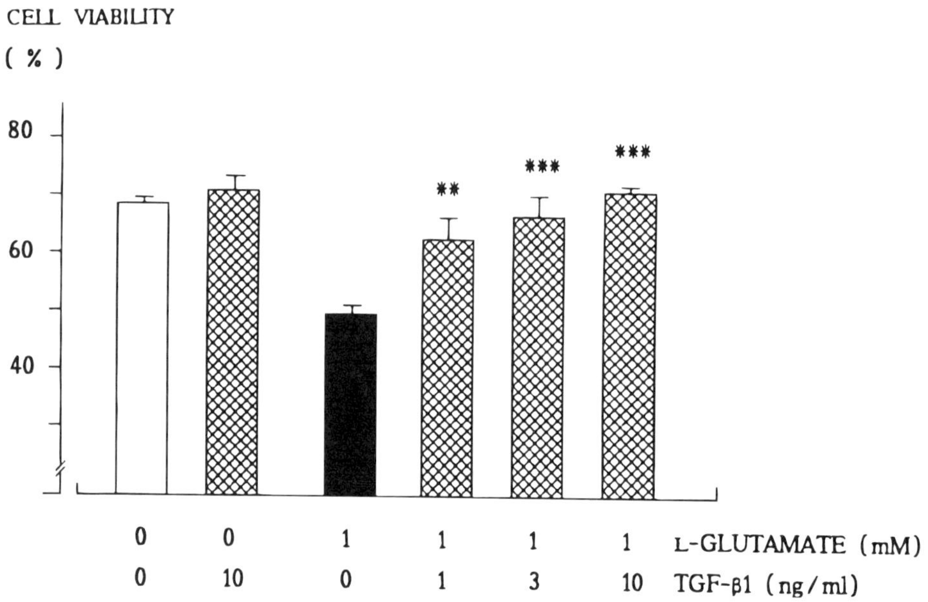

Fig. 3. Neuroprotection of cultured neurons by pretreatment with TGF-β1. Primary cultures of neurons obtained from embryonic chick telencephalon were exposed to 1 m*M* L-glutamate for a period of 60 min. Eighteen hours later, the percentage of viable neurons was determined by trypan blue exclusion. TGF-β1 was present in the medium from d 1 of culture up to d 5, including the period of glutamate exposure. The values are given as means ±SD from 4–5 cultures. Different from glutamate-treated controls: $^{**}p < 0.01$, $^{***}p < 0.001$ (analysis of variance, Duncan's test).

TGF-β1 is thought to induce angiogenesis and to inhibit the neutrophil adherence to endothelial cells injured by ischemia. These effects of TGF-β1 could facilitate the survival of neurons after ischemia.

4.1.2. *Neuroprotection by TGF-βs*

To protect neurons against damage is an urgent therapeutic task. Various drugs were experimentally and clinically tested for their capacity to protect neurons against ischemic or hypoxic injury. However, as indicated, neuronal damage results from many factors and none of these drugs acting against one of these damaging processes could completely prevent neuronal death. Thus, there is still need of a convincing neuroprotective strategy, and the question arises whether growth factors could be useful as neuroprotectants, because it is known that several of these peptides are critically important for the survival of neurons (Mattson and Furukawa, 1996).

In a first approach, we studied the potential neuroprotective effects of TGF-βs, which are multifunctional peptides acting on nearly every type of cell. We addressed these first experiments to the effects of three distinct TGF-β isoforms, 1, 2, and 3, on neuronal degeneration caused by energy deficiency of chick embryo telencephalic neurons in vitro (Prehn et al., 1993a; Henrich-Noack et al., 1994). The neuroprotective effects of the TGF-βs were indicated by histological and biochemical parameters. The cultured chick neurons were damaged with 1 m*M* cyanide or with 1 m*M* glutamate. Pretreatment and posttreatment with TGF-β1 prevented neuronal degeneration (Figs. 3 and 4). TGF-β1

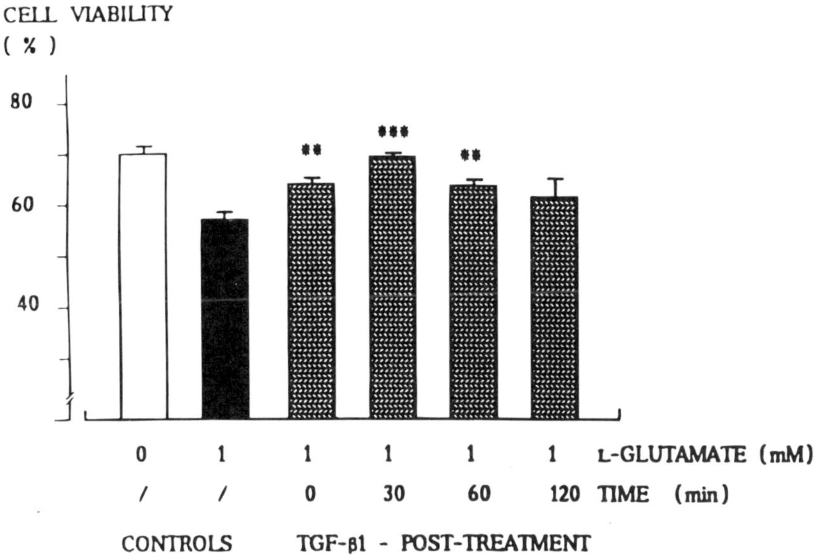

Fig. 4. Neuroprotection of cultured neurons by posttreatment with TGF-β1. Primary cultures of neurons obtained from embryonic chick telencephalon were exposed to 1 m*M* L-glutamate for 60 min. The percentage of viable neurons was determined by the Trypan blue exclusion method. TGF-β1 was added to the medium immediately after the glutamate treatment and remained in the cultures until the next day, when cell viability was assessed. The values are given as means ±SD from 4–5 cultures. Different from glutamate-treated controls: $**p < 0.01$, $***p < 0.001$ (analysis of variance, Duncan's test).

was found to exert the most pronounced neuroprotective activity. TGF-β2 did not cause any neuroprotection in these experiments, and the effect of TGF-β3 was restricted to a small dosage range. These experiments with primary cultures of chick neurons without glial cells demonstrate that TGF-β1 can act directly on neurons. To confirm its neuroprotective potency, TGF-β1 was also tested in primary cultures of rat cortical neurons, and similar neuroprotective effects were demonstrated (Prehn et al., 1993b). Furthermore, the protective effects of TGF-β1 have been examined in several other cell-culture models of neurodegeneration. These include death caused by glutamate receptor agonists (Prehn et al., 1995), X-irradiation (Jordan et al., 1997), β-amyloid (Prehn et al., 1996), and the AIDS virus coat protein gp120 (Meucci and Miller, 1997). TGF-β1 ameliorated cell death in all these cases. It should be noted that, in most of these experiments, death of neurons appears to occur primarily by apoptosis, although, in the case of excitotoxicity and gp120, necrosis also appears to be of importance.

Thus, it has become clear that TGF-β1, in particular, has the capacity to prevent degeneration of cultured neurons in a dose-dependent manner. However, the crucial question was whether this pharmacological potency of TGF-β1 could be demonstrated in vivo, as well. To this end, we initially used the mouse model of focal cerebral ischemia, injecting TGF-β1 into the cerebroventricle before occlusion of the middle cerebral artery. The cytokine reduced, not markedly, but statistically significantly, the infarct area on the mouse brain surface (Fig. 5; Prehn et al., 1993b).

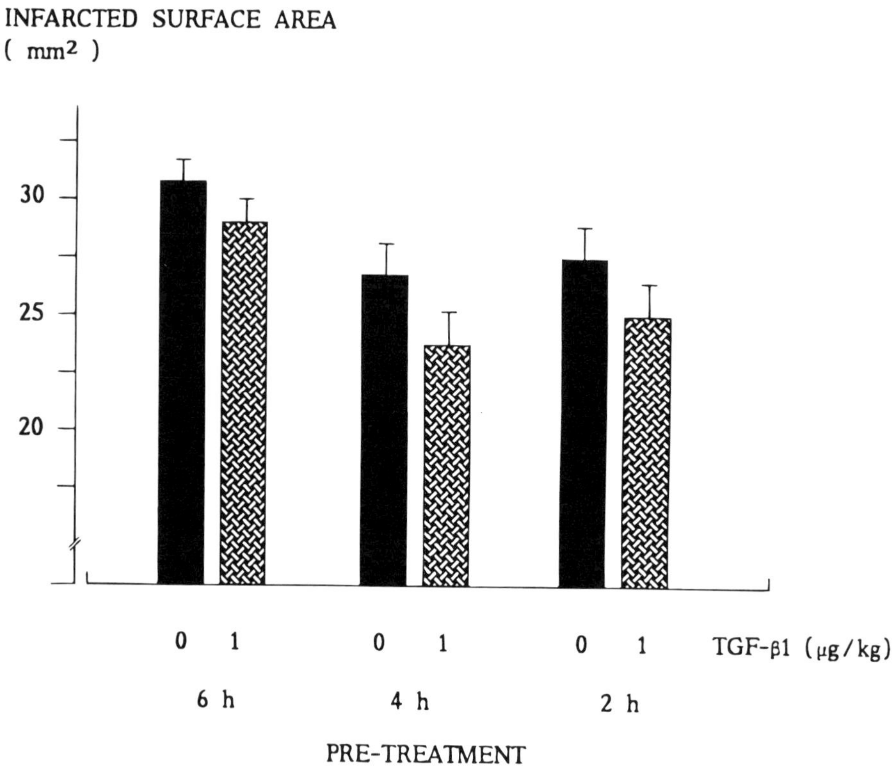

Fig. 5. Pretreatment of mice with TGF-β1 reduces the infarct after permanent occlusion of the middle cerebral artery. The area of ischemia was calculated planimetrically by means of an image-analyzing system. The animals received TGF-β1 intracerebroventricularly 6, 4, or 2 h prior to vessel occlusion. Values are means ±SD of 9–10 experiments. The overall effect of the TGF-β1 pretreatment was statistically significant ($p < 0.05$; two-way analysis of variance).

Furthermore, TGF-β1 caused significant protection of rat hippocampal neurons after transient global ischemia (Fig. 6; Henrich-Noack et al., 1996). In this model, TGF-β1 rescued hippocampal CA1 neurons at doses significantly lower than those of other growth factors. For instance, FGF-2 prevented the delayed death of hippocampal neurons in the gerbil model of global ischemia when given in a dose of 240 ng/d for a period of 4 d (Nakata et al., 1993). This treatment adds up to a total dose of approx 14 μ/kg FGF-2. NGF ameliorated neuronal degeneration after transient global cerebral ischemia in gerbils at a dose of 150 μ/kg (Shigeno et al., 1991). In contrast, 15 ng/kg TGF-β1 administered intracerebroventricularly protected the hippocampal CA1 neurons after transient global ischemia in the rat, i.e., around 1000–10,000 times less than for the above-mentioned factors (Henrich-Noack et al., 1996), and nearly as potent as the secreted form of β-amyloid precursor protein (Smith-Swintosky et al., 1994). TGF-β1, therefore, appears to be an extremely potent neuroprotective cytokine. Gross et al. (1993) used a rabbit model and caused cerebral ischemia by clot emboli. Four hours after embolization the infarct volume was found to be significantly reduced by 50 μg TGF-β1 administered intra-arterially. Probably, TGF-β1 could not be delivered adequately to brain tissue under this type of administration. When administered immediately after ischemia, TGF-β1

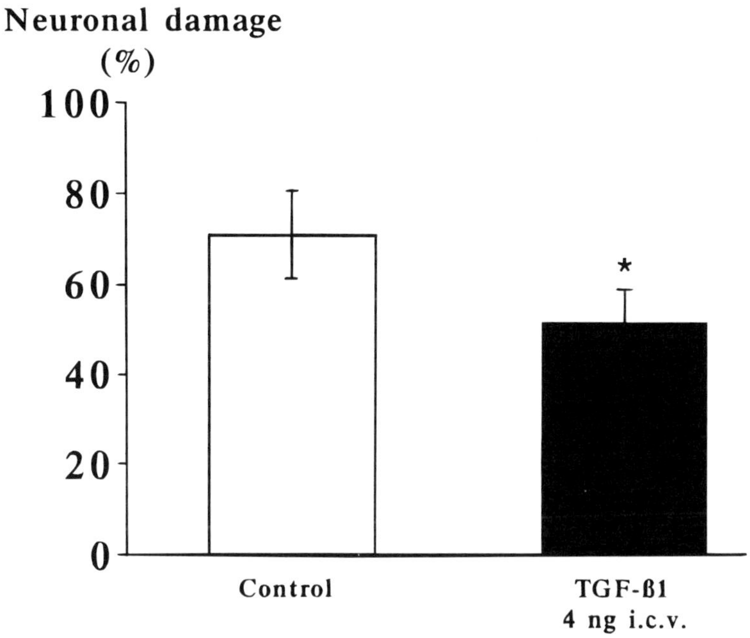

Fig. 6. TGF-β1 protects hippocampal neurons against damage caused by transient forebrain ischemia in the rat. TGF-β1 was administered intracerebroventricularly 1 h before ischemia. Neuronal damage was evaluated 7 d after ischemia. Values are means ±SD from 11–12 experiments. $^*p = 0.04$ (U-test).

was not effective at all. McNeill et al. (1994) used a rat model of hypoxic-ischemic brain injury, and reported that treatment of the rats with 10 ng TGF-β1 reduced the magnitude of neuronal loss and the area of cortical infarction. In addition, they observed reduced microglia reaction in brain tissue.

4.1.3. Possible Mechanisms of Action

As already indicated, TGF-β1 inhibits the activation of microglia and astrocytes caused by neuronal degeneration (Fig. 2), which might represent an important mechanism to limit the impact of inflammatory processes (Kuruvilla et al., 1991; Kiefer et al., 1993b). Exogenous TGF-β1 stimulates its own endogenous synthesis, as well as that of NGF, FGF, and the proto-oncogene c-*fos*, but not that of BDNF (Lindholm et al., 1990, 1992). TGF-β2 also increased NGF mRNA in rat hippocampus after ICV injection (Lindholm et al., 1990). Neuronal regeneration also may be inhibited by a scar produced by glial reactions, which forms a physical barrier to the growing axons. By inhibiting scar formation, TGF-β1 can contribute to repair mechanisms of the injured neurons (Lindholm et al., 1992). The TGF-β1 mimicks in some respects the beneficial effects of immunosuppresssive or anti-inflammatory drugs, but without discernible adverse effects (Kuruvilla et al., 1991).

Evidence is growing that TGF-β1 represents a control switch in many biological processes of various tissues. Considering these phenomena, it might be conceivable that TGF-β1 has a similar function in the injured (Prehn and Krieglstein, 1994). TGF-β1 has the capacity to improve the survival of neurons and, on the other hand, accelerate the removal of necrotic neurons (Prehn and Krieglstein, 1994). From a teleological point of view, both processes could be important for protecting brain tissue and function.

The remaining question is how TGF-βs could influence biochemical pathways to achieve their neuroprotective activity. Until now, two effects of TGF-βs that could be of significance for neuronal survival have been described by Prehn et al. (1994). TGF-β1 prevents Ca^{2+} of neurons and induces the oncogene *Bcl-2*. TGF-β1 reduced the increase in $[Ca^{2+}]_i$ of rat hippocampal neurons caused by the glutamate agonist *N*-methyl-D-aspartate or the Ca^{2+} ionophore 4-Br A 23187, and it elevated the *Bcl-2* content of cells. These two effects of TGF-β1 may protect against a wide range of toxic insults. The effect on $[Ca^{2+}]_i$ is achieved by an enhanced buffering and extrusion capacity of mitochondria. TGF-β1 and 2 produced a rapid and pronounced induction of Bcl-2, which is capable of protecting neurons against damage. The overexpression of Bcl-2 shifts the cellular redox potential to a more reduced state, without consistently affecting the major cellular antioxidant enzymes. Such a shift may be an appropriate cellular strategy to prevent damage from different insults, because the net cellular generation of reactive oxygen species is decreased (Ellerby et al., 1996). In addition, TGF-βs have been shown to reduce iNOS activity (Vodovotz et al., 1993) and, thereby, NO production is diminished. This effect of TGF-β could also contribute to its neuroprotective activity.

To summarize, TGF-β1 seems to be capable of protecting neurons against various insults by different mechanisms and by simultaneously orchestrating protective cellular and molecular events (Mattson et al., 1996).

4.2. *Alzheimer's Disease*

The histopathological hallmarks of Alzheimer's disease (AD) are neurofibrillary tangles and neuritic plaques (Tomlinson, 1980). Plaques and tangles are also found in normally aged brain, but to a far lesser degree. Immunochemical studies have shown that plaques contain several inflammatory markers, including complement, activated microglia and astrocytes, as well as various glia-derived cytokines (TGF-β1, TNFα, IL-1α, IL-1β, IL-6) (Patterson,1995; Sheng et al., 1995). The key question is whether this inflammatory reaction of AD brain tissue represents a defense mechanism and is the end stage of the amyloid cascade, or whether the cytokines contribute to the plaque development (Patterson, 1995).

Evidence increases that the TGF-βs are involved in the development of AD, but the specific role of the isoforms remains unclear. Flanders et al. (1995) analyzed brain tissue of AD patients and found TGF-β2 levels about three times increased, compared to controls. These authors did not find TGF-β1 and TGF-β3 selectively altered in any AD patient, and therefore they suggest that TGF-β2 is selective for neurofibrillary tangle-bearing neurons in AD; however, the nature of this association is unknown. TGF-β2 might have a function in tangle formation, or it could be a response to normalize the extracellular environment.

Peress and Perillo (1995) found TGF-β1IR limited to neuritic profiles within the senile plaques of AD patients. Neurofibrillary tangles, plaque neurites, microglia, astrocytes, and macrophages were shown to express TGF-β2IR. The TGF-β3 isoform appeared concentrated in Hirano bodies.

van der Wal et al. (1993) investigated brains of patients with AD and Down's syndrome (DS) and found TGF-β1-containing plaques mainly in the molecular layer of the dentate gyrus of AD patients; in DS patients TGF-β1 plaques were preferentially located in the entorhinal cortex. In AD, TGF-β1IR was also shown in neurofibrillary tangles. The authors suggest that TGF-β1 may play a role in the cascade of events leading to plaque formation (van der Wal et al., 1993).

Gray and Patel (1993) provided evidence that astrocytes are involved in the development of plaques in AD and DS. Astrocytes surround plaques throughout their development (Wisniewski et al., 1989) and synthesize TGF-β1, which is found in the plaques and appears to be increased in the surrounding cells (van der Wal et al., 1993). In cultured astrocytes, TGF-β1 produced an about sixfold increase in total mRNA of amyloid precursor protein (APP), but IL-1β had only a small effect. These results suggest that the formation of plaques is dependent on the ability of TGF-β1 to induce APP synthesis and matrix deposition (Gray and Patel, 1993).

It was shown in transgenic mice that astroglial overexpression of TGF-β1 induced an increased deposition of extracellular matrix components in the CNS, particularly around blood vessels along the glial limitans, as well as on neuronal cell bodies and processes. These data strongly suggest a direct cause–effect relationship between TGF-β1 and extracellular matrix formation, and it is argued that TGF-β1 could also contribute to the formation of amyloid plaques in AD (Wyss-Coray et al., 1995). On the other hand, TGF-β1 and -β2 isoforms were found to protect neurons dose-dependently against amyloid-induced injury (Chao et al., 1994).

These findings on glial reactions and glial-derived cytokines are consistent with the emerging concept of AD as an inflammatory disease that could be treated with anti-inflammatory drugs, and there is already some clinical evidence supporting that concept. On the other hand, it is argued that the enrichment of TGF-βs in the lesions of AD brains may dampen local inflammatory response. This might lead to a less effective clearance of toxic compounds in the damaged brain regions and allow a slow but progressive enlargement of lesions. In any case, the mechanisms of TGF-βs activities in AD have to be defined more precisely to provide insight into the potential therapeutic approaches.

4.3. Parkinson's Disease

Dopaminergic neurons of the nigrostriatal system are important in the control of motor performance and are a major focus of degeneration in Parkinson's disease. Movement disorders in this disease are primarily caused by loss of these dopaminergic neurons and substantial reduction of dopamine levels (Agid, 1991). Oxidative stress, increased iron, defects in enzymes with a capacity to cope with free radicals mediated by excitatory amino acids, and lack of availability of growth factors to balance the toxic influences have been implicated in the etiology of the idiopathic disease (Shoffner et al., 1991; Turski et al., 1991; Przedborski et al., 1992; Edwards, 1993; Gerlach et al., 1994). The classical therapy of chronic dopamine replacement (L-DOPA plus dopa decarboxylase inhibitor) has clear limitations that require supplementary or alternative strategies, including the application of neurotrophic factors (Unsicker, 1994; Unsicker al., 1996b). TGF-βs have been shown to promote, in a dose-dependent fashion, survival and differentiation of dopaminergic neurons isolated from E14 rat mesencephalon floor (Krieglstein and Unsicker, 1994; Poulsen et al., 1994; Krieglstein et al., 1995b). The effects of TGF-βs are not mediated by astroglial cells or any other numerical expansion of another cell population in these cultures. The effects of TGF-βs are more specific in terms of saturating amounts required (20 p*M*) than the effects of neurotrophins on these neurons (approx 200 p*M*), and are not dependent on initial exposure to serum (Krieglstein et al., 1996). The neurotoxin MPTP and MPP^+, its active metabolite, produce in vivo a syndrome with striking clinical and pathological resemblance to Parkinson's disease (Kopin and Markey, 1988; Forno et al., 1993). MPP^+ is also toxic to cultured dopaminergic neurons of the substantia nigra, therefore allowing the in vitro testing of putative beneficial agents that can sustain toxically impaired dopaminergic neurons. TGF-β1 reduced the MPP^+-mediated loss of tyrosine hydroxylase positive neurons from 62% to about 33%. Finally, TGF-β2 mRNA and protein occur in both the embryonic mesencephalon floor and the adult substantia nigra, raising the possibility that TGF-β is a physiologically relevant growth factor for midbrain dopaminergic neurons (Flanders et al., 1991; Unsicker et al., 1991; Krieglstein et al., 1995a).

REFERENCES

Agid, Y. (1991) Parkinson's disease: pathophysiology. *Lancet* **337,** 1321–1327.

Alexandrow, M. G., and Moses H. L. (1995) Transforming growth factor β and cell cycle regulation. *Cancer Res.* **55,** 1452–1457.

Anzano, M. A., Roberts, A. B., Smith, J. M., Sporn, M. B., and De Larco, J. E. (1983) Sarcoma growth factor from conditioned medium is composed of both type alpha and type beta transforming growth factors. *Proc. Natl. Acad. Sci. USA* **80,** 6264–6268.

Armstead, W. M., Mirro, R., Zuckerman, S. L., Shibata, M., and Leffler, C. W. (1993) Transforming growth factor-β attenuates ischemia-induced alterations in cerebrovascular responses. *Heart Circ. Physiol.* **33,** H381–H385.

Arora, K., Levine, M. S., and O'Connor, M. B. (1994) The screw gene encodes a ubiquitously expressed member of the TGF-beta family required for specification of dorsal cell fates in the Drosophila embryo. *Genes Dev.* **8,** 2588–2601.

Assoian, R. K., Frolik, C. A., Roberts, A. B., Miller, D. M., and Sporn, M. B. (1984) Transforming growth factor-beta controls receptor levels for epidermal growth factor in NRK fibroblasts. *Cell* **36,** 35–41.

Assoian, R. K., Komoiya, A., Meyers, C. A., Miller, D. M., and Sporn, M. B. (1983) Transforming growth factor-beta in human platelets. *J. Biol. Chem.* **258,** 7155–7160.

Baghdassarian, D., Toru-Delbauffe, D., Gavaret, J. M., and Pierre, M. (1993) Effects of transforming growth factor-β1 on the extracellular matrix and cytoskeleton of cultured astrocytes. *Glia* **7,** 193–202.

Blottner, D., Wolf, N., Lachmund, A., Flanders, K. C., and Unsicker, K. (1996) TGF-β rescues target-deprived preganglionic sympathetic neurons in the spinal cord. *Eur. J. Neurosci.* **8,** 202–210.

Bothwell, M. (1995) Functional interactions of neurotrophins and neurotrophin receptors. *Annu. Rev. Neurosci.* **18,** 223–253.

Buchman, V. L., Sporn, M., and Davies, A. M. (1994) Role of transforming growth factor-beta isoforms in regulating the expression of nerve growth factor and neurotrophin-3 mRNA levels in embryonic cutaneous cells at different stages of development. *Development* **120,** 1621–1629.

Burt, D. W. and Paton, I. R. (1992) Evolutionary origins of the transforming growth factor-beta gene family. *DNA Cell. Biol.* **11,** 497–510.

Celeste, A. and Wozney, J. M. (1994) *Eur. Patent Appl.* WO 94/26893, seq. id. no. 1.

Centrella, M., Horowitz, M. C., Wozney, J. M., and McCarthy, T. L. (1994) Transforming growth factor-β gene family members and bone. *Endocrine Rev.* **15,** 27–39.

Chalazonitis, A., Kalberg, J., Twardzik, D. R., Morrison, R. S., and Kaessler, J. A. (1992) Transforming growth factor beta has neurotrophic actions on sensory neurons in vitro and is synergistic with nerve growth factor. *Dev. Biol.* **152,** 121–132.

Chang, S. C., Hoang, B., Thoams, J. T., Vukicevic, S., Luyten, F. P., Ryba, N. J., et al. (1994) Cartilage-derived morphogenetic proteins. New members of the transforming growth factor-beta superfamily predominantly expressed in long bones during human embryonic development. *J. Biol. Chem.* **269,** 28,227–28,234.

Chao, C. C., Hu, S., Kravitz, F. H., Tsang, M., Anderson, W. R., and Peterson, P. K. (1994) Transforming growth factor-β protects human neurons against β-amyloid-induced injury. *Mol. Chem. Neuropathol.* **23,** 159–178.

Czarniecki, C. W., Chui, H. H., Wong, G. H. W., McCabe, S. M., and Palladino, M. A. (1988) Transforming growth factor-β1 modulates the expression of class II histocompatibility antigens on human cells. *J. Immunol.* **140,** 4217–4223.

Da Cunha, A. and Vitkovic, L. (1992) Transforming growth factor beta 1 (TGF-β1) expression and regulation in rat cortical astrocytes. *J. Neuroimmunol.* **36,** 157–169.

Da Cunha, A., Jefferson, J. A., Jackson, R. W., and Vitkovic, L. (1993) Glial cell specific mechanisms of TGF-β1 induction by IL-1 in cerebral cortex. *J. Neuroimmunol.* **42,** 71–86.

Delree, P., Ribbon, C., Martin, D., Rogister, B., Lefebvre, P. P., Rigo, J. M., et al. (1993) Plasticity of developing and adult dorsal root ganglion neurons as revealed in vitro. *Brain Res. Bull.* **30,** 231–237.

deMartin, R., Haendler, B., Hofer-Warbinek, R., Gaugitsch, H., Wrann, M., Schlusener, H., et al. (1987) Complementary DNA for human glioblastoma-derived T cell suppressor factor, a novel member of the transforming growth factor-β gene family. *EMBO J.* **6,** 3673–3677.

Derynck, R., Jarrett, J. A., Chen, E. Y., Eaton, D. H., Bell, J. R., Assoian, R. K., et al. (1985) Human transforming growth factor-beta cDNA sequence and expression in tumor cell lines. *Nature* **316,** 701–705.

Derynck, R., Lingquist, P. B., Lee, A., Wen, D., Tamm, J., Graycar, J. L., et al. (1988) A new type of transforming growth factor-β, TGF-β3. *EMBO J.* **7,** 3737–3743.

Derynck, R., Roberts, A. B., Winkler, M. E., Chen, E. Y., and Goeddel, D. V. (1984) Human transforming growth factor alpha: precursor structure and expression in E. coli. *Cell* **38,** 287–297.

Ding, A., Nathan, C. F., Graycar, D. J., Derynck, R., Stuehr, D.J., and Sirrmal, S. (1990) Macrophage deactivating factor and transforming growth factor-β1, β2, and β3 inhibit induction of macrophage nitrogen oxide synthesis by IFN-g. *J. Immunol.* **145,** 940–944.

Doapin, S., Cohen, G. H., and Davies, D. (1992) Structural similarity between transforming growth factor-β2 and nerve growth factor. *Science* **258,** 1160–1162.

Durbec, P., Marcos-Gutierrez, C. V., Kilkenny, C., Grigoriou, M., Wartiowaara, K., Suvanto, P., et al. (1996) GDNF signalling through the ret receptor tyrosine kinase. *Nature* **381,** 789–793.

Eccleston, P. A., Jessen, K. R., and Mirsky, R. (1989) Transforming growth factor-β and interferon-g have dual effects on growth of peripheral glia. *J. Neurosci. Res.* **24,** 524–530.

Ecochard, V., Cayrol, C., Foulquier, F., Zaraisky, A., and Duprat, A. M. (1995) A novel TGF-β-like gene, fugacin, specifically expresssed in the spemann organizer of xenopus. *Dev. Biol.* **172,** 699–703.

Edwards, R. H. (1993) Pathogenesis of Parkinson's disease. *Clinical Neurosci.* **1,** 36–44.

Elben, S. T., Fautsch, M. P., Burnette, R. J., Joshi, P., and Leof, E. B. (1994) Cell cycle-dependent inhibition of p34cdc2 synthesis by transforming growth factor beta 1 in cell cycle epithelial cells. *Cell Growth Differ.* **5,** 109–116.

Ellerby, L. M., Ellerby, H. M., Park, S. M., Holleran, A. L., Murphy, A. N., Fiskum, G., et al. (1996) Shift of the cellular oxidation-reduction potential in neural cells expressing Bcl-2. *J. Neurochem.* **67,** 1259–1267.

Ewen, M. E., Sluss, H. K., Whitehouse, L. L., and Livingston, D. M. (1993) TGF beta inhibition of Cdk4 synthesis is linked to cell cycle arrest. *Cell* **74,** 1009–1020.

Ferger, D. and Krieglstein, J. (1996) Cerebral ischemia: pharmacological bases of drug therapy. *Dementia* **7,** 161–168.

Finch, C. E., Laping, N. J., Morgan, T. E., Nichols, N. R., and Pasinetti, G. M. (1993) TGF-β1 is an organizer of responses to neurodegeneration. *J. Cell Biochem.* **53,** 314–322.

Flanders, K. C., Roberts, A. B., Ling, N., Fleudelys, B. E., and Sporn, M. B. (1988) Antibodies to peptide determinants of transforming growth factor-β and their applications. *Biochemistry* **27,** 739–746.

Flanders, K. C., Cissel, D. S., Mullen, L. T., Danielpour, D., Sporn, M. B., and Roberts, A. B. (1990) Antibodies to transforming growth factor-β2 peptides: specific detection of TGF-β2 in immunoassays. *Growth Factors* **3,** 45–52.

Flanders, K. C., Lüdecke, G., Engels, S., Cissel, D. S., Roberts, A. B., Kondaiah, P., et al. (1991) Immunhistochemical localization of transforming growth factor-βs in the nervous system of the mouse embryo. *Development* **113,** 183–191.

Flanders, K. C., Lüdecke, G., Renzig, J., Hamm, C., Cissel, D. S., and Unsicker, K. (1993) Effect of TGF-βs and bFGF on astroglial cell growth and gene expression in vitro. *Mol. Cell Neurosci.* **4,** 406–417.

Flanders, K. C., Lippa, C. F., Smith, T. W., Pollen, D. A., and Sporn, M. B. (1995) Altered expression of transforming growth factor-β in Alzheimer's disease. *Neurology* **45,** 1561–1569.

Forno, L. S., Delanney, L. E., Irwin, I., and Langston, J. W. (1993) Similarities and differences between MPTP-induced Parkinsonism and Parkinson's disease. *Adv. Neurol.* **60,** 600–608.

Franzen, P., ten Dijke, P., Ichijo, H., Yamashita, H., Schulz, P., Helsin, C.-H., and Miyazono, K. (1993) Cloning of a TGFβ type I receptor that forms a heteromeric complex with the TGFβ type II receptor. *Cell* **75,** 681–692.

Gehrmann, J., Banati, R. B., Wiessner, C., Hossmann, K. A., and Kreutzberg, G. W. (1995) Reactive microglia in cerebral ischaemia: an early mediator of tissue damage? *Neuropathol. Appl. Neurobiol.* **21,** 277–289.

Geng, Y. and Weinsberg, R. A. (1993) Transforming growth factor beta effects on expression of G1 cyclins and cyclin-dependent protein kinases. *Proc. Natl. Acad. Sci. USA* **90,** 10315–10319.

Gerlach, M., Ben-Shachar, D., Riederer, P., and Youdim, M. B. (1994) Altered brain metabolism of iron as a cause of neurodegenerative disease? *J. Neurochem.* **63,** 793–807.

Gray, C. W. and Patel, A. J. (1993) Regulation of β-amyloid precursor protein isoform mRNAs by transforming growth factor-β1 and interleukin-β1 in astrocytes. *Mol. Brain Res.* **19,** 251–256.

Griffith, D. L., Keck, P. C., Sampath, T. K., Rueger, D. C., and Carlson, W. D. (1996) Three-dimensional structure of recombinant human osteogenic protein 1: structural paradigm for the transforming growth factor β superfamily. *Proc. Natl. Acad. Sci. USA* **93,** 878–883.

Gross, C. E., Bednar, M. M., Howard, D. B., and Sporn, M. B. (1993) Transforming growth factor-β1 reduces infarct size after experimental cerebral ischemia in a rabbit model. *Stroke* **24,** 558–562.

Henrich-Noack, P., Prehn, J. H. M., and Krieglstein, J. (1994) Neuroprotective effects of TGF-β1. *J. Neural Transm.* **43,** 33–45.

Henrich-Noack, P., Prehn, J. H. M., and Krieglstein, J. (1996) TGF-β1 protects hippocampal neurons against degeneration caused by transient global ischemia: dose-response relationship and potential neuroprotective mechanisms. *Stroke* **27,** 1609–1615.

Hinck, A. P., Archer, S. J., Qian, S. W., Roberts, A. B., Sporn, M. B., Weatherbee, J. A., et al. (1996) Transforming growth factor β1: Three-dimensional structure in solution and comparison with the X-ray structure of transforming growth factor β2. *Biochemistry* **35,** 8517–8534.

Hötten, G., Neidhardt, H., Jacobowsky, B., and Pohl, J. (1994) Cloning and expression of recombinant human growth/differentiation factor 5. *Biochem. Biophys. Res. Commun.* **204,** 646–652.

Hötten, G., Neidhardt, H., Schneider, C., and Pohl, J. (1995) Cloning of a new member of the TGF-β family: a putative new activin βC chain. *Biochem. Biophys. Res. Commun.* **206,** 608–613.

Howe, P. H., Draetta, G., and Leof, E. B. (1991) Transforming growth factor β1 inhibition of p34cdc2 phosphorylation and histone H1 kinase ativity is associated with G1/S-phase growth arrest. *Mol. Cell Biol.* **11,** 1185–1194.

Hsu, S., Huang, F., Hafez, M., Winawer, S., and Friedman, E. (1994) Colon carcinoma cells switch their response to transforming growth factor beta 1 with tumor progression. *Cell Growth Differ.* **5,** 267–275.

Huang, S. S., O'Grady, P., and Huang, J. S. (1988) Human transforming growth factor-β: alpha2-macroglobulin complex is a latent form of transforming growth factor-β. *J. Biol. Chem.* **263,** 1535–1541.

Hunter, K. E., Sporn, M. B., and Davies, A. M. (1993) Transforming growth factor-βs inhibit mitogen-stimulated proliferation of astrocytes. *Glia* **7,** 203–211.

Ikada, M., Katagiri, T., Akimyama, S., Namiki, M., Komaki, M., Yamaguchi, A., et al. (1996) Bone morphogenetic protein-12 and –13 inhibit differentiation of myoblast, but do not induce their differentiation into osteoblasts. *Biochem. Biophys. Res. Commun.* **222,** 317–322.

Jakowlew, S. B., Dillard, P. J., Sporn, M. B., and Roberts, A. B. (1988) Complementary deoxyribonucleic acid cloning of an mRNA encoding transforming growth factor-beta 4 from chicken chondrocytes. *Mol. Endocrinol.* **2,** 1186–1195.

Jing, S., Wen, D., Yu, Y., Holst, P. L., Luo, Y., Fang, M., et al. (1996) GDNF-induced activation of the ret protein tyrosine kinase is mediated by GDNFR-α, a novel receptor for GDNF. *Cell* **85,** 1113–1124.

Jordan, J., Galindo, M. F., Prehn, J. H. M., Weichselbaum, R. R., Beckett, M., Ghadge, G. D., et al. (1997) p53 expression induces apoptosis in hippocampal pyramidal neuron cultures, *J. Neurosci.* **15**, 1397–1405.

Kiefer, R., Gold, R., Gehrmann, J., Lindholm, D., Wekerle, H., and Kreutzberg, G. W. (1993a) Transforming growth factor-β1 differentially regulates proliferation and MHC class-II antigen expression in forebrain and brainstem astrocyte primary cultures. *Brain Res.* **585,** 229–236.

Kiefer, R., Gold, R., Gehrmann, J., Lindholm, D., Wekerle, H., and Kreutzberg, G. W. (1993b) Transforming growth factor beta expression in reactive spinal cord microglia and meningeal inflammatory cells during experimental allergic neuritis. *J. Neurosci. Res.* **36,** 391–398.

Kiefer, R. (1994) Cytokine expression in the lesioned nervous system. *Neuropathol. Appl. Neurobiol.* **20,** 187–188.

Kiefer, R., Streit, W. J., Toyka, K. V., Kreutzberg, G. W., and Hartung, H.-P. (1995) Transforming growth factor-β1: a lesion-associated cytokine of the nervous system. *Int. J. Dev. Neurosci.* **13,** 331–339.

Kim, S. J., Angel, P., Lafyatis, R., Hattori, K., Kim, K. Y., Sporn, M. B., Karin, M., and Roberts, A. B. (1990) Autoinduction of transforming growth factor beta 1 is mediated by the AP-1 complex. *Mol. Cell Biol.* **10,** 1492–1497.

Kingsley, D. M. (1994) The TGF-β superfamily: new members, new receptors, and new genetic tests of function in different organisms. *Genes Dev.* **8,** 133–146.

Koff, A., Ohtsuki, M., Polyak, K., Roberts, J. M., and Massague, J. (1993) Negative regulation of G1 in mammalian cells: inhibition of cyclin E-dependent kinase by TGF-β. *Science* **260,** 536–539.

Kondaiah, P., Sands, M. J., Smith, J. M., Fields, A., Roberts, A. B., Sporn, M. B., and Melton, D. A. (1990) Identification of a novel transforming growth factor-β mRNA in Xenopus laevis. *J. Biol. Chem.* **265,** 1089–1093.

Kopin, I. J. and Markey, S. P. (1988) MPTP toxicity: implications for research in Parkinson's disease. *Annu. Rev. Neurosci.* **11,** 81–96.

Kotzbauer, P. T., Lampe, P. A., Heuckeroth, R. D., Golden, J. P., Creedon, D. J., Johnson, E. M., Jr., and Milbrandt, J. (1997) Neurturin, a relative of glial-cell-line-derived neurotrophic factor. *Nature* **384,** 467–470.

Kramer, I. M., Koornneef, I., de Laat, S. W., and van den Eijnden van Raaij, A. J. (1991) TGF-beta 1 induces phosphorylation of the cyclic AMP responsive element binding protein in ML-CC164 cells. *EMBO J.* **10,** 1083–1089.

Kreutzberg, G. W. (1996) Microglia: a sensor for pathological events in the CNS. *Trends Neurosci.* **19,** 312–318.

Krieglstein, J. (1996) *Pharmacology of Cerebral Ischemia 1996* medpharm, Scientific Publishers, Stuttgart.

Krieglstein, K. and Unsicker, K. (1994) Transforming growth factor beta promotes survival of dopaminergic neurons of the substantia nigra. *Neuroscience* **63,** 1189–1196.

Krieglstein, K., Rufer, M., Suter-Crazzolara, C., and Unsicker, K. (1995a) Neural functions of the transforming growth factor β. *Int. J. Dev. Neurosci.* **13,** 301–315.

Krieglstein, K., Suter-Crazzolara, C., Fischer, W. H., and Unsicker, K. (1995b) TGF-β superfamily members promote survival of midbrain dopaminergic neurons and protect them against MPP+ toxicity. *EMBO J.* **14,** 736–742.

Krieglstein, K., Maysinger, D., and Unsicker, K. (1996) The survival of mesencephalic dopaminergic neurons to the neurotrophins BDNF and NT-4 requires priming with serum: Comparison with members of the TGF-β superfamily and characterization of the serum-free culture system. *J. Neural Transm.* **47 (Suppl),** 247–258.

Krieglstein, K., and Unsicker, K. (1996) Distinct modulatory actions of TGF-β and LIF on neurotrophin-mediated survival of developing sensory neurons. *Neurochem. Res.* **21,** 843–850.

Krupinski, J., Kumar, P., Kumar, S., and Kaluza, J. (1996) Increased expression of TGF-β1 in brain tissue after ischemic stroke in humans. *Stroke* **27,** 852–857.

Kuruvilla, A. P., Shah, R., Hochwald, G. M., Liggitt, H. D., Palladino, M. A., and Thorbecke, G. J. (1991) Protective effect of transforming growth factor β1 on experimental autoimmune diseases in mice. *Proc. Natl. Acad. Sci. USA* **88,** 2918–2921.

Labourdette, G., Janet, T., Laeng, P., Perraud, F., Lawrence, D., and Pettman, B. (1990) Transforming growth factor type β1 modulates the effects of basic fibroblast growth factor and phenotypic expression of rat astroblasts in vivo. *J. Cell Physiol.* **144,** 473–488.

Laiho, M., DeCaprio, J. A., Ludlow, J. W., Livingston, D. M., and Massague, J. (1990) Growth inhibition by TGF-beta linked to suppression of retinoblastoma protein phosphorylation. *Cell* **62,** 175–185.

Laiho, M., Ronnstrand, L., Heino, J., Decaprio, J. A., Ludlow, J. W., Livingston, D. M., and Massague, J. (1991) Control of junB and extracellular matrix protein expression by transforming growth factor-beta 1 is independent of simian virus 40 T antigen-sensitive growth sensitive growth-inhibitory events. *Mol. Cell Biol.* **11,** 972–978.

Laping, N. J., Morgan, T. E., Nichols, N. R., Rozovsky, I., Young-Chan, C. S., Zarow, C., and Finch, C. E. (1994) Transforming growth factor-β1 induces neuronal and astrocyte genes: tubulin a1, glial fibrillary acidic protein and clusterin. *Neuroscience* **58,** 563–572.

Lefebvre, P. P., Martin, D., Staecker, H., Weber, T., Moonen, G., and Van De Water, T. R. (1992) TGFβ1 expression is initiated in adult auditory neurons by sectioning of the auditory nerve. *Neuroreport* **3,** 295–298.

Lehrmann, E., Kiefer, R., Finsen, B., Diemer, N. H., Zimmer, J., and Hartung, H. P. (1995) Cytokines in cerebral ischemia: expression of transforming growth factor beta-1 (TGF-β1) mRNA in the postischemic adult rat hippocampus. *Exp. Neurol.* **131,** 114–123.

Leof, E. B., Proper, J. A., Goustin, A. S., Shipley, G. D., DiCorlet, P. E., and Moses, H. L. (1986) Induction of c-sis mRNA and activity similar to platelet-derived growth factor by transforming growth factor beta: a proposed model for indirect mitogenesis involving autocrine activity. *Proc. Natl. Acad. Sci. USA* **83,** 2453–2457.

Lin, H. Y., Wang, X.-F., Ng-Eaton, E., Weinberg, R. A., and Lodish, H. F. (1992) Expression cloning of the TGF-β type II receptor, a functional transmembrane serine/threonine kinase. *Cell* **68,** 775–785.

Lindholm, D., Hengerer, B., Zafra, F., and Thoenen H. (1990) Transforming growth factor-β stimulates expression of nerve growth factor in the rat CNS. *Neuroreport* **1,** 9–12.

Lindholm, E., Castrén, E., Kiefer, R., Zafra, F., and Thoenen, H. (1992) Transforming growth factor-β1 in the rat brain: increase after injury and inhibition of astrocyte proliferation. *J. Cell Biol.* **117,** 395–400.

Logan, A. and Berry, M. (1993) Transforming growth factor in the injured CNS. *Trends Pharmacol. Sci.* **14,** 337–343.
Logan, A., Berry, M., Gonzalez, A. M., Frautschy, S. A., Sporn, M. B., and Baird, A. (1994) Effects of transforming growth factor β1 on scar production in the injured central nervous system of the rat. *Eur. J. Neurosci.* **3,** 355–363.
Logan, A., Frautschy, S. A., Gonzalez, A. M., Sporn, M. B., and Baird, A. (1992) Enhanced expression of transforming growth factor β1 in the rat brain after a localized cerebral injury. *Brain Res.* **587,** 216–225.
Lopez-Casillas, F., Cheifetz, S., Doody, J., Andres, J. L., Lane, W. S., and Massague, J. (1991) Structure and expression of the membrane proteoglycan betaglycan, a component of the TGF-beta receptor system. *Cell* **67,** 785–795.
Lyons, R. M., Gentry, L. E., Purchio, A. F., and Moses, H. L. (1990) Mechanisms of activation of latent recombinant transforming growth factor beta 1 by plasmin. *J. Cell Biol.* **110,** 1361–1367.
Madisen, L., Webb, N. R., Rose, T. M., Marquardt, H., Ikeda, T., Twardzik, D., Seyedin, S., and Purchio, A. F. (1988) Transforming growth factor-β2: cDNA cloning and sequence analysis. *DNA* **7,** 1–8.
Martinou, J.-C., le van Thai, A., Valette, A., and Weber, M. J. (1990) Transforming growth factor β_1 is a potent survival factor for rat embryo motoneurons in culture. *Dev. Brain Res.* **52,** 175–181.
Massague, J. (1992) Receptors for the TGF-beta family. *Cell* **69,** 1067–1070.
Massague, J., Attisano, L., and Wrana, J. L. (1994) The TGF-β family and its composite receptors. *Trends Cell Biol.* **4,** 172–178.
Massague, J. (1996) Crossing receptor bounderies. *Nature* **382,** 29, 30.
Mattson, M. P. and Furukawa, K. (1996) Programmed cell life: anti-apoptotic signaling and therapeutic strategies for neurodegenerative disorders. *Restorative Neurol. Neurosci.* **9,** 191–205.
Mattson, M. P., Barger, S. W., Furukawa, K., Bruce, A. J., Wyss-Coray, T., and Mucke, L. (1997) Cellular signaling roles of TGFβ, TNFα and βAPP in brain injury responses and Alzheimer's disease. *Brain Res. Rev.* **23,** 47–61.
McDonald, N. Q. and Hendrickson, W. A. (1993) A structural superfamily of growth factors containing a cystine knot motif. *Cell* **73,** 421–424.
McNeill, H., Williams, C., Guan, J., Dragunow, M., Lawlor, P., Sirimanne, E., Nikolics, K., and Gluckman, P. (1994) Neuronal rescue with transforming growth factor-β1 after hypoxic-ischaemic brain injury. *Neuroreport* **5,** 901–904.
Meno, C., Saijoh, Y., Fujii, H., Ikeda, M., Yokoyama, T., Yokoyama, M., Toyoda, Y., and Hamada, H. (1996) Left-right asymmetric expression of the TGFβ-family member lefty in mouse embryos. *Nature* **381,** 151–155.
Meucci, O. and Miller, R. J. (1996) Gp120 induced neurotoxicity in hippocampal pyramidal neuron cultures: protective effects of TGF-β1. *J. Neurosci.* **16,** 4080–4088.
Miyazono, K., Hellman, U., Wernstedt, C., and Heldin, C.-H. (1988) Latent high molecular weight complex of transforming growth factor β1. *J. Biol. Chem.* **263,** 6407–6415.
Miyazono, K., ten Dijke, P., Ichijo, H., and Heldin, C.-H. (1994) Receptors for transforming growth factor-beta. *Adv. Immunol.* **55,** 181–220.
Moos Jr. M., Wang, S., and Krinks, M. (1995) Anti-dorsalizing morphogenetic protein is a novel TGF-β homolog expressed in the spemann organizer. *Development* **121,** 4293–4301.
Morgan, T. E., Nichols, N. R., Pasinetti, G. M., and Finch, C. E. (1993) TGF-β1 mRNA increases in macrophage/microglial cells of the hippocampus in response to deafferentation and kainic acid-induced neurodegeneration. *Exp. Neurol.* **120,** 291–301.
Morganti-Kossman, M. C., Kossman, T., Brandes, M. E., Mergehagen, S. E., and Wahl, S. M. (1992) Autocrine and paracrine regulation of astrocyte function by transforming growth factor-β. *J. Neuroimmunol.* **39,** 163–174.
Mulder, K. M., and Brattain, M. G. (1988) Alterations in c-myc expression in relation to maturational status of human colon carcinoma cells. *Int. J. Cancer* **42,** 64–70.

Mulder, K. M., Levine, A. E., Hernandez, X., McKnight, M. K., Brattain, D. E., and Brattain, M. G. (1988) Modulation of c-myc by transforming growth factor-beta in human colon carcinoma cells. *Biochem. Biophys. Res. Commun.* **150,** 711–716.

Mulder, K. M., Humphrey, L. E., Choi, H. G., Childress-Fields, K. E., and Brattain, M. G. (1990) Evidence for c-myc in the signaling pathway for TGF-beta in well-differentiated human colon carcinoma cells. *J. Cell Physiol.* **145,** 501–507.

Mulder, K. M. and Morris, S. L. (1992) Activation of p21ras by transforming growth factor β in epithelial cells. *J. Biol. Chem.* **267,** 5029–5031.

Murzin, A. G. and Chothia, C. (1992) *Curr. Opinion Struct. Biol.* **2,** 895–903.

Nakata, N., Kato, H., and Kogure, K. (1993) Protective effects of basic fibroblast growth factor against hippocampal neuronal damage following cerebral ischemia in the gerbil. *Brain Res.* **605,** 353–356.

Nathan, C. and Sporn, M. (1991) Cytokines in context. *J. Cell Biol.* **113,** 981–986.

Nichols, N. R., Laping, N. J., Day, J. R., and Finch, C. E. (1991) Increase in transforming growth factor-β mRNA in hippocampus during responses to entorhinal cortex lesions in intact and adrenalectomized rats. *J. Neurosci. Res.* **28,** 134–139.

Noguti, T., Adachi-Yamada, T., Katagiri, T., Kawakami, A., Iwami, M., Ishibashi, J., Kataoka, H., Suzuki, A., Go, M., and Ishizaki, H. (1995) Insect prothoracicotropic hormone: a new member of the vertebrate growth factor superfamily. *FEBS Lett.* **376,** 251–256.

O'Connor-McCourt, M. D. and Wakefield, L. M. (1987) Latent transforming growth factor-β in serum. *J. Biol. Chem.* **262,** 14,090–14,099.

Oda, S., Nishimatsu, S., Murakami, K., and Ueno, N. (1995) Molecular cloning and functional analysis of a new activin β subunit: a dorsal mesoderm-inducing activity in xenopus. *Biochem. Biophys. Res. Commun.* **210,** 581–588.

Ohba, M., Shibanuma, M., Kuroki, T., and Nose, K. (1994) Production of hydrogen peroxide by transforming growth factor-beta 1 and its involvement in induction of egr-1 in mouse osteoblastic cells. *J. Cell Biol.* **126,** 1079–1088.

Özkaynak, E. and Oppermann, H. (1994) Eur. Patent Appl. WO 94/10203, seq. id. no. 1.

Pasinetti, G. M., Nichols, N. R., Tocco, G., Morgan, T., Laping, N., and Finch, C. E. (1993) Transforming growth factor β1 and fibronectin messenger RNA in rat brain: responses to injury and cell-type localization. *Neuroscience* **54,** 893–907.

Patterson, P. H. (1995) Cytokines in Alzheimer's disease and multiple sclerosis. *Curr. Opinion Neurobiol.* **5,** 642–646.

Peress, N. S. and Perillo, E. (1995) Differential expression of TGF-β1, 2 and 3 isotypes in Alzheimer's disease: a comparative immunohistochemical study with cerebral infarction, aged human and mouse control brains. *J. Neuropathol. Exp. Neurol.* **54,** 802–811.

Pertovaara, L., Sistonen, L., Bos, T. J., Vogt, P. K., Keski-Oja, J., and Alitalo, K. (1989) Enhanced jun gene expression is an early genomic response to transforming growth factor beta stimulation. *Mol. Cell Biol.* **9,** 1255–1262.

Peruche, B. and Krieglstein J. (1993) Mechanisms of drug actions against neuronal damage caused by ischemia—an overview. *Prog. Neuro-Psychopharmacol. and Biol. Psychiatry* **17,** 21–70.

Pietenpol, J. A., Holt, J. T., Stein, R. W., and Moses, H. L. (1990a) Transforming growth factor beta 1 suppression of c-myc gene transcription: role in inhibition of keratinocyte proliferation. *Proc. Natl. Acad. Sci. USA* **87,** 3758–3762.

Pietenpol, J. A., Stein, R. W., Moran, E., Yaciuk, P., Schlegel, R., Lyons, R. M., et al. (1990b) TGF-beta 1 inhibition of c-myc transcription and growth in keratinocytes is abrogated by viral transforming proteins with pRB binding domains. *Cell* **61,** 777–785.

Pircher, R., Jullien, P., and Lawrence, D. A. (1986) β-Transforming growth factor is stored in human blood platelets as a latent high molecular weight complex. *Biochem. Biophys. Res. Commun.* **136,** 30–37.

Polyak, K., Kato, J. Y., Solomon, M. J., Sherr, C. J., Massague, J., Roberts, J. M., and Koff, A. (1994) p27Kip1, a cyclin-Cdk inhibitor, links transforming growth factor-beta and contact inhibition to cell cycle arrest. *Genes Dev.* **8,** 9–22.

Poulsen, K. T., Armanini, M. P., Klein, R. D., Hynes, M. A., Phillips, H. S., and Rosenthal, A. (1994) TGFβ2 and TGFβ3 are potent survival factors for midbrain dopaminergic neurons. *Neuron* **13,** 1245–1252.

Prehn, J. H. M., Peruche, B., Unsicker, K., and Krieglstein, J. (1993a) Isoform-specific effects of transforming growth factors-β on degeneration of primary neuronal cultures induced by cytotoxic hypoxia or glutamate. *J. Neurochem.* **60,** 1665–1672.

Prehn, J. H. M., Backhauβ, C., and Krieglstein, J. (1993b) Transforming growth factor-β1 prevents glutamate neurotoxicity in rat neocortical cultures and protects mouse neocortex from ischemic injury in vivo. *J. Cereb. Blood Flow Metab.* **13,** 521–525.

Prehn, J. H. M. and Krieglstein, J. (1994) Opposing effects of transforming growth factor-1 on glutamate neurotoxicity. *Neuroscience* **60,** 7–10.

Prehn, J. H. M., Bindokas, V. P., Marcuccilli, C. J., Krajewski, S., Reed, J. C., and Miller, R. J. (1995) Regulation of neuronal bcl-2 protein expression and calcium homeostasis by transforming growth factor-β1 confers wide ranging protection on rat hippocampal neurons. *Proc. Natl. Acad. Sci. USA* **91,** 12,599–12,603.

Prehn, J. H. M., Bindokas, V., Jordan, J., Galindo, M. F., Ghadge, G. D., Roos, R. P., et al. (1996) Protective effect of transforming growth factor-β1 (TGF-β1) on β-amyloid neurotoxicity in rat hippocampal neurons. *Mol. Pharmacol.* **49,** 319–328.

Przedborski, S., Kostic, V., Jackson-Lewis, V., Naini, A. B., Simonetti, S., Fahn, S., et al. (1992) Transgenic mice with increased Cu/Zn-superoxide dismutase activity are resistant to N-methyl-4-phenyl-1,2,3,6-tetrahydropyridine-induced neurotoxicity. *J. Neurosci.* **12,** 1658–1667.

Roberts, A. B., Anzano, M. A., Lamb, L. C., Smith, J. M., and Sporn, M. B. (1981) New class of transforming growth factors potentiated by epidermal growth factor. *Proc. Natl. Acad. Sci. USA* **78,** 5339–5343.

Roberts, A. B., Anzano, M. A., Meyers, C. A., Wideman, J., Blacher, R., Pan, Y.-C., et al. (1983) Purification and properties of a type beta transforming growth factor from bovine kidney. *Biochemistry* **22,** 5692–5698.

Roberts, A. B. and Sporn, M. B. (1990) *Handbook of Experimental Pharmacology*, vol. 95. Springer Verlag, Heidelberg, pp. 419–472.

Rogister, B., Delree, P., Leprince, P., Martin, D., Sadzot, C., Malgrange, B., et al. (1993) Transforming growth factor β as a neuronoglial signal during peripheral nervous system response to injury. *J. Neurosci. Res.* **34,** 32–34.

Rufer, M., Flanders, K., and Unsicker, K. (1994) Presence and regulation of transforming growth factor beta mRNA and protein in the normal and lesioned rat sciatic nerve. *J. Neurosci. Res.* **39,** 412–423.

Saad, B., Constam, D. B., Ortmann, R., Moos, M., Fontana, A., and Schachner, M. (1991) Astrocyte-derived TGF-β2 and NGF differentially regulate neural recognition molecule expression by cultured astrocytes. *J. Cell Biol.* **115,** 473–484.

Satterwhite, D. J., Aakre, M. E., Gorska, A. E., and Moses, H. L. (1994) Inhibition of cell growth by TGF beta 1 is associated with inhibition of B-myb and cyclin A in both BALB/MK and Mv1Lu cells. *Cell Growth Differ.* **5,** 789–799.

Scherer, S. S., Kamholz, J., and Jakowlew, S. B. (1993) Axons modulate the expression of transforming growth factor-betas in Schwann cells. *Glia* **8,** 265–276.

Schlunegger, M. P. and Grütter, M. G. (1992) An unusual feature revealed by crystal structure at 2.2A resolution of human transforming growth factor-β2. *Nature* **358,** 430–434.

Sheng, J. G., Mrak, R. E., and Griffin, W. S. T. (1995) Microglial interleukin-1α expression in brain regions in Alzheimer's disease: correlation with neuritic plaque distribution. *Neuropathol. Appl. Neurobiol.* **21,** 290–301.

Shibanuma, M., Kuroki, T., and Nose, K. (1992) Isolation of a gene encoding a putative leucine zipper structure that is induced by transforming growth factor β1 and other growth factors. *J. Biol. Chem.* **267,** 10,219–10,224.

Shibanuma, M., Mashimo, J., Kuroki, T., and Nose, K. (1994) Characterization of the TGF-β1-inducible hic-5 gene that encodes a putative novel zinc finger protein and its possible involvement in cellular senescence. *J. Biol. Chem.* **269,** 26,767–26,774.

Shibuya, H., Yamaguchi, K., Shirakabe, K., Tonegawa, A., Gotoh, Y., Ueno, N., et al. (1996) TAB1: an activator of the TAK1 MAPKKK in TGF-β signal transduction. *Science* **272,** 1179–1182.

Shigeno, T., Mima, T., Takakura, K., Graham, D. I., Kato, G., Hashimoto, Y., Furukawa, S. (1991) Amelioration of delayed neuronal death in the hippocampus by nerve growth factor. *J. Neurosci.* **11,** 2914–2919.

Shoffner, J. M., Watts, R. L., Juncos, J. L., Torroni, A., and Wallace, D. C. (1991) Mitochondrial oxidative phosphorylation defects in Parkinson's disease. *Ann. Neurol.* **30,** 332–339.

Siesjö, B. K., Katsura, K. I., Zhao, Q., Folbergrová, J., Pahlmark, K., Siesjö, P., and Smith, M. L. (1995) Mechanisms of secondary brain damage in global and focal ischemia: a speculative synthesis. *J. Neurotrauma* **12,** 943–956.

Slingerland, J. M., Hengst, L., Pan, C. H., Alexander, D., Stampfer, M. R., and Reed, S. I. (1994) A novel inhibitor of cyclin-CDK activity detected in transforming growth factor beta-arrested epithelial cells. *Mol. Cell Biol.* **14,** 3683–3694.

Smith-Swintosky, V. L., Pettigrew, L. C., Craddock, S. D., Culwell, A. R., Rydel., R. E., and Mattson, M. P. (1994) Secreted forms of β-amyloid recursor protein protect against ischemic brain injury. *J. Neurochem.* **63,** 781–784.

Sriram, S., Flanders, K. C., Pendlebury, W., and Johns, L. (1991) Identification of TGFβ in brains of patients with multiple sclerosis. *Neurology* **41,** 219.

Stenzel, P., Angerer, L. M., Smith, B. J., Angerer, R. C., and Vale, W. W. (1994) The univin gene encodes a member of the transforming growth factor-beta superfamily with restricted expression in the sea urchin embryo. *Dev. Biol.* **166,** 149–158.

Storm, E. E., Huynh, T. V., Copeland, N. G., Jenkins, N. A., Kingsley, D. M., and Lee, S. J. (1994) Limb alterations in brachypodism mice due to mutations in a new member of the TGF beta superfamily. *Nature* **368,** 639–643.

Subramaniam, M., Harris, S. A., Oursler, M. J., Rasmussen, K., Riggs, B. L., and Spelsberg, T. C. (1995) Identification of a novel TGF-β-regulated gene encoding a putative zinc finger protein in human osteoblasts. *Nucleic Acid Res.* **23,** 4907–4912.

Suzumura, A., Sawada, M., Yamamoto, H., and Marunouchi, T. (1993) Transforming growth factor-β suppresses activation and proliferation of microglia in vitro. *J. Immunol.* **151,** 2150–2158.

Swindells, M. B. (1992) Structural similarity between transforming growth factor-beta 2 and nerve growth factor. *Science* **258,** 1160, 1161.

Takao, M., Hino, J., Takeshita, N., Konno, Y., Nishizawa, T., Matsuo, H., and Kangawa, K. (1996) Identification of rat bone morphogenetic protein 3b (BMP-3b), a new member of BMP-3. *Biochem. Biophys. Res. Commun.* **219,** 656–662.

ten Dijke, P., Miyazono, K., and Heldin, C.-H. (1996) Signaling via hetero-oligomeric complexes of type I and type II serine/threonine kinase receptors. *Curr. Opinion Cell Biol.* **8,** 139–145.

Todaro, G. J., Fryling, C., and De Larco, J. E. (1980) Transforming growth factors produced by certain human tumor cells: Polypeptides that interact with epidermal growth factor receptors. *Proc. Natl. Acad. Sci.* **77,** 5258–5262.

Tomlinson, B. E. (1980) *Biochemistry of Dementia* Wiley, Chichester, pp.15–51.

Treanor, J. J. S., Goodman, L., de Sauvage, F., Stone, D. M., Poulsen, K. T., Beck, C. D., et al. (1996) Characterization of a multicomponent receptor for GDNF. *Nature* **382,** 80–83.

Trupp, M., Arenas, E., Fainzilber, M., Nilsson, A.-S., Sieber, B.-A., Grigoriou, M., et al. (1996) Functional receptor for GDNF encoded by the c-ret proto-oncogene. *Nature* **381,** 785–789.

Tsunawaki, S., Sporn, M. B., Ding, A., and Nathan, C. (1988) Deactivation of macrophages by transforming growth factor-β. *Nature* **334,** 260.

Turski, L., Bressler, K., Rettig, K.-J., Löschmann, P.-A., and Wachtel, H. (1991) Protection of substantia nigra from MPP+ neurotoxicity by N-methyl-D-aspartate antagonists. *Nature* **349,** 414–418.

Unsicker, K., Flanders, K. C., Cissel, D. S., Lafyatis, R., and Sporn, M. B. (1991) Transforming growth factor beta isoforms in the adult rat central and peripheral nervous system. *Neuroscience* **44,** 613–625.

Unsicker, K. (1994) Growth factors in Parkinson's disease. *Prog. Growth Factor Res.* **5,** 73–87.

Unsicker, K., Meier, C., Krieglstein, K., Sartor, B. M., and Flanders, K. C. (1996a) Expression, localization, and function of transforming growth factor-βs in embryonic chick spinal cord, hindbrain, and dorsal root ganglia. *J. Neurobiol.* **29,** 262–276.

Unsicker, K., Suter-Crazzolara, C., and Krieglstein, K. (1996b) Functions of growth factors in the development and maintenance of midbrain dopaminergic neurons: concepts, facts, and prospects of the transforming growth factors β. In: CIBA Foundation *#196* (pp. 70–80).

van der Wal, E. A., Gómez-Pinella, F., and Cotman, C. W. (1993) Transforming growth factor-β1 is in plaques in Alzheimer und Down pathologies. *Neuroreport* **4,** 69–72.

Vodovotz, Y., Bogdan, C., Paik, J., Xie, Q. W., and Nathan, C. (1993) Mechanisms of suppression of macrophage nitric oxide release by transforming growth factor. *J. Exp. Med.* **178,** 605–613.

Wahl, S. M., Hunt, D. A., Wakefield, L., McCartney-Francis, N., Wahl, L., Roberts, A. B., and Sporn, M. B. (1987) Transforming growth factor beta (TGF-β) induces monocyte chemotaxis and growth factor production. *Proc. Natl. Acad. Sci. USA* **84,** 57,788–57,792.

Wahl, S. M., Allen, J. B., McCartney-Francis, N., Morganti-Kossman, M. C., Kossman, J. T., Ellington, L., et al. (1991) Macrophage and astrocyte-derived transforming growth factor β as a mediator of central nervous system dysfunction in aquired immune deficiency syndrome. *J. Exp. Med.* **173,** 981–991.

Wahl, S. M. (1992) Transforming growth factor beta (TGF-β) in inflammation: a cause and a cure. *J. Clin. Immunol.* **12,** 66–74.

Wakefield, L. M., Smith, D. M., Flanders, K. C., and Sporn, M. B. (1988) Latent transforming growth factor-β from human platelets. *J. Biol. Chem.* **263,** 7646–7654.

Wang, T., Danielson, P. D., Li, B.-Y., Shah, P. C., Kim, S. D., and Donahoe, P. K. (1996a) The p21ras farnesyltransferase a subunit in TGF-β and activin signaling. *Science* **271,** 1120–1122.

Wang, T., Li, B.-Y., Danielson, P. D., Shah, P. C., Rockwell, S., Lechleider, R. J., et al. (1996b) The immunophilin FKBP12 functions as a common inhibitor of the TGFβ family type I receptors. *Cell* **86,** 435–444.

Wang, X.-F., Lin, H. Y., Ng-Eaton, E., Downward, J., Lodish, H. F., and Weinberg, R. A. (1991) Expression cloning and characterization of the TGF-beta type III receptor. *Cell* **67,** 797–805.

Wang, X., Yue, T.-L., White, R. F., Barone, F. C., and Feuerstein, G. Z. (1995) Transforming growth factor-β1 exhibits delayed gene expression following focal cerebral ischemia. *Brain Res. Bull.* **36,** 606–609.

Wieser, R., Wrana, J. L., and Massague, J. (1995) GS domain mutations that constitutively activate TβR-I, the downstream signaling component in the TGF-β receptor complex. *EMBO J.* **14,** 2199–2208.

Wießner, C., Gehrmann, J., Lindholm, D., Töpper, R., Kreutzberg, G. W., and Hossmann, K. A. (1993) Expression of transforming growth factor-β1 and interleukin-1β mRNA in rat brain following transient forebrain ischemia. *Acta Neuropathol.* **86,** 439–446.

Wisniewski, H. M., Weigel, J., Wang, K. C., Kujawa, M., and Lack, B. (1989) Ultrastructural studies of the cells forming amyloid fibers in the classical plaques. *Can. J. Neurol. Sci.* **16,** 535–542.

Wozney, J. M. and Celeste, A. (1993) Eur. Patent App.l WO 93/00432, seq. id. no. 9.

Wrana, J. L., Attisano, L., Caramo, J., Zentela, A., Doody, J., Laiho, M., Wang, X.-F., and Massague, J. (1992) TGFβ signals through a heteromeric protein kinase receptor complex. *Cell* **71,** 1003–1014.

Wrana, J. L., Attisano, L., Wieser, R., Ventura, F., and Massague, J. (1994) Mechanism of activation of TGF-β receptor. *Nature* **370,** 341–347.

Wyss-Coray, T., Feng, L., Masliah, E., Ruppe, M. D., Lee, H. S., Toggas, S. M., Rockenstein, E. M., and Mucke, L. (1995) Increased central nervous system production of extracellular matrix components and development of hydrocephalus in transgenic mice overexpressing transforming growth factor-β1. *Am. J. Path.* **147,** 53–67.

Yamaguchi, K., Shirakabe, K., Shibuya, H., Irie, K., Oishi, I., Ueno, N., et al. (1995) Identification of a member of the MAPKKK family as a potential mediator of TGF-β signal transduction. *Science* **270,** 2008–2011.

Yao, J., Harvath, L., Gilbert, D. L., and Colton, C. A. (1990) Chemotaxis by a CNS macrophage, the microglia. *J. Neurosci. Res.* **27,** 36–42.

8
Protective and Regenerative Properties of GDNF in the Central Nervous System

Wayne A. Cass, Cecilia M. Kearns, and Don M. Gash

1. INTRODUCTION

Glial cell line-derived neurotrophic factor (GDNF) was isolated and cloned by Lin and colleagues (Lin et al., 1993, 1994) as part of a concerted effort to identify new trophic factors for midbrain dopamine (DA) neurons. Since the initial description of GDNF in 1993, studies from a number of laboratories have confirmed its potent effects on DA neurons. In addition, GDNF has been found to play an intricate role in the survival and development of a diverse set of neuronal and nonneuronal cells.

From the onset of its discovery, the potential therapeutic utilization of GDNF for treating parkinsonism was recognized (Lin et al., 1993). The basic rationale for such an application was straightforward. Parkinson's disease results from the progressive degeneration of DA cells in the substantia nigra, with a concomitant reduction of DA in the target region of these cells, the striatum (Jellinger, 1986; Hornykiewicz, 1993). Trophic factors for central DA neurons, it was reasoned, should be capable of upregulating residual DA neurons in the parkinsonian brain and increasing available DA levels in the striatum. Our studies have focused on testing this hypothesis and analyzing the effects of GDNF on the nigrostriatal dopaminergic system in rats and nonhuman primates. This chapter will first review much of the literature on GDNF and its effects on various cell types, and will conclude by delineating the specific effects of GDNF on the nigrostriatal dopaminergic system of nonhuman primates.

1.1. *Discovery and Characterization of GDNF*

GDNF was first described by Lin et al. (1993). It was isolated and purified from the conditioned medium of cultured rat glial cells from the B49 cell line. GDNF is a glycosylated, disulfide-bonded homodimer with an apparent molecular mass of 33–45 kDa (Lin et al., 1994). It contains seven conserved cysteine residues in the relative spacing of the members of the transforming growth factor-β (TGF-β) superfamily. However, it shares <20% homology with any of the known TGF-β proteins, and may represent a new subfamily of trophic factors.

From: Neuroprotective Signal Transduction *edited by M. P. Mattson*
Humana Press Inc., Totowa, NJ

1.2. Distribution of GDNF in the Brain and Body

GDNF mRNA is expressed in a developmentally and regionally specific manner in the central nervous system (CNS). Using *in situ* hybridization, Strömberg and colleagues found detectable levels of GDNF mRNA in embryonic and newborn rat striatum, but not in adult rats (Strömberg et al., 1993). These investigators also found GDNF mRNA-positive labeling in the thalamus, cerebellum, cortex, spinal cord, and pineal gland of embryonic or newborn rats. Using the more sensitive technique of reverse transcriptase-polymerase chain reaction (RT-PCR), GDNF mRNA expression has been found in substantia nigra and basal forebrain astrocytes, with the highest levels in early postnatal rat striatum (Schaar et al., 1993). Subsequent RT-PCR studies have demonstrated expression of GDNF mRNA in many brain regions, including the ventral mesencephalon and striatum, throughout development and in adult rats (Choi-Lundberg and Bohn, 1995), and in striatum, hippocampus, cortex, cerebellum, and spinal cord of adult rats (Springer et al., 1994). Additionally, Blum and Weickert (1995) have detected GDNF mRNA expression in the striatum throughout adult life in mice, using a nuclease protection assay. The presence of GDNF mRNA in the striatum throughout development and postnatal life suggests that endogenous GDNF may function as a target-derived trophic factor for midbrain DA neurons. The demonstration that GDNF is retrogradely transported by nigrostriatal DA neurons in adult rats further supports this idea (Tomac et al., 1995b). In the adult human, GDNF mRNA expression has been found in the striatum, hippocampus, cortex, and spinal cord (Springer et al., 1994).

Outside the CNS, GDNF mRNA expression has been found in almost every organ or structure examined. In embryonic rodents, these structures include limb bud, kidney, gut, skeletal muscle, lung, heart, adrenal gland, gonad, and spleen (Choi-Lundberg and Bohn, 1995; Trupp et al., 1995; Hellmich et al., 1996; Suvanto et al., 1996). In newborn rats the list includes all of the above, plus liver, bone, blood, sciatic nerve, and dorsal root ganglia (DRG) (Springer et al., 1994; Suter-Crazzolara and Unsicker, 1994; Choi-Lundberg and Bohn, 1995; Trupp et al., 1995). In adult rats, GDNF mRNA expression has been found in gut, skeletal muscle, liver, lung, heart, adrenal gland, gonad, and spleen (Choi-Lundberg and Bohn, 1995; Springer et al., 1995). The results of these studies suggest that GDNF may play an important role during the early development of many tissues and may have a trophic influence on peripheral neural, as well as nonneural, cells.

RT-PCR analysis of GDNF mRNA has indicated that there are two transcripts for GDNF in the brain and periphery (Schaar et al., 1993; Springer et al., 1994, 1995; Suter-Crazzolara and Unsicker, 1994; Cristina et al., 1995; Trupp et al., 1995). The larger transcript corresponds to the original 633-base-pair cDNA sequence published by Lin et al. (1993). The second transcript contains a deletion of 78 base pairs. The two transcripts appear to be present in most tissues in which GDNF mRNA is found; however, the relative abundance of the two forms can vary from tissue to tissue (Suter-Crazzolara and Unsicker, 1994). The biological significance of the two transcripts is unclear, because identical mature GDNF proteins appear to be produced by both mRNA isoforms (Trupp et al., 1995).

The gene coding for human GDNF has been mapped by two groups to the short arm of chromosome 5 (Bermingham et al., 1995; Schindelhauer et al., 1995). Although at present this locus does not appear to be the site of any known genetic disorders

(Bermingham et al., 1995), flanking sequences of the gene may be needed in future gene transfer experiments (Schindelhauer et al., 1995).

1.3. Multicomponent GDNF-Receptor and GDNF-Deficient Mice

Analysis of embryonic and newborn mice genetically modified to eliminate biologically active GDNF production have revealed a number of abnormalities. GDNF$^{-/-}$ mice die within 12–36 h of birth (Moore et al., 1996; Pichel et al., 1996; Sánchez et al., 1996). The entire enteric nervous system is absent and the kidneys and ureters are either missing or display severe developmental abnormalities. There are also significant decreases in the number of sensory and sympathetic neurons in the peripheral nervous system (PNS). The effects in the CNS at birth appear to be less dramatic than expected. There is an approximate 20% loss of trigeminal and lumbar motor neurons, but a significant reduction is not seen in the number of midbrain DA neurons and locus ceruleus noradrenergic neurons (Moore et al., 1996). Although the number of central catecholaminergic neurons at birth seems to be preserved in GDNF$^{-/-}$ mice, the effects of deleting GDNF on the postnatal survival and maturation of DA and norepinephrine neurons remains to be determined.

The resemblance between the phenotype of GDNF$^{-/-}$ mice and mice lacking the Ret receptor is striking, with the kidneys and enteric nervous system failing to develop in both mutants. Ret is a receptor-type tyrosine kinase whose ligand(s) had not been identified prior to the suggestion from the GDNF$^{-/-}$mice that it served as a functional receptor for GDNF. Four recently published papers indicate that this is indeed the case (Durbec et al., 1996; Jing et al., 1996; Treanor et al., 1996; Trupp et al., 1996). Two of these studies went further to demonstrate that a novel glycosyl-phosphatidylinositol-anchored membrane protein, designated as GDNFR-α, functioned with Ret to bind GDNF, and resulted in Ret receptor activation (Jing et al., 1996; Treanor et al., 1996; *see also* Massagué, 1996). At present, it is not clear whether there are other naturally occurring ligands for the Ret receptor or the Ret–GDNFR-α receptor complex.

Definitive mapping studies remain to be published, but the initial *in situ* hybridization studies have revealed GDNFR-α mRNA expression in numerous CNS sites, including the retina, hippocampus, cerebral cortex, cerebellum, lateral septum, and motor neurons in the spinal cord (Treanor et al., 1996). In the same study, GDNFR-α mRNA expression was also identified in numerous peripheral sites, including striated and smooth muscle adjacent to the enteric nervous system and developing nephrons of the kidney. Both GDNFR-α and Ret mRNA are expressed at high levels in the midbrain region containing DA neurons, the substantia nigra (Treanor et al., 1996; Trupp et al., 1996). However, GDNFR-α and Ret apparently are not always coexpressed in the same cells (Treanor et al., 1996). Schwann cells, which express GDNF mRNA, also express message for GDNFR-α, but not Ret. Ret is expressed in the enteric nervous system, but not in the surrounding muscle. Clearly, much remains to be learned about GDNF's interactions with target tissues.

2. GDNF EFFECTS IN CULTURE

2.1. Dopaminergic Neurons

GDNF was first identified by its ability to promote DA uptake in embryonic mesencephalic cells grown in culture (Lin et al., 1993, 1994). It also promoted the survival and morphological differentiation of dopaminergic neurons. The effects were relatively specific for DA neurons, because it did not increase the uptake of serotonin or GABA. In addition, GDNF did not lead to an increase in total astrocyte or oligodendrocyte number, suggesting that it is not a glial mitogen. It was these effects on dopaminergic cells in culture that led to the first suggestion that GDNF may be of potential therapeutic value in the treatment of Parkinson's disease (Lin et al., 1993).

Further studies have since confirmed the effects of GDNF in culture, i.e., promoting the survival and neurite outgrowth of embryonic dopamine cells without affecting glial growth (Krieglstein et al., 1995; Hou et al., 1996). Although the mechanism for GDNF's ability to prevent dopaminergic cell death in culture is not known, the prevention of apoptosis appears to be involved (Clarkson et al., 1995). GDNF was not able to prevent the reduction of dopaminergic cells and DA uptake in cultures exposed to 1-methyl-4-phenylpyridinium ion (MPP^+), the toxic metabolite of 1-methyl-4-phenyl-1,2,3,6-tetrahydropyridine (MPTP) (Hou et al., 1996). However, once the MPP^+ was removed from the cultures, GDNF protected against continuous cell death and promoted regrowth of damaged dopaminergic fibers (Hou et al., 1996).

2.2. *Other Types of Neurons*

Although most of the initial studies with GDNF have examined dopaminergic neurons or systems, more and more reports are appearing that demonstrate the trophic effects of GDNF on other types of neurons. Regarding culture systems, GDNF helps support the survival of many types of neurons in embryonic chicken sensory and autonomic ganglia (Buj-Bello et al., 1995; Ebendal et al., 1995; Trupp et al., 1995). Fiber outgrowth was also stimulated by GDNF in primary culture of sympathetic and ciliary ganglia (Ebendal et al., 1995). Similar to these peripheral neurons, GDNF also supports the survival and differentiation of Purkinje cells prepared from embryonic rats (Mount et al., 1995).

GDNF is a potent trophic factor for motor neurons. For example, GDNF supports the survival of cultured motor neurons from the embryonic rat spinal cord (Henderson et al., 1994). In cultures from embryonic rat spinal cord and ventral mesencephalon containing cranial nerve nuclei, GDNF increased the activity of choline acetyltransferase (Zurn et al., 1994), suggesting a trophic effect on these motor neurons. Further experiments confirmed this and also demonstrated that GDNF promotes motor neuron survival and increases neurite length (Zurn et al., 1996). Similarly, GDNF promotes the survival of cultured embryonic chick motor neurons (Oppenheim et al., 1995; Gouin et al., 1996). These results, as well as the in vivo results discussed in Section 3.2., have led to the suggestion by the above groups of investigators that GDNF may be a good candidate for the treatment of human motor neuron diseases, such as amyotrophic lateral sclerosis.

3. GDNF STUDIES IN RODENTS

The in vivo effects of GDNF on the nigrostriatal system of adult rodents are particularly dramatic. Hudson et al. (1995) have reported a three- to four-fold increase in nigral DA levels and increased DA turnover and tyrosine hydroxylase (TH) immunoreactivity in the ipsilateral striatum 3 wk after a single injection of GDNF into the substantia nigra of adult rats. They also noted an increase in TH-positive fiber density in the region of the injection site, suggesting that GDNF can induce neurite growth in adult DA neurons or that TH levels were upregulated in processes that previously contained undetectable levels. Behaviorally, the rats displayed increased locomotor activity, rearing, and stereotypies that lasted for at least 1 wk, but were gone by 3 wk post-GDNF injection. Beck et al. (1996) found that single injections of GDNF into the striatum or supranigral region in neonatal rats led to increased levels of DA, DA metabolites, and serotonin in both the striatum and mesencephalon. The GDNF-injected rat pups showed a slower rate of weight gain, hyperflexed forelimbs, and the appearance of clawed toes and kinked tails. However, these behavioral effects were transient and were no longer evident 26 d after injection. In contrast to the results by Hudson et al. (1995), Beck and colleagues found no change in nigral or striatal DA levels 3 wk after intrastriatal or intranigral GDNF injection in adult rats (Beck et al., 1996). The reasons for this discrepancy are unclear, but may be related to possible differences in the age of the animals, in injection sites, or in the GDNF preparation used in the study.

In oculo transplants have also been used to examine the in vivo effects of GDNF on DA neurons. Embryonic ventral mesencephalon, grafted into the anterior eye chamber of rats and treated weekly with GDNF, contained increased numbers of TH-positive neurons and increased DA fiber density (Strömberg et al., 1993). Similar results were found when the GDNF was delivered continuously by transplanting GDNF-containing fibrin glue balls along with the ventral mesencephalon pieces (Cheng et al., 1995). Further studies confirmed the effects of weekly GDNF injections *in oculo* and also demonstrated that GDNF treatment of developing grafts augmented potassium (K^+)-evoked overflow of DA (Johänsson et al., 1995). Johansson and colleagues also found that in transplants of mature ventral mesencephalon, GDNF did not alter the number of TH-positive neurons, but did increase the density of TH-positive nerve terminals in the region of the dopaminergic cell bodies (Johansson et al., 1995). The results of this latter study provide another line of evidence that adult dopaminergic neurons, like embryonic neurons, are responsive to the neurotrophic effects of GDNF.

3.1. Neuroprotective Effects on DA Neurons

GDNF exhibits potent neuroprotective properties when given prior to, or soon after, axotomy of nigral dopamine neurons or administration of dopaminergic toxins. Daily supranigral administration of GDNF significantly reduced degeneration of mesencephalic DA neurons following transection of the medial forebrain bundle (Beck et al., 1995). In mice given intranigral or intrastriatal GDNF 24 h prior to systemic MPTP, striatal and nigral (DA) levels are partially protected from depletion (Tomac et al., 1995a). When administered intranigrally 24 h prior to either an intranigral or intrastriatal 6-hydroxydopamine (6-OHDA) lesion (Kearns and Gash, 1995), or supranigrally 7 d after an intrastriatal 6-OHDA insult (Sauer et al., 1995), GDNF reduced the loss of nigral DA neurons. However, the number of surviving neurons are still below control numbers. Similarly, striatal DA levels were partially protected from depletion when GDNF was administered dorsal to the substantia nigra or into the lateral ventricle immediately prior to 6-OHDA (Opacka-Juffry et al., 1995).

In contrast to the partial neuroprotective effects in the above studies, more complete protection has been found using other experimental paradigms. Supranigral administration of GDNF every other day for 4 wk, starting on the day of an intrastriatal 6-OHDA lesion, completely prevented loss of cells in the substantia nigra (Sauer et al., 1995). Complete protection against nigral cell loss was also afforded by intranigral injection of

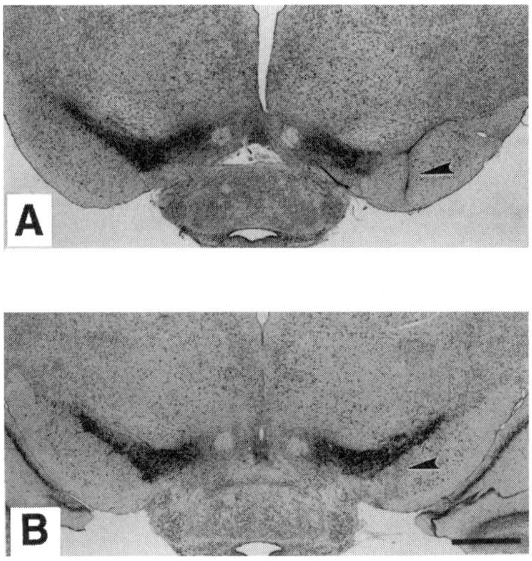

Fig. 1. Representative coronal sections through the midbrain of Fischer 344 rats processed for tyrosine hydroxylase immunocytochemistry. **(A)** The mesencephalon of animals treated with vehicle followed by 6-OHDA showed significant loss of substantia nigra DA neurons on the lesioned (right) side. Both vehicle and 6-OHDA were injected into the nigra. **(B)** In contrast, there was complete sparing of midbrain dopamine neurons in animals treated with 10 μg GDNF 6 h prior to an intranigral 6-OHDA lesion. Injection sites are shown by arrowheads. Scale bar = 1 mm.

GDNF 6 h prior to 6-OHDA (Fig. 1), but only partial protection if the GDNF was given at 1, 12, or 24 h prior to the 6-OHDA (Kearns et al., 1997). In addition, when GDNF was injected into the striatum 1 and 3 d before intrastriatal administration of 6-OHDA, and 1 and 4 d after, the density of tyrosine hydroxylase-positive fibers in the striatum was partially preserved and the number of tyrosine hydroxylase-positive neurons in the substantia nigra was almost completely preserved (Shults et al., 1996). When injected intrastriatally in rats 24 h prior to neurotoxic doses of methamphetamine, GDNF completely prevented reductions in striatal DA levels (Cass, 1996). Taken together, the above studies suggest that GDNF has profound neuroprotective effects on lesioned nigrostriatal DA neurons.

GDNF also provides protection against the effects of dopaminergic toxins on functional indices of dopamine dynamics and behavior. Using microdialysis, Opacka-Juffry et al. (1995) found that GDNF prevented 6-OHDA-induced decreases in basal extracellular DA levels in the striatum and that K^+-stimulated DA overflow was four times greater in GDNF-treated 6-OHDA-lesioned striatum, compared to control 6-OHDA-lesioned striatum. In addition, in vivo electrochemistry studies have demonstrated that GDNF administration can prevent the decrease in K^+-evoked overflow of DA normally observed in the striatum following treatment with neurotoxic doses of methamphetamine (Cass, 1996). Positron emission tomography studies of DA reuptake sites and behavioral studies of amphetamine-induced rotation have also indicated significant protection by GDNF against 6-OHDA toxicity (Opacka-Juffry et al., 1995). Shults et al. (1996) similarly found that GDNF could reduce amphetamine-induced rotational behavior in 6-OHDA-treated rats. Furthermore, locomotion, rearing, and motility, three behaviors thought to reflect DA transmission, are increased in mice treated with GDNF prior to MPTP, compared to mice treated with vehicle before MPTP (Tomac et al., 1995a). Thus, histological, neurochemical, functional, and behavioral studies have all shown that GDNF can protect dopaminergic neurons from the effects of various toxins and lesions. However, the type and the extent of the lesion, as well as the site of injection, the time-course for delivery, and the dose of GDNF all appear to be important for determining the extent of protection.

3.2. *Neuroprotective Effects on Other Neurons*

In addition to the trophic effects on cultured motor neurons discussed earlier, GDNF also has potent in vivo neuroprotective effects on motor neurons. GDNF rescues and prevents atrophy of facial motor neurons following axotomy in neonatal rats (Henderson et al., 1994; Zurn et al., 1994; Yan et al., 1995). Similar effects are found for spinal motor neurons following peripheral axotomy in chick embryos and neonatal mice (Oppenheim et al., 1995), and in adult mice (Li et al., 1995). In adult rats, local and systemic administration of GDNF reduces the attenuation of choline acetyltransferase immunoreactivity in the facial nucleus after transection of the facial nerve (Yan et al., 1995). However, in mutant mice that display progressive motor neuron degeneration, GDNF reduces loss of facial motor neurons, but does not increase their life span or prevent degeneration of myelinated axons in the facial nerve (Sagot et al., 1996). These researchers suggest that GDNF in combination with other trophic factors may be a more useful treatment for degenerative diseases of motor neurons (Sagot et al., 1996). In any case, the results of these studies provide further evidence that GDNF may prove useful in the treatment of lower motor neuron diseases.

In addition to its effects on cholinergic motor neurons, GDNF also has trophic effects on basal forebrain cholinergic neurons. When continuously infused for 2 wk into the lateral ventricles of rats with fimbria–fornix transections, GDNF reduced the loss of choline acetyltransferase-positive neurons (Williams et al., 1996). The potency of GDNF in preventing loss of cholinergic neurons was less than nerve growth factor but greater than brain-derived neurotrophic factor. Thus, GDNF may also be able to help in the treatment of Alzheimer's disease as well as the other neurodegenerative diseases already discussed.

Another type of neuron that can be affected by GDNF are norepinephrine neurons of the locus ceruleus. When fibroblasts genetically engineered to produce high amounts of GDNF were injected dorsolateral to the locus ceruleus 1 d prior to a 6-OHDA lesion of the dorsal catecholamine bundle, there was significant protection against loss of TH-positive neurons in the locus ceruleus (Arenas et al., 1995). In addition, GDNF led to sprouting of TH-positive fibers and increased TH activity and cell body size of the norepinephrine neurons on the lesioned side. GDNF thus appears to be a potent trophic factor for noradrenergic neurons, in addition to dopaminergic and cholinergic neurons.

GDNF may also have a role in protection against the effects of seizures or in recovery from seizures. Expression of GDNF mRNA is extremely low in the forebrain of adult rats. However, after inducing seizures by peripheral injection of pilocarpine or kainic acid, GDNF mRNA expression increases in striatum and neocortex (Schmidt-Kastner et al., 1994), and in hippocampus (Humpel et al., 1994; Schmidt-Kastner et al., 1994). Subseizure doses of kainic acid and *N*-methyl-D,L-aspartic acid also upregulate GDNF mRNA expression in striatum (Ho et al., 1995). Intracerebroventricular (icv) injection of GDNF prior to systemic administration of kainic acid suppresses seizure intensity and protects against neuronal cell loss in the hippocampus, amygdala, and thalamus (Martin et al., 1995). Together, these studies indicate that GDNF expression can be modulated by excitatory amino acids and suggest an additional therapeutic potential for GDNF in disorders involving acute excitotoxic processes.

3.3. *Neurorestorative Properties of GDNF on DA Neurons*

In addition to its neuroprotective effects, GDNF also exhibits neurorestorative properties. In mice, GDNF can partially reverse the effects of MPTP toxicity on nigrostriatal DA neurons and improve motor performance (Tomac et al., 1995a). In rats unilaterally lesioned with 6-OHDA, GDNF markedly reduced apomorphine-induced rotational behavior and increased nigral DA levels and TH-positive neurons on the lesioned side (Hoffer et al., 1994; Bowenkamp et al., 1995). These results provide additional evidence that GDNF may be useful in treating Parkinson's disease.

3.4. *GDNF's Effects on the Aging Nervous System*

Parkinson's disease is associated with aging, primarily affecting individuals over the age of 50. In addition, parkinsonian signs are frequently found in elderly people who do not have a clinical diagnosis of Parkinson's disease (Bennett et al., 1996). Thus, for GDNF to be effective in treating Parkinson's disease or those with parkinsonian signs, it will need to be able to have a positive influence on the aging nigrostriatal system.

The ability of GDNF to have a positive trophic effect on aged DA neurons in rats has recently been reported. Microdialysis experiments in aged (24 mo), normal rats have indicated that GDNF, injected intranigrally 3 wk prior, augments basal extracellular levels of DA and its two primary metabolites, homovanillic acid (HVA) and 3,4-dihydroxyphenylacetic acid (DOPAC) (Hebert and Gerhardt, 1997). Both K^+- and amphetamine-evoked overflow of DA were also increased, compared to vehicle-treated animals. Improvements in balance beam performance, which may be related to augmented dopaminergic functioning, have also been reported in aged rats that received intrastriatal GDNF (Bowenkamp et al., 1996).

GDNF is also neuroprotective in aged animals. Intranigral administration of GDNF 6 h prior to intranigral 6-OHDA significantly attenuated reduction of striatal and nigral DA content in 3-, 18-, and 24-month-old rats (Kearns et al., 1996, *Soc. Neurosci. Abstr.* **22,** 1953). Although the effectiveness of the GDNF appeared to be slightly less in the oldest animals, compared to the two younger groups, it was still neuroprotective. The results of the above studies in both normal and lesioned animals indicate that nigrostriatal DA neurons in aged animals are responsive to the neurotrophic effects of GDNF.

4. EFFECTS OF GDNF IN NONHUMAN PRIMATES

The positive dopaminergic effects of GDNF have led many investigators to suggest that GDNF may prove useful for the treatment of Parkinson's disease (Lin et al., 1993; Hoffer et al., 1994; Beck et al., 1995; Opacka-Juffry et al., 1995; Tomac et al., 1995a; and others). It may prevent further degeneration of nigrostriatal DA neurons and possibly restore some of the lost function. However, previous studies have used culture systems or rodents to characterize the effects of GDNF. The rodent brain is much smaller and less complex than the human brain, and rodent behaviors are not necessarily homologous to those of humans. In addition, adverse side effects were seen in the initial clinical trials using nerve growth factor and ciliary neurotrophic factor for treating other neurodegenerative diseases in humans (Petty et al., 1994; Lindsay, 1995). Thus, it is important to examine the effects of GDNF in nonhuman primates with a nervous system and behaviors closer to those of humans, before beginning clinical trials. Therefore, over the past few years, we have been examining the behavioral, anatomical, and neurochemical effects of GDNF in the normal and hemiparkinsonian rhesus monkey.

Adult female rhesus monkeys (*Macaca mulatta*) were used for all experiments. The animals ranged in age from 5 to 16 yr. All of the procedures performed on the animals were in strict accordance with the *NIH Guide for the Care and Use of Laboratory Animals* and were approved by institutional animal care and use committees. Details of many of the experimental procedures described can be found in recent publications (Gash et al., 1995, 1996).

4.1. Normal Animals

The behavioral effects of a single injection of GDNF or vehicle into the substantia nigra were examined in 12 normal, adult monkeys. The intranigral administration of human recombinant GDNF (150 μg) had little effect on the behavior of the recipient when monitored for 3 wk following administration (Gash et al., 1995). The most common effect observed was an increase in daytime and/or nocturnal home cage activity. After giving the animals 2 d to recover from surgery, there were no incidents of vomiting, dyskinesias, motor asymmetries, or stereotypic behaviors. The GDNF animals lost an average of 7% of total body weight over the 3-wk period after GDNF administration. The vehicle-treated animals gained an average 3% in body weight.

At the conclusion of the behavioral testing, the animals were deeply anesthetized (40 mg ketamine + 25 mg/kg pentobarbital, iv), perfused through the heart with ice-cold saline, and their brains removed. Some of the brains were used for immunohistochemistry and image analysis, and some of the brains were used for measuring monoamine and metabolites levels in the basal ganglia and midbrain. GDNF administration led to significant histological and neurochemical changes in the nigrostriatal DA system of these animals.

In normal monkeys given a single, unilateral injection of 150 μg GDNF into the substantia nigra and sacrificed 4 wk later, the perikaryal size of TH-positive neurons was increased by approx 8% on the injected side, compared to the contralateral side and compared to both sides of the vehicle recipients (Gash et al., 1995). In these same animals, TH-positive fiber area in the midbrain was increased by 78% on the injected side and 39% on the contralateral side (Fig. 2A,B). This increase in TH-positive fiber area on both the injected and contralateral sides of the ventral midbrain suggests that locally applied GDNF has the ability to travel through the brain parenchyma to target sites.

Postmortem monoamine levels were measured in various structures related to mesotelencephalic DA systems in these animals. For these studies, the animal brains were chilled and cut into 4-mm-thick coronal sections. Multiple tissue punches were taken from each region and analyzed for monoamines and monoamine metabolites, using high-performance liquid chromatography (HPLC), coupled with electrochemical detection, as described previously (Gash et al., 1995).

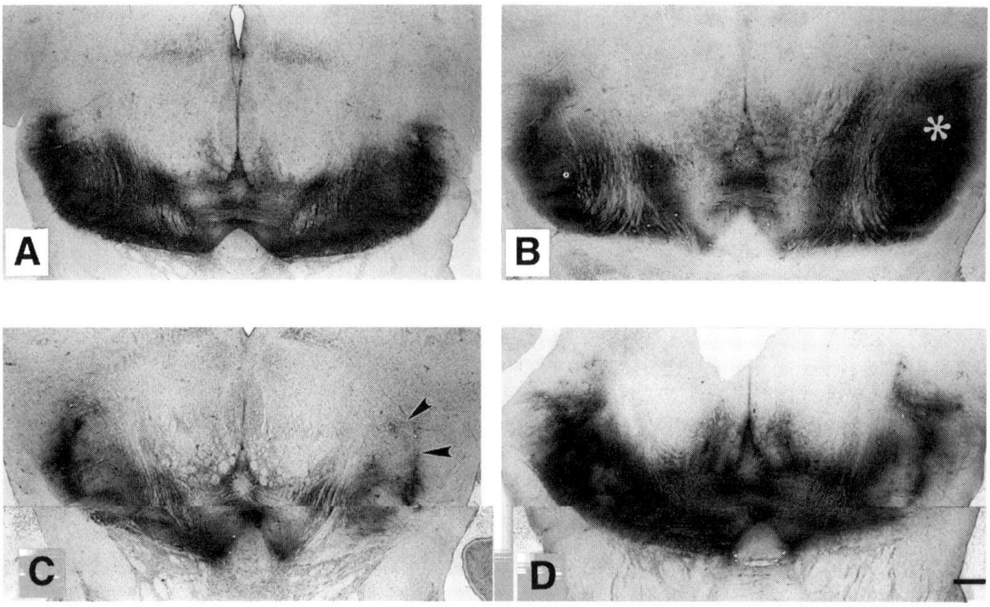

Fig. 2. DA neurons and their processes, as identified by immunocytochemical staining for tyrosine hydroxylase (TH), are shown in these coronal sections through the midbrain of four rhesus monkeys. **(A)** In normal animals, no changes could be seen in nigral DA neurons 4 wk after a vehicle injection into the right substantia nigra. **(B)** In contrast, at the same time point, a single intranigral injection of 150 μg GDNF significantly increased the number of neurites and perikaryal size on the right side (asterisk). **(C)** In animals receiving unilateral infusions of the neurotoxin MPTP through the right carotid artery, a marked reduction of TH-positive neurons was evident, especially on the right side (arrowheads). The left nigra was less severely lesioned. **(D)** Four weeks after an intranigral administration of 150 μg GDNF, significant bilateral increases in TH-positive fiber density and DA neuron cell size were measured in MPTP-lesioned animals. Scale bar = 1 mm.

Three weeks after a single injection of GDNF (150 μg) into the right substantia nigra of normal monkeys, DA levels in both the nigra and ventral tegmental area (VTA) were increased by approx 200% on the injected side (Gash et al., 1995). There were also significant increases in HVA and DOPAC in the same two regions. Trends for increased DA levels or DA metabolite levels were observed in the caudate nucleus, putamen, nucleus accumbens, and globus pallidus ipsilateral to the injection site. However, these increases did not usually reach statistical significance because of high intragroup variance. The GDNF injections had no significant effect on serotonin, 5-hydroxyindoleacetic acid, or norepinephrine levels in the same tissue samples.

Although DA levels were not elevated in the caudate nucleus and putamen in the normal animals following GDNF administration, releasable storage pools of DA in the striatum did appear augmented. In vivo electrochemistry was used to characterize potassium-evoked overflow of DA both before, and 3 wk after, intranigral administration of GDNF in three animals. GDNF administration led to a significant increase in evoked overflow of DA in both the caudate nucleus and putamen (Gash et al., 1995). There was no enhancement of release observed in three vehicle-treated animals. These results in normal monkeys correspond closely to effects seen in normal rats 3 wk after intranigral GDNF, i.e., increased nigral DA levels with no increase in striatal DA content (Hudson et al., 1995), and an increase in K^+-evoked overflow of DA in the striatum as assessed with in vivo electrochemistry (Hebert et al., 1997).

4.2. *Parkinsonian Animals*

The effects of intracerebral administration of GDNF were also examined in monkeys treated with MPTP. Unilateral MPTP lesions were made via intracarotid artery administration (0.4 mg/kg), as described previously (Bankiewicz et al., 1986; Kurlan et al., 1991; Ovadia et al., 1995). GDNF administration was made at least 3 mo post-MPTP treatment. At the time of GDNF injection, all of the monkeys displayed stable, moderate hemiparkinsonian signs. In contrast to the behavioral effects seen in normal monkeys, the behavioral effects of GDNF in the MPTP-lesioned animals were dramatic. Two weeks after a single injection of GDNF into the substantia nigra (150 µg), caudate nucleus (450 µg), or lateral ventricle (100 µg or 450 µg) on the lesioned side, there was significant improvement in parkinsonian features that lasted for at least 2 wk more (Gash et al., 1996). Notable effects were seen in reduction of bradykinesia and rigidity, and improvements in posture and balance. Tremor was the least affected of the cardinal signs of Parkinson's disease. In animals given icv injections of GDNF, the behavioral improvement was maintained by repeating GDNF administration at 4-wk intervals for a total of three injections.

The magnitude of the improvement in parkinsonian features was similar to that seen after treatment with levodopa. However, the effects of a single treatment with GDNF lasts for up to 4 wk (the longest time-point examined); improvement following a single levodopa treatment only lasts for a few hours (Greenamyre et al., 1994). Ongoing behavioral studies examining the interaction of GDNF with levodopa have indicated that the two treatments are additive, i.e., treatment with levodopa further improves the behavioral scores of GDNF-treated hemiparkinsonian monkeys. In addition, the adverse side effects observed following levodopa administration were dramatically reduced in animals that received GDNF (Miyoshi et al., 1997).

Similar to the normal animals, intracerebral administration of GDNF to MPTP-treated monkeys led to significant increases in midbrain DA neuron soma size and in TH-positive fiber area (Fig. 2C,D). MPTP-treated monkeys that received a single injection of GDNF or vehicle into the caudate nucleus or substantia nigra were killed 4 wk after trophic factor administration and their midbrains processed for quantitative image analysis, using unbiased stereological procedures (Gash et al., 1996). On the less-lesioned side, there were significant increases in nigral cell size (18%), nigral cell number (50%), and TH-positive fiber area (40%). On the more severely lesioned side, there were significant increases in nigral cell size (20%) and TH-positive fiber area (189%). Nigral cell number on the severely lesioned side tended to increase by an average of 76%; however, this increase was not statistically significant because of the variance of the data for this parameter. As with the normal monkeys, the bilateral effects observed following unilateral administration of trophic factor suggest that GDNF may be able to diffuse effectively throughout the brain. A preliminary report on the distribution of ^{125}I-GDNF following icv administration in rhesus monkeys indicates that it distributes predominantly to the substantia nigra and cerebral cortex of both hemispheres (Lapchak et al., 1996, *Am. Soc. Neural Transplant. Abstr.* **3,** 53).

The MPTP-treated monkeys that received three icv injections of GDNF or vehicle at 4-wk intervals were sacrificed 3 wk after the last dose, and multiple tissue punches were taken. In the GDNF-treated animals (100 or 450 μg/dose, $n = 3$ in each group), DA levels were increased by approx 200% on the lesioned side in the substantia nigra, VTA, and globus pallidus, compared to the lesioned side of three vehicle-treated animals (Gash et al., 1996). However, although DA levels in these three regions were increased, they were still significantly lower than DA levels on the less-lesioned side of the brain; the GDNF treatment improved DA levels in these animals, but it did not restore DA content to normal levels.

In the same three vehicle-treated and six GDNF-treated, MPTP-lesioned animals, tissue punches were taken from cortical areas, as well. For each side of each animal, 10 punches were taken from the dorsolateral surface of the prefrontal cortex, five from the cingulate gyrus, five from the lateral temporal cortex, and five from the lateral occipital cortex. DA levels were measured as described in Gash et al. (1995) and the values for all punches from a single region averaged to obtain a mean value for each region of each animal. The effects of the MPTP treatment itself on cortical DA levels were minor compared to the effects on midbrain and basal ganglia levels. In the vehicle-treated animals, DA levels were decreased by 38% in prefrontal cortex and 16% in cingulate cortex (Table 1). DA levels in the temporal and occipital cortices were not significantly altered. However, in the GDNF-treated animals, DA levels on the lesioned side in prefrontal cortex and cingulate cortex were not significantly reduced, suggesting that GDNF may have some restorative effect on MPTP-damaged mesocortical neurons. Mesocortical DA neurons in primates have been shown to be involved with cognitive abilities (Brozoski et al., 1979; Roberts et al., 1994; Arnsten et al., 1995). Thus, if GDNF proves useful for treating the motor dysfunctions in patients with Parkinson's disease, and if some of the cognitive deficits in these patients are caused by DA deficiencies, GDNF may also prove helpful in improving cognitive performance. Further studies are needed to more fully define the effects of GDNF on mesocortical DA systems.

5. CONCLUSIONS AND FUTURE DIRECTIONS

GDNF has been shown to have potent neurotrophic effects on many types of neurons, both in vitro and in vivo. The substantial neuroprotective and neurorestorative effects on rodent DA neurons are important properties for any new drug being considered for the treatment of Parkinson's disease. Further evidence that GDNF may be of significant therapeutic benefit comes from our studies of both normal and hemiparkinsonian monkeys. In these animals, intracerebral administration of GDNF has led to significant upregulation of the nigrostriatal DA system that lasts for weeks after a single administration. In normal animals, this upregulation included increases in nigral DA neuron soma size and tissue DA content on the injected side, and bilateral increases in TH-positive fiber density. Behavioral effects were minor and transient. In MPTP-treated animals, GDNF led to long-lasting bilateral increases in nigral cell size and TH-positive fiber density, an increase in the number of DA neurons in the nigra on the nonlesioned side, and a trend for increases in the number of nigral DA neurons on the MPTP-lesioned side. DA levels were elevated on the lesioned side in the substantia nigra, VTA, globus pallidus, prefrontal cortex, and cingulate cortex. GDNF substantially improved the parkinsonian behaviors in these monkeys and concomitant administration of levodopa further improves motor performance.

Table 1
DA Levels in Cortical Regions of MPTP-Treated Monkeys Following ICV Administration of GDNF or Vehicle[a]

	Left side	Right (lesioned) side
Prefrontal cortex		
Vehicle	21.9 ± 1.4	13.4 ± 1.2[b]
GDNF	19.4 ± 1.2	17.0 ± 1.8
Cingulate cortex		
Vehicle	29.5 ± 4.5	24.8 ± 5.4[b]
GDNF	26.3 ± 4.8	24.0 ± 4.8
Temporal cortex		
Vehicle	32.6 ± 7.4	29.0 ± 10.8
GDNF	26.6 ± 5.2	26.0 ± 7.2
Occipital cortex		
Vehicle	7.4 ± 2.4	8.1 ± 1.1
GDNF	8.0 ± 2.3	6.1 ± 1.5

[a] Levels are expressed as ng/g wet wt of tissue. Values are means ± SEM for three vehicle- and six GDNF-treated animals.

[b] $p \leq 0.05$ vs left side (paired *t*-test).

Taken together, the results of our studies provide evidence that GDNF is effective in upregulating the nigrostriatal DA system in nonhuman primates and that it may prove useful in therapeutic strategies for the treatment of Parkinson's disease. However, many questions remain to be answered and will require further experiments. Fundamental questions remain about GDNF-receptor interactions and the signaling pathways following receptor activation. What are the mechanisms by which GDNF provides neuroprotection for DA neurons and other neuronal populations? How does a single administration of GDNF stimulate changes in the nigrostriatal dopaminergic pathway that last for weeks? What are the long-term consequences of GDNF administration? How do the effects of GDNF vary with the age of the recipient? How does treatment with GDNF compare with other treatments? These questions and others will have to be examined and the answers carefully evaluated before the full potential of GDNF for treating neurodegenerative disorders is known.

ACKNOWLEDGMENTS

We thank Michael T. Dugan and M. Kyle Smoot for their technical assistance in these studies. This work was supported in part by grants from the United States Public Health Service, by contracts from Synergen and AMGEN, and by the University of Kentucky.

REFERENCES

Arenas, E., Trupp, M., Akerud, P., and Ibanez, C. F. (1995) GDNF prevents degeneration and promotes the phenotype of brain noradrenergic neurons in vivo. *Neuron* **15,** 1465–1473.

Arnsten, A. F. T., Cai, J. X., Steere, J. C., and Goldman-Rakic, P. S. (1995) Dopamine D2 receptor mechanisms contribute to age-related cognitive decline: the effects of quinpirole on memory and motor performance in monkeys. *J. Neurosci.* **15,** 3429–3439.

Bankiewicz, K. S., Oldfield, E. H., Chiueh, C. C., Doppman, J. L., Jacobowitz, D. M., and Kopin, I. J. (1986) Hemiparkinsonism in monkeys after unilateral internal carotid artery infusion of 1-methyl-4-phenyl-1,2,3,6-tetrahydropyridine (MPTP). *Life Sci.* **39,** 7–16.

Beck, K. D., Valverde, J., Alexi, T., Poulsen, K., Moffat, B., Vandlen, R. A., Rosenthal, A., and Hefti, F. (1995) Mesencephalic dopaminergic neurons protected by GDNF from axotomy-induced degeneration in the adult brain. *Nature* **373,** 339–341.

Beck, K. D., Irwin, I., Valverde, J., Brennan, T. J., Langston, J. W., and Hefti, F. (1996) GDNF induces a dystonia-like state in neonatal rats and stimulates dopamine and serotonin synthesis. *Neuron* **16,** 665–673.

Bennett, D. A., Beckett, L. A., Murray, A. M., Shannon, K. M., Goetz, C. G., Pilgrim, D. M., and Evans, D. A. (1996) Prevalence of parkinsonian signs and associated mortality in a community population of older people. *N. Engl. J. Med.* **334,** 71–76.

Bermingham, N., Hillermann, R., Gilmour, F., Martin, J. E., and Fisher, E. M. C. (1995) Human glial cell line-derived neurotrophic factor (GDNF) maps to chromosome 5. *Hum. Genet.* **96,** 671–673.

Blum, M., and Weickert, C. S. (1995) GDNF mRNA expression in normal postnatal development, aging, and in Weaver mutant mice. *Neurobiol. Aging* **16,** 925–929.

Bowenkamp, K. E., Hoffman, A. F., Gerhardt, G. A., Henry, M. A., Biddle, P., Hoffer, B. J., and Granholm, A.-C. E. (1995) Glial cell line-derived neurotrophic factor supports survival of injured midbrain dopaminergic neurons. *J. Comp. Neurol.* **355,** 479–489.

Bowenkamp, K. E., Lapchak, P. A., Hoffer, B. J., and Bickford, P. C. (1996) Glial cell line-derived neurotrophic factor reverses motor impairment in 16–17 month old rats. *Neurosci. Lett.* **211,** 81–84.

Brozoski, T. J., Brown, R. M., Rosvold, H. E., and Goldman, P. S. (1979) Cognitive deficit caused by regional depletion of dopamine in prefrontal cortex of rhesus monkey. *Science* **205,** 929–932.

Buj-Bello, A., Buchman, V. L., Horton, A., Rosenthal, A., and Davies, A. M. (1995) GDNF is an age-specific survival factor for sensory and autonomic neurons. *Neuron* **15,** 821–828.

Cass, W. A. (1996) GDNF selectively protects dopamine neurons over serotonin neurons against the neurotoxic effects of methamphetamine. *J. Neurosci.* **16,** 8132–8139.

Cheng, H., Hoffer, B., Strömberg, I., Russell, D., and Olson, L. (1995) The effects of glial cell line-derived neurotrophic factor in fibrin glue on developing dopamine neurons. *Exp. Brain Res.* **104,** 199–206.

Choi-Lundberg, D. L., and Bohn, M. C. (1995) Ontogeny and distribution of glial cell line-derived neurotrophic factor (GDNF) mRNA in rat. *Develop. Brain Res.* **85,** 80–88.

Clarkson, E. D., Zawada, W. M., and Freed, C. R. (1995) GDNF reduces apoptosis in dopaminergic neurons in vitro. *Neuroreport* **7,** 145–149.

Cristina, N., Chatellard-Causse, C., Manier, M., and Feuerstein, C. (1995) GDNF: existence of a second transcript in the brain. *Mol. Brain Res.* **32,** 354–357.

Durbec, P., Marcos-Gutierrez, C. V., Kilkenny, C., Grigoriou, M., Wartiowaara, K., Suvanto, P., et al. (1996) GDNF signalling through the Ret receptor tyrosine kinase. *Nature* **381,** 789–793.

Ebendal, T., Tomac, A., Hoffer, B. J., and Olson, L. (1995) Glial cell line-derived neurotrophic factor stimulates fiber formation and survival in cultured neurons from peripheral ganglia. *J. Neurosci. Res.* **40,** 276–284.

Gash, D. M., Zhang, Z., Cass, W. A., Ovadia, A., Simmerman, L., Martin, D., et al. (1995) Morphological and functional effects of intranigrally administered GDNF in normal rhesus monkeys. *J. Comp. Neurol.* **363,** 345–358.

Gash, D. M., Zhang, Z., Ovadia, A., Cass, W. A., Yi, A., Simmerman, L., et al. (1996) Functional recovery in GDNF-treated parkinsonian monkeys. *Nature* **380,** 252–255.

Gouin, A., Bloch-Gallego, E., Tanaka, H., Rosenthal, A., and Henderson, C. E. (1996) Transforming growth factor-beta 3, glial cell line-derived neurotrophic factor, and fibroblast

growth factor-2, act in different manners to promote motoneuron survival in vitro. *J. Neurosci. Res.* **43,** 454–464.
Greenamyre, J. T., Eller, R. V., Zhang, Z., Ovadia, A., Kurlan, R., and Gash, D. M. (1994) Antiparkinsonian effects of Remacemide hydrochloride, a glutamate antagonist, in rodent and primate models of Parkinson's disease. *Ann. Neurol.* **35,** 655–661.
Hebert, M. A. and Gerhardt, G. A. (1997) Behavioral and neurochemical effects of intranigral GDNF administration on aged Fischer 344 rats. *J. Pharmacol. Exp. Ther.*, in press.
Hebert, M. A., van Horne, C. G., Hoffer, B. J., and Gerhardt, G. A. (1996) Functional effects of GDNF in normal rat striatum: presynaptic studies using in vivo electrochemistry and microdialysis. *J. Pharmacol. Exp. Ther.* **279,** 1181–1190.
Hellmich, H. L., Kos, L., Cho, E. S., Mahon, K. A., and Zimmer, A. (1996) Embryonic expression of glial cell-line derived neurotrophic factor (GDNF) suggests multiple developmental roles in neural differentiation and epithelial-mesenchymal interactions. *Mech. Devel.* **54,** 95–105.
Henderson, C. E., Phillips, H. S., Pollock, R. A., Davies, A. N., Lemeulle, C., Armanini, M., et al. (1994) GDNF: a potent survival factor for motoneurons present in peripheral nerve and muscle. *Science* **266,** 1062–1064.
Ho, A., Gore, A. C., Weickert, C. S., and Blum, M. (1995) Glutamate regulation of GDNF gene expression in the striatum and primary striatal astrocytes. *Neuroreport* **6,** 1326–1330.
Hoffer, B. J., Hoffman, A., Bowenkamp, K., Huettl, P., Hudson, J., Martin, D., Lin, L.-F. H., and Gerhardt, G. A. (1994) Glial cell line-derived neurotrophic factor reverses toxin-induced injury to midbrain dopaminergic neurons in vivo. *Neurosci. Lett.* **182,** 107–111.
Hornykiewicz, O. (1993) Parkinson's disease and the adaptive capacity of the nigrostriatal dopamine system: possible neurochemical mechanisms. *Adv. Neurol.* **60,** 140–147.
Hou, J.-G. G., Lin, L.-F. H., and Mytilineou, C. (1996) Glial cell line-derived neurotrophic factor exerts neurotrophic effects on dopaminergic neurons in vitro and promotes their survival and regrowth after damage by 1-methyl-4-phenylpyridinium. *J. Neurochem.* **66,** 74–82.
Hudson, J., Granholm, A.-C., Gerhardt, G., Henry, M. A., Hoffman, A., Biddle, P., et al. (1995) Glial cell line-derived neurotrophic factor augments midbrain dopaminergic circuits in vivo. *Brain Res. Bull.* **36,** 425–432.
Humpel, C., Hoffer, B., Strömberg, I., Bektesh, S., Collins, F., and Olson, L. (1994) Neurons of the hippocampal formation express glial cell line-derived neurotrophic factor messenger RNA in response to kainate-induced excitation. *Neuroscience* **59,** 791–795.
Jellinger, K. (1986) Overview of morphological changes in Parkinson's disease. *Adv. Neurol.* **45,** 1–18.
Jing, S., Wen, D., Yu, Y., Holst, P. L., Luo, Y., Fang, M., et al. (1996) GDNF-induced activation of the ret protein tyrosine kinase is mediated by GDNFR-alpha, a novel receptor for GDNF. *Cell* **85,** 1113–1124.
Johansson, M., Friedemann, M., Hoffer, B., and Strömberg, I. (1995) Effects of glial cell line-derived neurotrophic factor on developing and mature ventral mesencephalic graphs in oculo. *Exp. Neurol.* **134,** 25–34.
Kearns, C. M. and Gash, D. M. (1995) GDNF protects nigral dopamine neurons against 6-hydroxydopamine in vivo. *Brain Res.* **672,** 104–111.
Kearns, C. M., Cass, W. A., Smoot, K., Kryscio, R., and Gash, D. M. (1997) GDNF protection against 6-OHDA: time dependence and requirement for protein synthesis. *J. Neurosci.*, in press.
Krieglstein, K., Suter-Crazzolara, C., Fischer, W. H., and Unsicker, K. (1995) TGF-β superfamily members promote survival of midbrain dopaminergic neurons and protect them against MPP^+ toxicity. *EMBO J.* **14,** 736–742.
Kurlan, R., Kim, M. H., and Gash, D. M. (1991) Oral levodopa dose-response study in MPTP-induced hemiparkinsonism monkeys: assessment with a new rating scale for monkey parkinsonism. *Mov. Disord.* **6,** 111–118.

Li, L., Wu, W., Lin, L.-F. H., Lei, M., Oppenheim, R. W., and Houenou, L. J. (1995) Rescue of adult mouse motoneurons from injury-induced cell death by glial cell line-derived neurotrophic factor. *Proc. Natl. Acad. Sci. USA* **92,** 9771–9775.

Lin, L.-F. H., Doherty, D. H., Lile, J. D., Bektesh, S., and Collins, F. (1993) GDNF: a glial cell line-derived neurotrophic factor for midbrain dopaminergic neurons. *Science* **260,** 1130–1132.

Lin, L.-F. H., Zhang, T. J., Collins, F., and Armes, L. G. (1994) Purification and initial characterization of rat B49 glial cell line-derived neurotrophic factor. *J. Neurochem.* **63,** 758–768.

Lindsay, R. M. (1995) Neuron saving schemes. *Nature* **373,** 289, 290.

Martin, D., Miller, G., Rosendahl, M., and Russell, D. A. (1995) Potent inhibitory effects of glial derived neurotrophic factor against kainic acid mediated seizures in the rat. *Brain Res.* **683,** 172–178.

Massagué, J. (1996) Crossing receptor boundaries. *Nature* **322,** 29, 30.

Miyoshi, Y., Zhang, Z., Ovadia, A., Lapchak, P. A., Collins, F., Hilt, D., Lebel, C., Kryscio, R., and Gash, M. (1997) Glial cell line-derived neurotrophic factor levodopa interactions and reduction of side effects in parkinsonian monkeys. *Ann. Neurol.*, in press.

Moore, M. W., Klein, R. D., Fariñas, I., Sauer, H., Armanini, M., Phillips, H., et al. (1996) Renal and neuronal abnormalities in mice lacking GDNF. *Nature* **382,** 76–79.

Mount, H. T. J., Dean, D. O., Alberch, J., Dreyfus, C. F., and Black, I. B. (1995) Glial cell lined-derived neurotrophic factor promotes the survival and morphological differentiation of Purkinje cells. *Proc. Natl. Acad. Sci. USA* **92,** 9092–9096.

Opacka-Juffry, J., Ashworth, S., Hume, S. P., Martin, D., Brooks, D. J., and Blunt, S. B. (1995) GDNF protects against 6-OHDA nigrostriatal lesion: in vivo study with microdialysis and PET. *Neuroreport* **7,** 348–352.

Oppenheim, R. W., Houenou, L. J., Johnson, J. E., Lin, L.-F. H., Li, L., Lo, A. C., et al. (1995) Developing motor neurons rescued from programmed and axotomy-induced cell death by GDNF. *Nature* **373,** 344–346.

Ovadia, A., Zhang, Z., and Gash, D. M. (1995) Increased susceptibility to MPTP toxicity in middle-aged rhesus monkeys. *Neurobiol. Aging* **16,** 931–937.

Petty, B. G., Cornblath, D. R., Adornato, B. T., Chaudhry, V., Flexner, C., Wachsman, M., Sinicropi, D., Burton, L. E., and Peroutka, S. J. (1994) The effects of systemically administered recombinant human nerve growth factor in healthy human subjects. *Ann. Neurol.* **36,** 244–246.

Pichel, J. G., Shen, L., Sheng, H. Z., Granholm, A.-C., Drago, J., Grinberg, A., et al. (1996) Defects in enteric innervation and kidney development in mice lacking GDNF. *Nature* **382,** 73–76.

Roberts, A. C., De Salvia, M. A., Wilkinson, L. S., Collins, P., Muir, J. L., Everitt, B. J., and Robbins, T. W. (1994) 6-Hydroxydopamine lesions of the prefrontal cortex in monkeys enhance performance on an analog of the Wisconsin Card Sort Test: possible interactions with subcortical dopamine. *J. Neurosci.* **14,** 2531–2544.

Sagot, Y., Tan, S. A., Hammang, J. P., Aebischer, P., and Kato, A. C. (1996) GDNF slows loss of motoneurons but not axonal degeneration or premature death of pmn/pmn mice. *J. Neurosci.* **16,** 2335–2341.

Sánchez, M. P., Silos-Santiago, I., Frisén, J., He, B., Lira, S. A., and Barbacid M. (1996) Renal agenesis and the absence of enteric neurons in mice lacking GDNF. *Nature* **382,** 70–73.

Sauer, H., Rosenblad, C., and Bjorklund, A. (1995) Glial cell line-derived neurotrophic factor but not transforming growth factor β3 prevents delayed degeneration of nigral dopaminergic neurons following striatal 6-hydroxydopamine lesions. *Proc. Natl. Acad. Sci. USA* **92,** 8935–8939.

Schaar, D. G., Sieber, B.-A., Dreyfus, C. F., and Black, I. B. (1993) Regional and cell-specific expression of GDNF in rat brain. *Exp. Neurol.* **124,** 368–371.

Schindelhauer, D., Schuffenhauer, S., Gasser, T., Steinkasserer, A., and Meitinger, T. (1995) The gene coding for glial cell line derived neurotrophic factor (GDNF) maps to chromosome 5p12–p13.1. *Genomics* **28,** 605–607.

Schmidt-Kastner, R., Tomac, A., Hoffer, B., Bektesh, S., Rosenzweig, B., and Olson, L. (1994) Glial cell line-derived neurotrophic factor (GDNF) mRNA upregulation in striatum and cortical areas after pilocarpine-induced status epilepticus in rats. *Mol. Brain Res.* **26,** 325–330.

Shults, C. W., Kimber, T., and Martin, D. (1996) Intrastriatal injection of GDNF attenuates the effects of 6-hydroxydopamine. *Neuroreport* **7,** 627–631.

Springer, J. E., Mu, X., Bergmann, L. W., and Trojanowski, J. Q. (1994) Expression of GDNF mRNA in rat and human nervous tissue. *Exp. Neurol.* **127,** 167–170.

Springer, J. E., Seeburger, J. L., He, J., Gabrea, A., Blankenhorn, E. P., and Bergman, L. W. (1995) cDNA sequence and differential mRNA regulation of two forms of glial cell line-derived neurotrophic factor in Schwann cells and rat skeletal muscle. *Exp. Neurol.* **131,** 47–52.

Strömberg, I., Björklund, L., Johansson, M., Tomac, A., Collins, F., Olson, L., Hoffer, B., and Humpel, C. (1993) Glial cell line-derived neurotrophic factor is expressed in the developing but not adult striatum and stimulates developing dopamine neurons in vivo. *Exp. Neurol.* **124,** 401–412.

Suter-Crazzolara, C., and Unsicker, K. (1994) GDNF is expressed in two forms in many tissues outside the CNS. *Neuroreport* **5,** 2486–2488.

Suvanto, P., Hiltunen, J. O., Arumae, U., Moshnyakov, M., Sariola, H., Sainio, K., and Saarma, M. (1996) Localization of glial cell line-derived neurotrophic factor (GDNF) mRNA in embryonic rat by in situ hybridization. *Eur. J. Neurosci.* **8,** 816–822.

Tomac, A., Lindqvist, E., Lin, L.-F. H., Ogren, S. O., Young, D., Hoffer, B. J., and Olson, L. (1995a) Protection and repair of the nigrostriatal dopaminergic system by GDNF in vivo. *Nature* **373,** 335–339.

Tomac, A., Widenfalk, J., Lin, L.-F. H., Kohno, T., Ebendal, T., Hoffer, B. J., and Olson, L. (1995b) Retrograde axonal transport of glial cell line-derived neurotrophic factor in the adult nigrostriatal system suggests a trophic role in the adult. *Proc. Natl. Acad. Sci. USA* **92,** 8274–8278.

Treanor, J. J. S., Goodman, L., de Sauvage, F., Stone, D. M., Poulsen, K. T., Beck, C. D., et al. (1996) Characterization of a multicomponent receptor for GDNF. *Nature* **382,** 80–83.

Trupp, M., Ryden, M., Jornvall, H., Funakoshi, H., Timmusk, T., Arenas, E., and Ibanez, C. F. (1995) Peripheral expression and biological activities of GDNF, a new neurotrophic factor for avian and mammalian peripheral neurons. *J. Cell Biol.* **130,** 137–148.

Trupp, M., Arenas, E., Fainzilber, M., Nilsson, A.-S., Sieber, B.-A., Grigoriou, M., et al. (1996) Functional receptor for GDNF encoded by the c-ret proto-onocogene. *Nature* **381,** 785–789.

Williams, L. R., Inouye, G., Cummins, V., and Pelleymounter, M. A. (1996) Glial cell line-derived neurotrophic factor sustains axotomized basal forebrain cholinergic neurons in vivo: dose-response comparison to nerve growth factor and brain-derived neurotrophic factor. *J. Pharmacol. Exp. Ther.* **277,** 1140–1151.

Yan, Q., Matheson, C., and Lopez, O. T. (1995) In vivo neurotrophic effects of GDNF on neonatal and adult facial motor neurons. *Nature* **373,** 341–344.

Zurn, A. D., Baetge, E. E., Hammang, J. P., Tan, S. A., and Aebischer, P. (1994) Glial cell line-derived neurotrophic factor (GDNF), a new neurotrophic factor for motoneurones. *Neuroreport* **6,** 113–118.

Zurn, A. D., Winkel, L., Menoud, A., Djabali, K., and Aebischer, P. (1996) Combined effects of GDNF, BDNF, and CNTF on motoneuron differentiation in vitro. *J. Neurosci. Res.* **44,** 133–141.

9
Tumor Necrosis Factor
The Good, the Bad, and the Umbra

Steven W. Barger

1. INTRODUCTION

The pace at which connections are being drawn between the nervous system and the immune system does not appear to be slowing. Although the relationship between these two systems is more commonly discussed in terms of concrete interactions modulating physiology, it can be expanded to include more abstract considerations, such as analogies between molecules and between basic principles of operation. Thus, we have come to see beyond the responsiveness of leukocytes to neuropeptides and neuromodulators like CRH and morphine, to recognize relationships exemplified by the compatibility between the principles of antigen presentation and Hebbian arguments about associative synaptic plasticity. Similarly, one can argue that innoculation is a form of conditioning and that neurons may be conditioned not only to strengthen a circuit required for information storage, but that they may also be conditioned to withstand greater metabolic demands or injurious conditions. The central nervous system (CNS) certainly maintains a degree of immune privilege; therefore, one must consider the role of resident leukocytes (i.e., microglia). That they help to combat infections is obvious, but microglia also respond to noninfectious pathologies with reactions more complex than the staid phagocytosis of debris; it is difficult to imagine that retention of the latter responses by evolution is accidental. This chapter focuses on a specific component of gliotic mediators, tumor necrosis factor (TNF). Some of the data discussed distinguishes between TNFα (cachectin) and TNFβ (lymphotoxin-α). However, their mechanisms of action and the responses of relevant target cells are similar enough that the more inclusive TNFs or TNF- will be used often to describe these shared actions. Emphasis is given to the role of TNFs in brain injury responses and to data that force consideration of the possibility that TNF expression in the CNS may have beneficial influences, particularly with respect to conditioning that enhances neuronal resistance to insults.

2. CHARACTERIZATION OF TNF

TNFα and TNFβ are homologous (28% in humans, 35% in mice) proteins with monomeric mol wt of 17 and 25 kDa, respectively. They are products of separate genes that lie on chromosome 6 in humans, in close proximity to one another and to

From: Neuroprotective Signal Transduction *edited by M. P. Mattson*
Humana Press Inc., Totowa, NJ

the more recent addition to the family: lymphotoxin-β (LT-β). They fall within the region often referred to as the major histocompatibility class III (MHC-III) group; nearby are genes relevant to other functions of the immune system. The close spatial relationship between the genes in this complex may lead to promoter overlap, and many of them are coordinately regulated. However, the expression patterns and regulation of TNFα and TNFβ differ considerably. TNFβ is almost restricted to activated T-lymphocytes, but TNFα is expressed primarily in cells of the mononuclear lineage (Pennica et al., 1984; Sawada et al., 1989) and may also be induced in B-lymphocytes (Sung et al., 1988), astrocytes (Sawada et al., 1989; Chung et al., 1991; Selmaj et al., 1991; Sun et al., 1995), and neurons (Gendron et al., 1991; Tchelingerian et al., 1994). TNFβ induction in microglia in certain disease states has also been reported (e.g., Selmaj et al., 1991).

TNFs are expressed basally in some cell types, but agents that activate T-cells or macrophages elevate expression dramatically. The mechanisms by which expression is regulated are complex, occurring at the levels of transcription, mRNA stability, and translation. For instance, TNFα mRNA is detectable in some cells that do not appear to express protein. TNFα transcription is controlled by the NFκB transcription factor (Shakhov et al., 1990) and is induced by concanavalin A and phorbol esters (English et al., 1991), among other activators. TNFα mRNA stability is regulated by a repeating AU-rich sequence in its 3′-untranslated region, the influence of which can be modulated by endotoxin (Han et al., 1990). Although TNFβ is also controlled by NFκB-dependent transcription (Messer et al., 1990; Paul et al., 1990), its induction is slower than that of TNFα and its mRNA is more stable (English et al., 1991).

The TNFs share a feature uncommon among peptide signaling molecules: a trimeric active structure. Although the predominant forms of TNFs are noncovalent homotrimers, LT-β appears to exist only as a heterotrimer with TNFβ polypeptide chains; this relationship may explain the apparent overlap in cell-type expression patterns between TNFβ and LT-β. Many of the peptide growth factors are active in a dimeric form. The dimer binds two otherwise independent tyrosine kinase receptor molecules, and it is thought that signal transduction is initiated when the two receptors are brought into close proximity by the ligand, so that they may phosphorylate one another. The TNF receptors are not kinases, but it is believed that TNF signal transduction involves recruitment of its receptors into homomultimeric complexes as well. Indeed, signal transduction can be elicited in chimeric TNF receptors that have binding domains for dimeric growth factors, or can spontaneously dimerize (Adam et al., 1995; Bazzoni et al., 1995).

An additional level of complexity is imparted to TNF structural considerations by the existence of forms integrally associated with the membrane. TNFα is released from its cells of synthesis by cleavage of an integral membrane "pro-TNF," perhaps by metalloproteases (Gearing, 1994; McGeehan et al., 1994). However, this cleavage is not requisite for activity (Peck et al., 1989; Macchia et al., 1993); in fact, some cells appear to respond preferentially to this membrane-associated form of TNF (Peck et al., 1989; Grell et al., 1995). LT-β appears to exist only in such an integral membrane form (Browning et al., 1993). This and related scenarios (such as the solubility of some cytokine receptors [see below]) have muddied the distinction between ligand and receptor in discussions of cytokine actions. These relationships also have strengthened the hypothesis that cytokines and growth factors evolved from cell adhesion molecules that originally functioned only as membrane-bound forms, an idea originally proposed after the discoveries of intracellular signal transduction pathways coupled to cell–cell adhesion molecules and integrins (receptors for extracellular matrix proteins). An additional implication of this nondiffusible TNF is that experimental analysis of CSF, culture-conditioned medium, or other fluids may not accurately detect changes in the expression of all the active TNF.

3. RECEPTORS AND SIGNAL TRANSDUCTION

The known high-affinity receptors for TNFα and TNFβ are the products of two separate but related genes. The largest mature form of one receptor is approx 55 kDa and is often termed TNF-R1; the other (TNF-R2) is approx 75 kDa in its largest form. They are homologous in their extracellular domains, and are members of a larger family that includes Fas and (distantly) the low-affinity NGF receptor. The cytoplasmic portions are usually short and divergent and predict no known enzymatic function. LT-β does not interact with either of the high-affinity TNF receptors, preferring instead a separate member of the receptor family (Crowe et al., 1994).

TNF-R1 and TNF-R2 have similar affinities for TNFα and TNFβ, and both are widely distributed. However, the two receptors differ in subcellular localization (Bradley et al., 1995), and some tissues appear deficient in TNF-R2. The receptors also appear to differ somewhat in the signals they convey intracellularly. TNF-R1 has been associated most closely with the ability of TNF to evoke cell death, but either receptor appears capable of killing cells alone (Heller et al., 1992; Bazzoni et al., 1995). For TNF-R1, this activity has been mapped to a "death domain" in the cytoplasmic tail (Tartaglia et al., 1993a), which is homologous to sequences present in other members of the receptor family capable of evoking apoptosis (Fas, CD30, and so on). TNF-R2 has been associated with the induction of various bioactivities, such as proliferation of thymocytes (Tartaglia et al., 1991, 1993b), migration of Langerhans cells (Wang et al., 1996), and participation in endotoxic shock (Pfeffer et al., 1993). In several bioassays, TNF-R2 seems to facilitate processes that more directly result from TNF-R1 activation, perhaps by increasing local concentrations of the ligand through an effect termed "ligand passing" (Tartaglia et al., 1993c; Bigda et al., 1994; Medvedev et al., 1996).

Both TNF-R1 and TNF-R2 can be processed to truncated forms that lack transmembrane regions and thus diffuse extracellularly, even humorally. In all paradigms tested to date, these soluble receptors can bind enough TNF to act as inhibitors of its actions (Van Zee et al., 1992; Howard et al., 1993). However, other cytokine receptors that are released can function as coligands (e.g., ciliary neurotrophic factor receptor-α), forming a complex with the conventional ligand (e.g., ciliary neurotrophic factor) to interact with cells in a way that evokes signal transduction (Davis et al., 1993). Such a scenario has never been documented with TNF receptors; but the integral membrane form of TNF-R1 can interact (perhaps indirectly) with at least one downstream effector enzyme that resides on the extracellular face of the plasma membrane (Wiegmann et al., 1994).

3.1. Receptor-Binding Proteins

Recent studies of protein–protein interactions between TNF receptors and a novel class of effector molecules has greatly enhanced our understanding of the sometimes confusing array of TNF actions (Fig. 1). In 1994, two proteins were cloned by virtue of their interaction with TNF-R2 in a yeast two-hybrid system (Rothe et al., 1994; Song

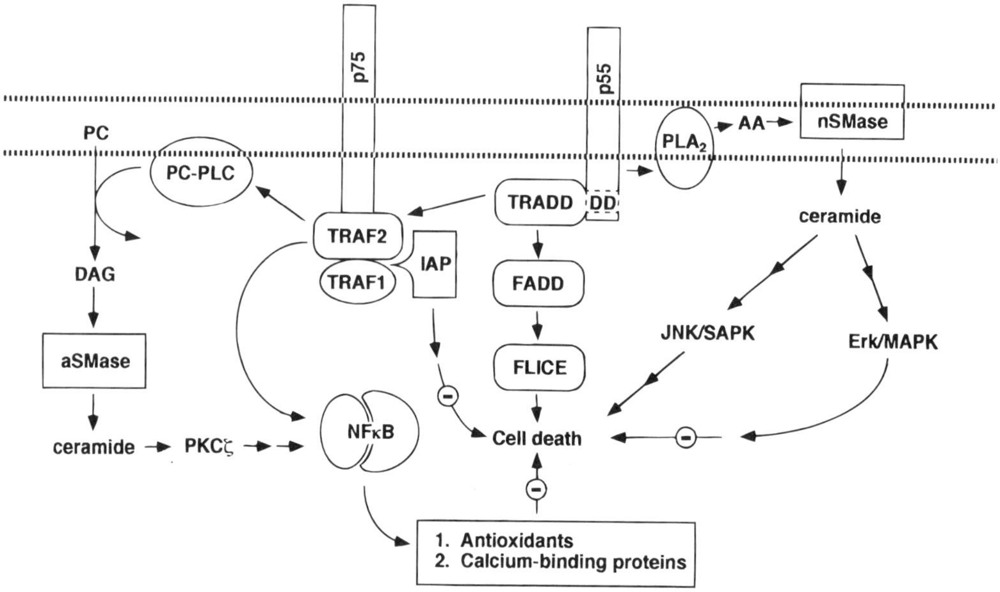

Fig. 1. Hypothetical connections of TNF signal transduction. TNFs bind two distinct receptors, TNF-R1 (p55) and TNF-R2 (p75). Intracellular proteins have recently been discovered that interact with the cytosolic domains of these receptors. TRAF2 appears to interact directly with TNF-R2; complexed with TRAF1, it can recruit a protein related to inhibitor of apoptosis (IAP). TRADD appears to interact preferentially with the TNF-R1's death domain, but it can also bind and transduce signals through FADD and TRAF2. Indeed, TRAF2 appears responsible for the ultimate stimulation of NFκB by TNF-R1 activation (Hsu et al., 1996). Also involved in NFκB stimulation is ceramide, produced exclusively by the acidic sphinomyelinase (aSMase) (Wiegmann et al., 1994). This enzyme can be activated by diacylglycerol (DAG) (Kolesnick, 1987). This is consistent with the fact that TNF stimulation of NFκB is blocked by an inhibitor of the phosphatidylcholine-specific phospholipase C (PC-PLC), which normally produces DAG in response to TNF (Schutze et al., 1992). Ceramide from aSMase activity can directly activate the ζ form of protein kinase C (PKC ζ), which then may stimulate NFκB indirectly (Muller et al., 1995). Nevertheless, some additional TRAF2-dependent signal may be required for NFκB stimulation, because Fas activation evokes aSMase activity without activating NFκB (Cifone et al., 1995). A region of TNF-R1 separate from the death domain can stimulate neutral SMase (nSMase; an enzyme of the outer membrane leaflet) to produce ceramide targeted separately from that produced by aSMase. TNF stimulates phospholipase A2 to produce arachidonic acid (AA), which can activate nSMase (Jayadev et al., 1994). The activation of Erk/MAPK by this arm of the signaling pathway is extrapolated from Fas mutants that lack the death domain, yet still stimulate nSMase and Erk/MAPK (Cifone et al., 1995). Recently, stimulation of survival or death by ceramide (or its metabolites) was distinguished by the stimulation of Erk/MAPK or JNK/SAPK, respectively (Cuvillier et al., 1996). Caveats and apparent inconsistencies of the model drawn here include the following: Wiegmann et al. (1994) did not observe stimulation of aSMase by TNF-R2, although others have demonstrated activation of NFκB by TNF-R2 (*see text*); TNF-R2 and TRADD probably compete for binding to the same site on TRAF2 (Hsu et al., 1996); PC-PLC and aSMase are involved in the ability of Fas to kill cells (Cifone et al., 1995).

and Donner, 1995). They were termed TRAF1 and TRAF2; only TRAF2 appears to interact directly with the receptor. TRAF2 is also a zinc-finger protein with a RING motif, suggesting the possibility that it is a transcription factor. Subsequently, the TRAF1–TRAF2 complex was found to interact with proteins homologous to a family of inhibitors of apoptosis (IAPs) originally isolated from a baculovirus (Rothe et al., 1995a).

A member of a second family of TNF receptor-associated proteins couples to TNF-R1. Similarly to TRAFs, TRADD (Hsu et al., 1995) was cloned by the yeast two-hybrid system, in this case, using the TNF-R1 "death domain" as the trap. The requisite interaction was a homophilic one that occurred through a region in TRADD that is itself homologous to the TNF-R1 "death domain" (in fact, many of the clones retrieved in this approach were TNF-R1 and other members of its family). The "death domain" of TRADD is somewhat homologous to FADD (a protein that seems to interact exclusively with Fas), RIP (which may interact with both TNF-R1 and Fas), and Reaper (a protein required for programmed cell death in *Drosophila*). Overexpression of either of these proteins induces apoptosis; interestingly, the *Drosophila* members of the IAP family (above) can block apoptosis triggered by Reaper, hinting at a possible functional control of TNF-R1-associated events by TNF-R2 signals. The downstream factors may include a cysteine protease termed FLICE (Muzio et al., 1996) or MACH (Boldin et al., 1996), which binds FADD and may be the relative of the interleukin-1β-converting enzyme (ICE) implicated by Enari et al. (1995) in cytokine-induced apoptosis.

3.2. *Enzymatic Signals*

Before the above receptor-binding proteins were even a twinkle in a cloner's eye, intracellular TNF signal transduction was linked to sundry enzymatic processes. TNFs have been shown to stimulate the activities of protein kinase C (PKC), phospholipase A2 (PLA2), phosphatidylcholine-specific phospholipase C (PC-PLC), sphingomyelinase (SMase), Erk/MAP kinases, and Jun nuclear kinase/stress-activated protein kinase (JNK/SAPK). Despite evidence linking these processes, their precise connections to one another and to the receptor-binding proteins (above) remain somewhat speculative (Fig. 1). Further, the role played by each of these pathways in downstream nuclear events and biological effects evoked by TNF (below) has been the subject of some controversy. For instance, several reports have implied a role for ceramide (a product of SMase activity) in the activation of NFκB, the most prominent transcription factor in TNF signaling (Schutze et al., 1992; Yang et al., 1993), but other data suggest that TNF induction of NFκB is independent of SMase activity (Dbaibo et al., 1993; Betts et al., 1994; Johns et al., 1994; Reddy et al., 1994; Westwick et al., 1995). Some discrepancies may reflect simple cell-type differences, but others clearly do not. Relevance of each signal transduction component will be clearer once the mechanistic chain of events connecting each of these signal transduction events to the TNF receptors has been elucidated.

3.3. *NFκB*

Most of the biological effects of TNFs have been attributed to influences on gene expression, and the chief mediator of such nuclear events is the transcription factor NFκB. First discovered and named for its impact on expression of the immunoglobulin κ chain, NFκB has since been documented to control several other genes, many of which are cytokines (including the TNFs themselves), or have another relationship with immune system function. Its active form is composed of two polypeptides: p50 and p65. The larger is a member of the Rel family of transcription factor proteins, which interact modularly to form various dimers with differing transcriptional actions. The regulation of this family is intriguing and appears unique in several ways. NFκB rests in the cytosol, complexed to an inhibitory subunit (IκB). Activating stimuli cause specific, ubiquitin-directed degradation of IκB. The remaining dimer then translocates to the nucleus and binds to κB enhancers. A particularly interesting facet of NFκB regulation is its activation by oxidative conditions (Schreck et al., 1991). This relationship has led to speculation that NFκB might play a role in promoting the damaging effects of reactive oxygen species (ROS). The activation of NFκB by TNFs, factors that can initiate cell death in some situations, has further promoted the speculative link of this transcription factor to cell death.

Since these circumstantial connections to cell death were reported, NFκB has been mechanistically dissociated from TNF-induced death. Although its stimulation by activated TNF-R1 is dependent on the so-called "death domain" of the receptor (Hsu et al., 1995), NFκB has been dissociated from the apoptotic pathway at the level of the TRADD receptor-binding protein (Hsu et al., 1995, 1996; Park and Baichwal, 1996). These studies suggest that TRADD heads a bifurcating signal that includes interactions with both TRAF2, which apparently transduces the activation of NFκB, and FADD, which apparently transduces the death signal. In cells expressing a mutant form of TRAF2, TNFα can no longer stimulate NFκB, yet can still elicit apoptosis. Furthermore, NFκB can be stimulated by activation of TNF-R2 (Medvedev et al., 1994; Rao et al., 1995; Rothe et al., 1995b), which lacks a "death domain." Other data suggest the possibility that the connection to ROS reveals a role for NFκB in sensing and responding to oxidative stress in a compensatory way, and the exquisite responsiveness and specificity apparent in the interaction between ROS and NFκB seems indicative of a role for mild oxidation in normal cellular signals (Schreck et al., 1992), perhaps including those initiated by TNFs. These normal signals may include factors that offer protection against oxidative stress by preemptively fortifying the cell's resistance. For instance, manganese superoxide dismutase (Mn-SOD) is induced by TNFα (Pang et al., 1992) and has NFκB consensus binding sites in its promoter. Large increases in intracellular ionic calcium ($[Ca^{2+}]_i$) can augment oxidative stress (Takasu et al., 1989; Oyama et al., 1994), but may be attenuated by TNFs' induction of the calcium (Ca)-binding protein calbindin (Cheng et al., 1994; Mattson et al., 1995) (incidentally, TRAF1 bears some homology to calbindin).

Recently, it was determined that κB-dependent transcription is activated by a neuroprotective agent: the secreted form of the Alzheimer's β-amyloid precursor protein (sAPP) (Barger and Mattson, 1996a). In addition, an antisense oligonucleotide-mediated decrease in IκBα, a manipulation designed to specifically activate NFκB, protects neuronal cells against insults such as excitotoxicity, Ca^{2+} ionophore, and amyloid β-peptide (Barger et al., 1995; Barger and Mattson, 1996a,b). NFκB also is activated by some neurotrophic or neuroprotective agents, such as nerve growth factor (Korner et al., 1994). These data expand the hypothesis that NFκB, at least in certain circumstances, can orchestrate a program of gene expression that is beneficial to the viability of some populations of cells. Indeed, the analogous argument can be made for TNFs themselves.

4. PHYSIOLOGICAL RESPONSES TO TNFs

From their discovery, the literature concerning TNFs has been focused on their negative attributes: cachexia, chronic inflammation, tissue necrosis, and cartilage and bone resorption. However, recent data suggest that TNFs may not have been selected by evolution for their harmful effects! Targeted gene disruption ("knockout") of TNF receptors has confirmed a role for TNFs in defense against certain infections, but the factors may have additional benefits that are less obvious in individual tissues. Many insults to the brain induce TNFα expression in the absence of any infection (below). Although it is true that the high concentrations of TNFα present at the center of an inflammatory reaction or other injury may eventually prove detrimental to that tissue, the umbra surrounding that focal lesion may gain from the factor's subtler influences.

One of the major roles of TNFs in vivo is coordination of the acute phase response (Beutler, 1995). TNFα is induced in cells of the monocyte lineage by endotoxin. When induced or injected systemically, TNFα exerts a dramatic influence on the liver, stimulating production of other cytokines, C-reactive protein, and serum amyloid A; it is also a major endogenous pyrogen. In addition, TNFs exert more localized effects: Expression within a given tissue (often by resident dendritic cells, for TNFα) causes changes in the endothelium of local capillaries, making them more attractive to circulating leukocytes. This results from a specific induction of appropriate cell adhesion proteins by TNFs and culminates in infiltration by neutrophils and T cells, with TNFs providing direct or indirect chemotactic signals.

Perhaps the most commonly cited activity of TNFs is the induction of cell death. Distinctions between necrotic and apoptotic death have generated much discussion, but similarities in the signals that can alternatively induce necrosis and apoptosis have blurred this difference (e.g., Bonfoco et al., 1995). In this regard, TNFα has been interpreted to induce both forms of cell death (Reid et al., 1989) or an atypical form of death that has characteristics of both apoptosis and necrosis (Fady et al., 1995). Of course, there are appropriate forms of cell death triggered by diffusible signals; activation-driven T-cell suicide has been considered an important role of Fas (Russell and Wang, 1993). However, TNF-receptor knockout mice have revealed no such roles in autoimmune suppression for TNFs. TNF cytotoxicity requires the presence of a macromolecular synthesis inhibitor, such as cycloheximide, in most cell types. This may reveal more about the cells being treated than about TNF. Many cell types appear programmed to undergo apoptosis if they receive an incomplete mitogenic stimulus; this scenario has been offered as a hypothetical defense against neoplasia and has been confirmed in cases in which cells are forced to overexpress a single oncogene, like *c-myc* (Evan et al., 1992; Shi et al., 1992). Therefore, it is not surprising that cells prevented from entering the cell division cycle by cycloheximide would enter a suicide program when they receive a TNF stimulus. The somewhat selective toxicity of TNFs has relevance to the original discoveries of tumor suppression by TNFα. In addition, it stands as an important caution in the interpretation of results obtained with TNFs in neuroblastoma cell lines or any other transformed cell.

TNFs and their receptors are expressed widely throughout development (Yamasu et al., 1989; Gendron et al., 1991; Kohchi et al., 1994; Pokholok et al., 1995), suggesting an important role in development of many tissues. However, functional ablation of TNFs in transgenic mice reveals no obvious developmental role (Kolls et al., 1994), and knockout of TNF-R1 causes only subtle effects on hematopoetic progenitor cell numbers (Zhang et al., 1995). The most obvious beneficial role for TNFs revealed in knockout mice is in defense against infection; in mice that lack TNF-R1, *Listeria monocytogenes* and *Mycobacterium tuberculosis* infections are uncontrolled (Rothe et al., 1993). These data suggest that TNFs function most importantly in the mature immune system. Indeed, an effect of TNFα on preimplantation embryo development (Tartakovsky and Ben-Yair, 1991) could be mediated by effects on the uterus rather than on the embryo itself (Hunt, 1993). Disruption of the TNFβ gene results in agenesis of the lymph nodes (De Togni et al., 1994). However, given the results of receptor knockout studies, this effect on lymph node development probably relies on the interaction between TNFβ and LTβ, a complex that binds a receptor distinct from TNF-R1 and -R2 (above).

4.1. TNF and Brain Injury Responses

Along with the many stress-related responses registered in the injured brain, TNFα can be increased as much as 10-fold in various injury conditions (Table 1). In most models, TNFα is one of the earliest cytokines induced, detectable as early as 1 h after the insult, and peaking between 4 and 12 h; however, it may be sustained as long as 5 d (Liu et al., 1994). Signals that inhibit the production of TNFα, such as prostaglandin E, IL-6 (Benveniste et al., 1995; Chao et al., 1995a), and TGFβ (Rimaniol et al., 1995; Wang et al., 1995), are also induced by brain injury. However, the kinetics are slower for these factors (on the order of days) and are generally consistent with the decline in TNFα levels, suggesting that such inhibitory signals are responsible for limiting the brain's exposure to a potentially harmful factor (below). In some cases, TNFα may be inducing its own negative regulators to effect a negative feedback loop (Aloisi et al., 1992; Chao et al., 1995b; Maerz et al., 1996). TNFα also has been reported to increase with aging (Siren et al., 1993) and aging-related diseases (Dickson et al., 1993; Mogi et al., 1994). However, its particular elevation in Alzheimer's disease appears to be highly variable (Selmaj et al., 1991; Cacabelos et al., 1994; Huberman et al., 1994).

The induction of TNFα in response to brain injury doubtlessly initiates secondary elevations of enzymes, structural proteins, additional cytokines, and other signaling molecules. One thing it probably does not contribute is an additional increase in glutamate (Zornow et al., 1993). Nevertheless, TNFα has been reported to exacerbate glutamate toxicity in human brain cultures (Gelbard et al., 1993; Chao and Hu, 1994). In some species, this effect appears to reflect the ability of TNFα to induce nitric oxide synthase (NOS) in microglia (Chao et al., 1993), causing increased production of neurotoxic nitric oxide (NO). However, NOS induction by cytokines is insubstantial in human glia (Peterson et al., 1994; Walker et al., 1995). Microglia have also been shown to elevate their production of the excitotoxin quinolinic acid in the presence of TNFα (Saito et al., 1993), but this only occurs when TNFα is combined with another stimulus, such as interferon-γ or endotoxin, factors not expected to be present in the absence of infection. The dramatic neurotoxicity of TNFα reported in the above studies may have more relevance in culture conditions than in vivo, as TNFα injections into rodent brains cause only temporary leukocyte invasion and no parenchymal damage or behavioral abnormalities (Wright and Merchant, 1992; Yamasaki et al., 1992).

Table 1
Brain Injury Conditions in Which TNFα Is Elevated

	Ref.
Clinical ischemia	Fassbender et al., 1994
Clinical trauma	Ross et al., 1994
Experimental ischemia	Liu et al., 1994; Szaflarski et al., 1995; Wang et al., 1994; Buttini et al., 1996; Saito et al., 1996
Experimental closed-head trauma	Shohami et al., 1994
Fluid percussion trauma	Taupin, 1993; Hayes et al., 1995; Fan et al., 1996
Ibotenic acid	Alvarez et al., 1994
Kainic acid	Minami et al., 1991
Radiation	Chiang and McBride, 1991
Surgical hippocampal lesion	Tchelingerian et al., 1993

If induction of TNFα by brain injury does have acute detrimental effects mediated by glia, they appear to be offset by the direct effects of TNFs on neurons. Cheng et al. (1994) first reported a positive influence of both TNFα and TNFβ on survival of cultured neurons. Hippocampal, cerebral cortical, and septal neurons survived the challenges of glucose deprivation and excitotoxicity at a higher rate after treatment with the TNFs. The suppression of glial growth in these cultures had no influence on the neuroprotective actions of TNFs, indicating direct action on the neurons themselves. Such neuroprotective effects may result in part from the elevation of expression of the Ca-binding protein calbindin. Calcium mediates a large component of the toxicity associated with both excitoxicity and glucose deprivation, and calbindin-expressing cells show greater survival rates in the face of calcium overloads (Mattson et al., 1991; Diop et al., 1995; Peterson et al., 1996). The TNFs induce calbindin expression in astrocytes (Mattson et al., 1995) and markedly increase the number of neurons with detectable expression (Cheng et al., 1994). Obviously, one mechanism by which agents can protect neurons against excitotoxic challenges is by decreasing the responses to glutamatergic neurotransmission. Long-term potentiation in hippocampal slices, a phenomenon exquisitely linked to glutamatergic transmission, is inhibited by TNFα when applied for greater than 50 min (Tancredi et al., 1992); this could be related to the ability of calbindin to inhibit postsynaptic potentiation (Chard et al., 1995).

Both TNFα and TNFβ also protect cultured neurons from the toxicity of the Alzheimer's amyloid β-peptide (Barger et al., 1995). This is significant because the toxicity of amyloid β-peptide involves ROS (Behl et al., 1994; Goodman and Mattson, 1996), as well as disruption of Ca regulation (Mattson et al., 1992; Mark et al., 1995). Consistent with this relationship, TNFs also reduce neuronal toxicity associated with iron treatment, and cellular peroxides evoked by iron are diminished by TNF treatment (Barger et al., 1995). This increased antioxidant capacity requires several hours of pretreatment with TNFs, suggesting a requirement for gene expression. It is likely that Mn-SOD and/or other antioxidants are induced in neurons by TNFs through activation of NFκB. This hypothesis is supported by studies using antisense oligonucleotides against IκBα to cause pharmacomolecular disinhibition of NFκB (Barger and Mattson, 1996b). Activation of κB-dependent transcription by this means results in protection of neurons against glutamate (Barger and Mattson, 1996b) and amyloid β-peptide (Barger et al., 1995), and it decreases peroxide production in neurons. IκBα antisense treatment also squelches the Ca^{2+} elevations evoked by glutamate and amyloid β-peptide (Barger et al., 1995; Barger and Mattson, 1996b). New data provides further support for the role of κB-dependent transcription in the neuroprotective actions of TNFα. Figure 2 depicts the ability of TNFα to ameliorate the death of cultured hippocampal neurons that resulted from glucose deprivation. Additional cultures were pretreated with double-stranded oligodeoxynucleotides corresponding to the κB enhancer sequence to act as a sink for any activated κB-binding transcription factors; this treatment has been determined empirically to markedly diminish the κB-binding activity activated by TNFα in these cultures. In the presence of this inhibitor, TNFα no longer promoted survival, consistent with the hypothesis that κB-dependent transcription plays a major role in this neuroprotective paradigm.

One possible explanation for the discrepancy between studies showing neurotoxicity of TNFs and those demonstrating neuroprotection is the species difference. Until recently, neuroprotective effects of TNFs were reported in studies utilizing only human TNFs on rodent cultures, but studies in which neurotoxicity was observed often utilized human TNF on human cultures. Human TNFα does not activate the rodent TNF-R2 efficiently. Therefore, the possibility remained that TNF's toxicity in human neuronal cultures resulted from more efficient activation of TNF-R2, although this would be counterintuitive, given the connections of TNF-R1 to cytotoxicity. Nevertheless, this possibility has been ruled out, because rat and mouse TNFαs have been demonstrated to confer neuroprotection in intraspecies cultures (Fig. 2; Bruce et al., 1996). This appears to leave the possibility that the diversity in reactions of neural cultures to TNFs results from variability in the numbers of glia present, especially given the hypothesis proposed by Chao et al. (1993) that microglia mediate TNF neurotoxicity.

Considering the direct neuroprotective actions of TNFs, contrasted to the apparently indirect microglia-mediated neurotoxicity reported in other studies, it would be difficult to predict the ultimate integration of these effects in vivo. Bruce et al. (1996) have now addressed this question empirically. Mice in which the TNF-R1 and TNF-R2 genes each had been independently disrupted were back-crossed to yield individuals devoid of receptors for the homotrimeric TNFs. No abnormalities in development or basic neurological behavior have been observed in these double knockouts. However, there are revealing alterations in responses to CNS injury in these mice. Transient occlusion of the middle cerebral artery to effect focal ischemia causes a larger cortical infarct area in the double knockout mice. In addition, kainic acid injections into the dorsal hippocampus also evokes more severe damage in the mice lacking TNF receptors, despite the fact that this lesion elevates TNFα levels to a similar degree in both knockout and wild-type mice. In contrast, kainic acid elevates Mn-SOD immunoreactivity only in the wild-type mice, consistent with the hypothesis that TNFα enhances antioxidant capacity (above). Systemically beneficial effects of TNFα may include the induction of vasodilatory C-type natriuretic peptide (CNP) (Suga et al., 1993) or a glucose transporter (Pekala

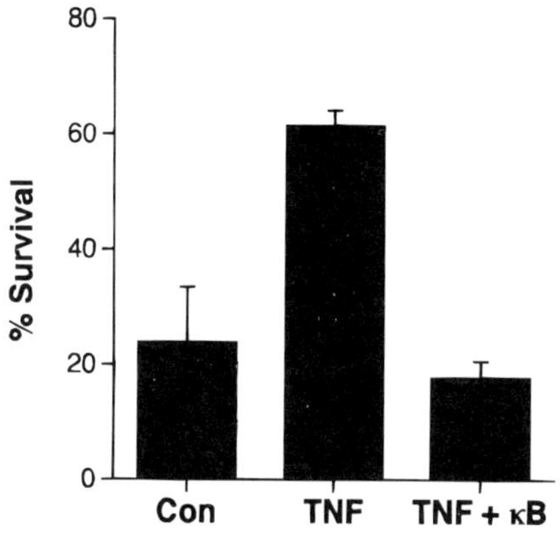

Fig. 2. TNF neuroprotection is dependent on the activity of κB-binding transcription factors. Rat primary hippocampal neurons (7 d in culture) were treated for 24 h with a control BSA solution (Con), 100 ng/mL rat TNFα (TNF; Peprotech), or rat TNFα added after a 3-h pretreatment with 10 μ*M* of a double-stranded oligonucleotide containing the κB enhancer sequence (TNF + κB). The cultures were then deprived of glucose for 18 h. The percentage of original neurons surviving the glucose deprivation was determined in triplicate as described (Cheng et al., 1994) and is depicted ± SEM. The κB enhancer sequence competes for the binding of κB-binding transcription factors to genomic DNA. No cell death resulted from nonspecific oligonucleotide toxicity (Barger and Mattson, 1996b).

et al., 1990; Boado et al., 1994) in vascular endothelial cells. However, the enhanced vulnerability resulting from disruption of TNF signaling persists in culture preparations from the TNF receptor-knockout mice (Bruce et al., 1996), demonstrating that at least much of the protection associated with TNF signaling results from a direct effect on CNS parenchyma. (Indeed, endogenous TNFα expression may play an important role in basal neuronal survival rates in culture, as anti-TNFα antibody dramatically reduced the survival of wild-type neuronal cultures to rates seen in cultures from the TNF-receptor knockout mice.)

The studies performed with the TNF-receptor knockout mice are compelling evidence that TNF ameliorates damage from at least acute CNS injury. However, Shohami et al. (1996) used a different model and reached the opposite conclusion. In the latter study, injections of either pentoxifylline or soluble TNF receptor lessened the cortical damage resulting from closed-head trauma. There are several possible explanations for the inconsistency of the two studies. First, pentoxifylline is of questionable utility in this application because it is cerebroprotective by virtue of prostaglandin-dependent vasodilatory activity (Noskovic et al., 1995; Tanahashi et al., 1995), inhibits the production of other cytokines, such as interleukin-1β (Weinberg et al., 1992), and has even been shown to elevate TNFα production (Reed and DeGowin, 1992). Alternatively, it is possible that closed-head trauma has a lower component of excitotoxic/oxidative damage than the ischemia and kainic acid models. Of course, the ever-present caveat of transgenic experiments applies to the TNF-receptor knockout study: Disruption of TNF signaling during development may have created a confounding aberrancy in adult injury responses in the knockouts. However, this seems unlikely because of the normal anatomy and behavior of the TNF-receptor knockout mice in examinations that were more than cursory (e.g., synaptophysin analysis, Morris water maze).

4.2. *Demyelination and TNF*

Regardless of any neuroprotective actions of TNF in acute brain injury, one must always avoid too much of a good thing. Chronic TNFα elevation has been proposed to play a role in demyelinating diseases. It is elevated in amyotrophic lateral sclerosis, and both TNFα and TNFβ are elevated in multiple sclerosis (Selmaj et al., 1991). Demyelination has been proposed to result from death of oligodendrocytes because they are sensitive to TNFα, at least in culture. Oligodendrocytes have been reported to express only TNF-R2 in vivo; it was postulated that their expression of TNF-R1 when cultured was responsible for their susceptibility to TNFα (Tchelingerian et al., 1995). Nevertheless, TNFα antibodies or soluble receptor can ameliorate experimental autoimmune encephalomyelitis, one of the leading animal models for multiple sclerosis (Selmaj and Raine, 1995). Furthermore, TNFα transgenic mice show a spontaneous demyelination (Probert et al., 1995). It is unclear whether these results stem from a direct effect of TNFα on oligodendrocytes or whether more elaborate immune interactions are responsible. TNFα can actually protect against demyelination in a virus-induced encephalomyelitis model (Paya et al., 1990).

It is important to consider that demyelination may not always be a negative event. Myelin seems to participate in the suppression of axonal regeneration in the CNS. Therefore, demyelination may be seen as simply another component of TNF's injury responses repertoire (Schneider-Schaulies et al., 1991). It has been noted that the TNF-R1 death domain is homologous to proteins involved in axonal guidance (Hofmann and Tschopp, 1995). A role for TNFα in the re-establishment of connections is supported by its influence on the outgrowth of neurites from primary hippocampal neurons (Fig. 3). Relatedly, PC12 cells adhere better to ex vivo sections of optic nerve if the nerves have been treated in vivo with TNFα (Schwartz et al., 1994). Given the dramatic influence of TNFs on the expression of cell adhesion molecules (CAMs) and other cell-surface proteins, it would not be surprising if TNFs induce factors that facilitate neurite attachment and outgrowth. Again, an excess of this activity could also be detrimental; in addition to the often mentioned possibility that TNF initiates or aggravates demyelination, aberrant sprouting and growth of neuritic processes could give rise to the dystrophic neurites present in Alzheimer's disease in particular.

5. CONCLUSIONS: THERAPEUTIC OPPORTUNITIES

TNFs are pleiotrophic factors expressed or inducible in many tissues, including the brain. Although maligned as the mediators of fever, shock, and necrotic tissue damage, they definitely can have beneficial effects on the survival of neurons. This role seems intuitively consistent with the well-characterized responsiveness of TNFα to brain injury. Nevertheless, it would be extremely risky to design therapies around treatment with TNFs themselves because of their diverse and sometimes unpredictable actions. The goal now should be to tease apart TNF signal transduction, so that the distinctions that dictate a cell's response to TNF can be identified and exploited for therapeutic advantage. For instance, the activation of transcription factors that bind κB enhancer

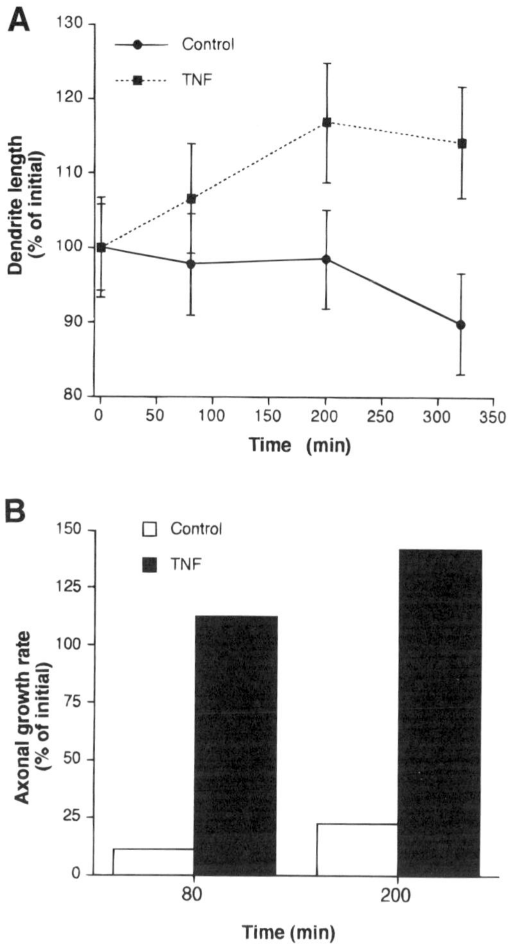

Fig. 3. Effects of TNFα on neurite outgrowth. **(A)** Rat primary hippocampal neurons (3 d in culture) were photographed at the intervals depicted after addition of control BSA or 30 ng/mL rat TNFα. Average dendrite length was measured in 15 cells/treatment and is depicted as percentage of the initial length ± SD. Note that control dendrites did not grow during the observation time, but TNFα-treated dendrites significantly elongated. **(B)** Axons were measured in the same neurons before treatment and at 80 and 200 min after treatment. Growth rate was calculated from elapsed time between photographs and is depicted as a percentage of the growth rate of the axons before treatment. Control axons slowed dramatically once photography began (perhaps because of the adverse effects of pH changes as the bicarbonate culture medium was exposed to ambient air). However, axon growth rate increased in the presence of TNFα.

sequences appears to account for most, if not all, of TNFα's neuroprotective activity. In addition, ceramide already has been shown to enhance neuronal survival (Goodman and Mattson, 1996); integration of this finding with a recent report of superior cell survival stimulated by metabolites of ceramide (Cuvillier et al., 1996) should lead pharmacologists down promising paths.

REFERENCES

Adam, D., Kessler, U., and Kronke, M. (1995) Cross-linking of the p55 tumor necrosis factor receptor cytoplasmic domain by a dimeric ligand induces nuclear factor-κB and mediates cell death. *J. Biol. Chem.* **270,** 17,482–17,487.

Aloisi, F., Care, A., Borsellino, G., Gallo, P., Rosa, S., Bassani, A., et al. (1992) Production of hemolymphopoietic cytokines (IL-6, IL-8, colony-stimulating factors) by normal human astrocytes in response to IL-1β and tumor necrosis factor-α. *J. Immunol.* **149,** 2358–2366.

Alvarez, X. A., Franco, A., Fernandez-Novoa, L., and Cacabelos, R. (1994) Effects of neurotoxic lesions in histaminergic neurons on brain tumor necrosis factor levels. *Agents & Actions* **41,** C70–72.

Barger, S. W. and Mattson, M. P. (1996a) Induction of neuroprotective κB-dependent transcription by secreted form of the Alzheimer's β-amyloid precursor. *Mol. Brain Res.* **40,** 116–126.

Barger, S. W. and Mattson, M. P. (1996b) Participation of gene expression in the protection against amyloid β-peptide toxicity by the β-amyloid precursor protein. *Ann. N.Y. Acad. Sci.* **777,** 303–308.

Barger, S. W., Hörster, D., Furukawa, K., Goodman, Y., Kriegelstein, J., and Mattson, M. P. (1995) Tumor necrosis factors α and β protect neurons against amyloid β-peptide toxicity: Evidence for involvement of a κB-binding factor and attenuation of peroxide and Ca^{2+} accumulation. *Proc. Natl. Acad. Sci. USA* **92,** 9328–9332.

Bazzoni, F., Alejos, E., and Beutler, B. (1995) Chimeric tumor necrosis factor receptors with constitutive signaling activity. *Proc. Natl. Acad. Sci. USA* **92,** 5376–5380.

Behl, C., Davis, J. B., Lesley, R., and Schubert, D. (1994) Hydrogen peroxide mediates amyloid β protein toxicity. *Cell* **77,** 817–827.

Benveniste, E. N., Tang, L. P., and Law, R. M. (1995) Differential regulation of astrocyte TNF-α expression by the cytokines TGF-β, IL-6 and IL-10. *Int. J. Dev. Neurosci.* **13,** 341–349.

Betts, J. C., Agranoff, A. B., Nabel, G. J., and Shayman, J. A. (1994) Dissociation of endogenous cellular ceramide from NF-κB activation. *J. Biol. Chem.* **269,** 8455–8458.

Beutler, B. (1995) TNF, immunity and inflammatory disease: lessons of the past decade. *J. Invest. Med.* **43,** 227–235.

Bigda, J., Beletsky, I., Brakebusch, C., Varfolomeev, Y., Engelmann, H., Bigda, J., Holtmann, H., and Wallach, D. (1994) Dual role of the p75 tumor necrosis factor (TNF) receptor in TNF cytotoxicity. *J. Exp. Med.* **180,** 445–460.

Boado, R. J., Wang, L., and Pardridge, W. M. (1994) Enhanced expression of the blood-brain barrier GLUT1 glucose transporter gene by brain-derived factors. *Mol. Brain Res.* **22,** 259–267.

Boldin, M. P., Goncharov, T. M., Goltsev, Y. V., and Wallach, D. (1996) Involvement of MACH, a novel MORT1/FADD-interacting protease, in Fas/APO-1- and TNF receptor-induced cell death. *Cell* **85,** 803–815.

Bonfoco, E., Krainc, D., Ankarcrona, M., Nicotera, P., and Lipton, S. A. (1995) Apoptosis and necrosis: two distinct events induced, respectively, by mild and intense insults with N-methyl-D-aspartate or nitric oxide/superoxide in cortical cell cultures. *Proc. Natl. Acad. Sci. USA* **92,** 7162–7166.

Bradley, J. R., Thiru, S., and Pober, J. S. (1995) Disparate localization of 55-kd and 75-kd tumor necrosis factor receptors in human endothelial cells. *Am. J. Pathol.* **146,** 27–32.

Browning, J. L., Ngam-ek, A., Lawton, P., DeMarinis, J., Tizard, R., Chow, E. P., et al. (1993) Lymphotoxin β, a novel member of the TNF family that forms a heteromeric complex with lymphotoxin on the cell surface. *Cell* **72,** 847–856.

Bruce, A. J., Boling, W., Kindy, M. S., Peschon, J., Kraemer, P. J., Carpenter, M. K., Holtsberg, F. W., and Mattson, M. P. (1996) Altered neuronal and microglial responses to excitotoxic and ischemic brain injury in mice lacking TNF receptors. *Nature Med.* **2,** 788–794.

Buttini M., Appel, K., Sauter, A., Gebicke-Haerter, P. J., and Boddeke, H. W. G. M. (1996) Expression of tumor necrosis factor α after focal cerebral ischaemia in the rat. *Neuroscience* **71,** 1–16.

Cacabelos, R., Alvarez, X. A., Franco-Maside, A., Fernandez-Novoa, L., and Caamano, J. (1994) Serum tumor necrosis factor (TNF) in Alzheimer's disease and multi-infarct dementia. *Methods Find. Exp. Clin. Pharmacol.* **16,** 25–35.

Chao, C. C. and Hu, S. (1994) Tumor necrosis factor-α potentiates glutamate neurotoxicity in human fetal brain cell cultures. *Dev. Neurosci.* **16,** 172–179.

Chao, C. C., Molitor, T. W., and Hu, S. (1993) Neuroprotective role of IL-4 against activated microglia. *J. Immunol.* **151,** 1473–1481.

Chao, C. C., Hu, S., Sheng, W. S., Peterson, P. K. (1995a) Tumor necrosis factor-α production by human fetal microglial cells: Regulation by other cytokines. *Devel. Neurosci.* **17,** 97–105.

Chao, C. C., Hu, S. X., Sheng, W. S., Tsang, M., and Peterson, P. K. (1995b) Tumor necrosis factor-α mediates the release of bioactive transforming growth factor-β in murine microglial cell cultures. *Clin. Immunol. Immunopath.* **77,** 358–365.

Chard, P. S., Jordan, J., Marcuccilli, C. J., Miller, R. J., Leiden, J. M., Roos, R. P, Ghadge, G. D. (1995) Regulation of excitatory transmission at hippocampal synapses by calbindin D28k. *Proc. Natl. Acad. Sci. USA* **92,** 5144–5148.

Cheng, B., Christakos, S., and Mattson, M. P. (1994) Tumor necrosis factors protect neurons against metabolic-excitotoxic insults and promote maintenance of calcium homeostasis. *Neuron* **12,** 139–153.

Chiang, C. S. and McBride, W. H. (1991) Radiation enhances tumor necrosis factor α production by murine brain cells. *Brain Res.* **566,** 265–269.

Chung, I. Y., Norris, J. G., and Benveniste, E. N. (1991) Differential tumor necrosis factor α expression by astrocytes from experimental allergic encephalomyelitis-susceptible and -resistant rat strains. *J. Exp. Med.* **173,** 801–811.

Cifone, M. G., Roncaioli, P., De Maria, R., Camqrda, G., Santoni, A., Ruberti, G., and Testi, R. (1995) Multiple pathways originate at the Fas/APO-1 (CD95) receptor: Sequential involvement of phosphatidylcholine-specific phospholipase C and acidic sphingomyelinase in the propagation of the apoptotic signal. *EMBO J.* **14,** 5859–5868.

Crowe, P. D., VanArsdale, T. L., Walter, B. N., Ware, C. F., Hession, C., Ehrenfels, B., et al. (1994) A lymphotoxin-—specific receptor. *Science* **264,** 707–710.

Cuvillier, O., Pirianov, G., Kleuser, B., Vanek, P. G., Coso, O. A., Gutkind, J. S., and Spiegel, S. (1996) Suppression of ceramide-mediated programmed cell death by sphingosine-1-phosphate. *Nature* **381,** 800–803.

Davis, S., Aldrich, T. H., Ip, N. Y., Stahl, N., Scherer, S., Farruggella, T., et al. (1993) Released form of CNTF receptor α component as a soluble mediator of CNTF responses. *Science* **259,** 1736–1739.

Dbaibo, G. S., Obeid, L. M., and Hannun, Y. A. (1993) Tumor necrosis factor-α (TNF-α) signal transduction through ceramide. Dissociation of growth inhibitory effects of TNF-α from activation of nuclear factor-κB. *J. Biol. Chem.* **268,** 17,762–17,766.

De Togni, P., Goellner, J., Ruddle, N. H., Streeter, P. R., Fick, A., Mariathasan, S., et al. (1994) Abnormal development of peripheral lymphoid organs in mice deficient in lymphotoxin. *Science* **264,** 703–707.

Dickson, D. W., Lee, S. C., Mattiace, L. A., Yen, S. H., and Brosnan, C. (1993) Microglia and cytokines in neurological disease, with special reference to AIDS and Alzheimer's disease. *Glia* **7,** 75–83.

Diop, A. G., Lesort, M., Esclaire, F., Dumas, M., and Hugon, J. (1995) Calbindin D28K-containing neurons, and not HSP70-expressing neurons, are more resistant to HIV-1 envelope (gp120) toxicity in cortical cell cultures. *J. Neurosci. Res.* **42,** 252–258.

Enari, M., Hug, H., and Nagata, S. (1995) Involvement of an ICE-like protease in Fas-mediated apoptosis. *Nature* **375,** 78–81.

English, B. K., Weaver, W. M., and Wilson, C. B. (1991) Differential regulation of lymphotoxin and tumor necrosis factor genes in human T lymphocytes. *J. Biol. Chem.* **266,** 7108–7113.

Evan, G. I., Wyllie, A. H., Gilbert, C. S., Littlewood, T. D., Land, H., Brooks, M., et al. (1992) Induction of apoptosis in fibroblasts by c-myc protein. *Cell* **69,** 119–128.

Fady, C., Gardner, A., Jacoby, F., Briskin, K., Tu, Y., Schmid, I., and Lichtenstein, A. (1995) Atypical apoptotic cell death induced in L929 targets by exposure to tumor necrosis factor. *J. Interferon Cytokine Res.* **15,** 71–80.

Fan, L., Young, P. R., Barone, F. C., Feuerstein, G. Z., Smith, D. H., and McIntosh, T. K. (1996) Experimental brain injury induces differential expression of tumor necrosis factor-α mRNA in the CNS. *Mol. Brain Res.* **36,** 287–291.

Fassbender, K., Rossol, S., Kammer, T., Daffertshofer, M., Wirth, S., Dollman, M., and Hennerici, M. (1994) Proinflammatory cytokines in serum of patients with acute cerebral ischemia: kinetics of secretion and relation to the extent of brain damage and outcome of disease. *J. Neurol. Sci.* **122,** 135–139.

Gearing, A. J., Beckett, P., Christodoulou, M., Churchill, M., Clements J., Davidson, A. H., J. L., et al. (1994) Processing of tumour necrosis factor-α precursor by metalloproteinases. *Nature* **370,** 555–557.

Gelbard, H. A., Dzenko, K. A., DiLoreto, D., del Cerro, C., del Cerro, M., and Epstein, L. G. (1993) Neurotoxic effects of tumor necrosis factor α in primary human neuronal cultures are mediated by activation of the glutamate AMPA receptor subtype: implications for AIDS neuropathogenesis. *Dev. Neurosci.* **15,** 417–422.

Gendron, R. L., Nestel, F. P., Lapp, W. S., and Baines, M. G. (1991) Expression of tumor necrosis factor α in the developing nervous system. *Int. J. Neurosci.* **60,** 129–136.

Goodman, Y. and Mattson M. P. (1996) Ceramide protects hippocampal neurons against excitotoxic and oxidative insults, and amyloid β-peptide toxicity. *J. Neurochem.* **66,** 869–872.

Grell, M., Douni, E., Wajant, H., Lohden, M., Clauss, M., Maxeiner, B., et al. (1995) The transmembrane form of tumor necrosis factor is the prime activating ligand of the 80 kDa tumor necrosis factor receptor. *Cell 83,* 793–802.

Han, J., Brown, T., and Beutler, G. (1990) Endotoxin-responsive sequences control cachectin/TNF biosynthesis at the translational level. *J. Exp. Med.* **171,** 465–475.

Hayes, R. L., Yang, K., Raghupathi, R., and McIntosh, T. K. (1995) Changes in gene expression following traumatic brain injury in the rat. *J. Neurotrauma* **12,** 779–790.

Heller, R. A., Song, K., Fan, N., and Chang, D. J. (1992) The p70 tumor necrosis factor receptor mediates cytotoxicity. *Cell* **70,** 47–56.

Hofmann, K. and Tschopp, J. (1995) The death domain motif found in Fas (Apo-1) and TNF receptor is present in proteins involved in apoptosis and axonal guidance. *FEBS Lett.* **371,** 321–323.

Howard, O. M., Clouse, K. A., Smith, C., Goodwin, R. G., and Farrar, W. L. (1993) Soluble tumor necrosis factor receptor: inhibition of human immunodeficiency virus activation. *Proc. Natl. Acad. Sci. USA* **90,** 2335–2339.

Hsu, H., Xiong, J., and Goeddel, D. V. (1995) The TNF receptor 1-associated protein TRADD signals cell death and NF-κB activation. *Cell* **81,** 495–504.

Hsu, H. L., Shu, H. B., Pan, M. G., and Goeddel, D. V. (1996) TRADD-TRAF2 and TRADD-FADD interactions define two distinct TNF receptor 1 signal transduction pathways. *Cell* **84,** 299–308.

Huberman, M., Shalit, F., Roth-Deri, I., Gutman, B., Brodie, C., Kott, E., and Sredni, B. (1994) Correlation of cytokine secretion by mononuclear cells of Alzheimer patients and their disease stage. *J. Neuroimmunol.* **52,** 147–152.

Hunt, J. S. (1993) Expression and regulation of the tumour necrosis factor-α gene in the female reproductive tract. *Reprod. Fertil. Dev.* **5,** 141–153.

Jayadev, S., Linardic, C. M., and Hannun, Y. A. (1994) Identification of arachidonic acid as a mediator of sphingomyelin hydrolysis in response to tumor necrosis factor α. *J. Biol. Chem.* **269,** 5757–5763.

Johns, L. D., Sarr, T., and Ranges, G. E. (1994) Inhibition of ceramide pathway does not affect ability of TNF-α to activate nuclear factor-κB. *J. Immunol.* **152,** 5877–5882.

Kohchi, C., Noguchi, K., Tanabe, Y., Mizuno, D., and Soma, G. (1994) Constitutive expression of TNF-α and -β genes in mouse embryo: roles of cytokines as regulator and effector on development. *Int. J. Biochem.* **26,** 111–119.

Kolesnick, R. N. (1987) 1,2-Diacylglycerols but not phorbol esters stimulate sphingomyelin hydrolysis in GH3 pituitary cells. *J. Biol. Chem.* **262,** 16,759–16,762.

Kolls, J., Peppel, K., Silva, M., and Beutler, B. (1994) Prolonged and effective blockade of tumor necrosis factor activity through adenovirus-mediated gene transfer. *Proc. Natl. Acad. Sci. USA* **91,** 215–219.

Korner, M., Tarantino, N., Pleskoff, O., Lee, L. M., and Debre, P. (1994) Activation of nuclear factor κB in human neuroblastoma cell lines. *J. Neurochem.* **62,** 1716–1726.

Liu, T., Clark, R. K., McDonnell, P. C., Young, P. R., White, R. F., Barone, F. C., and Feuerstein, G. Z. (1994) Tumor necrosis factor-α expression in ischemic neurons. *Stroke* **25,** 1481–1488.

Macchia, D., Almerigogna, F., Parronchi, P., Ravina, A., Maggi, E., and Romagnani, S. (1993) Membrane tumour necrosis factor-α is involved in the polyclonal B-cell activation induced by HIV-infected human T cells. *Nature* **363,** 464–466.

Maerz, P., Gadient, R. A., and Otten, U. (1996) Expression of interleukin-6 receptor (IL-6R) and gp130 mRNA in PC12 cells and sympathetic neurons: modulation by tumor necrosis factor α (TNF-α). *Brain Res.* **706,** 71–79.

Mark, R. J., Hensley, K., Butterfield, D. A., and Mattson, M. P. (1995) Amyloid β-peptide impairs ion-motive ATPase activities: evidence for a role in loss of neuronal Ca^{2+} homeostasis and cell death. *J. Neurosci.* **15,** 6239–6249.

Mattson, M. P., Rychlik, B., and Christakos, S. (1991) Evidence for calcium-reducing and excitoprotective roles for the calcium binding protein (calbindin-D28k) in cultured hippocampal neurons. *Neuron* **6,** 41–51.

Mattson, M. P., Cheng, B., Davis, D., Bryant, K., Lieberburg, I., and Rydel, R. E. (1992) β-Amyloid peptides destabilize calcium homeostasis and render human cortical neurons vulnerable to excitotoxicity. *J. Neurosci.* **12,** 379–389.

Mattson, M. P., Cheng, B., Baldwin, S. A., Smith-Swintosky, V. L., Keller, J., Geddes, J. W., Scheff, S. W., and Christakos, S. (1995) Brain injury and tumor necrosis factors induce calbindin D-28k in astrocytes: evidence for a cytoprotective response. *J. Neurosci. Res.* **42,** 357–370.

Medvedev, A. E., Espevik, T., Ranges, G., and Sundan, A. (1996) Distinct roles of the two tumor necrosis factor (TNF) receptors in modulating TNF and lymphotoxin α effects. *J. Biol. Chem.* **271,** 9778–9784.

Medvedev, A. E., Sundan, A., and Espevik, T. (1994) Involvement of the tumor necrosis factor receptor p75 in mediating cytotoxicity and gene regulating activities. *Eur. J. Immunol.* **24,** 2842–2849.

McGeehan, G. M., Becherer, J. D., Bast Jr., R. C, Boyer, C. M., Champion, B., Connolly, K. M., et al. (1994) Regulation of tumour necrosis factor-α processing by a metalloproteinase inhibitor. *Nature* **370,** 558–561.

Messer, G., Weiss, E. H., and Baeuerle, P. A. (1990) Tumor necrosis factor β (TNF-β) induces binding of the NF-κB transcription factor to a high-affinity κB element in the TNF-β promoter. *Cytokine* **2,** 389–397.

Minami, M., Kuraishi, Y., and Satoh, M. (1991) Effects of kainic acid on messenger RNA levels of IL-1β, IL-6, TNFα and LIF in the rat brain. *Biochem. Biophys. Res. Commun.* **176,** 593–598.

Mogi, M., Harada, M., Riederer, P., Narabayashi, H., Fujita, K., and Nagatsu, T. (1994) Tumor necrosis factor-α (TNF-α) increases both in the brain and in the cerebrospinal fluid from parkinsonian patients. *Neurosci. Lett.* **165,** 208–210.

Muller, G., Ayoub, M., Storz, P., Rennecke, J., Fabbro, D., and Pfizenmaier, K. (1995) PKCζ is a molecular switch in signal transduction of TNF-α, bifunctionally regulated by ceramide and arachidonic acid. *EMBO J.* **14,** 1961–1969.

Muzio, M., Chinnaiyan, A. M., Kischkel, F. C., O'Rourke, K., Shevchenko, A., Ni, J. (1996) FLICE, a novel FADD-homologous ICE/CED-3-like protease, is recruited to the CD95 (Fas/APO-1) death-inducing signaling complex. *Cell* **85,** 817–827.

Noskovic, P., Faberova, V., and Fabianova, M. (1995) Effect of a combination of pentoxifylline and nimodipine on lipid peroxidation in postischemic rat brain. *Mol. Chem. Neuropathol.* **25,** 97–102.

Oyama, Y., Furukawa, K., Chikahisa, L., and Hatakeyama, Y. (1994) Effect of N,N-diethyldithiocarbamate on ionomycin-induced increase in oxidation of cellular 2′,7′-dichlorofluorescin in dissociated cerebellar neurons. *Brain Res.* **660,** 158–161.

Pang, X. P., Ross, N. S., Park, M., Juillard, G. J., Stanley, T. M., and Hershman, J. M. (1992) Tumor necrosis factor-α activates nuclear factor κB and induces manganous superoxide dismutase and phosphodiesterase mRNA in human papillary thyroid carcinoma cells. *J. Biol. Chem.* **267,** 12,826–12830.

Park, A. and Baichwal, V. R. (1996) Systematic mutational analysis of the death domain of the tumor necrosis factor receptor 1-associated protein TRADD. *J. Biol. Chem.* **271,** 9858–9862.

Paul, N. L., Lenardo, M. J., Novak, K. D., Sarr, T., Tang, W. L., and Ruddle, N. H. (1990) Lymphotoxin activation by human T-cell leukemia virus type I-infected cell lines: role for NF-κB. *J. Virol.* **64,** 5412–5419.

Paya, C. V., Leibson, P. J., Patick, A. K., and Rodriguez, M. (1990) Inhibition of Theiler's virus-induced demyelination in vivo by tumor necrosis factor α. *Int. Immunol.* **2,** 909–913.

Peck, R., Brockhaus, M., and Frey, J. R. (1989) Cell surface tumor necrosis factor (TNF) accounts for monocyte- and lymphocyte-mediated killing of TNF-resistant target cells. *Cell. Immunol.* **122,** 1–10.

Pekala, P., Marlow, M., Heuvelman, D., Connolly, D. (1990) Regulation of hexose transport in aortic endothelial cells by vascular permeability factor and tumor necrosis factor-α, but not by insulin. *J. Biol. Chem.* **265,** 18,051–18,054.

Pennica, D., Nedwin, G. E., Hayflick, J. S., Seeburg, P. H., Derynck, R., Palladino, M. A., et al. (1984) Human tumour necrosis factor: precursor structure, expression and homology to lymphotoxin. *Nature* **312,** 724–729.

Peterson, D. A., Lucidi-Phillipi, C. A., Murphy, D. P., Ray, J., and Gage, F. H. (1996) Fibroblast growth factor-2 protects entorhinal layer II glutamatergic neurons from axotomy-induced death. *J. Neurosci.* **16,** 886–898.

Peterson, P. K., Hu, S., Anderson, W. R., and Chao, C. C. (1994) Nitric oxide production and neurotoxicity mediated by activated microglia from human versus mouse brain. *J. Infect. Dis.* **170,** 457–460.

Pfeffer, K., Matsuyama, T., Kundig, T. M., Wakeham, A., Kishihara, K., Shahinian, A., et al. (1993) Mice deficient for the 55 kd tumor necrosis factor receptor are resistant to endotoxic shock, yet succumb to L. monocytogenes infection. *Cell* **73,** 457–467.

Pokholok, D. K., Maroulakou, I. G., Kuprash, D. V., Alimzhanov, M. B., Kozlov, S. V., Novobrantseva, T. I., et al. (1995) Cloning and expression analysis of the murine lymphotoxin β gene. *Proc. Natl. Acad. Sci. USA* **92,** 674–678.

Probert, L., Akassoglou, K., Pasparakis, M., Kontogeorgos, G., and Kollias, G. (1995) Spontaneous inflammatory demyelinating disease in transgenic mice showing central nervous system-specific expression of tumor necrosis factor α. *Proc. Natl. Acad. Sci. USA* **92,** 11294–11298.

Rao, P., Hsu, K. C., and Chao, M. V. (1995) Upregulation of NF-κB-dependent gene expression mediated by the p75 tumor necrosis factor receptor. *J. Interferon Cytokine Res.* **15,** 171–177.

Reddy, S. A., Chaturvedi, M. M., Darnay, B. G., Chan, H., Higuchi, M., and Aggarwal, B. B. (1994) Reconstitution of nuclear factor κB activation induced by tumor necrosis factor requires membrane-associated components. Comparison with pathway activated by ceramide. *J. Biol. Chem.* **269,** 25,369–25,372.

Reed, W. R. and DeGowin, R. L. (1992) Suppressive effects of pentoxifylline on natural killer cell activity. *J. Lab. Clin. Med.* **119,** 763–771.

Reid, T. R., Torti, F. M., and Ringold, G. M. (1989) Evidence for two mechanisms by which tumor necrosis factor kills cells. *J. Biol. Chem.* **264,** 4583–4589.

Rimaniol, A. C., Lekieffre, D., Serrano, A., Masson, A., Benavides, J., and Zavala, F. (1995) Biphasic transforming growth factor-β production flanking the pro-inflammatory cytokine response in cerebral trauma. *Neuroreport* **7,** 133–136.

Ross, S. A., Halliday, M. I., Campbell, G. C., Byrnes, D. P., and Rowlands, B. J. (1994) The presence of tumour necrosis factor in CSF and plasma after severe head injury. *Br. J. Neurosurg.* **8,** 419–425.

Rothe, J., Lesslauer, W., Lotscher, H., Lang, Y., Koebel, P., Kontgen, F., (1993) Mice lacking the tumour necrosis factor receptor 1 are resistant to TNF-mediated toxicity but highly susceptible to infection by Listeria monocytogenes. *Nature* **364,** 798–802.

Rothe, M., Wong, S. C., Henzel, W. J., and Goeddel, D. V. (1994) A novel family of putative signal transducers associated with the cytoplasmic domain of the 75 kDa tumor necrosis factor receptor. *Cell* **78,** 681–692.

Rothe, M., Pan, M. G., Henzel, W. J., Ayres, T. M., and Goeddel, D. V. (1995a) The TNFR2-TRAF signaling complex contains two novel proteins related to baculoviral-inhibitor of apoptosis proteins. *Cell* **83,** 1243–1252.

Rothe, M., Sarma, V., Dixit, V. W., and Goeddel, D. V. (1995b) TRAF2-mediated activation of NF-κB by TNF receptor 2 and CD40. *Science* **269,** 1424–1427.

Russell, J. H., and Wang, R. (1993) Autoimmune gld mutation uncouples suicide and cytokine/proliferation pathways in activated, mature T cells. *Eur. J. Immunol.* **23,** 2379–2382.

Saito, K., Chen, C. Y., Masana, M., Crowley, J. S., Markey, S. P., and Heyes, M. P. (1993) 4-Chloro-3-hydroxyanthranilate, 6-chlorotryptophan and norharmane attenuate quinolinic acid formation by interferon-β-stimulated monocytes (THP-1 cells). *Biochem. J.* **291,** 11–14.

Saito, K., Suyama, K., Nishida, K., Sei, Y., and Basile, A. S. (1996) Early increases in TNF-α, IL-6 and IL-1β levels following transient cerebral ischemia in gerbil brain. *Neurosci. Lett.* **206,** 149–152.

Sawada, M., Kondo, N., Suzumura, A., and Marunouchi, T. (1989) Production of tumor necrosis factor-α by microglia and astrocytes in culture. *Brain Res.* **491,** 394–397.

Schneider-Schaulies, J., Kirchhoff, F., Archelos, J., and Schachner, M. (1991) Downregulation of myelin-associated glycoprotein on Schwann cells by interferon-γ and tumor necrosis factor-α affects neurite outgrowth. *Neuron* **7,** 995–1005.

Schreck, R., Rieber, P., and Baeuerle, P. A. (1991) Reactive oxygen intermediates as apparently widely used messengers in the activation of the NF-κB transcription factor and HIV-1. *EMBO J.* **10,** 2247–2258.

Schreck, R., Albermann, K., and Baeuerle, P. A. (1992) Nuclear factor κB: an oxidative stress-responsive transcription factor of eukaryotic cells (a review). *Free Radic. Res. Commun.* **17,** 221–237.

Schutze, S., Potthoff, K., Machleidt, T., Berkovic, D., Wiegmann, K., and Kronke, M. (1992) TNF activates NF-κB by phosphatidylcholine-specific phospholipase C-induced "acidic" sphingomyelin breakdown. *Cell* **71,** 765–776.

Schwartz, M., Sivron, T., Eitan, S., Hirschberg, D. L., Lotan, M., and Elman-Faber, A. (1994) Cytokines and cytokine-related substances regulating glial cell response to injury of the central nervous system. *Prog. Brain Res.* **103,** 331–341.

Selmaj, K. W., and Raine, C. S. (1995) Experimental autoimmune encephalomyelitis: immunotherapy with anti-tumor necrosis factor antibodies and soluble tumor necrosis factor receptors. *Neurology* **45,** S44–49.

Selmaj, K., Raine, C. S., Cannella, B., and Brosnan, C. F. (1991) Identification of lymphotoxin and tumor necrosis factor in multiple sclerosis lesions. *J. Clin. Invest.* **87,** 949–954.

Shakhov, A. N., Collart, P. V., Vassalli, P., Nedospasov, S. A., and Jongeneel, C. V. (1990) κB-type enhancers are involved in lipopolysaccharide-mediated transcriptional activation of the tumor necrosis factor α gene in primary macrophages. *J. Exp. Med.* **171,** 35–47.

Shi, Y., Glynn, J. M., Guilbert, L. J., Cotter, T. G., Bissonnette, R. P., and Green, D. R. (1992) Role for c-myc in activation-induced apoptotic cell death in T cell hybridomas. *Science* **257,** 212–214.

Shohami, E., Novikov, M., Bass, R., Yamin, A., and Gallily, R. (1994) Closed head injury triggers early production of TNFα and IL-6 by brain tissue. *J. Cereb. Blood Flow Metab.* **14,** 615–619.

Shohami, E., Bass, R., Wallach, D., Yamin, A., and Gallily, R. (1996) Inhibition of tumor necrosis factor α (TNFα) activity in rat brain is associated with cerebroprotection after closed head injury. *J. Cereb. Blood Flow Metab.* **16,** 378–384.

Siren, A. L., Liu, Y., Feuerstein, G., and Hallenbeck, J. M. (1993) Increased release of tumor necrosis factor-α into the cerebrospinal fluid and peripheral circulation of aged rats. *Stroke* **24,** 880–886.

Song, H. Y. and Donner, D. B. (1995) Association of a RING finger protein with the cytoplasmic domain of the human type-2 tumour necrosis factor receptor. *Biochem. J.* **309,** 825–829.

Suga, S., Itoh, H., Komatsu, Y., Ogawa, Y., Hama, N., Yoshimasa, T., and Nakao, K. (1993) Cytokine-induced C-type natriuretic peptide (CNP) secretion from vascular endothelial cells—evidence for CNP as a novel autocrine/paracrine regulator from endothelial cells. *Endocrinology* **133,** 3038–3041.

Sun, N., Grzybicki, D., Castro, R. F., Murphy, S., and Perlman, S. (1995) Activation of astrocytes in the spinal cord of mice chronically infected with a neurotropic coronavirus. *Virology* **213,** 482–493.

Sung, S. S., Jung, L. K., Walters, J. A., Chen, W., Wang, C. Y., and Fu, S. M. (1988) Production of tumor necrosis factor/cachectin by human B cell lines and tonsillar B cells. *J. Exp. Med.* **168,** 1539–1551.

Szaflarski, J., Burtrum, D., and Silverstein, F. S. (1995) Cerebral hypoxia-ischemia stimulates cytokine gene expression in perinatal rats. *Stroke* **26,** 1093–1100.

Takasu, N., Yamada, T., Shimizu, Y., Nagasawa, Y., and Komiya, I. (1989) Generation of hydrogen peroxide in cultured porcine thyroid cells: synergistic regulation by cytoplasmic free calcium and protein kinase C. *J. Endocrinol.* **120,** 503–508.

Tanahashi, N., Fukuuchi, Y., Tomita, M., Kobari, M., Takeda, H., and Yokoyama, M. (1995) Pentoxifylline ameliorates postischemic delayed hypoperfusion of the cerebral cortex following cardiac arrest in cats. *J. Neurol. Sci.* **132,** 105–109.

Tancredi, V., D'Arcangelo, G., Grassi, F., Tarroni, P., Palmieri, G., Santoni, A., and Eusebi, F. (1992) Tumor necrosis factor alters synaptic transmission in rat hippocampal slices. *Neurosci. Lett.* **146,** 176–178.

Tartaglia, L. A., Weber, R. F., Figari, I. S., Reynolds, C., Palladino, M. A, Jr., and Goeddel, D. V. (1991) The two different receptors for tumor necrosis factor mediate distinct cellular responses. *Proc. Natl. Acad. Sci. USA* **88,** 9292–9296.

Tartaglia, L. A., Ayres, T. M., Wong, G. H., and Goeddel, D. V. (1993a) A novel domain within the 55 kd TNF receptor signals cell death. *Cell* **74,** 845–853.

Tartaglia, L. A., Goeddel, D. V., Reynolds, C., Figari, I. S., Weber, R. F., Fendly, B. M., and Palladino, M. A, Jr. (1993b) Stimulation of human T-cell proliferation by specific activation of the 75-kDa tumor necrosis factor receptor. *J. Immunol.* **151,** 4637–

Tartaglia, L. A., Pennica, D., and Goeddel, D. V. (1993c) Ligand passing: the 75-kDa tumor necrosis factor (TNF) receptor recruits TNF for signaling by the 55-kDa TNF receptor. *J. Biol. Chem.* **268,** 18,542–18,548.

Tartakovsky, B. and Ben-Yair, E. (1991) Cytokines modulate preimplantation development and pregnancy. *Dev. Biol.* **146,** 345–352.

Taupin, V., Toulmond, S., Serrano, A., Benavides, J., and Zavala, F. (1993) Increase in IL-6, IL-1 and TNF levels in rat brain following traumatic lesion. Influence of pre- and post-traumatic treatment with Ro5 4864, a peripheral-type (p site) benzodiazepine ligand. *J. Neuroimmunol.* **42,** 177–185.

Tchelingerian, J. L., Quinonero, J., Booss, J., and Jacque, C. (1993) Localization of TNFα and IL-1α immunoreactivities in striatal neurons after surgical injury to the hippocampus. *Neuron* **10,** 213–224.

Tchelingerian, J. L., Vignais, L., and Jacque, C. (1994) TNFα gene expression is induced in neurones after a hippocampal lesion. *Neuroreport* **5,** 585–588.

Tchelingerian, J. L., Monge, M., Le Saux, F., Zalc, B., Jacque, C. (1995) Differential oligodendroglial expression of the tumor necrosis factor receptors in vivo and in vitro. *J. Neurochem.* **65,** 2377–2380.

Van Zee, K. J., Kohno, T., Fischer, E., Rock, C. S., Moldawer, L. L., and Lowry, S. F. (1992) Tumor necrosis factor soluble receptors circulate during experimental and clinical inflammation and can protect against excessive tumor necrosis factor α in vitro and in vivo. *Proc. Natl. Acad. Sci. USA* **89,** 4845–4849.

Walker, D. G., Kim, S. U., and McGeer, P. L. (1995) Complement and cytokine gene expression in cultured microglial derived from postmortem human brains. *J. Neurosci. Res.* **40,** 478–493.

Wang, B., Kondo, S., Shivji, G. M., Fujisawa, H., Mak, T. W., and Sauder, D. N. (1996) Tumour necrosis factor receptor II (p75) signalling is required for the migration of Langerhans' cells. *Immunology* **88,** 284–288.

Wang, X., Yue, T. L., Barone, F. C., White, R. F., Gagnon, R. C., and Feuerstein, G. Z. (1994) Concomitant cortical expression of TNF-α and IL-1β mRNAs follows early response gene expression in transient focal ischemia. *Mol. Chem. Neuropathol.* **23,** 103–114.

Wang, X., Yue, T. L., White, R. F., Barone, F. C., and Feuerstein, G. Z. (1995) Transforming growth factor-β 1 exhibits delayed gene expression following focal cerebral ischemia. *Brain Res. Bull.* **36,** 607–609.

Weinberg, J. B., Mason, S. N., and Wortham, T. S. (1992) Inhibition of tumor necrosis factor-α (TNF-α) and interleukin-1 β (IL-1β) messenger RNA (mRNA) expression in HL-60 leukemia cells by pentoxifylline and dexamethasone: dissociation of acivicin-induced TNF-α and IL-1β mRNA expression from acivicin-induced monocytoid differentiation. *Blood* **79,** 3337–3343.

Westwick, J. K., Bielawska, A. E., Dbaibo, G., Hannun, Y. A., and Brenner, D. A. (1995) Ceramide activates the stress-activated protein kinases. *J. Biol. Chem.* **270,** 22,689–22,692.

Wiegmann, K., Schutze, S., Machleidt, T., Witte, D., and Kronke, M. (1994) Functional dichotomy of neutral and acidic sphingomyelinases in tumor necrosis factor signaling. *Cell* **78,** 1005–1015.

Wright, J. L. and Merchant, R. E. (1992) Histopathological effects of intracerebral injections of human recombinant tumor necrosis factor-α in the rat. *Acta Neuropathol.* **85,** 93–100.

Yamasaki, T., Kikuchi, H., Moritake, K., Nagao, S., Iwasaki, K., Paine, J. T., Kagawa, T., and Namba Y. (1992) A morphological and ultrastructural investigation of normal mouse brain tissue after intracerebral injection of tumor necrosis factor. *J. Neurosurg.* **77,** 279–287.

Yamasu, K., Onoe, H., Soma, G., Oshima, H., and Mizuno, D. (1989) Secretion of tumor necrosis factor during fetal and neonatal development of the mouse: ontogenic inflammation. *J. Biol. Response Modif.* **8,** 644–655.

Yang, Z., Costanzo, M., Golde, D. W., and Kolesnick, R. N. (1993) Tumor necrosis factor activation of the sphingomyelin pathway signals nuclear factor κB translocation in intact HL-60 cells. *J. Biol. Chem.* **268,** 20,520–20,523.

Zhang, Y., Harada, A., Bluethmann, H., Wang, J. B., Nakao, S., Mukaida, N., and Matsushima, K. (1995) Tumor necrosis factor (TNF) is a physiologic regulator of hematopoietic progenitor cells: increase of early hematopoietic progenitor cells in TNF receptor p55-deficient mice in vivo and potent inhibition of progenitor cell proliferation by TNFα in vitro. *Blood* **86,** 2930–2937.

Zornow, M. H., Matsumoti, M., Wiley, C. A., and Strnat, M. A. (1993) Intracerebral injection of tumor necrosis factor-α does not increase brain concentrations of glutamate. *AIDS* **7,** 1689–1690.

10

Neuroprotective Strategies Based on Interleukin Signaling

David Martin, Gerald Miller, and Norman Fischer

1. CYTOKINES

Cytokines are soluble polypeptide mediators that control the growth, differentiation, and function of most cell types and play a particular role as mediators of inflammation. Cytokines form a network of interactive signals, some of which are stimulatory and others inhibitory. The physiologic roles of cytokines concern the coordination of processes between different cell types and the response to environmental stresses (Dinarello, 1996). With respect to the central nervous system (CNS), cytokines play a specific role and are particularly important in embryonic development, fever, neuroendocrine activation, and changes in behavior and mood. Furthermore, cytokines have been shown to be involved in numerous CNS disorders, such as cerebral ischemia, head trauma, apoptosis, regulation of the immune response to foreign antigens (HIV), and the development of cellular and humoral immunity and inflammatory responses (multiple sclerosis) (Rothwell and Hopkins, 1995). Thus, in the CNS, cytokines play a role in normal physiologic control, as well as being implicated in the pathogenesis of a number of neurologic diseases.

1.1. Interleukin-1

Interleukin-1 (IL-1) α and β are pleiotropic cytokines that mediate a wide range of biological activities, including immunoregulation, proinflammation, hematopoiesis, and neurodegeneration (Rothwell and Hopkins, 1995; Dinarello, 1996). These two cytokines share approx 26% identity in amino acid sequence, and bind to the same receptor on the surface of responsive cells (Dinarello and Thompson, 1992; Dinarello, 1996).

IL-1 can bind to at least two receptors, the 80-κDa (type I) and the 60-κDa (type II) receptor (Sims et al., 1989; McMahan et al., 1991). Analysis of the deduced amino acid sequences reveals that both receptors possess a single membrane-spanning segment, and the ligand-binding domains of both receptors possess three immunoglobulin (Ig)-like domains. A major difference between the type I and II receptors is the cytoplasmic region. The 215-amino acid cytoplasmic domain of type I receptor is large enough to have intrinsic enzyme activity or protein-binding activity, whereas type II receptor has a much shorter cytoplasmic domain of 29 amino acids. Thus, the type I receptor tranduces a signal, whereas the type II receptor can bind IL-1 but does not transduce a signal

From: Neuroprotective Signal Transduction *edited by M. P. Mattson*
Humana Press Inc., Totowa, NJ

(Sims et al., 1989; McMahan et al., 1991). In fact, the type II receptor acts as a sink for IL-1β, and has been termed a decoy receptor.

1.2. Interleukin-1 Receptor Antagonist (IL-1ra)

Since overproduction of cytokines can lead to tissue damage, it is not surprising that a number of endogenous molecules and mechanisms inhibit cytokine systems or action (Dinarello and Thompson, 1992; Rothwell and Hopkins, 1995; Dinarello, 1996). The binding of IL-1 to the type I IL-1 receptor and the subsequent signaling cascade are primary targets for potential drug intervention. The most widely studied of these is the inhibition of IL-1 by the IL-1 receptor antagonist (IL-1ra).

Identification and purification of a naturally occurring IL-1ra from medium conditioned by human monocytes have allowed the investigation of the physiological and pathophysiological actions of IL-1. IL-1ra is a 17-κDa polypeptide produced in response to the same stimuli as IL-1 itself. Recombinant human IL-1ra (rhIL-1ra) expressed in *Escherichia coli* appears to bind to more than one site on the IL-1 receptor, but unlike IL-1 itself, rhIL-1ra is not internalized and does not stimulate protein kinase activity, two early events associated with IL-1 activity (Eisenberg et al., 1990; Dripps et al., 1991; Dinarello and Thompson, 1992).

2. BRAIN ISCHEMIA

Brain injury resulting from stroke or cardiac arrest is a major public health problem in the industrialized world. For example, in the United States alone approx 500,000 strokes occur each year. Ischemia-related conditions are not only fatal, but leave thousands of patients mentally and physically disabled (Dobkin, 1995; Sacco, 1995).

There is considerable evidence supporting the involvement of cytokines in CNS injury. For example, data has been generated showing increases in IL-1β mRNA expression in rodent focal and global brain ischemia models (Minami et al., 1992b; Liu et al., 1993; Wang et al., 1994; Yabuchi et al., 1994). Furthermore, IL-1β levels were elevated in the gerbil hippocampus, striatum, and cortex after transient cerebral ischemia (Saito et al., 1996).

Inhibitors of IL-1, in particular IL-1ra, have been assessed in rodent models of focal ischemia. IL-1ra given centrally or peripherally at the time of occlusion or shortly after provided a remarkable degree of neuroprotection (Relton and Rothwell, 1992; Garcia et al., 1995; Relton et al., 1996). For example, the volume of cortical and striatal tissue damage in the rat and mouse was reduced by approx 50% (Relton and Rothwell, 1992; Garcia et al., 1995; Relton et al., 1996). Furthermore, improvements in neurologic outcome and a reduction in inflammatory cell infiltrate into the damaged areas were also observed in IL-1ra-treated animals (Garcia et al., 1995). Central administration of IL-1ra was also shown to be very effective in reducing neuronal cell death mediated by excitotoxins (Relton and Rothwell, 1992; Allan et al., 1995) in rodent models of temporal lobe epilepsy and traumatic brain injury (Toulmond and Rothwell, 1995). In addition, recent studies have shown that peripheral administration of IL-1ra also significantly inhibited neuronal cell death in rodent models of global ischemia (Martin et al., 1997) and perinatal anoxia-ischemia (Martin et al., 1994).

In vitro studies have demonstrated, using primary cortical cultures, that administration of IL-1β protected neurons against excitatory amino-acid-mediated neurodegeneration

(Strijbos and Rothwell, 1995). Therefore, data generated from these in vitro studies would conflict with recent in vivo studies suggesting that IL-1β contributes to neuronal cell death (Rothwell and Hopkins, 1995). Using mixed rat cortical cultures, we have shown that IL-1β significantly increases neuronal cell death in a dose-dependent manner. In addition, IL-1β stimulation of astrocyte-enriched cultures and subsequent exposure of mixed cultures to the media from astrocyte cultures also induced neuronal cell death. Addition of IL-1ra to the cultures in both of these situations blocked neuronal cell death in a dose-dependent manner (Martin et al., 1997). These data would suggest that IL-1β can act on microglial/astrocytes to produce/release cytotoxic factors that culminate in neuronal death.

3. MULTIPLE SCLEROSIS

Multiple sclerosis is a disease of the CNS that is characterized pathologically by areas of primary demyelination in the vicinity of a mononuclear cell perivascular inflammatory reaction. Although the etiology of the disease remains unknown, studies from various animal model systems strongly support the notion that cell-mediated immune reactions against myelin antigens contribute to the disease. However, the mechanism by which a cell-mediated immune response might lead to demyelination remains to be defined (Waksman and Reynolds, 1984; Hafler and Weiner, 1987; Paterson and Swanborg, 1988).

Recently, in vivo and in vitro studies have indicated a role for cytokines in the inflammatory demyelinating process. In particular, cytokines, such as tumor necrosis factor, interleukin-1, and interferon-γ, are produced by inflammatory cells in the periphery or within the CNS and may have potential to enhance inflammation in brain and spinal tissue by promoting myelin destruction by macrophages in inflamed CNS tissue, resulting in demyelination (Hafler and Weiner, 1987; Woodroofe, 1995). Furthermore, resident brain cells, microglia, and astrocytes also produce cytokines which may augment the inflammatory process (Chung and Benveniste, 1990; Woodroofe and Cuzner, 1993).

3.1. Interleukin-1 and MS

Increasing evidence indicates that IL-1β contributes to the inflammatory processes, in experimental allergic encephalomyelitis (EAE) (Woodroofe, 1995). EAE is a T-cell-mediated inflammatory demyelinating disorder of the CNS that serves as a prime animal model for multiple sclerosis (MS) (Paterson and Swanborg, 1988). Preclinical studies have demonstrated that IL-1β can augment the in vitro activation of encephalitogenic T-lymphocytes and enhance adoptive transfer of EAE (Jacobs et al., 1991). Using *in situ* hybridization procedures, IL-1 expression as well as other cytokines were examined in frozen sections of CNS tissue from MS cases (Woodroofe and Cuzner, 1993). IL-1 mRNA was detected predominately in perivascular inflammatory cuffs rather than in parenchymal cells. In addition, clinical studies have shown increased levels of IL-1 in cerebrospinal fluid (CSF) from MS patients (Beck et al., 1988; Hauser et al., 1990; Tsukada et al., 1991). In accordance with the clinical findings, EAE animals have increased IL-1 mRNA levels within the CNS (Symons et al., 1987; Bauer et al., 1993; Rubio and Capa, 1993; Okuda et al., 1995). Astrocytes and/or microglia are probably the main source of the increased IL-1 mRNA. The data collectively suggest that IL-1 may initiate or promote inflammation within the CNS.

3.2. Inhibitors of IL-1 in MS/EAE

There are a number of potential therapeutic targets for inhibiting IL-1 mediated responses. The one strategy that has been evaluated in many disease models is preventing interleukin-1 from activating the interleukin-1 receptor. Recent studies have addressed whether blocking the action of IL-1 attenuates the clinical and pathological sequelae associated with this CNS inflammatory disorder. These studies used two types of IL-1 inhibitors: IL-1 receptor antagonist (IL-1ra) (Martin and Near, 1995) and a soluble mouse recombinant IL-1 receptor (sIL-1R) (Jacobs et al., 1991). Jacobs et al. (1991) demonstrated that peripheral administration of IL-1 exacerbated EAE compared to vehicle-treated controls. Treatment of EAE rats with sIL-1r reduced the severity of paralysis and weight loss, and shortened the duration of the disease (Jacobs et al., 1991). Further support for the involvement of IL-1 in EAE has been shown by using IL-1ra. Treatment with IL-1ra daily, starting on d 9 postimmunization with myelin-basic protein during the effector phase of EAE, significantly inhibited clinical signs of EAE. IL-1ra delayed the onset, reduced the severity of paralysis and weight loss, and shortened the duration of disease (Martin and Near, 1995).

Therefore, based on the data generated from the ischemia- and multiple sclerosis-related studies, IL-1ra appears to be the first attractive and tangible option as a potential therapeutic agent for neurologic disorders in which IL-1 plays a predominant role in neuronal degenerative processes.

4. OTHER TARGETS: NUCLEAR FACTOR KAPPA B (NF-κB)

Transcription factors of the NF-κB family exert regulatory effects on the expression of a number of genes implicated in the cellular response to pathogen stimulation or injury (Baeuerle and Baltimore, 1996). First described as B-cell-specific DNA-binding proteins, they were later identified in a wide variety of cell types, including neurons and astrocytes (Sparacio et al., 1992, Bakalkin et al., 1993; Massa et al., 1993; Cauley and Verma, 1994). Several members of the family have been cloned in vertebrates (p65, p50, relB, c-rel, and p49); after homo- or heterodimerization of these subunits, the protein complexes formed bind to specific DNA sequences in the promotor of target genes and activate transcription. NF-κB complexes are present in a constitutively active form in the nucleus or an inactive form in the cytoplasm by virtue of their binding to inhibitory proteins (IκB) (Baeuerle and Baltimore, 1996). Induction of cytoplasmic NF-κB involves release from these proteins through phosphorylation or degradation of the inhibitory subunit, dimerization of active NF-κB subunits, and translocation of DNA-binding complexes to the nucleus (Baeuerle and Baltimore, 1996).

NF-κB activation occurs in response to various growth factors, mitogens, tumor promotors, DNA-damaging agents, or oxygen radicals. In particular, in cultured astrocytes, NF-κB is activated by inflammatory cytokines (IL-1) and other factors that are known to be released by activated microglia after CNS injuries, which suggests that these mediators play a role in changes in gene expression in reactive astrocytes (Sparacio et al., 1992; Friedman, 1993; Moynagh et al., 1993).

Systemic kainic acid, an excitotoxin, produces a consistent pattern of brain damage and reactive gliosis/astrocytosis. The vulnerable regions of the brain are hippocampal CA1 and CA3 regions, thalamus, amygdala, and the entorhinal and pyriform cortex

(Nadler, 1981; Ben-Ari, 1985). It has recently been shown that kainic acid-mediated neuronal cell death results in increases in NF-κB in nonneuronal and neuronal cells in areas of selective neuronal degeneration (Perez-Otano et al., 1996). The activated NF-κB also was shown to be localized in reactive astrocytes and coincident with gliosis. It is interesting that in vitro studies have shown that NF-κB is induced in glial cells in response to cytokines, such as IL-1β, where it regulates the expression of adhesion molecules, major histocompatibility complex class II molecules, proenkephalin, growth factors and other cytokines (IL-6) (Negro et al., 1992; Massa et al., 1993; Moynagh et al., 1993).

In the brain, microglial cells, activated by different brain lesions, constitute the major source for many cytokines (IL-1β, TNF-α, and TGFβ) (Woodroofe, 1995). Furthermore, central administration of kainic acid also induces expression of several cytokines in the hippocampus, particularly IL-1 (Minami et al., 1992a). In addition, administration of an IL-1 receptor antagonist attenuates kainic acid-mediated neuronal cell death and decreases astrocytosis and microgliosis (Martin et al., 1997). It would be interesting to know whether the inhibition of kainic acid-mediated cell death by IL-1ra also results in a decrease in NF-κB activity.

Thus, based on the above observations, it is conceivable that cytokines released by activated microglia are responsible for NF-κB induction. However, many factors other than cytokines can participate in acute and/or progressive neuronal injury and increase NF-κB activity (Baeuerle and Baltimore, 1996). Therefore, agents that can inhibit the triggering of NF-κB may be pivotal in preventing neuronal cell death, by hindering many multifactorial agents acting through a common denominator (NF-κB). However, recent studies by Goodman and Mattson would conflict with the latter statement, and suggest that NF-κB activation mediates neuroprotection (Goodman and Mattson, 1996).

Given the pleiotropic role of these transcription factors in controlling genes involved in the inflammatory response, immunoregulation, excitotoxicity, ischemia, and stress response, potential inhibitors of NF-κB promise to play an increasingly important role in suppressing pathological inflammatory responses in stroke, head trauma, AD, multiple sclerosis, and CNS infections (AIDS). The inhibition of NF-κB activation by steroid hormones, antioxidants, specific protease inhibitors, and other compounds may provide a pharmacological basis for interfering with many of the above diseases (Baeuerle and Henkel, 1994).

4.1. NO and NF-κB

Nitric oxide (NO) has received much attention recently by virtue of the wide range of its biological effects and its production by a variety of cell types and tissues (Moncada, 1991; Dawson and Snyder, 1994) NO is synthesized from guanidino-nitrogen of L-arginine and molecular oxygen by nitric oxide synthases (NOS), of which there are two major types. Constitutive NOS isoforms are activated by biological signals that transiently increase intracellular Ca^{2+}, and an inducible form (type II NOS) is Ca^{2+}-independent and is activated by a variety of agents, including the cytokines IL-1 and tumor necrosis factor (TNF). In the nervous system, activation of type II NOS results in the sustained release of large amounts of NO that have been shown to be implicated in the pathogenesis of several CNS diseases, such as multiple sclerosis, HIV encephalitis, and AD (Bukrinsky et al., 1995).

Substantial evidence demonstrates that IL-1 and/or TNF are involved in the neuropathological process leading to neuronal cell death (Rothwell and Hopkins, 1995). Similarly, there is evidence showing that IL-1-stimulated cells cause activation of NF-κB, which is crucial for the increase in iNOS gene expression and subsequent neuronal cell death (Kuno and Matsushima, 1994; Schini-Kerth et al., 1995; Beauparlant and Hiscott, 1996). Therefore, TNF, IL-1, NF-κB, or iNOS inhibitors have been shown to be potent inhibitors of neuronal cell loss in various models of ischemia, apoptosis, and multiple sclerosis (Hafler and Weiner, 1987; Jacobs et al., 1991; Relton and Rothwell, 1992; Martin et al., 1994; Allan et al., 1995; Garcia et al., 1995; Martin and Near, 1995; Rothwell and Hopkins, 1995; Toulmond and Rothwell, 1995; Woodroofe, 1995; Relton et al., 1996; Martin et al., 1997).

Studies have shown that IL-1 induces the synthesis of oxygen radicals in fibroblasts (Schreck et al., 1991). Schreck et al. (1991) showed that antioxidant reagents, such as *N*-acetyl-L-cysteine and pyrolidinedithiocarbamate, suppressed the activation of NF-κB by IL-1, demonstrating the essential role of reactive oxygen intermediates in intracellular signaling. Furthermore, our recent studies, using mixed rat cortical cultures, have demonstrated that IL-1β significantly increased neuronal cell death in a dose-dependent manner. Addition of inhibitors of nitric oxide, superoxide, or hydrogen peroxide attenuated IL-1β-mediated neuronal cell death in these cultures, as did IL-1ra (Martin et al., 1997). These results are consistent with Schreck studies, suggesting a role for reactive oxygen and NO intermediates in the cascade of events leading to neuronal cell death. Although other test systems have shown that inhibition of oxygen radicals and iNOS prevents NF-κB activation, we have not yet tested this hypothesis in our in vivo/in vitro cerebral ischemia and MS-related studies.

4.2. *MAP Kinases*

During mammalian development and in the mature organism, various extracellular stimuli act to promote survival and suppress cellular suicide programs (Oppenheim, 1991; Steller, 1995). Since there is competition for limiting amounts of target-derived survival factors in the nervous system, approximately half of all neurons produced by neurogenesis die during development (Cowan et al., 1984; Oppenheim, 1991; Steller, 1995). The cells that die typically undergo apoptosis, a morphologically and biochemically distinct form of programmed cell death. The processes of both cell survival and cell death involve highly regulated signaling pathways that are currently the subject of intense investigation. The intensity of this research is based on two aspects:

1. Defects in apoptosis may underlie various neurodegenerative disorders, such as AD, ischemia-induced apoptosis, and multiple sclerosis; and
2. Therapeutic approaches may be possible by regulating signaling pathways using small synthetic molecules.

To characterize molecular mechanisms that regulate neuronal apoptosis, several investigators have been examining the contributions to cell death by the MAP kinase family members, including ERK (extracellular signal-regulated kinase), JNK (c-JUN NH2-terminal protein kinase), and p38 (Gass et al., 1993; Hu and Wieloch, 1994; Xia et al., 1995). These MAP kinases are highly conserved proteins homologous to the HOG-1 stress gene in yeasts. In fact, when HOG-1 is deleted, yeasts fail to grow in hyperosmotic conditions; however, the mammalian gene coding for the IL-1-inducible p38 MAP kinase can reconstitute the ability of the yeast to grow in hyperosmotic conditions (Galcheva-Gargova et al., 1994; Sluss et al., 1994; Cano and Mahadevan, 1995).

A recent report demonstrated that withdrawal of nerve growth factor (NGF) from rat PC-12 pheochromocytoma cells led to sustained activation of the JNK and p38 enzymes and inhibition of ERKs. The effects of dominant-interfering or constitutively activated forms of various components of the JNK-p38 and ERK signaling pathways demonstrated that activation of JNK and p38 as well as concurrent inhibition of ERK are critical for induction of apoptosis in these cells. Therefore, the dynamic balance between growth factor-activated ERK and stress-activated JNK-p38 pathways may be important in determining whether a cell survives or undergoes apoptosis. The mechanisms involved in the removal of NGF and the subsequent changes in MAP kinase activation are not known (Hu and Wieloch, 1994). However, it has been suggested that NGF removal causes the release of IL-1 and/or TNF, which through the activation of their cognate receptors lead to the activation of the JNK-p38 signaling pathways (Rouse et al., 1994; Gupta et al., 1995; Raingeaud et al., 1995). Consequently, can the balance between ERK and JNK-p38 pathways be restored, by either inhibiting cytokine receptor activation or inhibiting JNK-p38 kinases directly, thus preventing or slowing neuronal apoptosis?

In the CNS, transient global ischemia causes tyrosine phosphorylation of MAP kinase. Tyrosine phosphorylation and activation of MAP kinase also occurs after electroconvulsive shock, but the mechanism of induction is unknown. Both ischemia and intense electrical activity can result in neuronal degeneration, raising the possibility that MAP kinase activation is an intermediate step in the process (Gass et al., 1993; Hu and Wieloch, 1994; Xia et al., 1995).

Cytokines, such as IL-1 and/or TNF, may play a role in the activation of MAP-kinases, which may contribute to the neuronal cell death observed in several neurologic disorders (Rouse et al., 1994; Gupta et al., 1995; Raingeaud et al., 1995). At least two members of the MAP kinase family (p38 and JNK) have been identified as part of an IL-1 phosphorylation pathway and are responsible for phosphorylating certain heat shock proteins that are elevated in ischemic-related disorders. Furthermore, activation of certain MAP kinases by IL-1 and/or TNF has been suggested to lead to neuronal degeneration through phosphorylation by MAP kinases of the microtubule-associated protein τ on sites that have been implicated in the etiology of neuronal loss in AD (Grundke-Iqbal et al., 1986; Udea et al., 1990; Drewes et al., 1992; Davis, 1993).

Therefore, the therapeutic effects of IL-1ra or TNF-binding protein (TNFbp) demonstrated in preclinical models of cerebral ischemia, head trauma, and multiple sclerosis may be caused by inhibition of IL-1- or TNF-mediated MAP kinase activation (Jacobs et al., 1991; Relton and Rothwell, 1992; Martin et al., 1994, 1995, 1997; Allan et al., 1995; Garcia et al., 1995; Martin and Near, 1995; Strijbos and Rothwell, 1995; Toulmond and Rothwell, 1995; Dawson, 1996; Relton et al., 1996). Specific inhibitors of MAP kinases potentially may be useful therapeutic agents for neurological disorders. A group of pyridinyl imidazole compounds has recently been found to be highly specific for inhibiting p38 MAP kinase (Saklatvala et al., 1996). They were discovered serendipitously as inhibitors of endotoxin-stimulated cytokine production. These inhibitors are potentially a powerful experimental probe to explore the role of p38 MAP kinases in neurophysiological and neuropathological conditions.

Neurobiology research of interleukin signaling systems is still in its infancy, but it is beginning to yield a wealth of information on many physiological functions in the CNS. This information may have many consequences. Foremost would be the growing number of proposals for many potential therapeutic indications for drugs interacting with these systems. If any drug modulating interleukin signaling systems proves to be useful in any neurodegenerative disorder, then neurobiological investigations of interleukin-signaling systems should contribute greatly to improved clinical treatment of patients, thereby reaching the ultimate goal of the biomedical researcher.

REFERENCES

Allan, S. M., Lawrence, C. B., and Rothwell, N. J. (1995) Mechanism of action of interleukin-1 (IL-1) in excitotoxic brain damage in the rat. *Neurosci. Abst.* **21,** 37.14.

Baeuerle, P. A. and Baltimore, D. (1996) NF-κB: ten years after. *Cell* **87,** 13–20.

Baeuerle, P. A. and Henkel, T. (1994) Function and activation of NF-κB in the immune system. *Ann. Rev. Immunol.* **12,** 141–179.

Bakalkin, G. Y., Yakovleva, A., and Terenius, L. (1993) NF-κappa B-like factors in the murine brain. Developmentally-regulated and tissue-specific expression. *Mol. Brain Res.* **20,** 137–146.

Bauer, J., Berkenbosch, F., Van Dam, A. M., and Dijkstra, C. D. (1993) Demonstration of interleukin-1 beta in Lewis rat brain during experimental allergic encephalomyelitis by immunocytochemistry at the light and ultrastructural level. *J. Neuroimmunol.* **48,** 13–21.

Beauparlant, P. and Hiscott, J. (1996) Biological and biochemical inhibitors of the NF-κB/Rel proteins and cytokine synthesis. *Cytokine Growth Factor Rev.* **7,** 175–190.

Beck, J., Rondot, P., Catinot, L., Falcot, E., Kircher, H., and Wietzerbin, J. (1988) Increased production of interferon gamma and tumor necrosis factor precedes clinical manifestation in multiple sclerosis: do cytokines trigger off exacerbations? *Acta Neurol. Scand.* **78,** 318–323.

Ben-Ari, Y. (1985) Limbic seizure and brain damage produced by kainic acid: mechanisms and relevance to human temporal lobe epilepsy. *Neuroscience* **10,** 375–403.

Bukrinsky, M. I., Nottet, H. S. L. M., Schmidtmayerova, H., Dubrovsky, L., Flanagan, C. R., Mullins, M. E., Lipton, S. A., and Gendelman, H. E. (1995) Regulation of nitric oxide synthase activity in human immunodeficiency virus type 1 (HIV-1)-infected monocytes: implications for HIV-associated neurological disease. *J. Exp. Med.* **181,** 735–745.

Cano, E. and Mahadevan, L. C. (1995) Parallel signal processing among mammalian MAPKs. *Trends Biochem. Sci.* **20,** 117–122.

Cauley, K. and Verma, I. M. (1994) Kappa B enhancer-binding complexes that do not contain NF-κappa B are developmentally regulated in mammalian brain. *Proc. Natl. Acad. Sci. USA* **91,** 390–394.

Chung, I. Y. and Benveniste, E. N. (1990) Tumor necrosis factor-α production by astrocytes. Induction by lipopolysaccharide, IFN-gamma, and IL-1β. *J. Immunol.* **144,** 2999–3007.

Cowan, W. M., Fawcett, J. W., O'Leary, D. D., and Stanfield, B. B. (1984) Regressive events in neurogenesis. *Science* **225,** 1258–1265.

Davis, R. J. (1993) The mitogen-activated protein kinase signal transduction pathway. *J. Biol. Chem.* **268,** 14,553–14,556.

Dawson, D. A., Martin, D., and Hallenbeck, J. M. (1996) Inhibition of tumor necrosis factor-alpha reduces focal cerebral ischemic injury in the spontaneously hypertensive rat. *Neurosci. Lett.* **218,** 41–44.

Dawson, T. M. and Snyder, S. H. (1994) Gases as biological messengers: nitric oxide and carbon monoxide in the brain. *J. Neurosci.* **14,** 5147–5159.

Dinarello, C. A. and Thompson, R. C. (1992) Blocking IL-1: interleukin-1 receptor antagonist in vivo and in vitro. *Immunol. Today* **12,** 404–410.

Dinarello, C. A. (1996) Biologic basis for interleukin-1 in disease. *Blood* **87,** 2095–2147.
Dobkin, B. The economic impact of stroke. (1995) *Neurology* **45,** S6–S9.
Drewes, G., Lichtenberg-Kraag, B., Doering, F., Mandelkow, E. M., Biennant, J., Goris, J., Doree, M., and Mandelkow, E. (1992) Mitogen activated protein (MAP) kinase transforms tau protein into an Alzheimer-like state. *EMBO J.* **11,** 2131–2138.
Dripps, D. J., Brandhuber, B. J., Thompson, R. C., and Eisenberg, S. P. (1991) Interleukin-1 (IL-1) receptor antagonist binds to the 80-κDa IL-1 receptor but does not initiate signal transduction. *J. Biol. Chem.* **226,** 10,331–10,336.
Eisenberg, S. P., Evans, R. J., Arend, N. P., et al. (1990) Primary structure and functional expression from complementary DNA of a known interleukin-1 receptor antagonist. *Nature* **343,** 341–346.
Friedman, W. J., Thakur, S., and Rabson, A. (1993) Induction of NF-κB transcription factors in rat hippocampal astrocytes by interleukin-1. Soc. *Neurosci. Abst.* **19,** 1307.
Galcheva-Gargova, Z., Derijard, B., Wu, I.-H., and Davis, R. J. (1994) An osmosensing signal transduction pathway in mammalian cells. *Science* **265,** 806–808.
Garcia, J. H., Liu, K. F., and Relton, J. K. (1995) Interleukin-1 receptor antagonist decreases the number of necrotic neurons in rats with middle cerebral artery occlusion. *Am. J. Pathol.* **147,** 1477–1485.
Gass, P., Kiessling. M., and Bading, H. (1993) Regionally selective stimulation of mitogen activated protein (MAP) kinase tyrosine phosphorylation after generalized seizures in the rat brain. *Neurosci. Lett.* **162,** 39–42.
Goodman, Y. and Mattson, M. P. (1996) Ceramide protects hippocampal neurons against excitotoxic and oxidative insults, and amyloid β-peptide toxicity. *J. Neurochem.* **66,** 869–872.
Grundke-Iqbal, I., Iqbal, K., Tung, Y., Quinlan, M., Wisniewski, H. M., and Binder, L. I. (1986) Abnormal phosphorylation of the microtubule-associated protein tau (tau) in Alzheimer cytoskeletal pathology. *Proc. Natl. Acad. Sci. USA* **83,** 4193–4197.
Gupta, S., Campbell, D., Derijard, B., and Davis, R. J. (1995) Transcription factor ATF2 regulation by the JNK signal transduction pathway. *Science* **267,** 389–393.
Hafler, D. A. and Weiner, H. L. (1987) T cells in multiple sclerosis and inflammatory central nervous disease. *Immunol. Rev.* **100,** 307–332.
Hauser, S. L., Doolittle, T. H., Lincoln, R., Brown, R. H., and Dinarello, C. A. (1990) Cytokine accumulation in CSF of multiple sclerosis patients. Frequent detection of interleukin-1 and tumor necrosis factor but not interleukin-6. *Neurology* **40,** 1735–1739.
Hu, B.-R. and Wieloch, T. (1994) Tyrosine phosphorylation and activation of mitogen-activated protein kinase in the rat brain following transient cerebral ischemia. *J. Neurochem.* **62,** 1357–1367.
Jacobs, C. A., Baker, P. E., Roux, E. R., et al. (1991) Experimental autoimmune encephalomyelitis is exacerbated by IL-1β and suppressed by soluble IL-1 receptor. *J. Immunol.* **146,** 2983–2989.
Kuno, K. and Matsushima, K. (1994) The IL-1 receptor signaling pathway. *J. Leukoc. Biol.* **56,** 542–547.
Liu, T., McDonnell, P. C., Young, P. R., White, R. F., Siren, A. L., Hallenbeck, J. M., Barone, F. C., and Feuerstein, G. Z. (1993) Interleukin-1β mRNA expression in ischemic rat cortex. *Stroke* **24,** 1746–1751.
Martin, D. and Near, S. L. (1995) Protective effect of the interleukin-1 receptor antagonist (IL-1ra) on experimental allergic encephalomyelitis in rat. *J. Neuroimmunol.* **61,** 241–245.
Martin, D., Chinookoswong, N., and Miller, G. (1994) The interleukin-1 receptor antagonist (rhIL-1ra) protects against cerebral infarction in a rat model of hypoxia-ischemia. *Exp. Neurol.* **130,** 362–367.
Martin, D., Miller, G., Neuberger, T., Relton, J., and Fischer, N. (1997) Role of IL-1 in neurodegeneration; Preclinical findings with IL-1ra and ICE inhibitors, in *Inflammatory Mechanisms and Management of Neurodegeneration* (Wood, P. L., ed) Humana, Totowa, NS, in press.

Martin, D., Near, S. L., Bendele, A., and Russell, D. (1995) Inhibition of tumor necrosis factor is protective against neurologic dysfunction after active immunization of Lewis rats with myelin basic protein. *Exp. Neurol.* **131,** 221–228.

Massa, P. T., Ozato, K., and McFarlin, D. E. (1993) Cell type-specific regulation of major histocompatibility complex (MHC) class I gene expression in astrocytes, oligodendrocytes, and neurons. *Glia* **8,** 201–207.

McMahan, C. J., Slack, J. L., Mosley, B., Cosman, D., et al. (1991) A novel IL-1 receptor, cloned from B cells by mammalian expression, is expressed in many cell types. *EMBO J.* **10,** 2821–2832.

Minami, M., Kuraishi, Y., and Yamaguchi, T. (1992a) Convulsants induce interleukin-1 β messenger RNA in rat brain. *Biochem. Biophys. Res. Commun.* **171,** 832–837.

Minami, M., Kuraishi, Y., Yabuuchi, K., Yamazaki, A., and Satoh, M. (1992b) Induction of interleukin-1 β mRNA in rat brain after transient forebrain ischemia. *J. Neurochem.* **58,** 390–392.

Moncada, S. (1991) Nitric oxide: physiology, pathophysiology, and pharmacology. *Pharmacol. Rev.* **43,** 109–142.

Moynagh, P. N., Williams, D. C., and O'Neill, L. A. (1993) Interleukin-1 activates transcription factor NF kappa B in glial cells. *Biochem. J.* **294,** 343–347.

Nadler, J. V. (1981) Kainic acid as a tool for the study of temporal lobe epilepsy. *Life Sci.* **24,** 2031–2042.

Negro, A., Tavella, A., Facci, L., Callegaro, L., and Skaper, S. D. (1992) Interleukin-1 beta regulates proenkephalin gene expression in astrocytes cultured from rat cortex. *Glia* **6,** 206–212.

Okuda, Y., Nakatsuji, Y., Fujimura, H., Esumi, H., Ogura, T., Yanagihara, T., and Sakoda, S. (1995) Expression of the inducible isoform of nitric oxide synthase in the central nervous system of mice correlates with the severity of actively induced experimental allergic encephalomyelitis. *J. Neuroimmunol.* **62,** 103–112.

Oppenheim, R. W. (1991) Cell death during development of the nervous system. *Ann. Rev. Neurosci.* **14,** 453–501.

Paterson, P. Y. and Swanborg, R. H. (1988) Demyelinating diseases of the central and peripheral nervous systems, in *Immunological Diseases*, 4th ed. (Samter, M., Talmage, D. W., Frank, M. M., Austen, K. F., and Claman, N. H., eds.), Little Brown and Co., Boston, MA, pp. 1877–1916.

Perez-Otano, I., McMillian, M. K., Chen, J., Bing, G., Hong, J. S., and Pennypacker, K. R. (1996) Induction of NF-κB-like transcription factors in brain areas susceptible to kainate toxicity. *Glia* **16,** 306–315.

Raingeaud, J., Gupta, S., Han, J., Rogers, J. S., Dickens, M., Ulevitch, R. J., and Davis, R. J. (1995) Pro-inflammatory cytokines and environmental stress cause p38 mitogen-activated protein kinase activation by dual phosphorylation on tyrosine and threonine. *J. Biol. Chem.* **270,** 7420–7426.

Relton, J. K. and Rothwell, N. J. (1992) Interleukin-1 receptor antagonist inhibits ischemic and excitotoxic neuronal damage in the rat. *Brain Res. Bull.* **29,** 243–246.

Relton, J. K., Martin, D., Thompson, R. C., and Russell, D. A. (1996) Peripheral administration of interleukin-1 receptor antagonist inhibits brain damage after focal cerebral ischemia in the rat. *Exp. Neurol.* **138,** 206–213.

Rothwell, N. J. and Hopkins, S. (1995) Cytokines and the nervous system II: actions and mechanisms of action. *Trends Neurosci.* **18,** 130–136.

Rouse, J., Cohen, P., Trigon, S., Morange, M., Alonso-Llamazares, A., Zamanillo, D., Hunt, T., and Nebreda, A. R. (1994) A novel kinase cascade triggered by stress and heat shock that stimulates MAPKAP kinase-2 and phosphorylation of the small heat shock proteins. *Cell* **78,** 1027–1037.

Rubio, N. and Capa, L. (1993) Differential IL-1 synthesis by astrocytes from Theiler's murine encephalomyelitis virus-susceptible and resistant strains of mice. *Cell Immunol.* **149,** 237–247.

Sacco, R. L. (1995) Risk factors and outcomes for ischemic stroke. *Neurology* **45,** S10–S14.

Saklatvala, J., Rawlinson, L., Waller, R. J., Sarsfield, S., Lee, J. C., Morton, L. F., Barnes, M. J., and Farndale, R. W. (1996) Role for p38 mitogen-activated protein kinase in platelet aggregation caused by collagen or a thromboxane analogue. *J. Biol. Chem.* **271,** 6586–6589.

Schini-Kerth, V. B., Mulsch, B. A., and Busse, R. (1995) Pyrrolidine dithiocarbamate selectively prevents the expression of inducible nitric oxide synthase in the rat aorta. *Eur. J. Pharmacol.* **265,** 83–87.

Schreck, R., Rieber, P., and Baeuerle, P. A. (1991) Reactive oxygen intermediates as apparently widely used messengers in the activation of the NF-κB transcription factor and HIV-1. *EMBO J.* **10,** 2247–2258.

Sims, J. E., Acres, B., Grubin, C. E., et al. (1989) Cloning the interleukin-1 receptor from human T cells. *Proc. Natl. Acad. Sci. USA* **86,** 8946–8950.

Sluss, H., Barrett, T., Derijard, B., and Davis, R. J. (1994) Signal transduction by tumor necrosis factor mediated by JNK protein kinases. *Mol. Cell. Biol.* **14,** 8376–8384.

Sparacio, S. M., Zhang, Y., Vilcek, J., and Benveniste, E. N. (1992) Cytokine regulation of interleukin-6 gene expression in astrocytes involves activation of an NF-κappa B-like nuclear protein. *J. Neuroimmunol.* **39,** 231–242.

Steller, H. (1995) Mechanisms and genes of cellular suicide. *Science* **267,** 1445–1449.

Strijbos, P. J. and Rothwell, N. J. (1995) Interleukin-1 beta attenuates excitatory amino acid induced neurodegeneration in vitro: involvement of nerve growth factor. *J. Neurosci.* **15,** 3468–3474.

Symons, J. A., Bundick, R. V., Suckling, A. J., et al. (1987) Cerebrospinal fluid interleukin-1 like activity during chronic relapsing experimental allergic encephalomyelitis. *Clin. Exp. Immunol.* **68,** 648–654.

Toulmond, S. and Rothwell, N. J. (1995) Interleukin-1 receptor antagonist inhibits neuronal damage caused by fluid percussion injury in the rat. *Brain Res.* **671,** 261–266.

Tsukada, N., Mijagi, K., Matsuda, M., Yanagisawa, N., and Yone, K. (1991) Tumor necrosis factor and interleukin-1 in the CSF and sera of patients with multiple sclerosis. *J. Neurol. Sci.* **102,** 230–234.

Udea, K., Masliah, E., Saitoh, T., Bakalis, S. L., Scoble, H., and Kosik, K. S. (1990) Alz-50 recognizes a phosphorylated epitope of tau protein. *J. Neurosci.* **10,** 3295–3304.

Waksman, B. H. and Reynolds, W. E. (1984) Multiple sclerosis as a disease of immune regulation. *Proc Soc. Exp. Biol. Med.* **175,** 282–294.

Wang, X., Yue, T. L., Barone, F. C., White, R. F., Gagnon, R. C., and Feuerstein, G. Z. (1994) Concomitant cortical expression of TNF-alpha and IL-1 beta mRNAs follows early response gene expression in transient focal ischemia. *Mol. Chem. Neuropathol.* **23,** 103–114.

Woodroofe, M. N. and Cuzner, M. L. (1993) Cytokine mRNA expression in inflammatory multiple sclerosis lesions: detection by non-radioactive in situ hybridization. *Cytokine* **5,** 583–588.

Woodroofe, M. N. (1995) Cytokine production in the central nervous system. *Neurology* **45,** S6–S10.

Xia, Z., Dickens, M., Raingeaud, J., Davis, R. J., and Greenberg, M. E. (1995) Opposing effects of ERK and JNK-p38 MAP kinases on apoptosis. *Science* **270,** 1326–1331.

Yabuchi, K., Minami, M., Katsumata, S., Yamazaki, A., and Satoh, M. (1994) An in situ hybridization study on interleukin-1 beta mRNA induced by transient forebrain ischemia in the rat brain. *Mol. Brain Res.* **26,** 135–142.

11
Signaling by β-Amyloid Precursor Protein

Katsutoshi Furukawa

1. INTRODUCTION

The β-amyloid precursor protein (βAPP) is widely expressed in neurons and glial cells throughout the mammalian nervous system. βAPP is a large transmembrane glycoprotein having a single transmembrane domain, a large extracelluar N-terminus, and a short cytoplasmic C-terminus (Kang et al., 1987). This protein is very important as a source of amyloid β peptide (Aβ), a 40–42 amino acid peptide that forms senile plaques in Alzheimer's disease (AD) patient brains. Senile plaques, along with neurofibrillary tangles, are the most characteristic pathological change in AD brains (Alzheimer, 1907). In normal physiological conditions, α-secretase cleaves βAPP between amino acids 16 and 17 of Aβ. The α-secretase cleavage prevents release of amyloidogenic Aβ and yields secreted forms of APP (sAPPα), which are released into the extracellular fluid (Weidemann et al., 1989; Esch et al., 1990; Sisodia et al., 1990). An alternative cleavage of βAPP within the endosomal–lysosomal pathway at the N-terminus of Aβ results in release of intact Aβ and secreted forms of APP without the Aβ1–16 domain (sAPPβ) (Haass et al., 1992; Seubert et al., 1992; Shoji et al., 1992; Busciglio et al., 1993). Previous studies paid great attention to the production and the functions of Aβ, but a regulated release of sAPPα and sAPPβ seems to be equally important (Furukawa et al., 1996b).

The first part of this chapter describes the metabolism and biochemical processing of βAPP. Several abnormalities in βAPP metabolism reportedly participate in the pathogenesis of AD. For example, Down's syndrome is caused by chromosome 21 trisomy, a chromosome that includes the βAPP gene; all persons with Down's syndrome develop AD-like neuropathology (St. George-Hyslop et al., 1987). Several genetic mutations in βAPP also have been discovered in familial AD patients whose brains have significant Aβ deposition (Goate et al., 1991; Mullan et al., 1992). In addition to the genetic abnormalities in βAPP, misprocessing of βAPP may play a role in sporadic AD.

The second part describes Aβ toxicity and its mechanisms. Aβ is cleaved from βAPP by β- and γ-secretases and is released into extracellular space. Aβ reportedly forms aggregated β-sheet structures, which show extreme toxicity to neurons and other cultured cells (Mattson et al., 1993a; Lorenzo and Yankner, 1994; Busciglio et al., 1995; Howlet et al., 1995). It has been suggested that Aβ production and aggregation are involved in

From: Neuroprotective Signal Transduction *edited by M. P. Mattson*
Humana Press Inc., Totowa, NJ

important pathomechanisms of AD. Therefore, this chapter reviews Aβ's toxicity, its aggregation, and free radical formation, with consideration of transgenic mouse models, and electrophysiological characteristics.

The third part considers the neuroprotective function of sAPPs. sAPPα is released from βAPP by α-secretase and has several functions, that are mostly beneficial to neurons (Saitoh et al., 1989; Mattson et al., 1993b; Furukawa et al., 1996b). Barger et al. (1995) recently reported that sAPPα significantly induces cGMP production in cultured neurons, and this cGMP induction mediates several neuronal functions mentioned below. Furukawa et al. (1996a) also reported that activation of K^+ channels by sAPPα stabilizes neuronal excitability and decreases $[Ca^{2+}]_i$. This K^+ channel activation by sAPP may participate in neuroprotection and neuroplasticity in the central nervous system (CNS). It has been reported that sAPP can protect cultured rat hippocampal neurons against glutamate toxicity and glucose deprivation (Mattson et al. 1993b; Furukawa et al., 1996b). Furthermore, Smith-Swintosky et al. (1994) showed that sAPPs protect hippocampal CA1 neurons in an in vivo ischemia model. These results suggest that sAPP could have strong neuroprotective effects in mammalian CNS neurons.

The last part of this chapter discusses several genes in which mutations cause hereditary forms of AD. Genes linked to AD have been found on chromosomes 1, 4, 14, and 21. The gene on chromosome 21 is βAPP, and its mutations reportedly cause a small percentage of cases of familial inherited AD (Levy et al., 1990; Goate et al., 1991; Wisniewski et al., 1991; Mullan et al., 1992). The other three genes are presenilin2 (PS2) (chromosome 1), apolipoprotein-E (Apo-E); (chromosome 4), and presenilinl (PS1) (chromosome 14). PS1 and PS2, which were found in 1995, have high homology with each other, including 7–9 transmembrane domains in their structure (Sherrington et al., 1995; Rogaev et al., 1995; Levy-Lehad et al., 1995a,b). Increased production of Aβ is reported in the fibroblasts from patients having PS1 and PS2 (Scheuner et al., 1996). ApoE, a component of lipoprotein complexes that regulate cellular cholesterol (Mahley, 1988), has three major alleles: E2, E3, and E4. Strittmatter et al. (1993) reported that people having ApoE4 allele show high frequency of late-onset AD, and AD patients with ApoE4 allele showed increased numbers of senile plaques (Rebeck et al., 1993; Schmechel et al., 1993). Therefore, these different genes may each cause AD by affecting βAPP processing.

2. PROCESSING OF βAPP

Senile plaque or amyloid plaque is one of the hallmarks of AD. Senile plaque is constituted of an insoluble peptide of 40–42 amino acids called Aβ (Glenner and Wong, 1984; Masters et al., 1985). Investigations of Aβ identified and clarified the molecular structure of its precursor, the βAPP (Kang et al., 1987) (Figs. 1 and 2). In order to make clear the pathogenesis of AD, many scientists have been investigating the physiopathological mechanisms of βAPP processing. The βAPP gene was localized to chromosome 21, also involved in Down's syndrome (chromosome 21 trisomy), a genetic disorder that shows AD-like neuropathology (St. George-Hyslop et al., 1987). The causal linkage of different βAPP mutations to familial hereditary AD provided a direct demonstration of the central role of βAPP in the pathogenesis of AD (Goate et al., 1991; Mullan et al., 1992).

Soluble Aβ is normally released from neuronal cells at low levels and circulates in cerebrospinal fluid at nanomolar levels (Haass et al., 1992). Several isoforms of βAPP can

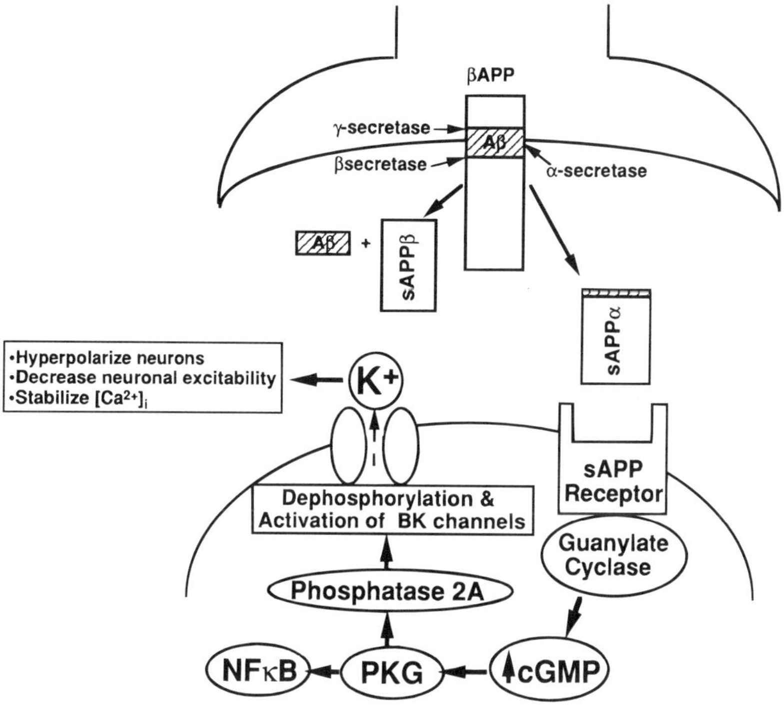

Fig. 1. Neuroprotective signal transduction pathways of sAPPs. sAPPα and sAPPβ are cleaved by α-secretase and β + γ-secretases, respectively. sAPPαs bind to receptors that may possess intrinsic guanylate cyclase activity. Cyclic GMP (cGMP) produced in response to sAPP activates cGMP-dependent protein kinase (PKG), which, through phosphorylation events, results in activation of K^+ channels, membrane hyperpolarization, and reduced Ca influx. PKG may also activate NFκB, resulting in delayed transcription-mediated neuroprotective responses.

Fig. 2. Structure of βAPP: sites of α-, β- and γ-secretase cleavages; heparin binding region (hbr) present in sAPPαs, but not in sAPPβs; pd, domain involved in promoting proliferation of nonneuronal cells.

be generated by alternative processing (*see* Mullan and Crawford, 1993, for review). βAPP is a transmembrane glycoprotein, and the Aβ sequence lies partly within the membrane domain. Two forms of βAPP contain a Kunitz-type protease inhibitor (KPI) domain near the extracellular N-terminus (βAPP751 and βAPP 770), but two forms do not have this domain (βAPP695 and βAPP714). An enzymatic cleavage of βAPP within the Aβ sequence by α-secretase results in release of the secreted form of βAPP (sAPPα) and prevents release of amyloidogenic Aβ into extracellular space. An alternative processing, which occurs exactly at the N-terminus of Aβ by β-secretase and at the C-terminus by γ-secretase, liberates intact Aβ and a form of sAPP trancated at the C-terminal (sAPPβ) (Haass et al., 1992; Seubert et al., 1992; Shoji et al., 1992; Busciglio et al., 1993).

Metabolism of βAPP can be influenced by environmental stimuli that activate Ca^{2+}-inositol phospholipid signaling pathways. For example, sAPPα is released by stimulation of muscarinic receptors (M1 and M3) that couple to G proteins, leading to inositol phospholipid turnover (Nitsch et al., 1992), and proteolytic processing of βAPP is mediated by activation of protein kinase C (Buxbaum et al., 1990). Furthermore, sAPPα can be released from presynaptic terminals by electrical stimulation or neuronal excitation (Nitsch et al., 1993).

Expression of βAPP is regulated during development of neurons (Ohyagi et al., 1990). For example, βAPP levels increase in parallel with the differentiation of embryonic carcinoma cells (Fukuchi et al., 1992). Some growth factors and cytokines upregulate βAPP expression, and neurotrophic factors may promote the release of sAPPα (Refolo et al., 1989). Furthermore, the expression of βAPP is stimulated by such neurotoxic insults as forebrain ischemia (Abe et al., 1991; Stephenson et al., 1992), brain trauma (Roberts et al., 1991), and excitotoxins (Siman et al., 1989).

3. NEUROTOXIC MECHANISMS OF Aβ

Many laboratories have been making an effort to establish transgenic mouse models of AD based upon overexpression of normal βAPP, expression of some fragment of βAPP, or expression of human mutant forms of βAPP (Quon et al., 1991; Sandhu et al., 1991; Kammesheidt et al., 1992; Lamb et al., 1993; Higgins et al., 1994; LaFerla et al., 1995). Most of these trials have failed to show reliable plaque formation or neuronal degeneration, but in 1995 Games and colleagues succeeded in producing transgenic mice showing remarkable thioflavin S-positive Aβ deposits, neuritic plaques, synaptic loss, astrocytosis, and microgliosis, by using a PDGF-β promoter and human mutant APP (V717F) as a a transgene. The deposits of Aβ began to appear in hippocampus and cortical regions from about 4 mo of age. By 10 mo of age, synapse loss and neuronal degeneration had occurred in brain regions where Aβ was deposited, but not in brain regions lacking Aβ. The Aβ deposition and neuronal loss in V717F mice strikingly indicate the probable correlation between βAPP processing and pathogenesis of AD.

Although Aβ is normally soluble, Aβ forms insoluble peptide aggregates in experimental conditions and in neural tissue by changes in environmental circumstances (e.g., Zn^{3+}, pH, oxidation, and so on) (Dyrks et al., 1992; Bush et al., 1994; Hensley et al., 1994). It has been reported that the neurotoxic potency of some batches of synthetic Aβ is related to the ability of the peptide to form aggregates (Busciglio et al., 1992; Mattson et al., 1993a; Pike et al., 1993). Chemical influences on the aggregation of Aβ and structure requirements of aggregated β-sheet have been investigated. For example, Aβ can aggregate at physiological pH, but it aggregates faster at more acidic pHs (Fraser et al., 1991; Hilbich et al., 1991; Burdick et al., 1992). Bush et al. (1994) reported that Aβ aggregates more rapidly in the presence of zinc. Aβ aggregation is accelerated by an oxidizing environment (Dyrks et al., 1992; Hensley et al., 1994), but antagonized by hexadecyl-*N*-methypiperidinium (Wood et al., 1996). The ability of Aβ to form aggregated sheet (β-sheet) may be critical for neurotoxicity, because Congo Red, which inhibits Aβ fibril formation, also prevented the Aβ toxicity, (Lorenzo and Yankner, 1994). Pike et al. (1995) reported that methionine 35 of Aβ plays a key role in both peptide fibril formation and Aβ toxicity based on studies in which they removed and/or replaced methionine 35 with other amino acids in Aβ and then investigated the aggregation and neurotoxicity of the resulting peptides.

The resting extracellular and intracellular free Ca^{2+} concentrations are 1–2 m*M* and 50–200 n*M*, respectively. This striking difference in the concentrations is conferred by several membrane-associated and cytosolic calcium (Ca)-regulating proteins (Clapham, 1995). One of the most important proteins regulating this free Ca^{2+} gradient is the plasma membrane calcium ATPase (PMCA), an energy-driven Ca pump. When neurons are stimulated by the binding of glutamate to its receptors or activation of voltage-dependent Ca^{2+} channels by membrane depolarization, the $[Ca^{2+}]_i$ rapidly increases to concentrations of well over 10 μ*M*. This increased $[Ca^{2+}]_i$ is rapidly removed by at least three pathways. One mechanism involves activity of the plasma membrane Na^+/Ca^{2+}-exchanger. The second mechanism involves activity of a smooth endoplasmic reticulum Ca^{2+}-ATPase, which pumps Ca^{2+} into the lumen of that organelle. A third mechanism involves cytoplasmic Ca^{2+}-binding proteins (for example, calbindin D_{28k}). In addition to these mechanisms, the plasma membrane Na^+/K^+-ATPase serves an indirect role in maintenance of low $[Ca^{2+}]_i$ by keeping the Na^+ gradient across the cell membrane and therefore discouraging Ca^{2+} influx through voltage-dependent Ca^{2+} channels. The major routes for Ca^{2+} influx are voltage-dependent Ca^{2+} channels, ligand-gated cation channels, and release from the intracellular Ca^{2+} store. Voltage-dependent Ca^{2+} channels are classified as T, L, N, P, Q, and R types. Representatives of ligand-gated cation channels are ionotropic glutamate receptors, which include NMDA, AMPA, and kainate receptors. In addition to channels and receptors in the plasma membrane, Ca^{2+} ions are released from intracellular Ca^{2+} stores (e.g., endoplasmic reticulum and mitochondria). The major route for Ca^{2+} release from endoplasmic reticulum is inositol trisphosphate (IP_3)-receptor channels. Muscarinic acetylcholine and metabotropic glutamate receptors are linked to a GTP-binding protein, which, in turn, can activate phospholipase C (PLC). The PLC cleaves phosphatidyl-inositol 4,5-*bis*-phosphate to release IP_3.

Several lines of evidence support the hypothesis that elevation in $[Ca^{2+}]_i$ is causally involved in neuronal degeneration in AD. Since data about the Ca^{2+} involvement in neurodegeneration has been summarized in a recent review (Mattson et al., 1996), this chapter will focus on the functions of Aβ and sAPP in relation to intracellular Ca^{2+} homeostasis.

Exposure of cultured human or rat neurons to Aβ25–35, Aβ1–38, or Aβ1–40 causes slow progressive elevation of $[Ca^{2+}]_i$ (Mattson et al., 1992, 1993a). Furthermore, Aβ greatly enhances $[Ca^{2+}]_i$ increase induced by glutamate. In addition to the receptor and channel functions of $[Ca^{2+}]_i$, Ca-binding proteins (calbindin or calretinin) may have important roles in $[Ca^{2+}]_i$ regulation in Alzheimer's brains (Sloviter, 1989; Iacopino et al., 1994). Neurons expressing calbindin in hippocampal cultures are relatively resistant to glutamate, Ca ionophore, and Aβ toxicity, and exhibit less $[Ca^{2+}]_i$ increase in response to these compounds (Mattson et al., 1993a; Cheng et al., 1994a; Barger et al., 1995). Therefore, Ca-binding proteins can afford protection to neurons, functioning as a buffer of intracellular free Ca^{2+} ions.

Application of Aβ to culture medium at concentrations of 1–50 μ*M* elicits significant neuronal degeneration (Yankner et al., 1990; Mattson et al., 1992; Pike et al., 1993). Behl et al. (1994) showed that Aβ can induce lipid peroxidation in cultured cells, and

Butterfield et al. (1994) reported lipid peroxidation in synapses using an electron paramagnetic resonance (EPR) spectroscopic method. The EPR method employs a spin-trapping compound that, on interaction with a free radical moiety, forms a spin adduct that is paramagnetic and can be detected in an EPR spectroscope. Their experiments indicate that Aβ may induce free radical production in neurons in an aggregation-related manner. Exposure of cultured hippocampal neurons to Aβ induced malondialdehyde accumulation (Goodman and Mattson, 1996a) and conjugation of 4-hydroxy-2,3-nonenal (HNE) to cellular proteins (Mark et al., 1996), indicating the induction of lipid peroxidation by Aβ. Intracellular H_2O_2 levels were increased in cultured rat hippocampal neurons by the application of Aβ (Goodman and Mattson, 1994). Other amyloidogenic peptides such as amylin may actuate a similar cytotoxic cascade of events involving oxidative stress and disruption of calcium homeostasis (May et al., 1993; Mattson and Goodman, 1995).

Production of free radicals is hypothesized to play a major role in the neurotoxicity of Aβ, because pretreatment with antioxidants, such as α-tocopherol (vitamin E) and propyl gallate, protects neurons against Aβ toxicity (Behl et al., 1994; Mark et al., 1995). Furthermore, glutathione also showed a strong neuroprotective effect against HNE and Aβ in primary cultured neurons (Mark et al., 1996). The most likely cellular site at which the oxidative reaction is initiated is plasma membrane, when Aβ is applied to neuronal cultures (Mattson et al., 1993a). Mark et al. (1995) reported that exposure of cultured rat hippocampal neurons to Aβ resulted in impairment of Na^+/K^+-ATPase activity (which normally protects neurons by reducing Na^+ influx) and resultant membrane depolarization contributed to subsequent elevation of $[Ca^{2+}]_i$ and cell death. Furthermore, ouabain, a selective Na^+/K^+-ATPase inhibitor, induced an elevation of $[Ca^{2+}]_i$ and neuronal apoptosis. The impairment of ion-motive ATPase activities was antagonized by antioxidants (vitamin E, propyl gallate, and PBN), suggesting that free radicals mediate impairment of pump activities. This neurotoxic mechanism of Aβ is likely to occur in human brains, because exposure of synaptosomes taken from neurologically normal aged humans to Aβ resulted in a highly significant impairment of both Na^+/K^+- and Ca^{2+}-ATPase activities (Fig. 3).

Aβ may also impair signal transduction pathways prior to cell death. Significantly impaired coupling of muscarinic receptors to M_1 muscarinic receptors to G proteins has been reported in postmortem brain tissue of AD patients (Flynn et al., 1991). They reported that the proportion of M_1 receptors existing in the high-affinity (G protein coupled) state is reduced in AD frontal cortex compared with age-matched controls. Furthermore, Aβ25–35 or Aβ1–40 attenuated the G protein-mediated Ca^{2+}-increase by the muscarinic agonist, carbachol (CCh) in the cultured rat cortical neurons (Kelly et al., 1996). These Aβs also diminished the membrane-associated GTPase activity and inositol phosphate release induced by CCh in cultured neurons. These effects by Aβ are antagonized by vitamin E, suggesting the involvement of free radicals. These results also indicate that Aβ may impair signal transduction mechanisms in the muscarinic pathway at an early stage of AD prior to neuronal degeneration.

Recently, several actions of Aβ have been investigated electrophysiologically. Arispe et al. (1993) introduced Aβ into phosphatidylserine liposomes and fused the liposomes with a planar lipid bilayer. The Aβ formed cation-selective channels capable of transporting calcium and monovalent cations. Furukawa et al. (1994) reported that

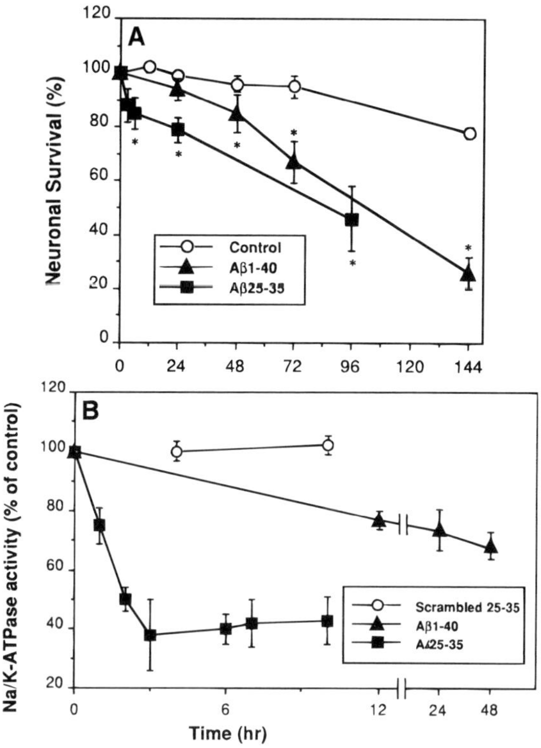

Fig. 3. Aβ inhibits Na^+/K^+-ATPase activity, which precedes neuronal death. **(A)** Cultures were exposed to vehicle (control), 50 μ*M* Aβ25–35, or 20 μ*M*, and neuronal survival was determined. Values represent the mean ±SD. $*p < 0.01$ compared to corresponding values in control. **(B)** Cultures were exposed to vehicle (control), 50 μ*M* Aβ25–35, 50 μ*M* scrambled Aβ 25–35, or 20 μ*M* Aβ 1–40, and Na^+/K^+-ATPase activity was determined. Values represent the mean ±SD. (Adapted with permission from Mark et al., 1996.)

Aβ elicits an irreversible nonselective ion current using dissociated rat cortical neurons. Such a nonselective inward current was also recorded from sympathetic ganglia of bullfrogs (Simmons and Schneider, 1993). On the other hand, potentiation of L-type voltage-activated Ca current by incubation with Aβ was reported (Davidson et al., 1994). These results suggest that Aβ may form ion channels capable of transporting Ca and other ions, and support the view that the channel activity of the polypeptide is involved in Aβ toxicity.

4. FUNCTIONS AND SIGNAL TRANSDUCTION PATHWAYS OF SAPP

Evidence that sAPP can protect neurons against neurotoxic insults relevant to the pathogenesis of cerebrovascular disease and AD comes from studies of excitotoxicity and Aβ-induced neuronal degeneration (Mattson et al., 1993b; Furukawa et al., 1996b). Pretreatment of cultured neurons from rat hippocampus with sAPPα before the application of either glutamate or Aβ resulted in a significant increase in neuronal survival. In vivo forebrain ischemia studies have shown that sAPPα (751 and 695 forms) can protect rat hippocampal CA1 neurons against transient ischemia-induced degeneration when

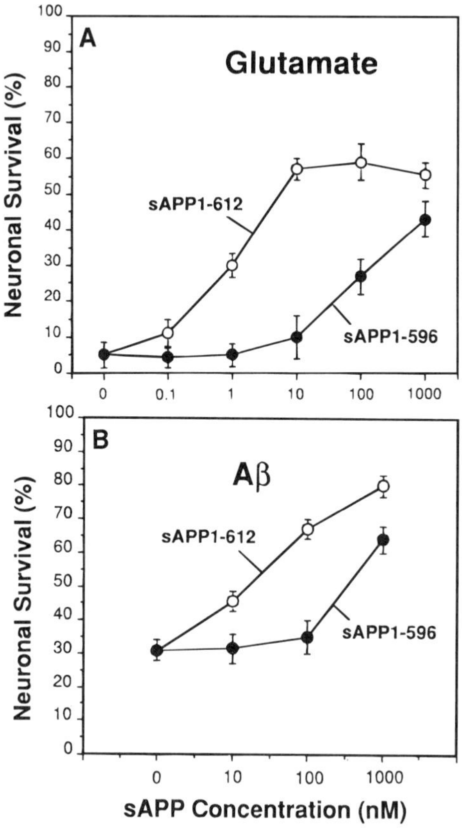

Fig. 4. Secreted APPs protect neurons in primary culture. Rat hippocampal cultures were exposed for 24 h to either 10 μ*M* glutamate (**A**) or 5 μ*M* Aβ25–35 after the pretreatment of sAPPs for 30 h. (**B**) Neuronal survival is expressed as percentage of initial number of neurons present prior to exposure to glutamate or Aβ. Values represent the mean and SEM of determinations made in 3–4 separate cultures. The secreted APPs employed included full-length sAPPα (APP1–612) and full-length sAPPβ (APP1–596). (Adapted with permission from Furukawa et al., 1996b.)

administered intracerebroventricularly immediately prior to the ischemic insult (Smith-Swintosky et al., 1994). In the case of spinal cord ischemia in rabbits, it was shown that sAPP peptides enhanced behavioral recovery after ischemic injury (Bowes et al., 1994). Further studies using primary culture have shown that sAPPα protects hippocampal neurons against oxidative stress and suppresses accumulation of reactive oxygen species (Goodman and Mattson, 1994). Pretreatment of cultured neurons with sAPPα confers resistance to oxidative insults, including exposure to glutamate, $FeSO_4$, and Aβ (Fig. 4); the hydrogen peroxide level in such pretreated cells, directly measured, is lowered.

In addition to protecting against neurotoxic insults, sAPPα can promote neurite outgrowth in cell culture and synaptogenesis in cortex in vivo (Ninomiya et al., 1994; Roch et al., 1994). Furthermore, the neuroprotective and neurotrophic roles of sAPP can be observed in a variety of brain injury paradigms that have been studied in vitro and in vivo. For example, levels of βAPP increase dramatically after ischemia in brain (Abe et al., 1991).

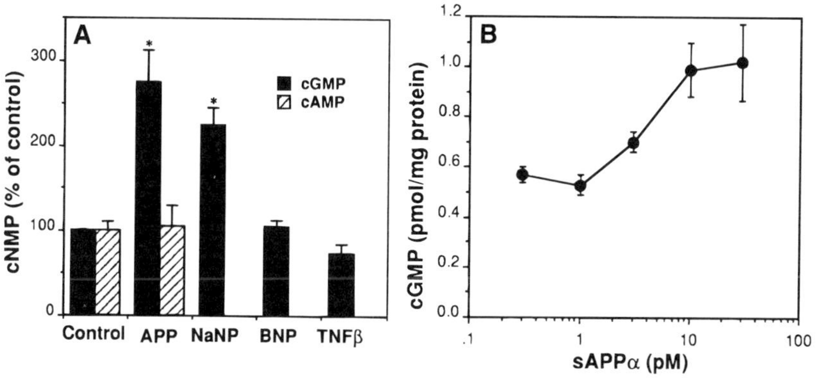

Fig. 5. Secreted APPα stimulates the levels of cGMP in neural cultures. Data are mean ±SEM (bars) value. **(A)** Hippocampal culture were treated with 500 p*M* sAPP695, 30 m*M* NaNP, 200 n*M* BNP, or 100 ng/mL of TNFβ for 5 min, and level of cGMP (closed bars) and cAMP (hatched bars) were determined. *$p < 0.05$ vs the control. **(B)** Hippocampal cultures were exposed to various concentrations of sAPPα695 for 5 min, and cGMP levels were determined. (Adapted with permission from Barger et al. 1995.)

sAPP released from synaptic terminals by electrical stimuli will bind to sAPP-specific receptors located on neuronal membrane, although the sAPP receptors have not yet been isolated. Barger et al. (1995) investigated change in cyclic nucleotides caused by sAPPα and reported that the application of sAPPα induced cGMP production, but did not increase cAMP production in cultured rat hippocampal neurons: At 5 min after the application of 500 p*M* sAPPα, cGMP increased approximately three-fold (Fig. 5). Because inhibition of the formation of nitric oxide or carbon monoxide did not affect the ability of sAPPα to rapidly lower intracellular Ca levels or elevate cGMP, they concluded that sAPPαs activate a membrane-associated guanylate cyclase rather than soluble (cytosolic) ones, perhaps similar to those present in the receptors for the natriuretic peptides (Barger and Mattson, 1995).

On the other hand, βAPP has a receptor-like structure and has a conserved cytoplasmic domain. Nishimoto and colleagues (1993) reported that βAPP forms a complex with G(o), a major GTP-binding protein in brain: The cytoplasmic βAPP sequence shows a specific G(o)-activating function that is necessary for complex formation. Their results suggest that abnormal βAPP-G(o) signaling is involved in the AD process.

sAPPαs can also activate transcription factors in neurons. Barger and Mattson (1996) indicated that sAPPα induced activation of a transcription factor, NFκB. They reported that transcription dependent on a κB element was enhanced by sAPPα in several cell lines, as measured by expression of a transfected chloramphenicol acetyltransferase reporter gene, and that sAPPα also induced an increase in κB DNA-binding activity in hippocampal neurons. Differentiated SH-SY5Y neuroblastoma cells, which were transfected with a chloramphenicol acetyltransferase (CAT) vector responsive to κB-binding transactivators, were used for a CAT assay. sAPPα increased CAT activity almost twice that of control, and Rp-8-Br-cGMPS (RGS), a selective inhibitor of cGMP-dependent kinase, antagonized this response. Therefore, we suggest that this NFκB induction by

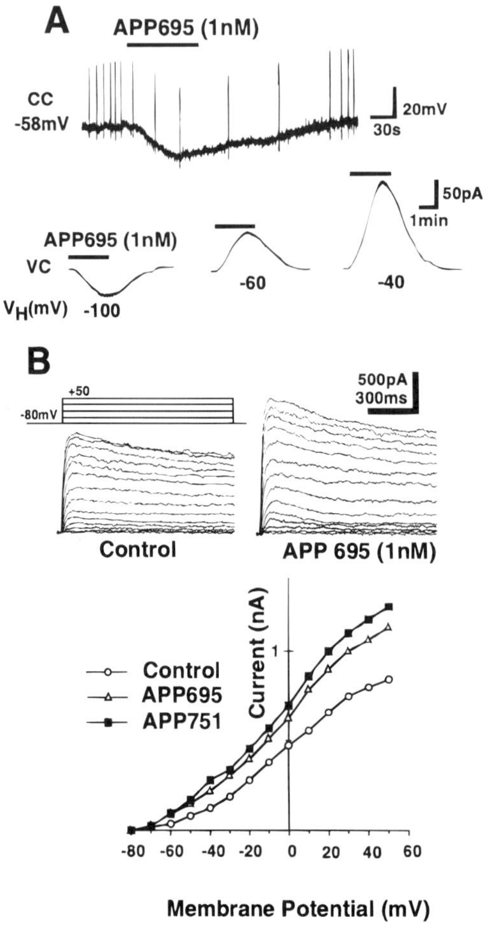

Fig. 6. Secreted APP reduces neuronal excitability by activating K^+ channels, and hyperpolarizes hippocampal neurons. **(A)** Upper record shows change in membrane potential recorded in current-clamp mode in neuron before, during, and following exposure to sAPPα695. The sAPPα elicited a transient hyperpolarization and greatly reduced action potential firing. The lower three traces show responses to sAPPα695 under voltage-clamp at three different potentials. **(B)** Top, representative current recordings in a neuron before (control) and 5 min after exposure to sAPPα751. Pulses were applied in voltage steps from –80 to +50 mV. Bottom, I/V relationships of individual recordings in the same cell before treatment, 5 min after exposure to sAPPα695, and 5 min after exposure to sAPPα751. (Adapted with permission from Furukawa et al., 1996a.)

sAPP can induce expression of some neuroprotective or neurotrophic genes, such as *bcl*-2 or antioxidant enzymes.

The first report on electrophysiological functions of sAPP was published in 1996, when Furukawa and co-workers (1996a) reported that sAPPαs activate K^+ channels, which are charybdotoxin (CHX)-sensitive through a cGMP pathway. Figure 6A shows the effect of sAPP on membrane potential under current-clamp conditions. The application of sAPPα hyperpolarizes a cultured rat hippocampal neuron and decreases spontaneous firing of action potentials. sAPPα also induced an outward current at holding potentials from –80 to –20 mV under the voltage-clamp condition. The reversal potential of the sAPPα-induced currents was about –85 mV, which is equal to the equilibrium potential of K^+ ion estimated from the Nernst equation. Therefore, the responses induced by sAPPα are mediated by the activation of K^+ channels. Three kinds of K^+ channel blockers (glibenclamide, apamine, and charybdotoxin) were examined in this study. Of these, only charybdotoxin antagonized the potentiation of K^+ current by sAPPα, indicating that the K^+ channel activated by sAPPαs is a large conductance, Ca^{2+}-dependent K^+ channel (BK channel).

The signal transduction pathways involved in the activation of K^+ channels by sAPPα were also examined. A specific cGMP-dependent kinase inhibitor, RGS, antagonized the K^+ channel potentiation by sAPPα. In addition, 8-Br-cGMP, a membrane-permeable derivative of cGMP, activated K^+ current by itself; this response may be activated by an increase in intracellular cGMP. Sodium nitroprosside, an inducer of nitric oxide formation, which activates soluble guanylate cyclase, also activated the K^+ current, an effect that was not further increased by sAPPs. Inhibitors of nitric oxide production, including methylene blue and *N*-methyl-arginine, did not block the sAPPs' effect, consistent with the activation of a membrane-associated guanylyl cyclase. Okadaic acid prevented activation of I_K by sAPPs, indicating the involvement of a protein phosphatase. The intracellular Ca^{2+}-chelating agent BAPTA-AM did not affect the sAPP response, and sAPP increased K^+ channel activity in single-channel recordings (outside-out configuration), using an internal solution containing BAPTA. These data suggest that sAPPs modulate BK channels by a Ca^{2+}-independent mechanism involving cGMP kinase and phosphatases. K^+ channel activation may stabilize neuronal excitability and intracellular levels of $[Ca^{2+}]_i$, because the application of sAPPα decreases $[Ca^{2+}]_i$ rapidly. From these results, sAPP is considered to stabilize neuronal excitability and intracellular Ca^{2+} homeostasis through a cGMP pathway.

In order to find out which sequence in sAPP has potent action for neuroprotection and activity-regulation, Furukawa et al. (1996b) employed several different types of sAPP preparations, including full-length recombinant sAPPs (sAPPα1–612 and sAPPβ1–596) purified from the culture supernatant of HEK293 cells transfected to express the corresponding cDNAs, N-terminally truncated GST-sAPP fusion proteins (sAPPα444–612 and sAPPβ444–596; sAPPα304–612 and sAPPβ304–596; sAPPα591–612 and sAPPβ591–596), and 20-amino acid synthetic sAPP peptides (sAPPα591–610 and sAPPβ576–595). Every sAPPα showed 100–1000 times stronger neuroprotective effect against glutamate and Aβ toxicities than those of sAPPβs. sAPPαs also significantly activated K^+ channels and attenuated Ca^{2+}-increase by glutamate, and their potencies were much higher than those of sAPPβs.

The fact that GST-sAPP591–612 and synthetic peptide sAPP591–610 exhibited neuroprotective and K^+-channel activating effects similar to those of full-length sAPP1–612 suggests that these actions of sAPPα reside in the sequences immediately surrounding the β-secretase cleavage site. Furthermore, because sAPPα contains a heparin-binding domain (VHHQK) near its carboxy-terminus (Brunden et al., 1993; Buee et al., 1993; Leveugle et al., 1994), it has been determined whether the heparin-binding domain played a role in sAPPα activities. Both heparinases I and II attenuated the activities of sAPPαs to protect neurons against excitotoxicity and Aβ toxicity, to activate K^+ channels, and to decrease the increase in $[Ca^{2+}]_i$ by glutamate. From these results, it is suggested that the heparin-binding domain at the C-terminus of sAPPαs is a potent receptor activator. Furthermore, mutations in βAPP previously shown to increase β-secretase cleavage (Cai et al., 1993; Citron et al., 1994) may promote neuronal degeneration as well, not only by increase in Aβ production, but also by increasing levels of less neuroprotective sAPPβs and reducing levels of more potent sAPPαs.

5. GENETIC MUTATIONS IN ALZHEIMER'S DISEASE

The inherited mutations of βAPP have been found in families with autosomal-dominant early-onset inheritance of AD. The first mutation was identified at βAPP codon 717, adjacent to the βAPP cleavage site that generates the Aβ C-terminus (Goate et al., 1991). A cell line expressing this mutant βAPP increases production of the 42-amino acid form of Aβ (Suzuki et al., 1994), the predominant form of Aβ in senile plaques (Roher et al., 1993). The second inherited βAPP mutation was found at the two amino acids near the Aβ N-terminus (Mullan et al., 1992), producing several times as much Aβ as normal genes. In addition to these mutations, Levy et al. (1990) reported that Dutch families having hereditary cerebral hemorrhage carry a mutation within Aβ that potentiates the rate of Aβ aggregation. These reports, which showed that families susceptible to AD carry mutations in βAPP that produce more Aβ or accelerate fibril formation, support the idea that Aβ production and its fibril formation are the causes of AD.

As mentioned in section 2, many labs have tried to establish a transgenic AD mouse based on human βAPP mutation (Quon et al., 1991; Sandhu et al., 1991; Kammesheidt et al., 1992; Lamb et al., 1993; Higgins et al., 1994; LaFerla et al., 1995). The transgenic mice obtained by Games and co-workers (1995), using PDGF-β and human mutant APP (V717F) as a promoter and a transgene, respectively, showed remarkable thioflavin S-positive Aβ deposits, neuritic plaques, synaptic loss, astrocytosis, and microgliosis. This success in transgenic animals also confirmed the theory that abnormal metabolism of βAPP is a cause of AD.

ApoE is a component of the very low-density lipoprotein (VLDL) and high-density lipoprotein complexes that contribute to the cellular uptake and regulation of cholesterol (Mahley, 1988). There are three alleles of ApoE (E2, E3, and E4), and Strittmatter et al. (1993) reported high frequency of ApoE4 allele in late-onset AD patients. ApoE–lipoprotein complex is taken up by neurons, leading to the intracellular free cholesterol release that is thought to be transported to nerve terminals, where it may be utilized for neurite outgrowth and synaptogenesis (Nathan et al., 1994). Although ApoE3 complexed with VLDL increases neurite outgrowth, ApoE4 complex inhibits the outgrowth. Therefore, it is suggested that ApoE4 allele may accelerate neuronal degeneration in aged brains by inhibiting a compensatory function for neurite outgrowth and synaptogenesis. Furthermore, brains of AD patients with ApoE4 allele show a significant increase in the number of senile plaques (Schmechel et al., 1993). In addition, Schmechel and coworkers reported that ApoE binds to soluble Aβ in vitro, indicating a strong correlation between ApoE and βAPP processing. ApoE also promotes amyloid fibril formation in an isoform-specific manner, that is, ApoE4 induces Aβ fibril formation more strongly than ApoE3 (Ma et al., 1994; Sanan et al., 1994). These in vitro data coincide with in vivo pathological results, which showed that ApoE is associated with amyloid plaque formation (Namba et al., 1991; Näslund et al., 1995). From these results, we can speculate that ApoE isoforms influence AD onset by affecting Aβ deposition and fibril formation.

Two novel genes inherited in familial AD patients were identified in 1995. One gene was found on chromosome 14, whose locus had been very common because it had been found in many early-onset familial AD families (Schellenberg et al., 1992). Mutations on chromosome 14 have been identified in a 467-amino acid protein having 7–9 transmembrane domains, and the protein was named S182 or PS1 (Sherrington et al., 1995). The other gene was identified on chromosome 1 as the genetic locus of familial AD in Volga German families (Levy-Lahad et al., 1995a,b; Li et al., 1995; Rogaev et al., 1995) and was named STM2 or PS2. The protein coded by PS2, which also has 7–9 transmembrane domain constructure, is highly homologous to PS1. PS1 and PS2 have significant homology to two proteins from *Caenorhabditis elegans*, SEL-12 and SPE-4. SEL-12 was identified as a protein that potentiates signaling through the Notch/LIN-12 family during development (Levitan and Greenwald, 1995). SPE-4 is located in Golgi, and functions by segregating membrane proteins during spermatogenesis (L'Hernault and Arduengo, 1992). Kovacs et al. (1996) reported that PS1 and PS2 are located on endoplasmic reticulum and Golgi complex in a transfected cell line. Therefore, the presenilins might regulate the transport or processing of proteins and intracellular Ca^{2+} release or uptake. The presenilins also affect βAPP processing and Aβ production (Scheuner et al., 1996). Scheuner and colleagues reported that fibroblasts obtained from people with PS1 mutations showed a high level of Aβ release, especially in the highly amyloidogenic Aβ42 form. Furthermore, Vito et al. (1996) reported the relationship between PS2 and apoptosis by identifying two cDNAs, ALG-2 and ALG-3, which modulate apoptosis. ALG-2 encoded a calcium-binding protein that participates in the induction of apoptosis by T-cell receptor, Fas, and glucocorticoid. On the other hand, ALG-3 had a high homology with PS2 and prevented T-cell receptor and Fas-induced apoptosis. From these results, it is proposed that PS1 and PS2 are involved in AD through regulation of βAPP metabolism and neuronal apoptosis.

Guo and colleagues reported that cells expressing PS1 mutant, L286V, exhibit increased vulnerability to amyloid β-peptide toxicity and apoptosis induced by nerve growth factor withdrawal (Guo et al., 1996). The cells expressing L286V also showed exaggerated Ca^{2+} response to bradykinin and carbachol, which induce Ca^{2+} release from endoplasmic reticulum. The endangering action of L286V involves oxidative stress and disruption of Ca homeostasis because antioxidants and Ca channel blockers counteract the adverse consequences of PS1 mutation. PS1 mutation may sensitize neurons by perturbing Ca^{2+} homeostasis, which is related to age-dependent accumulation of Aβ and reduction of trophic factor support.

6. FUTURE PROSPECTS FOR AD TREATMENT

This chapter assumes that βAPP is one of the key elements in the pathogenesis of AD. However, some neuroscientists insist that neither βAPP or Aβ is involved in AD pathogenesis, because strong Aβ deposition is sometimes observed in brains of nondemented people and no Aβ deposition is seen in some brains of severely demented (Morris et al., 1991a,b; Berg et al., 1993). Furthermore, there is an idea that Aβ deposition is not a real cause of dementia, but only a phenotype of AD. To counter this idea, I would like to cite four major evidences that βAPP and Aβ are causes of AD.

First, it is well known that amyloid plaques constituted by Aβ are richly found in AD brains (*see* Selkoe, 1994, for review). Although there are some exceptions, senile plaques are still an important pathological characteristic in AD (Giannakopoulos et al., 1994; Iwatsubo et al., 1994; Cummings and Cotman, 1995). Second, genetic mutations of βAPP are related to AD and AD-like diseases. Down's syndrome caused by chromosome 21 trisomy has AD-like neuropathology, that is, senile plaque-like conformation constituted by Aβ (St. George-Hyslop et al., 1987). Several mutations (e.g., codon 717) in βAPP have also been discovered in inherited familial AD patients (Goate et al., 1991; Mullan et al., 1992). Third, the transgenic mouse expressing human βAPP mutation showed similar neuropathology to human AD. The most reliable animal model is the transgenic mice established by Games et al. (1995). The mice having human mutant βAPP (V717F) display Aβ deposits, neuritic plaques, synaptic loss, astrocytosis, and microgliosis. Fourth, Aβ is very toxic to cultured and in vivo neurons (Yankner et al., 1990; Frautschy et al., 1991; Kowall et al., 1991; Mattson et al., 1992). Aβ applied to culture medium or injected into brain tissue deteriorates neurons significantly. The mechanisms of Aβ toxicity have been investigated by many neuroscientists, as is described in this chapter. Therefore, it is suggested that βAPP processing followed by Aβ deposition is the key mechanism of AD.

On the other hand, there is another argument that synaptic loss and neurofibrillary tangles indicate the severity of AD more accurately than the number of amyloid plaques (Morris et al., 1991b). However, Aβ shows toxicity without plaque formation in culture experiments and the number of plaques formed does not reflect total amount of Aβ in human brains. Furthermore, plaques of demented and nondemented patients consist of different protein compositions (Arai et al., 1990; Mesulam and Geula, 1994).

Several compounds offer potential for treatment of AD. First, estrogen is one of the strong candidates for AD therapy. Postmenopausal women taking estrogen replacement therapy have a reduced incidence of AD (Henderson et al., 1994). In vitro cell culture studies have shown that 17β-estradiol promotes long-term neuronal survival (Chowen et al., 1992), and in vivo animal model studies indicate that estrogen stimulate production of neurotrophic factors in the brain (Singh et al., 1995). Furthermore, estrogens, including 17β-estradiol and estriol, protect cultured rat hippocampal neurons against Aβ toxicity and oxidative insults (Goodman et al., 1996). Goodman and coworkers also reported that lipid peroxidation induced by Aβ was significantly inhibited in neurons pretreated with the estrogens, suggesting that the estrogens have antioxidant functions.

There are several sources of evidence that some inflammatory processes participate in neurodegeneration in AD. Breitner et al. (1994) reported that use of nonsteroidal anti-inflammatory agents may reduce the risk of developing of AD. In addition, inhibitors of the arachidonic acid cascade (a pathway activated in many inflammatory conditions) protect cultured rat neurons against Aβ toxicity (Goodman et al., 1994). Postmortem AD brains show the presence of a variety of inflammation-related molecules in association with senile plaques and degenerating neurons (Eikeleboom et al., 1994). Activated microglia are also associated with plaque, and some of their roles in inflammatory processes are well known, since they produce superoxide and hydrogen peroxide. Therefore, inflammatory responses must exist in AD brains, and anti-inflammatory compounds could be a key to the cure of AD.

The expression of neurotrophic factors and cytokines increases in response to neuronal injury. In experimental models of traumatic, ischemic, and excitotoxic neuronal injuries in rats and mice, the following trophic factors and cytokines are highly expressed: basic fibloblast growth factor (bFGF), nerve growth factor (NGF), brain-derived neurotrophic factor (BDNF), insulin-like growth factor-1 (IGF-1), transforming growth factor-β (TGFβ), sAPPs, tumor necrosis factor (TNF), interleukin-1β, and protease nexin-1 (PN-1) (*see* Mattson and Scheff, 1994, for review). In support of functions for expression of these compounds in protecting neurons and promoting recovery, many in vitro and in vivo studies have indicated that application of the trophic factors or cytokines can prevent neuronal death and/or promote neuronal outgrowth. Basic FGF reportedly has very potent neuroprotective and neurotrophic effects; for example, bFGF promoted neurite outgrowth and long-term survival of cultured neurons and protected neurons against the excitotoxicity, Aβ toxicity, and glucose deprivation (Mattson et al., 1989; Mattson and Rychlik, 1990; Cheng and Mattson, 1992; Mattson et al., 1993a). Other neurotrophic factors and cytokines also have been reported to have significant trophic and protective effects (*see* Mattson et al., 1996, for review). The mechanisms of neuroprotective functions of these compounds may be enhancement of intracellular Ca^{2+} stabilization and reduction of free radicals. For example, bFGF can attenuate increase in $[Ca^{2+}]_i$ induced by Aβ or glutamate through increasing expression of calbindin (Ca^{2+}-binding protein) (Collazo et al., 1992) and reducing expression of NMDA receptor protein (Mattson et al., 1993c). Furthermore, bFGF also induces Cu/Zn-SOD and glutathione reductase expression (Mattson et al., 1995), suggesting that induction of such antioxidant enzymes may underlie its ability to protect neurons against Aβ toxicity and oxidative insults.

βAPP mRNA and protein expression increase when a brain is injured (Abe et al., 1991; McKenzie et al., 1994; Banati et al., 1995), and we have shown that the sAPPs have potent neuroprotective effects in cultured rat and human neurons against glutamate toxicity, Aβ toxicity, and glucose deprivation (Mattson et al., 1993b; Furukawa et al., 1996b). The sAPPs also protected hippocampal neurons in in vivo rat transient-ischemia model (Smith-Swintosky et al., 1994). The sAPPs or related compounds must be major candidates for a drug to cure AD.

Signal transduction pathways activated by sAPPs include cGMP-mediated, sAPP-induced activation of K^+ channels and suppression of neuronal excitability. The neuroprotective effects of sAPPs may be mediated by this acute signal transduction mechanism. In addition to the acute mechanism, the sAPPs can activate transcription factor NFκB, which may lead to further induction of neuroprotective genes and proteins. For example, sAPPs may increase resistance of neurons to oxidative insults by inducing the expression of antioxidant enzymes. In rat neuronal cultures, sAPPs activate NFκB and induce expression of Cu/Zn-SOD and catalase (Barger and Mattson, 1996). The sAPPs may also activate MAP-kinases, which are well-known as components of neurotrophic factor signaling pathways (Greenberg et al., 1994). Cheng and coworkers (1994b) reported that TNFα and TNFβ strongly induce expression of a calcium-binding protein, calbindin-D_{28k}, apparently through NFκB induction. sAPPα may therefore also have a signal transduction pathway similar to that of TNFs (Bruce et al., 1996).

The differential potencies of sAPPαs and sAPPβs, reported in a recent study (Furukawa et al., 1996b), indicate that physiological and pathophysiological factors that influence α- and β-secretase cleavages of βAPP may affect neuronal plasticity and the pathogenesis of AD. It has been reported that sAPPs play roles in the regulation of neurite outgrowth, synaptogenesis, and synaptic plasticity (Milward et al., 1992; Ninomiya et al., 1994). We can propose that some βAPP mutations, which function by altering the regulation of βAPP, so that β-secretase cleavage is increased and/or α-secretase cleavage reduced, are causally linked to AD.

Furthermore, Goodman and Mattson (1996b) have recently reported that K^+ channel openers, such as diazoxide, levochromakalim, and pinacidil, protect rat hippocampal neurons against oxidative injury and Aβ toxicity. K^+ channel openers suppressed intracellular peroxide and the generation of peroxides induced by $FeSO_4$ and Aβ. Therefore, it is suggest that sAPPα protects neurons not only through stabilizing $[Ca^{2+}]_i$ but also by suppressing free radical formation.

ACKNOWLEDGMENTS

I want to thank S. W. Barger, V. L. Smith-Swintosky, B. Cheng, R. J. Mark, and M. P. Mattson for contributions to work from this laboratory.

REFERENCES

Abe, K., Tanzi, R. E., and Kogure, K. (1991) Selective induction of Kunitz-type protease inhibitor domain-containing amyloid precuorsor protein mRNA after persistent focal ischemia in rat cerebral cortex. *Neurosci. Lett.* **125,** 172–174.

Alzheimer, A. (1907) Uber eine eigenartige Erkrangkung der Hirmrinde. *All. Z. Psychiatr.* **64,** 146–148.

Arai, H., Lee, V. M.-Y., Otvos, L. Jr., Greenberg, B. D., Lowery, D. E., Sharma, S. K., Schmidt, M. L., and Trojanowski, J. Q. (1990) Defined neurofilament, τ, and β-amyloid precursor protein epitopes distinguish Alzheimer from non-Alzheimer senile plaques. *Proc. Natl. Acad. Sci. USA* **87,** 2249–2253.

Arispe, N., Pollard, H. B., and Rojas, E. (1993) Giant multilevel cation channels formed by Alzheimer disease amyloid β-protein [AβP-(1–40)] in bilayer membranes. *Proc. Natl. Acad. Sci. USA* **90,** 10,573–10,577.

Banati, R. B., Gehrmann, J., Wiessner, C., Hossmann, K. A., and Kreutzberg, W. (1995) Glial expression of the beta-amyloid precursor protein APP in global ischemia. *J. Cereb. Blood Flow Metab.* **15,** 647–654.

Barger, S. W. and Mattson, M. P. (1995) Secreted form of the Alzheimer's amyloid precursor protein stimulates a membrane-associated guanylate cyclase. *Biochem. J.* **311,** 45–47.

Barger, S. W., Fiscus, R. R., Ruth, P., Hofmann, F., and Mattson, M. P. (1995) Role of cyclic GMP in the regulation of neuronal calcium and survival by secreted forms of β amyloid precursor. *J. Neurochem.* **64,** 2087–2096.

Barger, S. W. and Mattson, M. P. (1996) Secreted form of the Alzheimer's β-amyloid precursor induces neuroprotective κB-dependent transcription. *Mol. Brain Res.* **40,** 116–126.

Behl, C., Davis, J. B., Lesley, R., and Schubert, D. (1994) Hydrogen peroxide mediates amyloid β protein toxicity. *Cell* **77,** 817–827.

Berg, L., McKeel Jr., D. W., Miller, J. P., Baty, J., and Morris, J. C. (1993) Neuropathological indexes of Alzheimer's disease in demented and nondemented persons aged 80 years and older. *Arch. Neurol.* **50,** 349–358.

Bowes, M. P., Masliah, E., Otero, D. A. C., Zivin, J. A., and Saitoh, T. (1994) Reduction of neurobiological damage by a peptide segment of the amyloid β-A4 protein precursor in a rabbit spinal cord ischemia model. *Exp. Neurol.* **129,** 112–119.

Breitner, J. C. S., Gau, B. A., Welsh, K. A., Plassman, B. L., McDonald, W. M., Helms, M. J., and Anthony, J. C. (1994) Inverse association of anti-inflammatory treatments and Alzheimer's disease. *Neurology* **44,** 227–232.

Bruce, A. J., Boling, W., Kindy, M. S., Peschon, J., Kramer, P. J., Carpenter, M. K., and Holtberg, F. W. (1996) Altered neuronal and microglial responses to excitotoxic and ischemic brain injury in mice lacking TNF receptors. *Nature Med.* **2,** 788–794.

Brunden, K. R., Richter-Cook, N. J., Chaturvedi, N., and Frederickson, R. C. A. (1993) pH-dependent binding of synthetic β-amyloid peptides to glycosaminoglycans. *J. Neurochem.* **61,** 2147–2154.

Buee, L., Ding, W., Delcourte, A., and Fillint, H. (1993) Binding of secreted human neuroblastoma proteoglycans to the Alzheimer's amyloid A4 peptide. *Brain Res.* **601,** 154–163.

Burdick, D., Soregham, B., Kwon, M., Kosmoski, J., Knauer, M., Henschen, A., Yates, J., et al. (1992) Assembly and aggregation properties of synthetic Alzheimer's A4/Beta amyloid peptide analogs. *J. Biol. Chem.* **267,** 546–554.

Busciglio, J., Lorenzo, A., and Yanker, B. A. (1992) Methodological variables in the assessment of β amyloid neurotoxicity. *Neurobiol. Aging* **13,** 609–612.

Busciglio, J., Yeh, J., and Yanker, B. A. (1993) β-Amyloid neurotoxicity in human cortical culture is not mediated by excitotoxins. *J. Neurochem.* **61,** 1565–1568.

Busciglio, J., Lorenzo, A., Yeh, J., and Yankner, B. A. (1995) β-Amyloid fibrils induce tau phosphorylation and loss of microtubule binding. *Neuron.* **14,** 879–888.

Bush, A. I., Pettingell, W. H., Multhaup, G., deParadis, M., Vonsattel, J.-P., Gusella, J. F., et al. (1994) Rapid induction of Alzheimer Aβ amyloid formation by zinc. *Science* **259,** 514–516.

Butterfield, D. A., Hensley, K., Harris, M., Mattson, M. P., and Carney, J. (1994) β-amyloid peptide free radical fragments initiate synaptosomal lipoperoxidation in a sequence-specific fashion: implications to Alzheimer's disease. *Biochem. Biophys. Res. Commun.* **200,** 710–715.

Buxbaum, J. D., Gandy, S. E., Cicchetti, P., Ehrlich, M. E., Czemik, A. J., Francasso, R. P., et al. (1990) Processing Alzheimer β/A4 amyloid precursor protein: modulation by agents that regulate protein phosphorylation. *Proc. Natl. Acad. Sci. USA* **87,** 6003–9006.

Cai, X. D., Golde, T. E., and Younkin, S. G. (1993) Release of excess amyloid beta protein from a mutant amyloid beta protein precursor. *Science* **259,** 514–516.

Cheng, B. and Mattson, M. P. (1992) Glucose deprivation elicits neurofibrillary tangle-like antigenic changes in hippocampal neurons: prevention by NGF and bFGF. *Exp. Neurol.* **117,** 114–123.

Cheng, B., Barger, S. W., and Mattson, M. P. (1994a) Staurosporine, K-252a and K-252b stabilize calcium homeostasis and promote survival of CNS neurons in the absence of glucose. *J. Neurochem.* **62,** 1319–1329.

Cheng, B., Christakos, S., and Mattson, M. P. (1994b) Tumor necrosis factor protect neurons against metabolic excitotoxic insults and promote maintenance of calcium homeostasis. *Neuron.* **12,** 139–153.

Chowen, J. A., Torres-Aleman, I., and Garcia-Segura, L. M. (1992) Trophic effects of estradiol on fetal rat hypothalamic neurons. *Neuroendocrinology* **56,** 895–901.

Citron, M., Vigo-Pelfrey, C., Teblow, D. B., Miller, C., Schenk, D., Johnston, J., et al. (1994) Excessive production of amyloid beta-protein by peripheral cells of symptomatic and presymptomatic patients carrying the Swedish familial Alzheimer disease mutation. *Proc. Natl. Acad. Sci. USA* **91,** 11,993–11,997.

Clapham, D. E. (1995) Calcium signaling. *Cell* **80,** 259–268.

Collazo, D., Takahashi, H., and McKay, R. D. (1992) Cellular targets and trophic functions of neurotrophin-3 in the developing rat hippocampus. *Neuron.* **9,** 643–656.

Cummings, B. J. and Cotman, C. W. (1995) Image analysis of β-amyloid load in Alzheimer's disease and relation to dementia severity. *Lancet* **346,** 1524–1528.

Davidson, R. M., Shajenko, L., Donta, T. S. (1994) Amyloid β-peptide (AβP) potentiate a nimodipine-sensitive L-type barium condactance in N1E-115 neuroblastoma cells. *Brain Res.* **643,** 324–327.

Dyrks, T., Dyrks, E., Hartmann, T., Masters, C., and Beyreuter, K. E. (1992) Amyloidogenicity of beta A4 and beta A4-bearing amyloid protein precursor fragments by metal-catalyzed oxidation. *J. Biol. Chem.* **267,** 18,210–18,217.

Eikeleboom, P., Zhan, S. S., van Gool, W. A., and Allsop, D. (1994) Inflammatory mechanisms in Alzheimer's disease. *Trends Pharmacol. Sci.* **15,** 447–450.

Esch, F. S., Keim, P. S., Beattie, E. C., Blacher, R. W., Culwell, A. R., Oltersdorf, T., McClure, D., and Ward, P. J. (1990) Cleavage of amyloid β-peptide during constitutive processing of its precursor. *Science* **248,** 1122–1124.

Flynn, D. D., Weinstein, D. A., and Mash, D. C. (1991) Loss of high-affinity agonist binding to M1 muscarinic receptor in Alzheimer's disease: implications for failure of cholinergic replacement therapies. *Ann. Neurol.* **29,** 256–262.

Fraser, P. E., Nguyen, J. T., Surewics, W. K., and Kirschner, D. A. (1991) pH-dependent stractural transitions of Alzheimer amyloid peptides. *Biophys. J.* **60,** 1190–1201.

Frautschy, S. A., Baird, A., and Cole, G. M. (1991) Effects of injected Alzheimer β-amyloid cores in rat brain. *Proc. Natl. Acad. Sci. USA* **92,** 369–371.

Fukuchi, K., Deeb, S. S., Kamino, K., Ogburn, C. E., Snow, A. D., Sekiguchi, R. T., et al. (1992) Increased expression of the β-amyloid protein precursor and microtubule-associated protein tau during the differentiation of murine embryonal acrcinoma cells. *J. Neurochem.* **58,** 1863–1873.

Furukawa, K., Abe, Y., and Akaike, N. (1994) Amyloid β-protein-induced irreversible current in rat cortical neurones. *Neuroreport* **5,** 2016–2018.

Furukawa, K., Barger, S. W., Blalock , E., and Mattson, M. P. (1996a) Activation of K^+ channels and suppression of neuronal activity by secreted β-amyloid precursor protein. *Nature* **379,** 74–78.

Furukawa, K., Sopher, B., Rydel, R. E., Begley, J. G., Martin, G. M., and Mattson, M. P. (1996b) Increased activity-regulating and neuroprotective efficacy of α-secretase derived secreted APP is conferred by the heparin-binding domain of Aβ. *J. Neurochem.*, **63,** 1882–1896.

Games, D., Adams, D., Alessandrini, R., Berthelette, P., Blackwell, C., Carr, T., et al. (1995) Alzheimer-type neuropathology in transgenic mice overexpressing V717F β-amyloid precursor protein. *Nature* **373,** 523–527.

Giannakopoulos, P., Hof, P. R., Giannakopoulos, A. S., Buee-Scherrer, V., Surini, M., Delacourte, A., and Bouras, C. (1994) Dementia in the oldest-old: quantitative analysis of 12 cases from psychiatric hospital. *Dementia* **5,** 348–356.

Glenner, G. G. and Wong, C. W., (1984) Alzheimer's disease: initial report of the purification and characterization of a novel cerebrovascular amyloid protein. *Biochem. Biophys. Res. Commun.* **120,** 885–890.

Goate, A., Chartier-Harlin, M.-C., Mullan, M., Brown, J., Crawford, F., Fidani, L., et al. (1991) Segregation of a missense mutation in the amyloid precursor protein gene with familial Alzheimer's disease. *Nature* **349,** 704–706.

Goodman, Y. and Mattson, M. P. (1994) Secreted forms of β-amyloid precursor protein protect hippocampal neurons against amyloid β-peptide-induced oxidative injury. *Exp. Neurol.* **128,** 1–12.

Goodman, Y. and Mattson, M. P. (1996a) Ceramide protects hippocampal neurons against excitotoxic and oxidative insults, and amyloid β-peptide toxicity. *J. Neurochem.* **66,** 869–872.

Goodman, Y. and Mattson, M. P. (1996b) K^+ channel openers protect hippocampal neurons against oxidative injury and amyloid β-peptide toxicity. *Brain Res.* **706,** 328–332.

Goodman, Y., Steiner, M. R., Steiner, S. M., and Mattson, M. P. (1994) Nordihydroguaiaretic acid protects hippocampal neurons against amyloid β-toxicity, and attenuates free radical and calcium accumulation. *Brain Res.* **654,** 171–176.

Goodman, Y., Bruce, A. J., Cheng, B., and Mattson, M. P. (1996) Estrogens attenuate and corticosterone exacerbates excitotoxicity, oxidative injury and amyloid β-peptide toxicity in hippocampal neurons. *J. Neurochem.* **66,** 1836–1844.

Greenberg, S. M., Koo, E. H., Selkoe, D. J., Qiu, W. Q., and Kosik, K. S. (1994) Secreted beta-amyloid precursor protein stimulates mitogen-activated protein kinase and enhances tau phosphorylation. *Proc. Natl. Acad. Sci. USA* **91,** 7104–7108.

Guo, Q., Furukawa, K., Sopher, B. L., Pham, D. G., Robinson, N., Martin, G. M., and Mattson, M. P. (1996) Alzheimer's PS-1 mutation perturbs calcium homeostasis and sensitizes PC12 cells to death induced by amyloid β-peptide. *Neuroreport* **8,** 379–383.

Haass, C., Schlossmacher, M. G., Hung, A. Y., Vigo-Pelfrey, C., Mellon, A., Ostaszewski, B., et al. (1992) Amyloid β-peptide is produced by cultured cells during normal metabolism. *Nature* **359,** 322–325.

Henderson, V. W., Paganini-Hill, A., Emanuel, C. K., Dunn, M. E., and Buckwalter, J. G. (1994) Estrogen replacement therapy in older women. Comparisons between Alzheimer's disease cases and nondemented control subjects. *Arch. Neurol.* **51,** 869–900.

Hensley, K., Carney, J. M., Mattson, M. P., Aksenova, M., Wu, J. F., Floyd, R., and Butterfield, D. A. (1994) A model for β-amyloid aggregation and neurotoxicity based on free radical generation by the peptide: relevance to Alzheimer's disease. *Proc. Natl. Acad. Sci. USA* **91,** 3270–3274

Higgins, L. S., Holtzman, D. M., Rabin, J., Mobley, W. C., and Cordell, B. (1994) Transgenic mouse brain histopathology resembles early Alzheimer's disease. *Ann. Neurol.* **35,** 598–607.

Hilbich, C., Kisters-Woike, B., Reed, J., Masters, C. L., Beyreuther, K. (1991) Aggregation and secondary structure of synthetic amyloid beta A4 peptides of Alzheimer's disease. *J. Mol. Biol.* **218,** 149–163.

Howlet, D. R., Jennings, K. H., Lee, D. C., Clark, M. S. G., Brown, F., Wetzel, R., Wood, S. J., Camilleri, P., and Roberts, G. W. (1995) Aggregation state and neurotoxic properties of Alzheimer β-amyloid peptide. *Neurodegeneration* **4,** 23–32.

Iacopino, A. M., Quintero, E. M., and Miller, E. K. (1994) Calbindin-D_{28K}: a potential neuroprotective protein. *Neurodegeneration* **3,** 1–20.

Iwatsubo, T., Odaka, A., Suzuki, N., Mizusawa, H., Nukina, N., and Ihara, Y. (1994) Visualization of Aβ42(43) and Aβ40 in senile plaques with end-specific Aβ monoclonals: evidence that an initially deposited species is Aβ42(43). *Neuron* **13,** 45–53.

Kammesheidt, A., Boyce, F. M., Spanoyannis, A. F., Cummings, B. J., Oretegón, M., Cotman, C., et al. (1992) Deposition of β/A4 immunoreactivity and neuronal pathology in transgenic mice expressing the carboxy-terminal fragment of the Alzheimer amyloid precursor protein. *Proc. Natl. Acad. Sci. USA* **89,** 10,857–10,861.

Kang J., Lemaire, H. G., Unterbeck, A., Salbaum, J. M., Masters, C. L., Gresesckik, K.-H., et al. (1987) The presence of Alzheimer's disease amyloid A4 protein resembles a cell-surface receptor. *Nature* **325,** 733–736.

Kelly, J. F., Furukawa, K., Barger, S. W., Rengen, M. R., Mark, R. J., Blanc, E. M., Roth, G. S., and Mattson, M. P. (1996) Amyloid β-peptide disrupts carbachol-induced muscarinic cholinergic signal transduction in cortical neurons. *Proc. Natl. Acad. Sci. USA* **93,** 6753–6758.

Kovacs, D. M., Fausett, H. J., Page, K. J., Kim, T.-W., Moir, R. D., Merriam, D. E., et al. (1996) Alzheimer-associated presenilins 1 and 2, neuronal expressin in brain and localization to intracellular membranes in mammalian cells. *Nature Med.* **2,** 224–229.

Kowall, N. W., Beal, M. F., Busciglio, J., Duffy, L. K., and Yankner, B. A. (1991) An in vivo model for the neurodegenerative effects of β amyloid and protection by substance P. *Proc. Natl. Acad. Sci. USA* **88,** 7247–7251.

LaFerla, F. M., Tinkel, B. T., Bieberich, C. J., Haudenschild, C. C., and Jay, G. (1995) The Alzheimer's Aβ peptide induces neurodegeneration and apoptotic cell death in transgenic mice. *Nature Genet.* **9,** 21–30.

Lamb, B. T., Sisodia, S. S., Lawler, A. M., Slunt, H. H., Kitt, C. A., Kearns, W. G. (1993) Introduction and expression of the 400 kilobase amyloid precursor protein gene in transgenic mice. *Nature Genet.* **5,** 22–30.

Leveugle, B., Scanameo, A., Ding, W., and Fillit, H. (1994) Binding of heparin sulfate glycosaminoglycan to β-amyloid peptide: inhibition by potentially therapeutic polysulfated compounds. *Neuroreport* **5,** 1389–1392.

Levine, E. S., Dreyfus, C. Y., Black, I. B., and Plummer, M. R. (1995) Brain-derived neurotrophic factor rapidly enhances synaptic transmission in hippocampal neurons via post-synaptic tyrosine kinase receptor. *Proc. Natl. Acad. Sci. USA* **92,** 8074–8077.

Levitan, D. and Greenwood, I. (1995) Facilitation of *lin-12*-mediated signaling by *sel-12,* a *Caenorhabditis elegans* S182 Alzheimer's disease gene. *Nature* **377,** 351–354.

Levy, E., Carman, M. D., Fernadez-Madrid, I. J., Power, M. D., Lieberburg, I., van Duinen, S. G., et al. (1990) Mutation of the Alzheimer's disease amyloid gene in hereditary cerebral hemorrhage, Dutch-type. *Science* **246,** 1124–1126.

Levy-Lehad, E., Wijsman, E. M., Nemans, E., Anderson, L., Goddard, K. A. B., Weber, J. L., Bird, T. D., and Schellenberg, G. D. (1995a) A familial Alzheimer's disease locus on chromosome 1. *Science* **269,** 970–973.

Levy-Lahad, E., Wasco, W., Poorkaj, P., Romano, D. M., Oshima, J., Pettingell, W. H., et al. (1995b) Candidate gene for the chromosome 1 familial Alzheimer's disease locus. *Science* **269,** 973–977.

L'Hernault, S. W., and Arduengo, P. M. (1992) Mutation of a putative sperm membrane protein in *Caenorhabditis eleganse* prevents sperm differentiation but not its associated meiotic divisions. *J. Cell Biol.* **119,** 55–68.

Li, J., Ma, J., and Potter, H. (1995) Identification and expression analysis of a potential familiar Alzheimer disease gene on chromosome 1 related to AD3. *Proc. Natl. Acad. Sci. USA* **90,** 7951–7955.

Lorenzo A. and Yankner, B. A. (1994) β-amyloid neurotoxicity requires fibril formation and is inhibited by Congo red. *Proc. Natl. Acad. Sci. USA* **92,** 8074–8077.

Ma, J., Yee, A., Brewer Jr., H. B., Das, S., and Potter, H. (1994) Amyloid-associated proteins α_1-antichymotrypsin and apolipoprotein E promote assembly of Alzheimer β-protein into filaments. *Nature* **372,** 92–94.

Mahley, R. W. (1988) Apolipoprotein E: cholesterol transport protein with expanding role in cell biology. *Science* **240,** 622–630.

Mark, R. J., Hensley, K., Butterfield, D. A., and Mattson, M. P. (1995) Amyloid β-peptide impairs ion-motive ATPase activities: evidence for a role in loss of neuronal Ca^{2+} homeostasis and cell death. *J. Neurosci.* **25,** 6239–6249.

Mark, R. J., Blanc, E. M., and Mattson, M. P. (1996) Amyloid β-peptide and oxidative cell injury in Alzheimer's disease. *Mol. Neurobiol.* **12,** 211–224.

Masters, C. L., Simms, G., Weinman, N. A., Multhaup, G., McDonald, B. L., and Beyreuther, K. (1985) Amyloid plaque core protein in Alzheimer disease and Down syndrome. *Proc. Natl. Acad. Sci. USA* **82,** 4245–4249.

Mattson, M. P. and Rychlik, B. (1990) Glia protect hippocampal neurons against excitatory amino-acid induced degeneration: involvement of fibroblast growth factor. *Int. J. Dev. Neurosci.* **8,** 399–415.

Mattson, M. P. and Scheff, S. W. (1994) Endogenous neuroprotection factors and traumatic brain injury: mechanisms of action and implications for therapies. *J. Neurotrauma* **11,** 3–33.

Mattson, M. P. and Goodman, Y. (1995) Different amyloidogenic peptides share a similar mechanism of neurotoxicity involving reactive oxygen species and calcium. *Brain Res.* **676,** 219–224.

Mattson, M. P., Murrain, M., Guthrie, P. B., and Kater, S. B. (1989) Fibroblast growth factor and glutamate: opposing actions in the generation and degeneration of hippocampal neuroarchitecture. *J. Neurosci.* **9,** 3728–3740.

Mattson, M. P., Cheng, B., Davis, D., Bryant, K., Lieberburg, I., and Rydel, R. E. (1992) β-amyloid peptides destabilize calcium homeostasis and render human cortical neurons vulnerable to excitotoxicity. *J. Neurosci.* **9,** 3728–3740.

Mattson, M. P., Tomaselli, K., and Rydel, R. E. (1993a) Calcium-destabilizing and neurodegenerative effects of aggregated β-amyloid peptide are attenuated by basic FGF. *Brain Res.* **621,** 35–49.

Mattson, M. P., Cheng, B., Culwell, A., Esch, F., Lieberburg, I., and Rydel, R. E. (1993b) Evidence for excitoprotective and intraneuronal calcium-regulating roles for secreted forms of β-amyloid precursor protein. *Neuron* **10,** 243–254.

Mattson, M. P., Kumar, K., Cheng, B., Wang, H., and Michaelis, E. K. (1993c) Basic FGF regulates the expression of a functional 71 kDa NMDA receptor protein that mediates calcium influx and neurotoxicity in cultured hippocampal neurons. *J. Neurosci.* **13,** 4575–4588.

Mattson, M. P., Lovell, M. A., Furukawa, K., and Markesbery, W. R. (1995) Neurotrophic factors attenuate glutamate-induced accumulation of peroxides, elevation of $[Ca^{2+}]_i$ and neurotoxicity, and increase antioxidant enzyme activities in hippocampal neurons. *J. Neurochem.* **65,** 1740–1751.

Mattson, M. P., Furukawa, K., Bruce, A. J., Mark, R. J., and Blanc, E. (1997) Calcium homeostasis and free radical metabolism as convergence points in the pathophysiology of dementia, in *Molecular Mechanisms of Dementia* (Wasco, W. and Tanzi, R. E., eds), Humana, Totowa, NJ, pp. 103–143.

May, P. C., Boggs, L. N., and Fuson, K. S. (1993) Neurotoxicity of human amylin in rat primary hippocampal cultures: similarity to Alzheimer's disease amyloid-beta neurotoxicity. *J. Neurochem.* **61,** 2330–2333.

McKenzie, J. E., Gentleman, S. M., Roberts, G. W., Graham, D. I., and Royston, M. C. (1994) Increased numbers of beta APP-immunoreactive neurones in the entorhinal cortex after head injury. *Neuroreport* **6,** 161–164.

Mesulam, M.-M. and Geula, C. (1994) Butyrylcholineesterase reactivity differentiates the amyloid plaques of aging from those of dementia. *Ann. Neurol.* **36,** 722–724.

Milward, E. A., Papadopoulos, R., Fuller, S. J., Moir, R. D., Small, D., Beyreuther, K., and Masters, C. L. (1992) The amyloid protein precursor of Alzheimer's disease is a mediator of the effects of nerve growth factor on neurite outgrowth. *Neuron* **9,** 129–137.

Morris, J., Mirra, S., Davis, P., Gearing, M., Mobs, R., Hughes, J. P., and Heyman, A. (1991a) Correlatin of dementia severity with plaque and tangle freuqencies in midfrontal cortex in CERAD cases with Alzheimer's disease. *Ann. Neurol.* **30,** 249.

Morris, J. C., McKeel, D. W., Jr., Storandt, M., Rubin, E. H., Price, J. L., Grant, E. A., et al. (1991b) Very mild Alzheimer's disease: informant-based clinical, psychometric, and pathologic distinction from normal aging. *Neurology* **41,** 469–478.

Mullan, M., Crawford, F., Axelman, K., Houlden, H., Lilius, L., Winblad, B., and Lannfelt, L. (1992) A pathogenic muation for probable Alzheimer's disease in the APP gene at the N-terminus of β-amyloid. *Nature Genet.* **1,** 345–347.

Mullan, M. and Crawford, F. (1993) Genetic and molecular advances in Alzheimer's disease. *Trends Neuosci.* **16,** 398–403.

Namba, Y., Tomonaga, M., Kawasaki, H., Otomo, E., and Ikeda, K. (1991) Apolipoprotein E immunoreactivity in cerebral amyloid deposits and neurofibrillary tangles in Alzheimer's disease and kuru plaque amyloid in Creuszfeldt-Jacob disease. *Brain Res.* **541,** 163–166.

Näslund, J., Thyberg, J., Tjernberg, L. O., Wernstedt, C., Karlström, A. R., Bogdanovic, N. (1995) Characterization of stable complexes involving apolipoprotein E and the amyloid β peptide in Alzheimer's disease brain. *Neuron* **15,** 219–228.

Nathan, B. P., Bellosta, S., Sanan, D. A., Weisgraber, K. H., Mahley, R. W., and Pitas, R. E. (1994) Differential effects of apolipoproteins E3 and E4 on neuronal growth in vitro. *Science* **264,** 850–852.

Ninomiya, H., Roch J.-M., and Saitoh T. (1994) Secreted form of amyloid b/A4 protein precursor (APP) binds to two distinct APP binding sites on rat B103 neuron-like cells through two different domains, but only one site is involved in neuritotropic activity. *J. Neurochem.* **63,** 495–500.

Nishimoto, I., Okamoto, T., Matsuura, Y., Takahashi, S., Okamoto, T., Murayama, Y., and Ogata, E. (1993) Alzheimer amyloid protein precursor complexes with brain GTP-binding protein Go. *Nature* **362,** 75–79.

Nitsch, R. M., Slack, B. E., Wurtman, R. J., and Growdon, J. W. (1992) Release of Alzheimer amyloid precursor derivatives stimulated by activation of muscarinic acetylcholine receptors. *Science* **258,** 304–307.

Nitsch, R. M., Farber, S. A., Growdon, J. H., and Wurtman, R. J. (1993) Release of amyloid beta-protein precursor derivatives by electrical depolarization of rat hippocampal slices. *Proc. Natl. Acad. Sci. USA* **90,** 5191–5193.

Ohyagi, Y., Takahashi, K., Kamegai, M., and Tabira, R. (1990) Developmental and differential expression of beta amyloid protein precursor mRNA in mouse brain. *Biochem. Biophys. Res. Commun.* **167,** 54–60.

Ostrowski, J., Sims, J. E., Sibley, C. H., Valentine, M. A., Dower, S. K., Meier, K. E. and Bomsztyk, K. (1991) A serine/threonine kinase activity is closely associated with a 65-kDa phosphoprotein specifically recognized by the κB enhancer element. *J. Biol. Chem.* **226,** 12,722–12,733.

Pike, C., Burdick, D., Walencewicz, A., Glabe, C., and Cotman, C. (1993) Neurodegeneration induced by β-amyloid peptides in vitro: the role of peptide assembly state. *J. Neurochem.* **13,** 1676–1686.

Pike, C. J., Walencewicz-Wasserman, A. J., Kosmoski, J., Cribbs, D. H., Glabe, C. G., and Cotman, C. W. (1995) Structure-activity analyses of β-amyloid peptides: contributions of the β25–35 region to aggregation and neurotoxicity. *J. Neurochem.* **64,** 253–265.

Quon, D., Wong, Y., Catalano, R., Scardina, J. M., Murakami, K., and Codell, B. (1991) Formation of β-amyloid protein deposits in brains of transgenic mice. *Nature* **35,** 239–241.

Rebeck, G. W., Reiter, J. S., Stickland, D. K., and Hyman, B. T. (1993) Apolipoprotein E in sporadic Alzheimer's disease: allelic variation and receptor interactions. *Neuron* **11,** 575–580.

Roberts, G., Gentleman, S., Lynch, A., and Graham, D. (1991) βA4 amyloid protein deposition in brain after head trauma. *Lancet* **338,** 1422,1423.

Refolo, L. M., Salton, S. R. J., Anderson, J. P., Mehta, P., and Robakis, N. K. (1989) Nerve and epiderma growth factors induce the release of the Alzheimer amyloid precursor from PC12 cell cultures. *Biochem. Biophys. Res. Commun.* **164,** 664–670.

Roch, J. M., Masliah, E., Roch-Levecq, A. C., Sundsmo, M. P., Otero, D. A., Veinbergs, I., and Saitoh, T. (1994) Increase of synaptic density and memory retention by a peptide representing the trophic domain of the amyloid beta/A4 protein precursor. *Proc. Natl. Acad. Sci. USA* **91,** 7450–7454.

Rogaev, E. I., Sherrington, R., Rogaev, E. A., Levesque, G., Ikeda, M., Liang, Y., et al. (1995) Familial Alzheimer's disease in kindred with missense mutations in a gene on chromosome 1 related to the Alzheimer's disease type 3 gene. *Nature* **376,** 775–778.

Roher, A. E., Lowenson, J. D., Clarke, S., Wolkow, C., Wang, R., Cotter, R. J., et al. (1993) Structural alterations in the peptide backbone of β-amyloid core protein may account for its deposition and stability in Alzheimer's disease. *J. Biol. Chem.* **268,** 3072–3083.

Saitoh, T., Sundsmo, M., Roch, J.-M., Kimura, N., Cole, G., Schubert, D., Oltersdorf, T., and Schenk, D. B. (1989) Secreted form of amyloid β protein precursor is involved in the growth regulation of fibroblasts. *Cell* **58,** 615–622.

Sandhu, F. A., Salim, M., and Zain, S. B. (1991) Expression of the human beta-amyloid protein of Alzheimer's disease specifically in the brains of transgenic mice. *J. Biol. Chem.* **266,** 21, 331–21,334.

Sanan, D. A., Weisgraber, K., Huang, D. Y., Saunders, A., Schmechel, D., Wisniewski, T., et al. (1994) Apolipoprotein E associates with Aβ amyloid peptide to form novel monofibrils: isoform apoE4 associates more efficiently than apo E3. *J. Clin. Invest.* **94,** 880–889.

Schellenberg, G. D., Bird, T. D., Wijsman, E. M., Orr, H. T., Anderson, L., Nemens, E., et al. (1992) Genetic linkage evidance for a familial Alzheimer's disease locus on chromosome 14. *Science* **258,** 668–671.

Scheuner, D., Eckman, C., Jensen, M., Song, X., Citron, M., Suzuki, N., et al. (1996) Secreted amyloid β-protein similar to that in the senile plaques of Alzheimer's disease is increased in vivo by the presenilin 1 and 2 and *APP* mutaions linked to familial Alzheimer's disease. *Nature Med.* **2,** 864–870.

Schmechel, D. E., Saunders, A. M., Strittmatter, W. J., Crain, B. J., Hulette, C. M., Joo, S. H., et al. (1993) Increased amyloid β-peptide deposition in cerebral cortex as a consequence of apolipoprotein E genotype in late-onset Alzheimer disease. *Proc. Natl. Acad. Sci. USA* **90,** 9649–9653.

Selkoe, D. J. (1994) Cell biology of the amyloid β-protein precursor and the mechanism of Alzheimer's disease. *Ann. Rev. Cell. Biol.* **10,** 373–403.

Seubert, P., Vigo-Pelfrey, C., Esch, F., Lee, M., Dovey, H., Davis, D., et al. (1992) Isolation and quantitation of soluble Alzheimer's β-peptide from biological fluids. *Nature* **359,** 325–327.

Sherrington, R., Rogaev, E. I., Liang, Y., Rogaeva, E. A., Levesque, G., Ikeda, M., et al. (1995) Cloning of a gene bearing missense mutaions in early-onset familial Alzheimer's disease. *Nature* **375,** 754–760.

Shoji, M., Golde, T. E., Ghiso, J., Cheung, T. T., Estus, S., Shaffer, L. M., et al. (1992) Production of the Alzheimer amyloid β protein by normal proteolytic processing. *Science* **258,** 126–129.

Siman, R., Card, J. P., Nelson, R. B., and Davis, L. G. (1989) Expression of β-amyloid precursor protein in reactive astrocyte following neuronal damage. *Neuron* **3,** 275–285.

Simmons, M. A. and Schneider, C. R. (1993) Amyloid β peptides act directly on single neurons. *Neuroscci. Lett.* **150,** 133–136.

Singh, M., Meyer, E. M., and Simpkins, J. W. (1995) The effect of ovariectomy and estradiol replacement on brain-derivered neurotrophic factor messenger ribonucleic acid expression in cortical and hippocampal brain regions of female Sprague-Dawley rats. *Endocrinology* **136,** 2320–2324.

Sisodia, S. S., Koo, E. H., Beyreuther, K., Unterbeck, A., and Price, D. L. (1990) Evidence that β-amyloid protein in Alzheimer's disease is not derived by normal processing. *Science* **248,** 492–495.

Sloviter, R. S. (1989) Calcium-binding protein (calbindin-D28k) and parvalbumin immunocytochemistry: localization in the rat hippocampal neurons to seizure activity. *J. Comp. Neurol.* **280,** 183–196.

Smith-Swintosky, V. L., Pettigrew, L. C., Craddock, S. D., Culwell, A. R., Rydel, R. E., and Mattson, M. P. (1994) Secreted forms of β-amyloid precursor protein protect against ischemic brain injury. *J. Neurochem.* **63,** 781–784.

Spitz, D., Malcolm, R. R., and Roberts, R. J. (1990) Cyototoxicity and matabolism of 4-hydroxy-2-nonenal and 2-nonenal in H_2O_2-resistant cell lines. *Biochem. J.* **265,** 453–459.

St. George-Hyslop, P. H., Tanzi, R. E., Polinsky, R. J., Haines, J. L., Nee, L., Watkins, P. C., et al. (1987) The genetic defect causing familial Alzheimer's disease maps on chromosome 21. *Science* **235,** 885–890.

Stephenson, D. T., Rash, K., and Clemens, J. A. (1992) Amyloid precursor protein accumulates in regions of neurodegeneration following focal cerebral ischemia in the rat. *Brain Res.* **593,** 128–135.

Strittmatter, W. J., Saunders, A. M., Schmechel, D., Pericak-Vance, M., Enghild, J., Salvesen, G. S., and Roses, A. D. (1993) Apolipoprotin E: high avidity binding to β-amyloid and increased frequency of type 4 allele in late-onset familial Alzheimer disease. *Proc. Natl. Acad. Sci. USA* **90,** 1977–1981.

Suzuki, N., Cheung, T. T., Cai, X.-D., Odaka, A., Otvos Jr., L., Eckman, C., et al. (1994) An increased percentage of long amyloid β protein secreted by familial amyloid β protein precursor (βAPP_{717}) mutants. *Science* **264,** 1336–1340.

Vito, P., Lacana, E., and D'Adamio, L. (1996) Interfering with apoptosis: Ca^{2+}-binding protein ALG-2 and Alzheimer's disease gene ALG-3. *Science* **271,** 521–525.

Weidemann, A., König, G., Bunke, D., Fischer, P., Salbaum, J. M., Masters, C. L., and Beyreuther, K. (1989) Identification, biogenesis, and localization of precursors of Alzheimer's disease A4 amyloid protein. *Cell* **57,** 115–126.

Wisniewski, T., Ghiso, J., and Frangione, B. (1991) Peptides homologous to the amyloid protein of Alzheimer's disease containing a glutamine for glutamic acid substitution have accelerated amyloid fibril formation. *Biochem. Biophys. Res. Commun.* **179,** 1247–1254.

Wood, S. J., MacKenzie, L., Maleeff, B., Hurle, M. R., and Wetzel, R. (1996) Selective inhibition of Aβ fibril formation. *J. Biol. Chem.* **271,** 4086–4092.

Yankner, B. A., Duffy, L. K., and Kirschner, D. A. (1990) Neurotrophic and neurotoxic effects of amyloid β protein: reversal by tachykinin neuropeptides. *Science* **250,** 279–282.

12

Clinical Potential of Agents That Affect Thrombin Signaling in Degenerative and Traumatic Neurologic Disorders

Barry W. Festoff

1. INTRODUCTION

1.1. *Background on Classical Concepts of Thrombin's Role in Coagulation*

Historically, thrombin has been a favorite enzyme protease to study because of its activity towards large substrates, particularly those that are zymogens for other enzymes involved in coagulation. The most well known, in this context, is Factor XIII activation to XIIIa. Thrombin catalyzes this reaction, activating a powerful Ca^{2+}-dependent enzyme for crosslinking fibrin via the γ-carboxamide group of glutamine residues and the ε-amino group of lysine. However, in order for Factor XIIIa to utilize the fibrin substrate, this molecule must result from specific and selective cleavage by thrombin of the plasma protein, fibrinogen (Fenton, 1986; Fenton et al., 1993).

Somewhat less appreciated is thrombin's unique ability to affect its own regulation by forming complexes on cell surfaces with the integral membrane proteoglycan, thrombomodulin, to activate protein C (Esmon, 1989,a,b, 1995). Activated protein C then becomes a potent anticoagulant, regulating the cascade by proteolyzing Factors V and X. In addition, the chondroitin sulfate in thrombomodulin binds to a basic surface on thrombin that is also involved in heparin interaction. The chondroitin sulfate enhances the affinity of thrombin for thrombomodulin approx 10- to 20-fold, making thrombomodulin a more potent inhibitor of coagulation, altering thrombin's conformation and specificity, and accelerating thrombin inhibition by the serpin, antithrombin (Esmon, 1995).

1.2. *Activation of Thrombin*

Thrombin itself, like all other coagulation serine proteases, is synthesized as a zymogen, prothrombin. Prothrombin is transcribed primarily in the liver from a roughly 2.3 kb mRNA (Degen, 1992). In the mouse, it is quite abundant, representing 0.07% of all mRNA (Degen et al., 1990). Prothrombin is activated in the plasma and on the surfaces of cells by a prothrombinase complex consisting of Factors Va and Xa, as well as Ca^{2+} and phospholipids (Fenton et al., 1991). However, not much is known of the transcriptional regulation of the prothrombin gene (Bancroft et al., 1992). Prothrombin is synthesized as a preproprotein consisting of a putative Gla domain, two kringle domains,

From: Neuroprotective Signal Transduction *edited by M. P. Mattson*
Humana Press Inc., Totowa, NJ

and a two-chain protease domain. Amino-acid sequence analysis from a number of species, from hagfish to human, suggest that the thrombin B-chain and the propeptide-Gla domain are the regions most required for the common function(s) of vertebrate prothrombins (Banfield et al., 1994).

1.3. Binding to Receptors

Thrombin binds to a number of receptors, which include both high- and low-affinity binding sites (Greco and Jamieson, 1991). The high-affinity sites are activated by binding of thrombin, and not by proteolytic cleavage (Carney et al., 1992). The most well known are thrombomodulin (TM), discussed below, and platelet protein GPIb.

1.3.1. Thrombomodulin (TM)

As noted, TM is an integral membrane proteoglycan, with chondroitin sulfate side chains (Esmon, 1989a,b, 1995); it consists of six epidermal growth factor (EGF)-like domains. TM has been found not only in the vascular environment (vascular endothelial cells, syncytiotrophoblasts, platelets), but also in some types of tumor cells (Dittman et al., 1994). In this regard, fetomodulin (FM), a differentiation-induced glycoprotein, was found to be expressed in mesenchyme and in the neuroepithelium of the mesencephalic flexure early in murine embryogenesis (Imada et al., 1987). This suggested important roles for FM in the central nervous system (CNS). Subsequently, by protein analysis, as well as by gene-sequencing analysis, FM was shown to be identical to TM (Imada et al., 1990). In fact, a recent report indicates that TMnull mice, after homologous recombination to "knockout" the gene for TM, died before E9.5 *in utero*; during neural tube formation in the early embryonic period, and well before the formation of the vascular system (Healy et al., 1995; Rosenberg, 1995). These reports suggest that TM may play important roles in CNS development. This has been supported by studies showing developmental regulation of this receptor, or acceptor, for thrombin in postimplantation mouse embryos (Weiler-Guettler et al., 1996).

1.3.2. Platelet Glycoprotein GPIb

This glycoprotein on the surface of platelets binds thrombin with high affinity (Colman, 1991), after which aggregation is induced. GPIb also binds von Willebrand factor at a different domain from where thrombin binds. If platelets are treated with chymotrypsin, they are less sensitive to thrombin. In addition, platelets from patients with the Bernard-Soulier syndrome, which lack GPIb, are also less responsive to thrombin than platelets from normal donors. However, Bernard-Soulier platelets respond normally to the peptide SFLLRN, a synthetic peptide of yet another receptor for thrombin (Kinlough-Rathbone et al., 1995).

1.3.3. Proteolytically Activated Receptor for Thrombin (PAR-1)

With the cloning of the PAR-1 (Vu et al., 1991a), questions concerning the disparate effects of thrombin, both intra- and extravascularly, have started to be answered. PAR-1 belongs to the seven-transmembrane domain receptor family, and contains within its proposed exodomain a thrombin cleavage site. Although this varies slightly from species to species, the cleavage site, which is at R42–S43 in the human PAR-1, is largely conserved. The tetrapeptide just amino-terminal to this cleavage is identical to the thrombin cleavage site in human protein C, LDPR (single-letter amino acid code) (Esmon, 1995). Of equal interest, carboxyl to this site in the PAR-1, is a domain similar to the C-terminus of the leech thrombin inhibitor, hirudin. It is to this site that thrombin probably binds prior to cleavage of the exodomain (Brass et al., 1992; Seiler et al., 1992). The KYEPF sequence is known to bind thrombin's anion exosite I (Vu et al., 1991b), just as the DEEFI sequence at the C-terminus of hirudin does (Liu et al., 1991). Thrombin-receptor-active peptides (TRAPs), such as SFLLRN, activate the receptor (Vu et al., 1991a), as mentioned above. Following activation, various second-messenger-mediated events take place, such as adenylate cyclase inhibition, inositol triphosphate (IP_3) elevation, and $[Ca^{2+}]_i$ enhancement (Grand et al., 1996).

Thus, although PAR-1 is but one of several thrombin-binding sites found on platelets and other cells (Carney et al., 1986), it is the only one capable of eliciting a second-messenger response. It is coupled to two heterotrimeric G proteins, G_i, for adenylate cyclase inhibition, and G_q, for phospholipase C activation with subsequent IP_3 generation, $[Ca^{2+}]_i$ mobilization, and protein kinase C (PKC) activation (Grand et al., 1996). A second proteolytically activated receptor, PAR-2 (Nystedt et al., 1994), responds to trypsin as well as to TRAPs, such as the hexamer SFLLRN, but not to thrombin; it also enhances $[Ca^{2+}]_i$ mobilization (Mirza et al., 1996). PAR-2 also has a specific active peptide, SLIGRL, but the physiological ligand (protease) for PAR-2 is not yet known. Although it has not yet been reported in the CNS (Nystedt et al., 1995), recent evidence suggests that both protein and gene expression for PAR-2 will be detected there.

In addition, thrombin also interacts with tyrosine kinase (Trk) receptors and low molecular mass monomeric G proteins, such as the small Ras protein Rho. Rho is involved in actin stress fiber formation, and within neurons may mediate thrombin's effect on neurite retraction (Jalink and Moolenaar, 1992) and, possibly, on neuronal death. In fact, altering the interaction of thrombin with Rho may be an area for possible therapeutic exploitation by regulating thrombin signal transduction in neurodegeneration and trauma. These strategies are further explored below.

Another potentially interesting relationship between G protein and Trk receptors in thrombin signaling (Delafontaine et al., 1996; Du et al., 1996), may be the basis for neuroprotective effects of insulin-like growth factor I and II (IGF-I, II) (Cheng and Mattson, 1992). All of these actions depend on the protease function of thrombin, since anything that alters proteolytic activity also causes biological inactivity (Brass et al., 1994). Thrombin inhibitors, such as hirudin, protease nexin I (PNI), and others, oppose these actions on cells.

2. POSTTRANSLATIONAL REGULATION BY INHIBITORS

Molecules that inhibit thrombin are either active site or anion exosite I or II inhibitors.

2.1. *D-Phe-Pro-Arg Chloromethyl Ketone (PPACK)*

Included in the first group are the molecules originally synthesized by Elliot Shaw and his colleagues at Brookhaven National Laboratories. The most used of these is PPACK (Hansen and Harker, 1988). Like phosphofluoridation of the S1 residue with diisopropylfluorophosphate (DFP), PPACK causes irreversible inactivation of thrombin proteolytic activity.

2.2. *Warfarin*

Warfarin sodium acts by inhibiting syntheisis of vitamin K-dependent coagulation factors, which results in sequential decreases in Factors VII, IX, X, and II (thrombin).

2.3. *Argatroban*

This is one of several synthetic active site thrombin inhibitors, with considerable specificity (Winn et al., 1993). Originally MD-805, argatroban is an agrinine derivative with a K_i in the range of 20 n*M* and may have greater potential to inhibit thrombin bound to fibrin or cell surfaces (Berry et al., 1994). This latter attribute may provide greater appeal for argtroban-like drugs in bound thrombin on neurons or their processes.

2.4. *Invertebrate and Recombinant*

2.4.1. *Hirudin*

Hirudin is a 65-amino-acid polypeptide isolated from the leech *Hirudo medicinalis* that acts as a thrombin-specific inhibitor (Fenton et al., 1991). It is unique in that inhibition is two-step or bivalent, since two binding domains are present for thrombin on the molecule. In the first step, thrombin binds to negatively charged residues at the COOH-terminal, residues 54–65, and in the second, at the active site NH2-terminal, residues 1–49 (Fenton et al., 1991). The former site interacts ionically with the anion exosite I, the same site that PART interacts with (Liu et al., 1991).

2.4.2. *Hirulogs*

Recombinant molecules, so-called hirulogs, such as hirulog-1 (Witting et al., 1992), have been designed that have higher affinity than hirudin for the catalytic and anion-binding exosite I, respectively. In addition, chimeric molecules have been synthesized that link the COOH-terminal tail of hirudin to RGDS cell-adhesion sequences (disintegrins) (Knapp et al., 1992), and block both antithrombin and antiplatelet aggregation. Such chimeric antithrombins may be attractive in approaches to extravascular thrombin effects, especially within the CNS.

2.5. *Endogenous Mammalian Inhibitors (Serpins)*

Serpins are serine protease inhibitors that generally form 1:1 molar ratio complexes with their cognate target proteases (Carrell and Travis, 1985; Carrell and Boswell, 1987). They are relatively large proteins, 40–65 kDa, and are synthesized without a proinhibitor form (Potempa et al., 1994). Many are synthesized in the liver, and may be acute phase reactants. Increasing evidence suggests that local synthesis, including that within the CNS, may regulate local protease excess (Potter et al., 1992).

2.5.1. *Antithrombin (AT)*

Also known as ATIII, this is the principal regulator of thrombin in the vascular space. The glycosaminoglycan (GAG) heparin effects anticoagulation by enhancing the interaction of AT with thrombin, because both molecules possess heparin binding sites, forming thrombin–AT (TAT) complexes.

2.5.2. *Heparin Cofactor II (HC-II)*

Just as heparin enhances TAT complexes and interaction, another GAG, dermatan sulfate, catalyzes the inhibition of thrombin by HCII (Liu et al., 1995). The concentrations of HCII are higher in the plasmas of individuals with congenital ATIII deficiency, and in those of pregnant women, than in controls. However, the role of HCII as a physiologic thrombin inhibitor is unknown.

2.5.3. Protease Nexin I (PNI)

PNI, originally PN, was first isolated from human foreskin fibroblast culture conditioned media by Baker and colleagues in 1980 (Baker et al., 1980). That same year, a hypothesis was presented concerning localization of a protease inhibitor to the synaptic neuromuscular junction (NMJ) as being implicated in the pathogenesis of amyotrophic lateral sclerosis (ALS) (Festoff, 1980). PNI was subsequently localized to the NMJ (Festoff and Hantaï, 1987; Hantaï et al., 1988; Festoff et al., 1991), and was found to be an effective inhibitor of trypsin, plasmin, and urokinase-type plasminogen activator (uPA), as well as thrombin. By sequence homology, it was placed into the serpin class (Baker et al., 1986).

In parallel with these studies, a protein was detected in rat C6 glioma media that enhanced neurite outgrowth for neuroblastoma and other neuronal cells, but was different from NGF (Monard et al., 1975); later it was found to possess antiprotease activity toward the same proteases inhibited by PNI (Guenther et al., 1985). In that same year, after a meeting on PAs in Italy, Baker and Monard exchanged reagents and determined that the glioma factor and PNI were essentially the same, if not identical (Gloor et al., 1986). This subsequently was confirmed by protein and nucleic acid sequencing (McGrogan et al., 1988, 1990; Sommer et al., 1987).

PNI is the most potent tissue inhibitor of thrombin (Stone et al., 1987) and has an interesting target protease specificity shift when it is bound to the cell or basement membrane surface (Farrell and Cunningham, 1987; Farrell et al., 1988; Wagner et al., 1989b; Donovan et al., 1994). This fact probably explains the important role that PNI plays in containing thrombin's effects in tissues, particularly in the CNS.

2.5.4. Placental Thrombin Inhibitor (PTI)

PTI, originally termed cytoplasmic antiprotease (CAP) is relatively unique in that it is an intracellularly acting serpin (Coughlin et al., 1993; Morgenstern et al., 1993; Riewald et al., 1996). Although PNI has been detected in human placenta (White et al., 1993), it is likely that PTI is the physiological regulator for thrombin within this tissue. Another thrombin inhibitor was detected in extracts of rat 9L gliosarcoma cells when implanted in rat brain (Rao et al., 1990). This novel inhibitor was clearly distinct from PNI; it was present only in the tumor growing in rat brain and undetectable either in the normal brain tissue or in vitro, and was in a latent, but sodium dodecyl sulfate-activatable, state, but did not bind uPA. It formed 88 kDa complexes with radiolabeled thrombin (Rao et al., 1990), but has not yet been further characterized. As an indication of the potential binding sites of thrombin with various molecules discussed above, Table 1 may be helpful.

2.6. *Background of Extravascular Thrombin Action*

Although thrombin's original effects were thought to be restricted to platelets, cultured cells have been shown to respond dramatically to active thrombin. It must be active, although minor controversy still remains, since prothrombin does not appear to have direct effects (Turnell et al., 1995). Thus, prothrombin activation is a key step in its action, so controlling activation is critical. Prothrombinase is associated with coagulation factors X and V (Fenton, 1986; Fenton et al., 1991) and Factor X, at least, which may be effectively controlled with recombinant tick anticoagulant protein (rTAP),

Table 1
Thrombin Interactions with Various Ligands

Ligand	Catalytic site	Anion binding exosite I	Anion binding exosite II
Fibrinogen		+	
Hirudin	+	+	
Thrombin receptor (PAR-1)		+	
Thrombomodulin			+
GPIb		+	
Heparin			+
Serpins (?)	+		+

recently cloned and expressed (Ripka et al., 1995). Other nonconventional activators of prothrombin have been detected on macrophages (Lindahl et al., 1989), tumor cells, and cells of neural origin (Sekiya et al., 1994).

Thrombin is not only the key enzyme in hemostasis, but it also aggregates platelets, is a mitogen for vascular smooth muscle cells (SMC) and others, and elicits various responses from cells, including those of the nervous system. Fibroblasts and endothelial cells have been studied most, but thrombin has similar effects on bone, tumor, skin, uterine and other cell types (Carney and Cunningham, 1978). In general, thrombin has mitogenic effects on these proliferatively competent cells. The concentration of prothrombin in human plasma is 1.37 μmol (Fenton, 1986). As will be seen in Section 2.8.1., this is ample for the nanomolar amounts that cause deleterious effects on neurons: neurite retraction, cell rounding, and cell death.

Several studies have attempted to study activation inhibition from the perspective of metastasis. This has been in the context of lung, bone, and other secondary sites, and with various primary cancers, including prostate. Historically, clotting has long been associated with cancer, calling to mind the Trousseau phenomenon (Trousseau, 1865; Rao et al., 1990). More recent studies show that thrombin may increase the metastatic potential of tumor cells (Wojtukiewicz et al., 1993). On the other hand, other authors report that thrombin inhibitors promote metastasis (Gasic, 1991), indicating a gap in our knowledge relative to thrombin, its inhibition, and metastasis. Thrombin increases platelet adhesiveness to tumor cells in vitro (Nierodzik et al., 1991), and treatment of tumor cells with thrombin enhanced pulmonary metastasis 10–160-fold (Nierodzik et al., 1992). Several tumor cell lines are known to express the PAR-1 (Wojtukiewicz et al., 1995; Nierodzik et al., 1996). However, even if metastasis was inhibited, this did not necessarily translate into increased survival (Merriman et al., 1989).

Thus, if we are to extrapolate from these studies on metastasis to the nervous system, and to both neurodegenerative and neurotraumatic diseases, we are not on very solid ground. On the one hand, it would be simple to conclude that the goal would be to limit thrombin's action, or more precisely, its coagulative proteolytic action. Implied in this is that we would aim to block thrombin's activation of the PAR-1. On the other hand, if thrombin inhibition may promote metastasis, and if even blocking metastasis does not enhance survival, perhaps we would want to limit thrombin inhibition. It is clear that more basic investigations of the effects of both of these diametrically opposite approaches are needed, at the molecular, cellular, and organismic levels. Finally, although most studies suggest the involvement of thrombin proteolytic activity, in metastasis, Bar-Shavit and colleagues (1995) have been interested in the RGD domain in thrombin.

2.8. *Background of Thrombin Action in Nervous System*

2.8.1. *Cells in Culture*

Studies of the effects of thrombin on neural cells began in the 1980s, with the work of Snider in Richelsen's group at the Mayo Clinic and Means and colleagues at Cincinnati. Thrombin was shown to increase cyclic GMP (cGMP) production initially in N1E-115 mouse neuroblastoma cells (Snider and Richelson, 1983; Snider, 1986; Snider et al., 1986). The coagulation protease was then proposed to have toxic effects, as might occur in the brain after hemorrhage, with wider implications suggesting mediation through prostacyclins (Means and Anderson, 1986). This theme, of toxic effects of thrombin on CNS cells, although speculative, has received considerable support in the past 10 years.

In fact, thrombin has deleterious effects on neural cells,with the exception of its effects on astrocytes (which can be divided into two different categories, proliferative and shape-changing effects). Our own recent work implicated thrombin in synapse retraction and elimination (Zoubine et al., 1996). Thrombin's action in neurite retraction has been studied by several groups. This effect is thrombin-specific, not shared by closely related members of the serine protease superfamily (such as plasmin and urokinase). Thrombin retracts neuroblastoma neurites at low concentrations (Gurwitz and Cunningham, 1988). PNI, known also as glia-derived nexin (GDN), promotes neurite outgrowth in these cells (Guenther et al., 1985; Gloor et al., 1986) by inhibiting thrombin (Gurwitz and Cunningham, 1990). Thrombin elicits retraction by activating PAR-1 (Suidan et al., 1992), with participation of Rho of the Ras superfamily of monomeric G proteins, involved in actin stress-fiber formation. Other members in the family, Rac and cdc42, are concerned with axon formation and guidance, while Rho, like thrombin, is involved in retraction. Inhibition of Rho, by ADP-ribosylation with the C3 exoenzyme of *Clostridium botulinum*, by the TK inhibitor genistein, or by inhibition of isoprenylation of Rho's C-terminal (with lovastatin or compactin), completely blocks thrombin's ability to retract neurites (Jalink and Moolenaar, 1992). Missing from these reports was evidence of PAR-1s cleavage, an area that we are actively studying at this time. A recent study (Shea, 1996) suggests that PAR-1 may be topographically localized in neurons to the axon terminal, as well as to that portion of the soma nearest the axon hillock or main portion of the process. Immunocytochemical confirmation may be possible.

Rho alteration has not been used to interfere with thrombin's effects on neural cell killing. In this regard, our recent experiments with mouse NSC19 cells, a hybrid of N18TG2 murine neuroblastoma and mouse spinal cord motor neurons (Cashman et al., 1992), are of interest.

We incubated subconfluent NSC19 cells in 96-well plates in N1 supplemented DMEM with addition of the following: pertussis toxin (PTX; inhibits G_o/G_i), 400 ng/mL;

Table 2
Effects of G Protein Modulators

Modulator	G protein effect
Pertussis toxin (PTX)	Inhibits G_o/G_i
Cholera toxin (CTX)	Stimulates G_s
Mastoparan (MAS)	Stimulates G_o/G_i
Lovastatin (LOV)	Inhibits Rho isoprenylation

cholera toxin (CTX; stimulates G_s), 10 μg/mL; mastoparan (Mas; stimulates G_o/G_i), 40 μ*M*; and lovastatin (Lov; blocks 21 kDa Ras protein Rho isoprenylation), 40 μ*M* (*see* Table 2). After 24 h no morphological changes were observed when compared to control. We then added α-thrombin (1 m*M*) and incubated another 24 h. Following this, we scored neurons with processes, and calculated the percentage of process-bearing cells (arrowheads) from total number of cells. Pretreatment of cells with CTX or with LOV prevented α-thrombin-mediated neurite retraction effect on motor neurons, whereas both PTX and mastoparan exacerbated thrombin's action (Fig. 1).

Addition of either lysophosphatidic acid (LPA) or thrombin receptor active peptide (TRAP) to serum-starved neurons causes rapid growth cone collapse, neurite retraction, and transient rounding of the cell body (Jalink et al., 1993, 1994). These effects of either LPA or thrombin are all blocked by ADP-ribosylation, using the C3 exoenzyme. These results show that LPA or thrombin-induced neurite retraction and rounding of the cell body, all occurring prior to cell death, are mediated via Rho. Studies of neurite retraction by others found that serum caused retraction, accounted for by LPA, and involved both [Ca^{2+}] mobilization and Rho (Dyer et al., 1992a,b; Tigyi et al., 1996a,b). Thrombin may also increase intracellular LPA levels. As yet, no reports of LPA-mediated neuronal cell death or of its prevention with ADP ribosylation of Rho have appeared.

However, evidence is rapidly accumulating on thrombin's role as a mediator of neuronal cell death. In the last 1–2 yr, some degree of controversy has developed as to whether low concentrations of thrombin actually protected rat primary astrocytes and hippocampal neurons grown under hypoxic or hypoglycemic conditions (Vaughan et al., 1995). However, slightly higher thrombin concentrations clearly induced cell death (Smith-Swintosky et al., 1995; Vaughan et al., 1995). In fact, low thrombin levels exacerbated glucose deprivation-induced cell death of rat hippocampal neurons, implicating calcium mobilization and PNI neuroprotection (Smith-Swintosky et al., 1995). Our data with mouse spinal cord neurons, under activity-dependent conditions, confirm PNI's neuroprotection of thrombin-induced killing, but also suggest a biphasic "shoulder" for survival at lower concentrations of thrombin (Festoff et al., 1996). Despite these advances, it is still unclear whether thrombin induces neuronal death through apoptosis or necrosis. If the former, thrombin may be an apoptogen for neurons, because it is a mitogen for dividing cells.

Mention was made of thrombin's stimulation of Trk activity (Grand et al., 1996). Crosstalk between thrombin's G-protein-coupled signal transduction pathway and traditional Trk receptors, such as insulin and IGF-I, may also be possible (Delafontaine et al., 1996; Du et al., 1996). Although we have not analyzed data on such an action enhancing

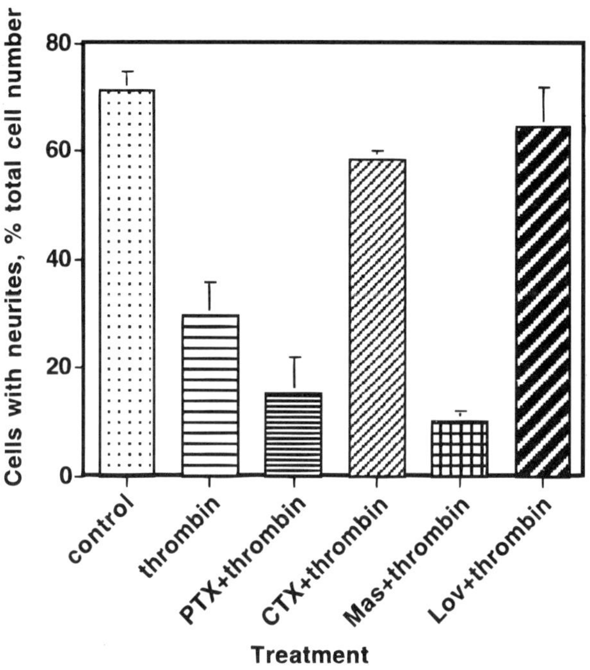

Fig. 1. Effects of several G protein modulators on motor neuron survival and neurite outgrowth. NSC19 cells were cultured in growth medium and switched to differentiation medium and finally serum-free for 24 h. Modulators were then preincubated for an additional 24 h and, finally, 100 n*M* α-thrombin was added for 48 h. Cells were viewed under phase microscopy and cells with neurites, along with total cells, were counted.

the neuroprotective effect of IGF-I on neurons, we have data on a similar effect in Schwann cells. We incubated a mouse Schwann cell line, MSC80 (Boutry et al., 1992), determined whether it was responsive to thrombin, and, if so, if IGF-I addition altered that response. Varying thrombin from p*M* to μ*M*, we found proliferation at low concentrations, but cell loss and aggregation at higher μ*M* levels of thrombin. We then fixed the concentration at μ*M*, and found that adding IGF-I in the MTT growth assay enhanced survival (Table 3). This suggests that protection may be afforded and that cross-talk between the Trk IGFR and the G-coupled PAR-1 may result. We are conducting similar experiments with NSC19, and, subsequently, with primary neuronal cultures. Since Trk and G-protein-coupled receptor crosstalk has been detected in vascular smooth muscle cells (SMC) (Delafontaine et al., 1996; Du et al., 1996), and in the clonal MSC80 Schwann cell line, promoting this crosstalk may be another strategy to prevent neuronal cell death by thrombin. Studies in mitotically active cells, such as SMC and the MSC80 line, show that thrombin's mitotic effects are mediated via the IGFR Trk. It may also be possible that IGFR phosphorylation by thrombin, with added IGF-I, could allow for neuroprotection in neurons.

An interesting aspect of thrombin signal transduction influences on neuroprotection comes from studies of the amyloid beta peptide (Aβ). Mattson and colleagues showed that Aβ can destabilize calcium homeostasis in human cortical neurons (Barger et al.,

Table 3
Viability of Mouse Schwann (MSC80) Cells in Culture Exposed to Thrombin and IGF-I

Treatment	D 1	D 2	D 3
Control	0.042 ± 0.004[a]	0.050 ± 0.006	0.004 ± 0.002
Thrombin	0.050 ± 0.005	0.014 ± 0.002	0.012 ± 0.003
Thrombin + IGF-I	0.061 ± 0.006	0.023 ± 0.001	0.015 ± 0.001

[a] Absorbance values at 590 nm in MTT assay (Mosmann, 1983) are shown with 2.7 µ*M* thrombin and 1.07 µ*M* IGF-I.

1993; Mattson, 1994). The interplay with excitotoxins was also shown, since such treatment increased neuronal vulnerability to glutamate. In a subsequent study, these authors demonstrated that, in human neuroblastoma cells, $[Ca^{2+}]_i$ increase is induced by thrombin through cleavage of the PAR-1 and short-term treatment with Aβ increased thrombin-mediated enhancement of $[Ca^{2+}]_i$ (Mattson and Begley, 1996). Downregulation of PAR-1, using antisense oligonucleotides (ODNs), decreased $[Ca^{2+}]_i$ after thrombin addition. Chronic treatment of these neuronal cells with Aβ actually blunted thrombin-mediated $[Ca^{2+}]_i$. Addition of Aβ also resulted in lipid peroxidation, demonstrated by thiobarituric acid fluorescence (TBAR), suggesting the connection of thrombin to the thrombin–PNI balance and free radical generation (Hensley et al., 1994). Oxidative damage to PNI has also been shown to occur in the presence of Aβ, interfering with PNI's ability to inactivate thrombin (Bolkenius and Monard, 1995). The use of certain anti-oxidants, such as water-soluble α tocopherol analogs, effectively prevented such damage and might be useful in Alzheimer's disease (AD) therapeutic design (Bolkenius and Monard, 1995). A recent study suggested that Aβ may mediate these damaging effects on neurons through a specific cell surface receptor, the receptor for advanced glycation end products (RAGE) (Yan et al., 1996). This RAGE was shown to mediate toxic effects of the peptide on both neurons and microglia, which are activated by Aβ.

2.8.2. *In Vivo and in Human Disease*

Much of the recent interest in a role for thrombin in human neurologic disease, outside of its function in hemorrhagic stroke, has come from concepts of mechanisms underlying AD (Ho et al., 1994). Thrombin was shown by immunostaining to be present in senile plaques, in neurofibrillary tangles, and associated with diffuse amyloid in AD brain, but it was only detected in blood vessels and plasma in the brain tissue of age-matched controls (Akiyama et al., 1992). Preapsorbtion of antisera by prothrombin did not eliminate immunoreactivity, suggesting that prothrombin is more rapidly converted into active thrombin in AD brain. Thrombin can proteolytically cleave the β amyloid protein precursor (β-APP_{695}) in vitro at its C-terminal, the Arg510–Ile511 bond, to generate a 28 kDa fragment, possibly an intermediate in production (Igarashi et al., 1992). Thus, thrombin can process βAPP to generate an amyloidogenic fragment. In AD brains, the normally regulating influence of PNI on thrombin might be diminished, resulting in the enhancement of such abnormal processing pathway for β-APP.

In contrast to thrombin, extractable PNI inhibitory activity was reduced in AD brain compared to control individuals, with PNI activity in hippocampus as low as in whole brain (Wagner et al., 1989a). However, although this activity was reduced sevenfold, Western immunoblotting showed an increase in endogenous PNI–protease complexes, which comigrated with authentic PNI–thrombin complexes (Wagner et al., 1989a), suggesting that increased complex formation had occurred in vivo, possibly in an attempt to neutralize elevated thrombin activity, in the AD brain. Monard and his colleagues showed that GDN–PNI is enriched in the olfactory system: bulb, tract, and lobe (Reinhard et al., 1988)—an important finding, since the olfactory system undergoes almost continuous degeneration and regeneration in the adult mammalian nervous system. Our finding of high levels of functionally active PNI in normal human cerebrospinal fluid (CSF) (Festoff et al., 1992), coupled with the inability to detect PNI in plasma (Baker et al., 1986), certainly suggests its local synthesis within the nervous system by brain and spinal tissue cells. However, the relative amount of PNI in the CNS vs the peripheral system is still controversial. Using a monoclonal neutralizing antibody, Cunningham and co-workers studied the codistribution of PNI and PNII–APP in different tissues of nonhuman primates immunochemically and by complex formation (Wagner et al., 1993) and reported detection of PNI only in brain. Using this same antibody, Cunningham failed to detect PNI in CSF (Cunningham, 1992). In contrast, both PNI mRNA levels and protease inhibition referable to PNI in murine tissues were the highest in seminal vesicles, with the brain content being only about 11% (Vassali et al., 1993). At first, one might suspect that perhaps a different gene, or alternative splicing, might account for the disparity. This seems unlikely, since we found that the sequence of the murine PNI gene (*Spi4*) is identical in brain and seminal vesicle (Citron et al., 1996). The chromosomal assignment of the human PNI gene (*PI7*) to 2q33–35, is syntentic with mouse chromosome 1 (Carter et al., 1995), where it lies between the AChR γ-subunit and villin (Voss et al., 1996).

Interactions between thrombin and growth factors, as described above for IGF-I, may also be of interest. For example, thrombin was shown to enhance secretion of nerve growth factor (NGF) from cultured astrocytes (Debeir et al., 1996). Of possible relevance in approaches to AD is the use of certain nootropic compounds (Nicholson, 1990). Nootropics, such as piracetam, facilitate learning and retrieval of information, and may have neuroprotective effects. In this regard, the hypoxanthine agents, pentoxyfylline and the related propentofylline, are gaining attention for possible treatment of both AD and vascular dementia (Rother et al., 1996). Propentofylline has been shown to increase synthesis of NGF in astrocytes (Shinoda et al., 1990; Fuji et al., 1993; Nabeshima et al., 1994) and antithrombotic action (Saletu et al., 1990). Propentofylline also enhances adenosine release, and adenosine has been shown to have a neuroprotective effect. Propentofylline may also depress free radical formation in activated microglia and influence astrocyte reactions (Schubert et al., 1994).

Since neuroprotection afforded by neurtrophins and other growth factors is now seen as a very real approach to therapy in neurodegeneration and after neurotrauma, the signal transduction interactions with thrombin may offer additional aspects for drug discovery. Thus, many of the same agents suggested to have roles in neurodegeneration may equally have effects in neurotrauma.

3. CLINICAL STRATEGIES IN THROMBIN-SIGNALING INTERVENTION

In general, one might envisage the therapeutic potential of agents that alter thrombin signal transduction and, hence, offer neuroprotection in one of several ways. They might promote coagulation potential for thrombin, enhancing its action as a procoagulant. Alternatively, they might promote thrombin's anticoagulation potential. Since thrombin accumulates at sites of vascular injury and interacts with a myriad of procoagulant substrates, interruption of this by enhancing its complex with TM, which instead activates protein C and attenuates clot formation, would seem desirable. In this regard, protein engineering has produced a thrombin variant by mutagenesis (a single substitution at E229A) that markedly shifted thrombin to anticoagulant activity (Gibbs et al., 1995). Or, they may effect neither, and, instead, act downstream by affording some interruption of thrombin's signal transduction pathway. Although we are primarily concerned with neuroprotection, it must be kept in mind that positive or negative effects on astrocytes may have profound actions on neuronal health and survival.

4. SUMMARY AND NEW HORIZONS

4.1. Agonists and Antagonists of Thrombin Binding

Certainly, in this regard, we must consider the TRAPs, as well as peptides that are directed against specific domains, or antibodies to these domains, of the PAR-1. For example, peptides or antibodies that bridge the cleavage domain and hirudin-like binding site may be offered to prevent thrombin binding and cleavage activation of the PAR-1. In a similar manner, low-dose activation of PAR-1 by TRAPs may desensitize the PAR-1 to cleavage activation by thrombin.

4.2. Small Molecules that Regulate Prothrombin Expression

Although very little is known about transcriptional regulation for prothrombin in liver, much less in other tissues (Bancroft et al., 1990, 1992, Chow et al., 1991), this would appear to be an area for exploration.

4.3. Control of Prothrombin Activation

Activation of prothrombin is probably the most effective means for controlling thrombin's activity (Andrew et al., 1990). In this regard, agents to control activation of Factor X to Xa, such as the tick protein (rTAP) (Ripka et al., 1995), may have efficacy. For nonconventional prothrombinase activities, such as found on certain neural cells (Sekiya et al., 1994), inhbition with EDTA and inhibitors of Ca^{2+}-activated proteases, but not with inhibitors of serine proteases, may be useful.

4.4. Small Molecules that Regulate Serpin Induction

Since the balance of serine proteases and their inhibitors is likely to be critical in neurdegenerative and traumatic disorders of the CNS (Festoff and Hantaï, 1987; Festoff, 1990; Monard et al., 1992; Cunningham et al., 1993), efforts should be toward reestablishing this balance after thrombin accumulates at sites of injury or disease. In this regard, a serpin isolated from *Xenopus laevis* is upregulated by estrogen (Holland et al., 1992). Similarly, PNI, found concentrated in mouse seminal vesicles (Vassalli et al., 1993), and expressed from the same gene there as in brain (Citron et al., 1996), is androgen-dependent at the level of transcription. Likewise, retinoic acid has been shown to upregulate the synthesis of J6 (identical to heat-shock protein HSP47) by inducing a nuclear protein similar or identical to GATA-2 or GATA-3 (Wang, 1994). Other molecules may have similar induction capabilities for serpins.

4.5. *Antisense Oligonucleotide Therapies*

Mention already was made of the use of antisense ODNs to the PAR-1 in human neuroblastoma cells to modulate the effects of thrombin on [Ca^{2+}] and neuronal cell death (Mattson and Begley, 1996). One might consider similar studies to determine the effects of antisense ODNs to PNI to alter the balance, or to prothrombin, for that matter, if local synthesis rather than transudation from the vascular space is responsible for increased thrombin activity.

4.6. *Modulation of PAR-1 Signal Transduction Pathway*

In addition to direct approaches to thrombin interactions with PAR-1, downstream strategies may be equally or more effective. Mention was already made of preliminary studies to modulate thrombin's interaction with heterotrimeric G proteins, such as G_s or G_q, and some of our own preliminary studies in this regard. The recent availability of PAR-1 knockout mice (Connolly et al., 1996; Trejo et al., 1996; Darrow et al., 1996) should allow for additional progress on approaches to PAR-1 signaling in neural cells.

4.7. *Interruption by TM Fragments*

Another approach, similar to the use of TRAPs, is blocking of thrombin's procoagulant action by interrupting its interaction with PAR-1 by applying EGF-domain fragments of TM. Disulfide-bonded TM fragments have been expressed in *Pischia pastoris* (White et al., 1994), and such peptides have significant activity in blocking thrombin–hirudin and thrombin–fibrinogen interactions (Srinivasan et al., 1994). Peptides corresponding to the fourth, fifth, and sixth EGF domains are the most active (Lougheed et al., 1995). We have studies underway to determine the effects of pretreatment of neurons with these TM fragments on thrombin-induced [Ca^{2+}] influx and neuronal cell death.

4.8. *Cell and Gene Therapies*

Finally, consideration must be given to cell and gene therapy approaches to the problem of thrombin's negative effects in the CNS. In this regard, approaches to degenerative and traumatic CNS disorders by this strategy are emerging (Doering, 1994). From many of the studies enumerated above, such approaches may also be entertained for reestablishing the thrombin–protease nexin I balance in the CNS, presumed to be dysregulated in Alzheimer's disease, ALS, and brain and spinal cord injury. Cautious optimism for success in one or more of these approaches is certainly warranted.

5. SUMMARY

In this chapter, a brief review of the biochemical interactions of thrombin with several of its substrates, acceptor molecules, and inhibitors is presented. The role of these interactions in coagulation on extravascular cells, and specifically regarding the CNS, is discussed. Emphasis has been placed on interactions with the low-affinity, proteolytically activated receptor for thrombin and its unique signal transduction pathway. From consideration of thrombin-signaling agents, strategies to alter thrombin's negative effects have emerged that may ultimately prove useful in the pharmacological arsenal for neurodegenerative and traumatic neurological disorders.

ACKNOWLEDGMENTS

Support for the studies reported came, in part, from the Medical Research Service, Department of Veterans Affairs and the Defense and Veterans Head Injury Project, Jackson Foundation, USUHS. The author gratefully acknowledges the participation of Bruce A. Citron, Mikhail N. Zoubine, Chi Yong, and Sherri Zhang in some of the studies discussed, and, especially, Irina V. Smirnova, who also critically reviewed the manuscript.

REFERENCES

Akiyama, H., Ikeda, K., Kondo, H., and McGeer, P. L. (1992) Thrombin accumulation in brains of patients with Alzheimer's disease. *Neurosci. Lett.* **146,** 152–154.

Andrew, M., Schmidt, B., Mitchell, L., Paes, B., and Ofosu, F. (1990) Thrombin generation in newborn plasma is critically dependent on the concentration of prothrombin. *Thromb. Haemost.* **63,** 27–30.

Baker, J. B., Knauer, D. J., and Cunningham, D. D. (1986) Protease nexins: secreted protease inhibitors that regulate protease actions at and near the cell surface, in *The Receptors* (Conn, P. M., ed.). Academic Press, New York, pp. 153–172.

Baker, J. B., Low, D. A., Simmer, R. L., and Cunningham, D. D. (1980) Protease-nexin: a cellular component that links thrombin and plasminogen activator and mediates their binding to cells. *Cell* **21,** 37–45.

Bancroft, J. D., McDowell, S. A., and Degen, S. J. F. (1992) The human prothrombin gene: transcriptional regulation in HepG2 cells. *Biochemistry* **31,** 12,469–12,476.

Bancroft, J. D., Schaefer, L. A., and Degen, S. J. (1990) Characterization of the Alu-rich 5′-flanking region of the human prothrombin-encoding gene: identification of a positive cis-acting element that regulates liver-specific expression. *Gene* **95,** 253–260.

Banfield, D. K., Irwin, D. M., Walz, D. A., and MacGillivray, R. T. (1994) Evolution of prothrombin: isolation and characterization of the cDNAs encoding chicken and hagfish prothrombin. *J. Mol. Evol.* **38,** 177–187.

Bar-Shavit, R., Ginzburg, Y., Maoz, M., Vlodavsky, I., and Peretz, T. (1995) The involvement of thrombin RGD in metastasis: characterization of a cryptic adhesive site. *Isr. J. Med. Sci.* **31,** 86–94.

Barger, S. W., Smith-Swintosky, V. L., Rydel, R. E., and Mattson, M. P. (1993) beta-Amyloid precursor protein mismetabolism and loss of calcium homeostasis in Alzheimer's disease. *Ann. NY Acad. Sci.* **695,** 158–164.

Berry, C. N., Girardot, C., Lecoffre, C., and Lunven, C. (1994) Effects of the synthetic thrombin inhibitor argatroban on fibrin-or clot-incorporated thrombin: comparison with heparin and recombinant hirudin. *Thromb. Haemost.* **72,** 381–386.

Bolkenius, F. N. and Monard, D. (1995) Inactivation of protease nexin-1 by xanthine oxidase-derived free radicals. *Neurochem. Int.* **26,** 587–592.

Boutry, J. M., Hauw, J. J., Gansmuller, A., Di Bert, N., Pouchelet, M., and Baron Van Evercooren, A. (1992) Establishment and characterization of a mouse Schwann cell line which produces myelin in vivo. *J. Neurosci. Res.* **32,** 15–26.

Brass, L. F., Pizarro, S., Ahuja, M., Belmonte, E., Blanchard, N., Stadel, J. M., and Hoxie, J. A. (1994) Changes in the structure and function of the human thrombin receptor during receptor activation, internalization, and recycling. *J. Biol. Chem.* **269,** 2943–2952.

Brass, L. F., Vassallo Jr., R. R., Belmonte, E., Ahuja, M., Cichowski, K., and Hoxie, J. A. (1992) Structure and function of the human platelet thrombin receptor. Studies using monoclonal antibodies directed against a defined domain within the receptor N terminus. *J. Biol. Chem.* **267,** 13,795–13,798.

Carney, D. H. and Cunningham, D. D. (1978) Cell surface action of thrombin is sufficient to initiate division of chick cells. *Cell* **14,** 811–823.

Carney, D. H., Redin, W., and McCroskey, L. (1992) Role of high-affinity thrombin receptors in postclotting cellular effects of thrombin. *Semin. Thromb. Hemost.* **18,** 91–103.

Carrell, R. W. and Boswell, D. R. (1987) Serpins: The super family of serine protease inhibitors, in *Protease Inhibitors* (Barret, A. and Salvesen, G., eds.). Elsevier, Amsterdam, pp. 403–418.

Carrell, R. W. and Travis, J. (1985) Alpha1-antitrypsin and the serpins: variation and countervariation. *Trends Biochem. Sci.* **10,** 20–24.

Carter, R. E., Cerosaletti, K. M., Burkin, D. J., Fournier, R. E. K., Jones, C., Greenberg, B. D., Citron, B. A., and Festoff, B. W. (1995) The gene for the serpin thrombin inhibitor (P17), protease nexin I, is located on human chromosome 2q33–q35 and on syntenic regions in the mouse and sheep genomes. *Genomics* **27,** 196–199.

Cashman, N. R., Durham, H. D., Blusztajn, J. K., Oda, K., Tabira, T., Shaw, I. T., Dahrouge, S., and Antel, J. P. (1992) Neuroblastoma x spinal cord (NSC) hybrid cell lines resemble developing motor neurons. *Dev. Dynamics* **194,** 209–221.

Cheng, B. and Mattson, M. P. (1992) IGF-I and IGF-II protect cultured hippocampal and septal neurons against calcium-mediated hypoglycemic damage. *J. Neurosci.* **12,** 1558–1566.

Chow, B. K., Ting, V., Tufaro, F., and MacGillivray, R. T. (1991) Characterization of a novel liver-specific enhancer in the human prothrombin gene. *J. Biol. Chem.* **266,** 18,927–18,933.

Citron, B. A., Smirnova, I. V., Ratzlaff, K., and Festoff, B. W. (1996) Protease nexin I in mouse brain is expressed from the same gene that is expressed in seminal vesicle. *J. Mol. Neurosci.* **7,** 183–191.

Colman, R. W. (1991) Receptors that activate platelets. *Proc. Soc. Exp. Biol. Med.* **197,** 242–248.

Connolly, A. J., Ishihara, H., Kahn, M. L., Fraese Jr., R. V., and Coughlin, S. R. (1996) Role of the thrombin receptor in development and evidence for a second receptor. *Nature* **381,** 516–519.

Coughlin, P. B., Tetaz, T., and Salem, H. H. (1993) Identification and purification of a novel serine proteinase inhibitor. *J. Biol. Chem.* **268,** 9541–9547.

Cunningham, D. D. (1992) Regulation of neuronal cells and astrocytes by protease nexin-1 and thrombin. *Ann. NY Acad. Sci.* **674,** 228–236.

Cunningham, D. D., Pulliam, L., and Vaughan, P. J. (1993) Protease nexin-1 and thrombin: injury-related processes in the brain. *Thromb. Haemost.* **70,** 168–171.

Darrow, A. D., Fung-Leuong, W.-P., Ye, R. D., Santulli, R. J., Cheung, W.-I., Derian, C. K., et al. (1996) Biological consequences of thrombin receptor deficiency in mice. *Thromb. Haemostasis* **76,** 860–866.

Debeir, T., Benavides, J., and Vige, X. (1996) Dual effects of thrombin and a 14-amino acid peptide agonist of the thrombin receptor on septal cholinergic neurons. *Brain Res.* **708,** 159–166.

Degen, S. J., Schaefer, L. A., Jamison, C. S., Grant, S. G., Fitzgibbon, J. J., Pai, J. A., and Elliott, R. W. (1990) Characterization of the cDNA coding for mouse prothrombin and localization of the gene on mouse chromosome 2. *DNA Cell Biol.* **9,** 487–498.

Degen, S. J. F. (1992) The prothrombin gene and its liver-specific expression. *Semin. Thromb. Hemost.* **18,** 230–242.

Delafontaine, P., Anwar, A., Lou, H., and Ku, L. (1996) G-protein coupled and tyrosine kinase receptors: evidence that activation of the insulin-like growth factor I receptor is required for thrombin-induced mitogenesis of rat aortic smooth muscle cells. *J. Clin. Invest.* **97,** 139–145.

Dittman, W. A., Nelson, S. C., Greer, P. K., Horton, E. T., Palomba, M. L., and McCachren, S. S. (1994) Characterization of thrombomodulin expression in response to retinoic acid and identification of a retinoic acid response element in the human thrombomodulin gene. *J. Biol. Chem.* **269,** 16,925–16,932.

Doering, L. C. (1994) Nervous system modification by transplants and gene transfer. *Bioessays* **16,** 825–831.
Donovan, F. M., Vaughan, P. J., and Cunningham, D. D. (1994) Regulation of protease nexin-1 target protease specificity by collagen type IV. *J. Biol. Chem.* **269,** 17,199–17,205.
Du, J., Sperling, L. S., Marrero, M. B., Phillips, L., and Delafontaine, P. (1996) G-protein and tyrosine kinase receptor cross-talk in rat aortic smooth muscle cells: thrombin- and angiotensin II-induced tyrosine phosphorylation of insulin receptor substrate-1 and insulin-like growth factor 1 receptor. *Biochem. Biophys. Res. Commun.* **218,** 934–939.
Dyer, D., Tigyi, G., and Miledi, R. (1992a) The effect of active serum albumin on PC12 cells: I. Neurite retraction and activation of the phosphoinositide second messenger system. *Brain Res. Mol. Brain Res.* **14,** 293–301.
Dyer, D., Tigyi, G., and Miledi, R. (1992b) The effect of active serum albumin on PC12 cells: II. Intracellular Ca^{2+} transients and their role in neurite retraction. *Brain Res. Mol. Brain Res.* **14,** 302–9.
Esmon, C. T. (1995) Thrombomodulin as a model of molecular mechanisms that modulate protease specificity and function at the vessel surface. *FASEB J.* **9,** 946–955.
Esmon, C. T. (1989a) The roles of protein C and thrombomodulin in the regulation of blood coagulation. *J. Biol. Chem.* **264,** 4743–4746.
Esmon, N. L. (1989b) Thrombomodulin. *Prog. Hemost. Thromb.* **9,** 29–55.
Farrell, D. H. and Cunningham, D. D. (1987) Glycosaminoglycans on fibroblasts accelerate thrombin inhibition by protease nexin-1. *Biochem. J.* **245,** 543–550.
Farrell, D. H., Wagner, S. L., Yuan, R. H., and Cunningham, D. D. (1988) Localization of protease nexin-1 on the fibroblast extracellular matrix. *J. Cell. Physiol.* **134,** 179–188.
Fenton, J. W. (1986) Thrombin. *Ann. NY Acad. Sci.* **485,** 5–15.
Fenton, J. W., Ofosu, F. A., Brezniak, D. V., and Hassouna, H. I. (1993) Understanding thrombin and hemostasis. *Hematol. Oncol. Clin. North. Am.* **7,** 1107–1119.
Fenton, J. W., Ofosu, F. A., Moon, D. G., and Maraganore, J. M. (1991) Thrombin structure and function: why thrombin is the primary target for antithrombotics. *Blood Coagulation Fibrinolysis* **2,** 69–75.
Fenton, J. W., Villanueva, G. B., Ofosu, F. A., and Maraganore, J. M. (1991) Thrombin inhibition by hirudin: how hirudin inhibits thrombin. *Haemostasis* **21(Suppl. 1)**, 27–31.
Festoff, B. W. (1980) Neuromuscular junction macromolecules in the pathogenesis of amyotrophic lateral sclerosis. *Med. Hypotheses* **6,** 121–131.
Festoff, B. W. (1990) Serine proteases and serpins in the nervous system: roles in development and in malignant and degenerative disease. In NATA ASI Series Plenum, New York, pp. 360.
Festoff, B. W. and Hantaï, D. (1987) Plasminogen activators and inhibitors: roles in muscle and neuromuscular regeneration. *Prog. Brain Res.* **71,** 423–431.
Festoff, B. W., Nelson, P. G., and Brenneman, D. E. (1996) Prevention of activity-dependent neuronal death: vasoactive intestinal polypeptide stimulates astrocytes to secrete the thrombin-inhibiting, neurotrophic serpin, protease nexin I. *J. Neurobiol.* **30,** 255–266.
Festoff, B. W., Rao, J. S., and Chen, M. (1992) Protease nexin I, thrombin- and urokinase-inhibiting serpin, concentrated in normal human cerebrospinal fluid. *Neurology* **42,** 1361–1366.
Festoff, B. W., Rao, J. S., and Hantaï, D. (1991) Plasminogen activators and inhibitors in the neuromuscular system: III. The serpin protease nexin I is synthesized by muscle and localized at neuromuscular synapses. *J. Cell. Physiol.* **147,** 76–86.
Fuji, K., Hiramatsu, M., Hayashi, S., Kameyama, T., and Nabeshima, T. (1993) Effects of propentofylline, a NGF synthesis stimulator, on alterations in muscarinic cholinergic receptors induced by basal forebrain lesion in rats. *Neurosci. Lett.* **150,** 99–102.
Gasic, G. J. (1991) Thrombin inhibitors promote metastasis. *Period. Biol.* **93,** 633–640.
Gibbs, C. S., Coutré, S. E., Tsiang, M., Li, W.-X., Jain, A. K., Dunn, K. E., et al. (1995) Conversion of thrombin into an anticoagulant by protein engineering. *Nature* **378,** 413–416.

Gloor, S., Odink, K., Guenther, J., Nick, H., and Monard, D. (1986) A glia-derived neurite promoting factor with protease inhibitory activity belongs to the protease nexins. *Cell* **47,** 687–693.

Grand, R. J. A., Turnell, A. S., and Grabham, P. W. (1996) Cellular consequences of thrombin-receptor activation. *Biochem. J.* **313,** 353–368.

Greco, N. J. and Jamieson, G. A. (1991) High and moderate affinity pathways for alpha-thrombin-induced platelet activation. *Proc. Soc. Exp. Biol. Med.* **198,** 792–799.

Guenther, J., Nick, H., and Monard, D. (1985) A glia-derived neurite-promoting factor with protease inhibitory activity. *EMBO J.* **4,** 1963–1966.

Gurwitz, D. and Cunningham, D. D. (1990) Neurite outgrowth activity of protease nexin-1 on neuroblastoma cells requires thrombin inhibition. *J. Cell. Physiol.* **142,** 155–162.

Gurwitz, D. and Cunningham, D. D. (1988) Thrombin modulates and reverses neuroblastoma neurite outgrowth. *Proc. Natl. Acad. Sci. USA* **85,** 3440–3444.

Hansen, S. R. and Harker, L. A. (1988) Interrruption of acute platelet dependent thrombosis by the synthetic antithrombin D-phenylalanyl-L-prolyl-L-arginil chloromethyl ketone. *Proc. Natl. Acad. Sci. USA* **85,** 3184–3188.

Hantaï, D., Rao, J. S., and Festoff, B. W. (1988) Serine proteases and serpins: their possible roles in the motor system. *Revue Neurologique* **144,** 680–687.

Healy, A. M., Rayburn, H. B., Rosenberg, R. D., and Weiler, H. (1995) Absence of the blood-clotting regulator thrombomodulin causes embryonic lethality in mice before development of a functional cardiovascular system. *Proc. Natl. Acad. Sci. USA* **92,** 850–854.

Hensley, K., Carney, J. M., Mattson, M. P., Aksenova, M., Harris, M., Wu, J. F., Floyd, R. A., and Butterfield, D. A. (1994) A model for β-amyloid aggregation and neurotoxicity based on free radical generation by the peptide: relevance to Alzheimer disease. *Proc. Natl. Acad. Sci. USA* **91,** 3270–3274.

Ho, G. J., Smirnova, I. V., Akaaboune, M., Hantai, D., and Festoff, B. W. (1994) Serine proteases and their serpin inhibitors in Alzheimer's disease. *Biomed. Pharmacother.* **48,** 296–304.

Holland, L. J., Suksang, C., Wall, A. A., Roberts, L. R., Moser, D. R., and Bhattacharya, A. (1992) A major estrogen-regulated protein secreted from the liver of Xenopus laevis is a member of the serpin superfamily. Nucleotide sequence of cDNA and hormonal induction of mRNA. *J. Biol. Chem.* **267,** 7053–7059.

Igarashi, K., Murai, H., and Asaka, J. (1992) Proteolytic processing of amyloid beta protein precursor (APP) by thrombin. *Biochem. Biophys. Res. Comm.* **185,** 1000–1004.

Imada, M., Imada, S., Iwasaki, H., Kime, A., Yamaguchi, H., and Moore, E. E. (1987) Fetomodulin: marker surface protein of fetal development which is modulatable by cyclic AMP. *Dev. Biol.* **122,** 483–491.

Imada, S., Yamaguchi, H., Nagumo, M., Katayanagi, S., Iwasaki, H., and Imada, M. (1990) Identification of fetomodulin, a surface marker protein of fetal development, as thrombomodulin by gene cloning and functional assays. *Dev. Biol.* **140,** 113–122.

Jalink, K., Eichholtz, T., Postma, F. R., van Corven, E. J., and Moolenaar, W. H. (1993) Lysophosphatidic acid induces neuronal shape changes via a novel, receptor-mediated signaling pathway: similarity to thrombin action. *Cell Growth Differ.* **4,** 247–255.

Jalink, K. and Moolenaar, W. H. (1992) Thrombin receptor activation causes rapid neural cell rounding and neurite retraction independent of classic second messengers. *J. Cell Biol.* **118,** 411–419.

Jalink, K., van Corven, E. J., Hengeveld, T., Morii, N., Narumiya, S., and Moolenaar, W. H. (1994) Inhibition of lysophosphatidate- and thrombin-induced neurite retraction and neuronal cell rounding by ADP ribosylation of the small GTP-binding protein Rho. *J. Cell Biol.* **126,** 801–810.

Kinlough-Rathbone, R. L., Perry, D. W., and Packham, M. A. (1995) Contrasting effects of thrombin and the thrombin receptor peptide, SFLLRN, on aggregation and release of 14C-serotonin by human platelets pretreated with chymotrypsin or serratia marcescens protease. *Thromb. Haemost.* **73,** 122–125.

Knapp, A., Degenhardt, T., and Dodt, J. (1992) Hirudisins. Hirudin-derived thrombin inhibitors with disintegrin activity. *J. Biol. Chem.* **267,** 24,230–24,234.

Lindahl, U., Pejler, G., Bogwald, J., and Seljelid, R. (1989) A prothrombinase complex of mouse peritoneal macrophages. *Arch. Biochem. Biophys.* **273,** 180–188.

Liu, L., Dewar, L., Song, Y., Kulczycky, M., Blajchman, M. A., Fenton II, J. W., et al. (1995) Inhibition of thrombin by antithrombin III and heparin cofactor II in vivo. *Thromb. Haemost.* **73,** 405–412.

Liu, L. W., Vu, T. K., Esmon, C. T., and Coughlin, S. R. (1991) The region of the thrombin receptor resembling hirudin binds to thrombin and alters enzyme specificity. *J. Biol. Chem.* **266,** 16,977–16,980.

Lougheed, J. C., Bowman, C. L., Meininger, D. P., and Komives, E. A. (1995) Thrombin inhibition by cyclic peptides from thrombomodulin. *Protein Sci.* **4,** 773–780.

Mattson, M. P. (1994) Calcium and neuronal injury in Alzheimer's disease. Contributions of beta-amyloid precursor protein mismetabolism, free radicals, and metabolic compromise. *Ann. NY Acad. Sci.* **747,** 50–76.

Mattson, M. P. and Begley, J. G. (1996) Amyloid beta peptide alters thrombin induced calcium responses in cultured human neural cells. *Amyloid Int. J. Exp. Clin. Invest.* **3,** 28–40.

McGrogan, M., Kennedy, J., Golini, F., Ashton, N., Dunn, F., Bell, K., et al. (1990) Structure of the human protease nexin gene and expression of recombinant forms of PNI, in *Serine Proteases and Their Serpin Inhibitors in the Nervous System* (Festoff, B. W., ed.). Elsevier, Amsterdam, pp. 147–161.

McGrogan, M., Kennedy, J., Li, M., Hsu, C., Scott, R., Simonson, C. W., and Baker, J. (1988) Molecular cloning and expression of two forms of human protease nexin 1. *Biotechnology* **6,** 172–177.

Means, E. D. and Anderson, D. K. (1986) Thrombin interactions with central nervous system tissue and implications of these interactions. *Ann. NY Acad. Sci.* **485,** 314–322.

Merriman, R. L., Shackelford, K. A., Tanzer, L. R., Campbell, J. B., Bemis, K. G., and Matsumoto, K. (1989) Drug treatments for metastasis of the Lewis lung carcinoma: lack of correlation between inhibition of lung metastasis and survival. *Cancer Res.* **49,** 4509–4516.

Mirza, H., Yatsula, V., and Bahou, W. F. (1996) The proteinase activated receptor-2 (PAR-2) mediates mitogenic responses in human vascular endothelial cells. *J. Clin. Invest.* **97,** 1705–1714.

Monard, D., Suidan, H. S., and Nitsch, C. (1992) Relevance of the balance between glia-derived nexin and thrombin following lesion in the nervous system. *Ann. NY Acad. Sci.* **674,** 237–242.

Monard, D. D., Stockel, K., Goodman, R., and Thoenen, H. (1975) Distinction between nerve growth factors and glial factors. *Nature* **258,** 444,445.

Morgenstern, K. A., Henzel, W. J., Baker, J. B., Wong, S., Pastuszyn, A., and Kisiel, W. (1993) Isolation and characterization of an intracellular serine proteinase inhibitor from a monkey kidney epithelial cell line. *J. Biol. Chem.* **268,** 21,560–21,568.

Mosmann, T. (1983) Rapid colorimetric assay for cellular growth and survival: application to proliferation and cytotoxicity assays. *J. Immunol. Meth.* **65,** 55–63.

Nabeshima, T., Nitta, A., Fuji, K., Kameyama, T., and Hasegawa, T. (1994) Oral administration of NGF synthesis stimulators recovers reduced brain NGF content in aged rats and cognitive dysfunction in basal-forebrain-lesioned rats. *Gerontology* **2,** 46–56.

Nicholson, C. D. (1990) Pharmacology of nootropics and metabolically active compounds in relation to their use in dementia. *Psychopharmacology* (Berlin) **101,** 147–159.

Nierodzik, M. L., Kajumo, F., and Karpatkin, S. (1992) Effect of thrombin treatment of tumor cells on adhesion of tumor cells to platelets in vitro and tumor metastasis in vivo. *Cancer Res.* **52,** 3267–3272.

Nierodzik, M. L., Plotkin, A., Kajumo, F., and Karpatkin, S. (1991) Thrombin stimulates tumor-platelet adhesion in vitro and metastasis in vivo. *J. Clin. Inves.* **87,** 229–236.

Nierodzik, M. L. R., Bain, R. M., Liu, L. X., Shivji, M., Takeshita, K., and Karpatkin, S. (1996) Presence of the seven transmembrane thrombin receptor on human tumour cells: effect of activation on tumour adhesion to platelets and tumour tyrosine phosphorylation. *B. J. Haematol.* **92,** 452–457.

Nystedt, S., Emilsson, K., Larsson, A. K., Strîmbeck, B., and Sundelin, J. (1995) Molecular cloning and functional expression of the gene encoding the human proteinase-activated receptor 2. *Eur. J. Biochem.* **232,** 84–89.

Nystedt, S., Emilsson, K., Wahlestedt, C., and Sundelin, J. (1994) Molecular cloning of a potential proteinase activated receptor. *Proc. Natl. Acad. Sci. USA* **91,** 9208–9212.

Potempa, J., Korzus, E., and Travis, J. (1994) The serpin superfamily of proteinase inhibitors: structure, function, and regulation. *J. Biol. Chem.* **269,** 15,957–15,960.

Potter, H., Nelson, R. B., Das, S., Siman, R., Kayyali, U. S., and Dressler, D. (1992) The involvement of proteases, protease inhibitors, and an acute phase response in Alzheimer's disease. *Ann. NY Acad. Sci.* **674,** 161–173.

Rao, J. S., Baker, J. B., Morantz, R. A., Kimler, B., Evans, R., and Festoff, B. W. (1990) Serpin inhibitors of urokinase and thrombin in normal rat brain and the 9L brain tumor: evidence for elevated expression of protease nexin I-like inhibitor and a novel sodium dodecyl sulfate-activated tumor antithrombin. *Cancer Res.* **50,** 5039–5044.

Reinhard, E., Meier, R., Halfter, W., Rovelli, G., and Monard, D. (1988) Detection of glia-derived nexin in the olfactory system of the rat. *Neuron* **1,** 387–394.

Riewald, M., Morgenstern, K. A., and Schleef, R. R. (1996) Identification and characterization of the cytoplasmic antiproteinase (CAP) in human platelets. Evidence for the interaction of CAP with endogenous platelet proteins. *J. Biol. Chem.* **271,** 7160–7167.

Ripka, W., Brunck, T., Stanssens, P., LaRoche, Y., Lauwereys, M., Lambeir, A. M., et al. (1995) Strategies in the design of inhibitors of serine proteases of the coagulation cascade—Factor Xa. *Eur. J. Med. Chem.* **30 (Suppl.),** 87S–100S.

Rosenberg, R. D. (1995) The absence of the blood clotting regulator thrombomodulin causes embryonic lethality in mice before development of a functional cardiovascular system. *Thromb. Haemost.* **74,** 52–57.

Rother, M., Kittner, B., Rudolphi, K., Rossner, M., and Labs, K. H. (1996) HWA 285 (propentofylline)—a new compound for the treatment of both vascular dementia and dementia of the Alzheimer type. *Ann. NY Acad. Sci.* **777,** 404–409.

Saletu, B., Moller, H. J., Grunberger, J., Deutsch, H., and Rossner, M. (1990) Propentofylline in adult-onset cognitive disorders: double-blind, placebo-controlled, clinical, psychometric and brain mapping studies. *Neuropsychobiology* **24,** 173–184.

Schubert, P., Rudolphi, K. A., Fredholm, B. B., and Nakamura, Y. (1994) Modulation of nerve and glial function by adenosine—role in the development of ischemic damage. *Int. J. Biochem.* **26,** 1227–1236.

Seiler, S. M., Michel, I. M., and Fenton, J. W. (1992) Involvement of the "tethered-ligand" receptor in thrombin inhibition of platelet adenylate cyclase. *Biochem. Biophys. Res. Commun.* **182,** 1296–1302.

Sekiya, F., Usui, H., Inoue, K., Fukudome, K., and Morita, T. (1994) Activation of prothrombin by a novel membrane-associated protease. An alternative pathway for thrombin generation independent of the coagulation cascade. *J. Biol. Chem.* **269,** 32,441–32,445.

Shea, T. B. (1996) The influence of surface protease activity on neuritogenesis and neurite retraction is dependent upon mode of neurite induction: inhibition of thrombin-mediated surface protcasc activity is crucial for ncuritc outgrowth but not for mainenance of existng neurites. *Neuroscience Res. Comm.* **19,** 57–60.

Shinoda, I., Furukawa, Y., and Furukawa, S. (1990) Stimulation of nerve growth factor synthesis/secretion by propentofylline in cultured mouse astroglial cells. *Biochem. Pharmacol.* **39,** 1813–1816.

Smith-Swintosky, V. L., Zimmer, S., Fenton, J. W., II, and Mattson, M. P. (1995) Protease nexin-1 and thrombin modulate neuronal Ca^{2+} homeostasis and sensitivity to glucose deprivation-induced injury. *J. Neurosci.* **15,** 5840–5850.

Snider, R. M. (1986) Thrombin effects on cultured nerve cells: clinical implications and evidence for a novel mechanism of neuronal activation. *Ann. NY Acad. Sci.* **485,** 310–313.

Snider, R. M., McKinney, M., and Richelson, E. (1986) Thrombin binding and stimulation of cyclic guanosine monophosphate formation in neuroblastoma cells. *Semin. Thromb. Hemost.* **12,** 253–262.

Snider, R. M., and Richelson, E. (1983) Thrombin stimulation of guanosine 3′,5′-monophosphate formation in murine neuroblastoma cells (clone N1E-115) Science **221,** 566–568.

Sommer, J., Gloor, S. M., Rovelli, G. F., Hofsteenge, J., Nick, H., Meier, R., and Monard, D. (1987) cDNA sequence coding for a rat glia-derived nexin and its homology to members of the serpin superfamily. *Biochemistry* **26,** 6407–6410.

Srinivasan, J., Hu, S., Hrabal, R., Zhu, Y., Komives, E. A., and Ni, F. (1994) Thrombin-bound structure of an EGF subdomain from human thrombomodulin determined by transferred nuclear Overhauser effects. *Biochemistry* **33,** 13,553–13,560.

Stone, S. R., Nick, H., Hofsteenge, J., and Monard, D. (1987) Glial-derived neurite-promoting factor is a slow-binding inhibitor of trypsin, thrombin, and urokinase. *Arch. Biochem. Biophys.* **252,** 237–244.

Suidan, H. S., Stone, S. R., Hemmings, B. A., and Monard, D. (1992) Thrombin causes neurite retraction in neuronal cells through activation of cell surface receptors. *Neuron* **8,** 363–375.

Tigyi, G., Fischer, D. J., Sebok, A., Marshall, F., Dyer, D. L., and Miledi, R. (1996a) Lysophosphatidic acid-induced neurite retraction in PC12 cells: neurite-protective effects of cyclic AMP signaling. *J. Neurochem.* **66,** 549–558.

Tigyi, G., Fischer, D. J., Sebok, A., Yang, C., Dyer, D. L., and Miledi, R. (1996b) Lysophosphatidic acid-induced neurite retraction in PC12 cells: control by phosphoinositide-Ca^{2+} signaling and Rho. *J. Neurochem.* **66,** 537–548.

Trejo, J.-A., Connolly, A. J., and Coughlin, S. R. (1996) The cloned thrombin receptor is necessary and sufficient for activation of mitogen-activated protein kinase and mitogenesis in mouse lung fibroblasts. *J. Biol. Chem.* **271,** 21536–21541.

Trousseau, A. (1865) *Plegmasia Alba Dolens, Clinique Medicale de l'Hotel Dieu de Paris.* Balliere, Paris, pp. 94–96; 654–1062.

Turnell, A. S., Brant, D. P., Brown, G. R., Finney, M., Gallimore, P. H., Kirk, C. J., et al. (1995) Regulation of neurite outgrowth from differentiated human neuroepithelial cells: a comparison of the activities of prothrombin and thrombin. *Biochem. J.* **308,** 965–973.

Vassalli, J. D., Huarte, J., Bosco, D., Sappino, A. P., Sappino, N., Velardi, A., et al. (1993) Protease-nexin I as an androgen-dependent secretory product of the murine seminal vesicle. *EMBO J.* **12,** 1871–1878.

Vaughan, P. J., Pike, C. J., Cotman, C. W., and Cunningham, D. D. (1995) Thrombin receptor activation protects neurons and astrocytes from cell death produced by environmental insults. *J. Neurosci.* **15,** 5389–5401.

Voss, G. C., Botteri, F., Monard, D., and Jockusch, H. (1996) Localization of protease nexin 1 gene spi4 between villin and acetylcholine receptor gamma genes on mouse chromosome 1. *Mammal. Genome* **7,** 167–168.

Vu, T. K., Hung, D. T., Wheaton, V. I., and Coughlin, S. R. (1991a) Molecular cloning of a functional thrombin receptor reveals a novel proteolytic mechanism of receptor activation. *Cell* **64,** 1057–1068.

Vu, T. K., Wheaton, V. I., Hung, D. T., Charo, I., and Coughlin, S. R. (1991b) Domains specifying thrombin-receptor interaction. *Nature* **353,** 674–677.

Wagner, S. L., Geddes, J. W., Cotman, C. W., Lau, A. L., Gurwitz, D., Isackson, P. J., and Cunningham, D. D. (1989a) Protease nexin-1, an antithrombin with neurite outgrowth activity, is reduced in Alzheimer disease. *Proc. Natl. Acad. Sci. USA* **86,** 8284–8288.

Wagner, S. L., Lau, A. L., and Cunningham, D. D. (1989b) Binding of protease nexin-1 to the fibroblast surface alters its target proteinase specificity. *J. Biol. Chem.* **264,** 611–615.

Wagner, S. L., Van Nostrand, W. E., Lau, A. L., Farrow, J. S., Suzuki, M., Bartus, R. T., et al. (1993) Co-distribution of protease nexin-1 and protease nexin-2 in brains of non-human primates. *Brain Res.* **626,** 90–98.

Wang, S. Y. (1994) A retinoic acid-inducible GATA-binding protein binds to the regulatory region of J6 serpin gene. *J. Biol. Chem.* **269,** 607–613.

Weiler-Guettler, H., Aird, W. C., Rayburn, H., Husain, M., and Rosenberg, R. D. (1996) Developmentally regulated gene expression of thrombomodulin in postimplantation mouse embryos. *Development* **122,** 2271–2281.

White, C. E., Kempi, N. M., and Komives, E. A. (1994) Expression of highly disulfide-bonded proteins in Pichia pastoris. *Structure* **2,** 1003–1005.

White, E. A., Baker, J. B., McGrogan, M., and Kitos, P. A. (1993) Protease nexin 1 is expressed in the human placenta. *Thromb. Haemost.* **69,** 119–123.

Winn, M. J., Jain, K., and Ku, D. D. (1993) Argatroban and inhibition of the vasomotor actions of thrombin. *J. Cardiovasc. Pharmacol.* **22,** 754–760.

Witting, J. I., Bourdon, P., Brezniak, D. V., Maraganore, J. M., and Fenton, J. W. (1992) Thrombin-specific inhibition by and slow cleavage of hirulog-1. *Biochem. J.* **283,** 737–743.

Wojtukiewicz, M. Z., Tang, D. G., Ben-Josef, E., Renaud, C., Walz, D. A., and Honn, K. V. (1995) Solid tumor cells express functional "tethered ligand" thrombin receptor. *Cancer Res.* **55,** 698–704.

Wojtukiewicz, M. Z., Tang, D. G., Ciarelli, J. J., Nelson, K. K., Walz, D. A., Diglio, C. A., Mammen, E. F., and Honn, K. V. (1993) Thrombin increases the metastatic potential of tumor cells. *Int. J. Cancer* **54,** 793–806.

Yan, S. D., Chen, X., Fu, J., Chen, M., Zhu, H. J., Roher, A., et al. (1996) Rage and amyloid beta peptide neurotoxicity in Alzheimer's disease. *Nature* **382,** 685–691.

Zoubine, M. N., Ma, J. Y., Smirnova, I. V., Citron, B. A., and Festoff, B. W. (1996) A molecular mechanism for synapse elimination: novel inhibition of locally-generated thrombin delays synapse loss in neonatal mouse muscle. *Dev. Biol.* **179,** 447–457.

13
Thrombin and Thrombin Inhibitors

Roles in Acute and Chronic Neurodegenerative Disorders

Virginia L. Smith-Swintosky

1. INTRODUCTION

Thrombin is a 37 kDa glycosylated trypsin-like serine protease that has multiple biological functions. It is formed from a precursor protein, the zymogen prothrombin, by a complex of Factor Xa, Va, phospholipid, and calcium (Tracy et al., 1981). Prothrombin is synthesized by the liver and secreted into the blood stream (Fenton, 1986), where it is cleaved to mature thrombin as part of a cascade of proteolytic events (Davie et al., 1991). Recently, prothrombin mRNA has been reported to be expressed in brain tissue (Dihanich et al., 1991).

Thrombin exerts its actions via both soluble target proteins and cell surface receptors. Its classical role in hemostasis includes cleavage of fibrinogen to clottable fibrin and the activation of platelets to cause their degranulation and aggregation (Davey and Luscher, 1967; Berndt and Phillips, 1981; Davie et al., 1991). In vitro studies have shown that thrombin can activate cell surface receptors on many cell types, including lymphocytes (Chen and Buchanan, 1975; Bizios et al., 1985; Naldini et al., 1991; Tordai et al., 1993), osteoblasts (Jenkins et al., 1993, 1994), tumor cells (Wojtukiewicz et al., 1993), endothelial cells (Weksler et al., 1978; Prescott et al., 1984; Daniel et al., 1986; Gelehrter and Szncycer-Laszuk, 1986; Goligorsky et al., 1989; Tiruppathi et al., 1992), vascular smooth muscle cells (McNamara et al., 1993), fibroblasts (Chen and Buchanan, 1975), epithelial cells (Reddan et al., 1982), neurons (Gurwitz and Cunningham, 1988, 1990; Zurn et al., 1988; Grabham et al., 1991; Jalink and Moolenar, 1992; Suidan et al., 1992; Smith-Swintosky et al., 1995a,b; Vaughan et al., 1995), and astrocytes (Perraud et al., 1987; Loret et al., 1989; Cavanaugh et al., 1990; Neveu et al., 1993; Beecher et al., 1994; Vaughan et al., 1995). Thrombin's versatility allows it a role in a wide variety of biological functions, such as hemostasis, artherosclerosis/restenosis, wound repair, inflammation, bone remodeling, tumor growth, and neural development/disease.

Thrombin is inhibited by several potent endogenous serine protease inhibitors, such as protease nexin-1, anti-thrombin III, and heparin cofactor 2, which are found circulating in the blood or are associated with the cell surface. Much of the research regarding thrombin's actions in the CNS stresses the importance of the delicate balance that exists between thrombin and its inhibitors in CNS function.

From: Neuroprotective Signal Transduction *edited by M. P. Mattson*
Humana Press Inc., Totowa, NJ

2. THROMBIN RECEPTOR

In 1991, Vu et al. cloned the human thrombin receptor and found it to be a member of the seven transmembrane G protein-coupled receptor family, with a novel mechanism of activation. They discovered that within the long 100-amino-acid extracellular N-terminus there was a putative thrombin cleavage site (Arg41–Ser42). When this site was mutated to an uncleavable Arg/Pro site, the receptor could no longer be activated by thrombin. Thus, Coughlin's group proposed that upon cleavage by thrombin, a new N-terminus was generated that functioned as a "tethered ligand," which binds intramolecularly to an undefined receptor site effecting receptor activation. The tethered ligand hypothesis was substantiated by studies utilizing synthetic peptides corresponding to the first six amino acids of this new amino terminus, SFLLRN, or thrombin receptor-activated peptide (TRAP-6), which activated the receptor. Such studies have been replicated in several cell types, indicating TRAP peptides can duplicate the actions of thrombin, although in general much higher concentrations of peptide are required to mimic thrombin's actions (e.g., neurite retraction; Jalink and Moolenaar, 1992; Suidan et al., 1992). Within 1 min of thrombin receptor activation, the receptor is inactivated by phosphorylation of serine and threonine residues in the cytoplasmic tail by an unknown receptor kinase (Brass et al., 1992; Ishii et al., 1994). The receptor is internalized by coated pits and endosomes, and either recycled to the cell surface or degraded in the lysosomes (Hoxie et al., 1993; Hein et al., 1994; Brass et al., 1994).

2.1. Cellular Responses to Thrombin in the CNS

McKinney et al. (1983) provided the first evidence for thrombin receptor in the CNS. They reported that radiolabeled human α-thrombin bound rapidly, reversibly, and with high affinity to postmortem human brain ($K_D = 20.6\ nM$) and spinal cord ($K_D = 6\ nM$) tissue. However, it was not until 1991 that Coughlin's group isolated and cloned a human thrombin receptor reportedly expressed in platelets and endothelial cells (Vu et al., 1991). In general, it appears that this cloned receptor mediates the majority of thrombin's effects on different cell types. In 1994, Niclou et al. used primers to the rat thrombin receptor (Zhong et al., 1992) to characterize the developmental expression of the receptor in rat brain (embryonic to PN28), utilizing *in situ* hybridization and Northern blotting. They found thrombin receptor mRNA in the brain and spinal cord, especially neocortex, cingulate/resplenial cortex, hypothalamus, thalamus, pretectum, ventral mesencephalon, hippocampus (hilus, CA2 > CA1, DG, subiculum), cerebellum, and olfactory bulb, so that mRNA levels were highest in embryonic stages, but then decreased with age. Furthermore, recent evidence suggests that thrombin receptor expression is localized in neuronal, glial, and ependymal cells, although white matter and most cerebrovascular endothelial cells were negative (Weinstein et al., 1995).

Until 1991, the expression of thrombin or its precursor prothrombin in tissues other than liver was not demonstrated conclusively, because of difficulty in avoiding serum contamination. Using sensitive mRNA detection methods, such as PCR, Northern blotting and *in situ* hybridization, Dihanich et al. demonstrated that prothrombin is expressed in the brain throughout development (E13–PN21), especially in the olfactory bulb, cortex, cerebellum, and hippocampus. They also demonstrated that prothrombin is present in several neuronal and glial cell lines (Dihanich et al., 1991). Taking their findings together with recent evidence, it appears that prothrombin is co-localized with the thrombin receptor in rat brain, indicating that locally synthesized thrombin could act in an autocrine or paracrine manner to regulate specific processes (Weinstein et al., 1995). However, it is unknown whether the factors required for prothrombin conversion to thrombin in the blood are present in brain or are even required.

2.2. *Normal Functions of Thrombin*

The normal physiological role for thrombin in the CNS is unknown; however, the developmental expression of prothrombin and thrombin receptor in the brain and synaptogenesis studies in the peripheral nervous system (PNS) suggest it may play an important role in neural development. Prothrombin, thrombin receptor, and endogenous thrombin inhibitors (such as protease nexin-1 [PN-1]), appear to be coexpressed at high levels in the CNS during early development, tapering off to a basal level in the adult brain (Choi et al., 1990; Wagner et al., 1993; Coughlin, 1994; Niclou et al., 1994; Weinstein et al., 1995). The early stages of CNS development are marked with tremendous synaptic reorganization. Thrombin has been shown to have a direct effect on the morphology of embryonic cells in culture, so that low concentrations of thrombin (p*M*) induce neurite retraction in both astrocytes and neurons, resulting in synapse loss. PN-1, a potent cell-surface thrombin inhibitor, reverses the thrombin effect (Zurn et al., 1988; Cunningham and Gurwitz, 1989; Cavanaugh et al., 1990; Farmer et al., 1990; Jalink and Moolenaar, 1992; Suidan et al., 1992; Beecher et al., 1994; Brewer, 1995; Grabham and Cunningham, 1995; Turnell et al., 1995). Elegant studies utilizing an in vitro model of the mammalian neuromuscular junction (NMJ) have demonstrated that thrombin is required for polyneuronal synapse elimination that is followed by PN-1-induced stabilization of the NMJ in mouse skeletal muscle (Liu et al., 1994; Akaaboune et al., 1995; and *see* Chapter 12 of this book). Similar studies have not been performed in CNS tissue; however, the evidence suggests that an intricate interplay exists between thrombin and its inhibitors, particularly PN-1, in regulating synaptic organization in the CNS.

3. THROMBIN AND NEURONAL INJURY

Several lines of evidence suggest thrombin levels are altered in acute and chronic neurodegenerative conditions (Table 1). Acute ischemic stroke appears to involve thrombin on several levels. First, nearly 80% of all strokes are thrombotic (or embolic) in origin, in that thrombin induces blood clotting and platelet aggregation leading to thrombus formation, and eventually to arterial occlusion and ischemia. However, thrombin receptor is also present in the brain and may directly contribute to neuronal death that occurs as a result of ischemic stroke. Nitsch et al. (1993) demonstrated that, after transient forebrain ischemia, PN-1 is re-expressed in the adult rat brain for up to 1 yr after the insult, suggesting indirectly that endogenous thrombin is increased during ischemic stroke. Similarly, thrombin has been implicated in traumatic brain injury. In 1994, Suzuki et al. reported that patients who had suffered severe head injury, such as contusion or intracerebral hemorrhage, displayed diphasic increases in thrombin–antithrombin III (AT3) complexes in blood sampled from the internal jugular vein (the main draining vessel of the brain), compared to a peripheral vein. They found an initial peak of complex formation at 3 h postinjury, which decreased significantly by 12 h, and remained depressed until another highly significant peak at 72 h postinjury. The authors proposed that the early phase reflected the systemic coagulation and fibrinolytic

Table 1
Thrombin Implicated in Neurodegenerative Conditions

Thrombin-induced effect	Injury/disease	Reference
Acute		
Thrombin is neurotoxic	Ischemic stroke, epilepsy, traumtic brain injury	Smith-Swintosky et al., 1995; Vaughan et al., 1995
PN-1 increased in rat transient forebrain ischemia	Ischemic stroke	Nitsch et al., 1993
Thrombin exacerbates glucose deprivation-induced damage in hippocampal neuronal cultures	Ischemic stroke	Smith-Swintosky et al., 1995
Thrombin protects cultured hippocampal neurons from metabolic/oxidative insults	Ischemic stroke	Vaughan et al., 1995
Thrombin, thrombin-AT3 complexes increase in the brain after severe head trauma	Traumatic brain injury	Suzuki et al., 1994
Thrombin induces reactive gliosis, inflammation, and edema traumatic brain injury	Ischemic stroke,	Nishino et al., 1993, 1994
Chronic		
Thrombin localized in senile plaques and neurofibrillary tangles	Alzheimer's disease	Akiyama et al., 1992
PN-1 Prod decreased and PN-1-thrombin complexes increased and associated with senile plaques	Alzheimer's disease	Wagner et al., 1989
Thrombin-induced APP processing in vitro	Alzheimer's disease	Igarashi et al., 1992; Chong et al., 1994; Davis-Salinas et al., 1994; Brewer, 1996
Thrombin exacerbates amyloid β-peptide toxicity in hippocampal cultures	Alzheimer's disease	Smith-Swintosky et al., 1995
Thrombin protects against amyloid β-peptide toxicity in hippocampal cultures	Alzheimer's disease	Pike et al., 1996
PN-1 sensitivity to oxidation	Alzheimer's disease	Bolkenius and Monard, 1995
Thrombin degrades tau protein	Alzheimer's disease	Olesen, 1994

abnormalities because of traumatic shock and hypoxia, and that the late phase increase indicated specific thrombin generation in the CNS. The thrombin–AT3 complex levels in the jugular vein of patients who died were relatively higher than the survivors in both the initial and later stages (Suzuki et al., 1994).

Recently, animal studies have demonstrated that direct infusions of α-thrombin into the brain lead to infiltration of inflammatory cells, proliferation of mesenchymal cells, induction of angiogenesis, an increase in reactive astrocytes, and development of pale tissue foci caused by spongiosis. The experimenters concluded that thrombin may significantly contribute to the inflammation, brain edema, and reactive gliosis that occurs following CNS injury, including stroke (Nishino et al., 1993, 1994).

Among these emerging reports are studies from several research groups utilizing cell culture technology to examine the effects of thrombin on neural tissue. Their results demonstrated that α-thrombin is toxic to cultured rat hippocampal and septal cholinergic neurons and cortical astrocytes in a concentration-dependent manner (Smith-Swintosky et al., 1995a; Vaughan et al., 1995; Debeir et al., 1996; Fig. 1). Moreover, additional studies were performed to examine the interactive effects between thrombin and neurological insults associated with aging and ischemic/traumatic neuronal injury (e.g., oxidative stress, excitotoxicity, decreased glucose utilization/transport) on neuronal cell survival. The evidence suggests that subtoxic levels of thrombin (<30 n*M*) potentiate neuronal damage induced by a variety of insults, such as glucose deprivation and glutamate toxicity (Smith-Swintosky et al., 1995a), but higher concentrations of thrombin (>100 n*M*) appear to protect neurons from such insults (Vaughan et al., 1995). The mechanism by which thrombin protects cells is unknown, although it has been shown that thrombin can induce synthesis/secretion of growth factors in culture (Neveu et al., 1993; Pike et al., 1996). Growth factors have been shown to afford cultured neurons significant protection against excitotoxic/metabolic injury (for review, *see* Mattson et al., 1993b). Moreover, PN-1 protects neurons from glucose-deprivation-induced damage and blocks thrombin-induced neurotoxicity (Smith-Swintosky et al., 1995a). It is unclear whether thrombin can directly induce increased expression of PN-1 in response to injury. However, thrombin has been shown to induce reactive gliosis in the brain and inflammation in the periphery, which results in cytokine release (Nishino et al., 1993, 1994; Shin et al., 1995). Cytokines and other injury-related factors have been shown to stimulate PN-1 secretion in cultured brain cells (Vaughan and Cunningham, 1993), suggesting that brain injury-induced cytokine cascades may counteract thrombin-induced neuronal damage. These experimental studies suggest that thrombin and PN-1 are likely to play important roles in acute neurological insults, including stroke and traumatic brain injury.

Chronic neurodegenerative disorders such as Alzheimer's disease (AD) also appear to involve thrombin activation. In 1992, Akiyama et al. demonstrated with immunohistochemistry that thrombin was present in AD brains. Positive staining was restricted to vessels and residual plasma in controls, but was also present in senile plaques, some diffuse amyloid deposits, and neurofibrillary tangles (NFTs) in the brains of AD patients. In addition, Wagner et al. (1989) demonstrated that PN-1–thrombin complexes were present in senile plaques and that PN-1 levels were decreased sevenfold in AD brains, compared with age-matched control brains. More recently, in vitro studies have demonstrated direct effects of thrombin on amyloid precursor protein (APP) and on the

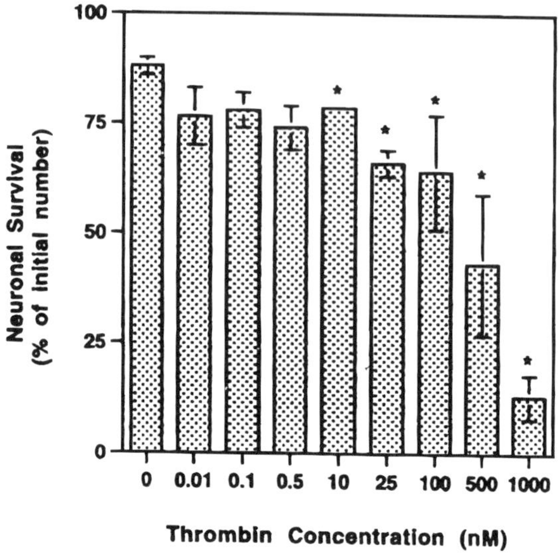

Fig. 1. Dose-response curve of rat hippocampal cultures treated with control (saline) or α-thrombin. Cell viability was assessed 48 h posttreatment. Cell survival is expressed as percent of initial number of cells present pretreatment. Values represent mean + SEM (n = 8 fields/treatment in two separate experiments). (Adapted with permission from Smith-Swintosky et al., 1995a.)

microtubule-associated protein tau, both of which are pathologically altered and accumulate in the AD brain. Reports regarding thrombin's effect on APP vary. In bacteria (Igarashi et al., 1992) and platelets (Chong et al., 1994), thrombin appears to cleave APP, thereby generating the amyloidogenic fragment, amyloid β-peptide (Aβ), which aggregates in senile plaques and is neurotoxic (Koh et al., 1990; Busciglio et al., 1992; Mattson et al., 1992). However, Davis-Salinas et al. (1994) demonstrated in glioblastoma cultures that thrombin-receptor activation induced secretion and nonamyloidogenic processing of APP, yielding secreted forms of the large precursor protein (e.g., APP695), previously shown to be neuroprotective against a variety of neurological insults (Mattson et al., 1993a; Barger et al., 1994; Goodman and Mattson, 1994; Smith-Swintosky et al., 1994; Furukawa et al., 1996). Most recently, Brewer (1996) reported that thrombin does not appear to induce APP processing in neurons cultured in serum-free medium. Altogether, it is unclear how thrombin affects APP processing in the AD brain (e.g., neurotoxic neurotrophic). Nevertheless, several studies suggest that thrombin can influence Aβ-toxicity in neuronal cultures. The results are consistent with previous reports regarding thrombin's effects on acute neuronal injury (e.g., glucose deprivation), in that low concentrations of thrombin (≤30 nM) appear to exacerbate Aβ-induced neurotoxicity via increased intracellular calcium levels and oxidative stress (Smith-Swintosky et al., 1995b), but high concentrations of thrombin (≥100 nM) protect neurons (Pike et al., 1996). Notably, PN-1 significantly protects cultured hippocampal neurons from Aβ-toxicity (Smith-Swintosky et al., 1995b), further demonstrating the balance between thrombin and PN-1 in regulating neuronal survival in neurodegenerative conditions. Recent evidence suggests that PN-1 is quite sensitive to oxidative inactivation

(Bolkenius and Monard, 1995). Aβ may be implicated in decreased PN-1 activity in AD brain, because it has been shown to induce protein oxidation and inactivation of several endogenous enzymes associated with AD (Smith et al., 1991; Hensley et al., 1994; Mark et al., 1995). The dual action of thrombin on neurons in response to Aβ may adversely affect neuronal survival and plasticity, because both trophic and toxic actions of thrombin in neurons have been reported (Smith-Swintosky et al., 1995b; Vaughan et al., 1995; Pike et al., 1996; Mattson and Begley, 1996).

Although much attention is focused on the role of amyloid in AD, NFTs are another unique histological hallmark of the disease. The most prominent structural element of NFTs is the paired helical filament (PHF), which consists primarily of the microtubule-associated protein tau (Delacourte and Defossez, 1986; Grundke-Iqbal et al., 1986; Goedert et al., 1988). Six different isoforms of tau exist in the brain, all of which are present in PHF (Goedert et al., 1989). Functionally, however, PHF tau differs from soluble tau, as it has a much lower capacity to assemble microtubules. Olesen (1994) demonstrated that thrombin proteolytically degrades tau protein in vitro, thereby possibly reducing its capacity to promote microtubule assembly compared with full-length tau; it may therefore contribute to PHF tau formation in AD.

Taking these lines of evidence together, it seems reasonable to conclude that thrombin may contribute to the degeneration that occurs in both acute and chronic neurodegenerative disorders on several levels (Fig. 2). Moreover, the balance that exists between thrombin and endogenous thrombin inhibitors, such as PN-1, appears to be critical for neural cells during development and injury/disease.

3.1. *Thrombin Inhibitors Are Neuroprotective*

Compelling evidence suggesting that endogenous thrombin inhibitors, such as PN-1, can reverse the deleterious effects of thrombin and provide neuroprotection against a variety of metabolic/excitotoxic insults has spurred a growing interest in the development of thrombin inhibitors as therapeutic agents for neurodegenerative disorders. Several thrombin inhibitors currently under investigation appear promising; however, there have been few reports of their effects in the CNS. Hirudin, a potent thrombin inhibitor isolated from leeches, has been used in several in vitro assays to demonstrate thrombin-specific effects on neurons. For example, hirudin has been shown to reverse thrombin-induced neurite retraction in astrocytes and neurons (Suidan et al., 1992; Cunningham and Gurwitz, 1989; Grabham and Cunningham, 1995), and reverse thrombin-induced neuroprotection against Aβ toxicity (Pike et al., 1996). In addition, recombinant hirudin reportedly inhibits thrombin-induced platelet aggregation (Maraganore et al., 1989) and can prolong the time required for thrombotic occlusion of lesioned arteries in vivo (Takiguchi et al., 1995), suggesting a potential value for hirudin–hirudin analogs in treating ischemic stroke. Argatroban, a synthetic selective thrombin inhibitor, may also have beneficial effects related to the CNS. Studies indicate that argatroban treatment prevents reocclusion following tissue plasminogen activator-induced thrombolysis in a canine model of femoral arterial thrombosis (Mellott et al., 1990) and significantly prolongs the time required for thrombotic occlusion to occur in a middle cerebral artery occlusion model of focal ischemia in the rat (Umemura et al., 1995). However, initial studies with argatroban in humans have not demonstrated significant therapeutic efficacy. Stroke patients treated with argatroban showed decreased

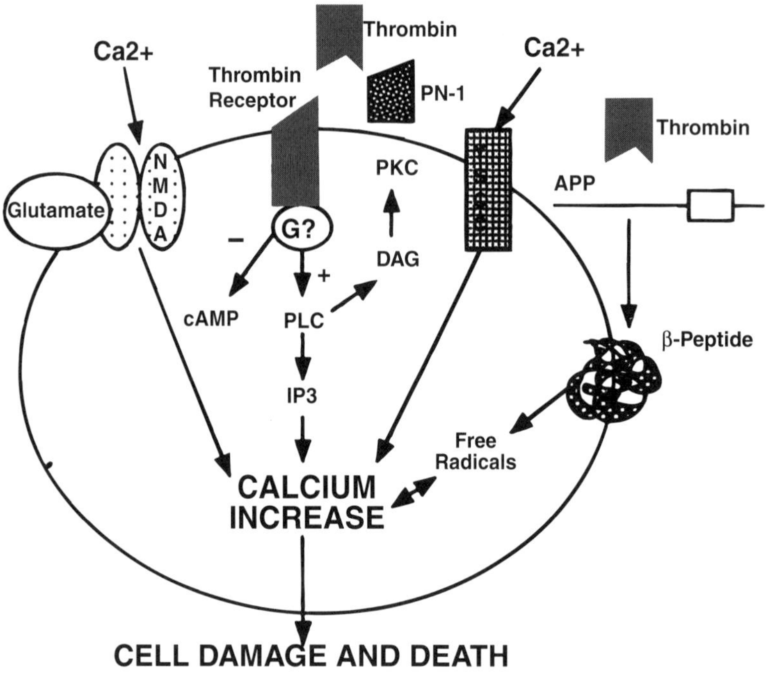

Fig. 2. Summary diagram of proposed thrombin-signaling pathways involved in neurodegeneration. Excitotoxic and metabolic insults involve excessive stimulation of glutamate receptors, such as the NMDA receptor, leading to an influx of Ca^{2+} into the cell. Change in membrane potential can activate other voltage-sensitive Ca^{2+} channels (VSCC), thereby further increasing intracellular Ca^{2+} levels. Thrombin has been shown in nonneuronal cells to stimulate IP_3 formation, which could lead to release of Ca^{2+} from intracellular stores. Taken together, thrombin can exacerbate neural injury by potentiating the rise in intracellular Ca^{2+}. If not compensated for, sustained increases in Ca^{2+} (μM) can activate various lipases, kinases, and proteases that damage the cell and can eventually lead to cell death. Protease nexin-1 (PN-1), is a potent cell surface thrombin inhibitor. It has been shown to completely block thrombin's neurotoxic effects and provide neuroprotection against a variety of insults, including glucose deprivation and glutamate toxicity. In addition, there exists an important balance between thrombin and PN-1 in APP processing and Aβ-peptide-induced neurotoxicity, so that thrombin has been shown to generate β-peptide in vitro and therefore contribute to the neuronal death that occurs in AD. On the other hand, PN-1 has been shown to be neuroprotective against β-amyloid toxicity.

plasma levels of D-dimer (an indicator of hypercoagulability) relative to control patients; however, the decrease did not correspond to a significant improvement in neurological scores (Kario et al., 1995).

3.2. *Thrombin Receptor Signaling in the CNS*

Another approach to therapeutic intervention for thrombin-induced neurotoxicity is to develop agents that interfere with thrombin signaling events in neural cells. Thrombin receptor signaling utilizes multiple transduction pathways, including phosphatidylinositol (PI) hydrolysis, protein kinase C (PKC), phospholipase A_2, calcium, cyclic nucleotides, tyrosine kinase-mediated pathways, Ras-mediated pathways, and possibly other protease-activated receptors (for an excellent review, *see* Grand et al., 1996). However, little is known about thrombin receptor signaling in the CNS. Thrombin has been shown to significantly increase intracellular calcium levels $[Ca^{2+}]_i$ in cultured hippocampal neuron, which appear to be dependent on calcium influx, since the calcium increase did not occur in calcium-deficient media (Smith-Swintosky et al., 1995a). In other cell types, thrombin-induced elevation of $[Ca^{2+}]_1$ involves PI hydrolysis, leading to the generation of $InsP_3$-mediated release of calcium from intracellular stores (e.g., endoplasmic reticulum) and PKC activation (Baffy et al., 1994). However, preliminary evidence suggests that thrombin and TRAP-6 do not significantly stimulate PI hydrolysis in cultured rat hippocampal neurons (unpublished observation). Such results would support previous evidence indicating that thrombin-induced increases in $[Ca^{2+}]_i$ are dependent on calcium influx. In addition, thrombin has been shown to induce the formation of free radicals in hippocampal neuronal cultures, as measured by the oxidation indicator dye, 2,7 dichlorofluoroscein (Smith-Swintosky et al., 1995b). Several studies suggest that free radicals can contribute to increases in $[Ca^{2+}]_1$, thus providing thrombin with another possible means of elevating intracellular calcium levels (Mattson et al., 1996). In neurons, this is quite important, as a result of a large literature indicating that sustained increases in $[Ca^{2+}]_1$ can be lethal to cells (for review, *see* Siesjo, et al., 1989; Mattson, 1992).

Thrombin has been shown to affect the formation of cyclic nucleotides. Snider and Richelson (1983) demonstrated that thrombin induces cGMP formation that is dependent on extracellular calcium in N1E-115 neuroblastoma cells. Similar studies in primary neuronal cultures have not been reported. More is known about the effects of thrombin on the generation of cAMP. In general, thrombin has been shown to inhibit agonist-stimulated adenylate cyclase activity in platelets, megakaryoblastic HEL cell membranes, vascular smooth muscle cells, and fibroblasts (at concentrations <0.1 n*M*), thereby reducing intracellular cAMP levels (Magnaldo et al., 1988; Brass and Woolkasis, 1992; Seiler et al., 1992; Kanthou et al., 1996). This effect of thrombin is thought to be mediated by G_1 since thrombin-induced inhibition of adenylate cyclase can be abolished by treatment with pertussis-toxin (Hung et al., 1992). However, thrombin also signals through $G_{\alpha z}$, which is found predominantly in neuronal tissue and platelets, and has 65% amino acid sequence similarity with $G_{\alpha i}$, but is insensitive to pertussis-toxin and inhibits adenylate cyclase activity (Carlson et al., 1989; Gagnon et al., 1991; Kozasa and Giman, 1995). It is not known if thrombin inhibits adenylate cyclase in neurons. Considerable evidence suggests that elevation of cyclic nucleotides in neurons is neuroprotective (for review, *see* Mattson and Furukawa, 1996). For example, agents that elevate cAMP and cGMP levels have been shown to protect cultured hippocampal neurons against excitotoxic and metabolic insults (Mattson and Kater, 1988; Barger et al., 1994; *see* Chapter 10). Therefore, thrombin-induced inhibition of cyclic nucleotide formation in neurons may contribute to its neurotoxic effects (Fig. 1). Alternate thrombin signaling pathways reported in other cell types have not been explored in the CNS (e.g., phosphorylation of growth factor tyrosine kinase receptors; van Obberghen-Schilling and Pouyssegur, 1993; Rao et al., 1995). Further examination of thrombin receptor signaling in neurons and glia may prove critical to developing brain-selective thrombin receptor antagonists.

4. FUTURE DIRECTIONS

There is still much we do not know about thrombin effects on the CNS. Preliminary evidence in neuronal cultures suggests that thrombin receptor activation in the CNS may be pharmacologically distinct from other cell types (e.g., dose-response, signaling cascades). Thrombin receptor knockout (KO) mice could prove useful in determining thrombin's role in the CNS. Recently, two groups of researchers have generated KOs against the murine-cloned thrombin receptor gene (Connelly et al., 1996; Darrow et al., 1996). Knocking out this gene generates a partial lethality, in that approx 50% of the embryos die between embryonic d 9 and 10. Those that survive appear to be normal upon gross inspection. Platelets respond to thrombin, but their fibroblasts do not. These results support the idea that a separate platelet receptor for thrombin exists (Connelly et al., 1996; Darrow et al., 1996). There have been no reports of CNS studies in these mice. An examination of thrombin's effects on neuronal cultures from the KO mice should prove useful in determining if a brain-specific thrombin receptor exists. Furthermore, injury studies in vivo should provide valuable information as to how much of the cellular damage that occurs is caused by thrombin receptor activation (e.g., ischemic stroke, head trauma).

With the discovery of the thrombin receptor came the question as to whether there were other receptors that could be activated by proteases. In 1994, Sundelin's group provided the first evidence for a new thrombin receptor family member, which they called protease-activated receptor 2 (PAR-2) (Nystedt et al., 1995). The receptor was cloned from a mouse genomic library, expressed in Xenopus oocytes, and found to be activated by trypsin at low concentrations (p*M*). Sequence alignment with the thrombin receptor suggested a possible cleavage site analogous to that used for activation of the thrombin receptor, so that mutation of this site from Arg/Ser to Arg/Pro prevented activation of PAR-2 by trypsin. As with the thrombin receptor, a synthetic hexapeptide (SLIGRL) representing the first six amino acids following the putative PAR-2 cleavage site was an agonist for PAR-2 activation. However, the native physiological activating protease for the PAR-2 receptor remains unknown. Initial studies using Northern blot analysis suggested that PAR-2 mRNA is expressed in highly vascularized tissues, such as kidney, small intestine, and stomach (Nystedt et al., 1995). Subsequent studies have demonstrated functional receptor in several other cell types, including keratinocytes (Santulli et al., 1995), endothelial cells (Mirza et al., 1996), vascular and gastric smooth muscle cells (Al-Ani et al., 1995), Jurkat T leukemic cells (Mari et al., 1996), and neurons (unpublished observation). Although the specific role of this receptor in the CNS is unclear, recent evidence indicates it may play a role in neurodegeneration (Smith-Swintosky, et al., 1997). In addition, it is interesting to note that there may be crosstalk between the thrombin receptor and PAR-2, controlling receptor activation (Mirza et al., 1996). Further investigation of the activation and transduction mechanisms of these receptors should provide fundamental information necessary for understanding their in CNS development and disease, ultimately leading to the generation of novel therapeutic agents for neurodegenerative conditions.

ACKNOWLEDGMENTS

I thank Michael Beblavy and Cathy Cheo-Issacs for their work on the phosphatidylinositol hydrolysis studies. In addition, I am grateful for my colleagues, Patricia Andrade-Gordon, Claudia Derian, and Richard Shank, for their helpful comments on this manuscript.

REFERENCES

Akiyama, H., Skeda, K., Kondo, H., and McGreer P. L. (1992) Thrombin accumulation in brains of patients with Alzheimer's disease. *Neurosci. Lett.* **146,** 152–154.

Al-Ani, B., Saifeddine, M., and Hollenberg, M. D. (1995) Detection of functional receptors for the proteinase-activated-receptor-2-activating polypeptide, SLIGRL-NH2, in rat vascular and gastric smooth muscle. *Can. J. Physiol. Pharmacol.* **73,** 1203–1207.

Baffy, G., Yang, L., Raj, S., Manning, D. R., and Williamson, J. R. (1994) G protein coupling to the thrombin receptor in chinese hamster lung fibroblasts. *J. Biol. Chem.* **269,** 8483–8487.

Barger, S. W., Fiscus, R. R., Ruth, P., Hofmann, F., and Mattson, M. P. (1994) Role of cyclic GMP in the regulation of neuronal calcium and survival by secreted forms of β amyloid precursor. *J. Neurochem.* **64,** 2087–2096.

Beecher, K. L., Anderson, T. T., Fenton II, J. W., and Festoff, B. W. (1994) Thrombin receptor peptide induces shape changes in neonatal murine astrocytes in culture. *J. Neurosci. Res.* **37,** 108–115.

Berndt, M. C. and Phillips, D. R. (1981) *Platelets in Biology and Pathology*, 2nd ed. Elsevier, Amsterdam, pp. 43–75.

Bizios, R., Lai, L., Fenton, J., Sonder, S., and Malik, A. (1985) Thrombin-induced aggregation of lymphocytes: non-enzymatic induction by a hirudin-blocked thrombin exosite. *Thromb. Res.* **38,** 425–431.

Bolkenius, F. N. and Monard, D. (1995) Inactivation of protease nexin-1 by xanthine oxidase-derived free radicals. *Neurochem. Int.* **26,** 587–592.

Brass, L. F. and Woolkalis, M. J. (1992) Dual regulation of cyclic AMP formation by thrombin in HEL cells, a leukemic cell line with megararyocytic properties. *Biochem. J.* **281,** 73–80.

Brass, L. F., Vassallo, R. R., Belmonte, E., Abuja, M., Cichowski, L., and Hoxie, J. A. (1992) Structure and function of the human platelet thrombin receptor. Studies using monoclonal antibodies directed against a defined domain within the receptor N terminus. *J. Biol. Chem.* **267,** 1469–1477.

Brass, L. F., Pizarro, S., Ahuja, M., Belmonte, E., Blanchard, N., Stadel, J. M., and Hoxie, J. A. (1994) Changes in the structure and function of the human thrombin receptor during receptor activation, internalization and recycling. *J. Biol. Chem.* **269,** 2943–2052.

Brewer, G. J. (1995) Exogenous thrombin inhibits neuritogenesis in cultured neuroblastoma cells but not in rat hippocampal neurons. *Brain Res.* **683,** 258–263.

Brewer, G. J. (1996) Thrombin causes cell spreading and redistribution of β-amyloid immunoreactivity in cultured hippocampal neurons. *J. Neurochem.* **67,** 119–130.

Busciglio, J., Lorenzo, As., and Yankner, B. A. (1992) Methodological variables in the assessment of beta amyloid neurotoxicity. *Neurobiol. Aging* **13,** 609–612.

Carlson, L. L., Weaver, D. R., and Reppert, S. M. (1989) Melatonin signal transduction in hamster brain: Inhibition of adenlyl cyclase by a pertussis toxin-sensitive G protein. *Endocrinology (Baltimore)* **125,** 2670–2676.

Cavanaugh, K. P., Gurwitz, D., Cunningham, D. D., and Bradshaw, R. A. (1990) Reciprocal modulation of astrocyte stellation by thrombin and protease nexin-1. *J. Neurochem.* **54,** 1735–1743.

Chen, L. B. and Buchanan, J. M. (1975) Mitogenic activity of blood components. I. Thrombin and Prothrombin. *Proc. Natl. Acad. Sci.* **72,** 131–135.

Choi, B. H., Suzuki, M., Kim, T., Wagner, S. L., and Cunningham, D. D. (1990) Protease nexin-1: localization in the human brain suggests a protective role against extravasated serine proteases. *Am. J. Pathol.* **137,** 741–747.

Chong, Y. H., Jung, J. M., Choi, W., Park, C. W., Choi, K. S., and Suh, Y. H. (1994) Bacterial expression, purification of full length and carboxyl terminal fragment of Alzheimer amyloid precursor protein and their proteolytic processing by thrombin. *Life Sci.* **54,** 1259–1268.

Connelly, A. J., Ishihara, H., Kahn, M. L., Farese, R. V., and Coughlin, S. R. (1996) Role of the thrombin receptor in development and evidence for a second receptor. *Nature* **381,** 516–519.

Coughlin, S. R. (1994) Molecular mechanisms of thrombin signaling. *Semin. Hematol.* **31,** 270–277.

Cunningham, D. D. and Gurwitz, D. (1989) Proteolytic regulation of neurite outgrowth from neuroblastoma cells by thrombin and protease nexin-1. *J. Cell. Biochem.* **39,** 55–64.

Daniel, T. O., Gibbs, V. C., Milfay, D. F., Garovoy, M. R., and Williams, L. T. (1986) Thrombin stimulates c-sis gene expression in microvascular endothelial cells. *J. Bio. Chem.* **261,** 9579–9582.

Darrow, A. L., Fung-Leung, W.-P., Ye, R. D., Santulli, R. J., Cheung, W.-M., Derian, C. K., et al. (1996) Biological consequences of thrombin receptor deficiency in mice, submitted.

Davey, M. G. and Luscher, E. F. (1967) Actions of thrombin and other coagulant and proteolytic enzymes on blood platelets. *Nature* **216,** 857–858.

Davie, E. W., Fujikawa, K., and Kisiel, W. (1991) The coagulation cascade: interaction, maintenance and regulation. *Biochemistry* **30,** 10,363–10,370.

Davis-Salinas, J., Saporito-Irwin, S. M., Donovan, F. M., Cunningham, D. D., and Van Nostrand, W. E. (1994) Thrombin receptor activation induces secretion and nonamyloidogenic processing of amyloid beta-protein precursor. *J. Biol. Chem.* **269,** 22,623–22,627.

Debeir, T., Benavides, J., and Vige, X. (1996) Dual effects of thrombin and a 14-amino acid peptide agonist of the thrombin receptor on septal cholinergic neurons. *Brain Res.* **708,** 159–166.

Delacourte, A. and Defossez, A. (1986) Alzheimer's disease: tau proteins, the promoting factors of microtubule assembly, are major components of paired helical filaments. *J. Neurol. Sci.* **76,** 173–186.

Dihanich, M., Kaser, M., Reinhard, E., Cunningham, D., and Monard, D. (1991) Prothrombin mRNA is expressed by cells of the nervous system. *Neuron* **6,** 575–581.

Farmer, L., Sommer, J., and Monard, D. (1990) Glia-derived nexin potentiates neurite extension in hippocampal pyramidal cells in vitro. *Dev. Neurosci.* **12,** 73–80.

Fenton II, J. W. (1986) Thrombin. *Ann NY Acad Sci.* **485,** 5–15.

Gagnon, A. W., Manning, D. R., Catani, I., Gerwitz, A., Poncz, M., and Brass, L. F. (1991) Identification of Gzα as a pertussis-toxin insensitive G protein in human platelets and megakaryocytes. *Blood* **78,** 1247–1253.

Gelehrter, T. D. and Szncycer-Laszuk, R. (1986) Thrombin induction of plasminogen activator-inhibitor in cultured human endothelial cells. *J. Clin. Invest.* **77,** 165–169.

Goedert, M., Wischik, C. M., Crowther, R. A., Walker, J. E., and Klug, A. (1988) Cloning and sequencing of the cDNA encoding a core protein of the paired helical filament of Alzheimer's disease: identification as the microtubule-associated protein tau. *Proc. Natl. Acad. Sci. USA* **85,** 4051–4055.

Goedert, M., Spillantine, M. G., Jakes, R., Rutherford, D., and Crowther, R. A. (1989) Multiple isoforms of human microtubule-associated protein tau: sequences and localization in neurofibrillary tangles of Alzheimer's disease. *Neuron* **3,** 519–526.

Goligorsky, M. S., Menton, D. N., Laszlo, A., and Lum, H. (1989) Nature of thrombin-induced sustained increase in cytosolic calcium concentration in cultured endothelial cells. *J. Biol. Chem.* **264,** 16,771–16,775.

Goodman, Y. and Mattson, M. P. (1994) Secreted forms of β-amyloid precursor protein protect hippocampal neurons against amyloid β-peptide-induced oxidative injury. *Exp. Neurol.* **128,** 1–12.

Grabham, P. W., Monard, D., Gallimore, P. H., and Grand, R. J. A. (1991) Modulation of human neurite outgrowth by serine proteases: a comparison of the interaction of thrombin and prothrombin with glia-derived nexin. *Eur. J. Neurosci.* **3,** 663–668.

Grabham, P. and Cunningham, D. D. (1995) Thrombin receptor activation stimulates astrocyte proliferation and reversal of stellation by distinct pathways: involvement of tyrosine phosphorylation. *J. Neurochem.* **64,** 583–591.

Grand, R. J. A, Turnell, A. S., and Grabham, P. W. (1996) Cellular consequences of thrombin-receptor activation. *Biochem. J.* **313,** 353–368.

Grundke-Iqbal, I., Iqbal, K., Quinlan, M., Tung, Y.-C., Zaidi, M. S., and Wisnievski, H. M. (1986) Microtubule-associated protein tau. A component of Alzheimer's paired helical filaments. *J. Biol. Chem.* **261,** 6084–6089.

Gurwitz, D. and Cunningham, D. D. (1988) Thrombin modulates and reverses neuroblastoma neurite outgrowth. *Proc. Natl. Acad. Sci. USA* ***85*** , 3440–3444.

Hein, L., Ishii, K., Coughlin, S. R., and Kobilka, B. K. (1994) Intracellular targeting and trafficking of thrombin receptors. A novel mechanism for resensitization of a G protein-coupled receptor. *J. Biol. Chem.* **269,** 27,719–27,729.

Hensley, K., Carney, J. M., Mattson, M. P., Aksenova, M., Harris, M., Wu, J. F., Floyd, R., and Butterfield, D. A. (1994) A model for β-amyloid aggregation and neurotoxicity based on free radical generation by the peptide: relevance to Alzheimer's disease. *Proc. Natl. Acad. Sci. USA* **91,** 3270–3274.

Hoxie, J. A., Ahuja, M., Belmonte, E., Pizarro., S., Parton, R., and Brass, L. F. (1993) Internalization and recycling of activated thrombin receptors. *J. Biol. Chem.* **268,** 13,756–13,763.

Hung, D. T., Wong, Y.-T., Vu, T.-K. H., and Coughlin, S. R. (1992) The cloned platelet thrombin receptor couples to at least two distinct effectors to stimulate phophoinositide hydrolysis and inhibit adenlyl cyclase. *J. Biol. Chem.* **267,** 20,831–20,834.

Igarashi, K., Murai, H., and Asaka, J.-I. (1992) Proteolytic processing of amyloid β protein precursor (APP) by thrombin. *Biochem. Biophys. Res. Commun.* **185,** 1000–1004.

Ishii, K., Chen, J., Ishii, M., Koch, W. J., Freedman, N. J., Lefkowitz, R. J., and Coughlin, S. R. (1994) Inhibition of thrombin receptor signaling by a G-protein coupled receptor kinase. Functional specificity among G-protein coupled receptor kinases. *J. Biol. Chem.* **269,** 1125–1130.

Jalink, K. and Moolenaar, W. H. (1992) Thrombin receptor activation causes rapid neural cell rounding and neurite retraction independent of classic second messengers. *J. Cell. Biol.* **118,** 411–419.

Jenkins, A. L., Bootman, M. D., Taylor, C. W., Mackie, E. J., and Stone, S. R. (1993) Characterization of the receptor responsible for thrombin-induced intracellular calcium responses in osteoblast-like cells. *J. Biol. Chem.* **268,** 21,433–21,437.

Jenkins, A. L., Bootman, M. D., Berridge, M. J., and Stone, S. R. (1994) Differences in intracellular calcium signaling after activation of the thrombin receptor by thrombin and agonist peptide in osteoblast-like cells. *J. Biol. Chem.* **269,** 17,104–17,110.

Kanthou, C., Kanse, S. M., Kakkar, V. V., and Benzakour, O. (1996) Involvement of pertussis toxin-sensitive and -insensitive G proteins in α-thrombin signaling on cultured human vascular smooth muscle cells. *Cell Signal.* **8,** 59–66.

Kario, K., Kodama, K., Koide, M., and Matsuo, T. (1995) Thrombin inhibition in the acute phase of ischemic stroke using argatroban. *Blood Coagul. Fibrinol.* **6,** 423–427.

Koh, J.-H., Yang, L. L., and Cotman, C. W. (1990) β-amyloid protein increases the vulnerability of cultured cortical neurons to excitotoxic damage. *Brain Res.* **533,** 315–320.

Kozasa, T. and Giman, A. G. (1995) Purification of recombinant G proteins from Sf9 cells by hexahistidine tagging of associated subunits. Characterization of α12 and inhibition of adenlyl cyclase by αz. *J. Biol. Chem.* **270,** 1734–1741.

Liu, Y., Fields, D., Festoff, B. W., and Nelson, P. G. (1994) Proteolytic action of thrombin is required for electrical activity-dependent synapse reduction. *Proc. Natl. Acad. Sci. USA* **91,** 10,300–10,304.

Loret, C., Sensenbrenner, M., and Labourdette, G. (1989) Differential phenotypic expression induced in cultured rat astroblasts by acidic fibroblast growth factor, epidermal growth factor and thrombin. *J. Biol. Chem.* **264,** 8319–8327.

Magnaldo, I., Pouyssegur, J. and Paris, S. (1988) Thrombin exerts a dual effect on stimulated adenylate cycalse in hamster fibroblasts, an inhibition via a GTP-binding protein and a potentiation via activation of protein kinase C. *Biochem. J.* **253,** 711–719.

Maraganore, J. M., Chao, B., Joseph, M. L., et al. (1989) Anticoagulant activity of synthetic hirudin peptides. *J. Biol. Chem.* **264,** 8692–8698.

Mari, B., Guerin, S., Far, D. F., Breitmayer, L.-P., Belhacene, N., Peyron, J.-F., Rossi, B., and Auberger, P. (1996) Thrombin and trypsin-induced Ca^{2+} mobilization in human T-cell lines through activation with different protease-activated receptors. *FASEB J.* **10,** 309–316.

Mark, R. J., Hensley, K., Butterfield, D. A., and Mattson, M. P. (1995) Amyloid β-peptide impairs ion-motive ATPase activites: evidence for a role in loss of neuronal Ca^{2+} homeostasis and cell death. *J. Neurosci.* **15,** 6239–6249.

Mattson, M. P. and Kater, S. B. (1988) Isolated hippocampal neurons in cryopreserved long-term cultures: development of neuroarchitecture and sensitivity to NMDA. *Int. Dev. Neurosci.* **6,** 439–452.

Mattson, M. P. (1992) Calcium as sculptor and destroyer of neural circuitry. *Exp. Gerontol.* **27,** 29–49.

Mattson, M. P., Cheng, B., Davis, D., Bryant, K., Lieberburg, I., and Rydel, R. E. (1992) β-amyloid peptides destabilize calcium homeostasis and render human cortical neurons vulnerable to excitotoxicity. *J. Neurosci.* **12,** 376–389.

Mattson, M. P., Cheng B., Culwell, A. R., Esch, F. S., Lieberburg, I., and Rydel, R. E. (1993a) Evidence for excitoprotective and intraneuronal calcium-regulating roles for secreted forms of the β-amyloid precursor protein. *Neuron* **10,** 243–254.

Mattson, M. P., Cheng, B., and Smith-Swintosky, V. L. (1993b) Neurotrophic factor mediated protection from excitotoxicity and disturbances in calcium and free radical metabolism. *Sem. Neurosci.* **5,** 295–307.

Mattson, M. P., Barger, S. W., Furakawa, K., Mark, R. J., Smith-Swintosky, V. L., Pettigrew, L. C., and Bruce, A. J. (1996) *Neurodegenerative Diseases.* Plenum, New York.

Mattson, M. P. and Begley, J. G. (1996) Amyloid β-peptide alters thrombin-induced calcium responses in cultured human neural cells. *Amyloid: Int. J. Exp. Clin. Invest.* **2,** 28–40.

Mattson, M. P. and Furakawa, K. (1996) Programmed cell life: anti-apoptotic signaling and therapeutic strategies for neurodegenrative disorders. *Restor. Neurol. Neurosci.* **9,** 191–205.

McKinney, M., Snider. R. M., and Richelson, E. (1983) Thrombin binding to human brain and spinal cord. *Mayo Clin. Proc.* **58,** 829–831.

McNamara, C. A., Sarembock, J. J., Gimple, L. W., Fenton, J. W., Coughlin, S. R., and Owens, G. R. (1993) Thrombin stimulates proliferation of cultured rat aortic smooth muscle cells by a proteolytically activated receptor. *J. Clin. Invest.* **91,** 94–98.

Mellott, M. J., Connoly, T. M., York, S. J., and Bush, L. R. (1990) Prevention of reocclusion by MCI-9038, a thombin inhibitor, following t-PA-induced thrombolysis in a canine model of femoral arterial thrombosis. *Thromb. Haemost.* **64,** 526–534.

Mirza, H., Yatsula, V., and Bahou, W. F. (1996) The protease activated receptor-2 (PAR-2) mediates mitogenic responses in human vascular endothelial cells: molecular characterization and evidence for functional coupling to the thrombin receptor. *J. Clin. Invest.* **97,** 1705–1714.

Naldini, A., Carney, D. H., Bocci, V., and Kimpel, G. R. (1991) Thrombin induced T-cell activation. *FASEB J.* **5,** 1378A.

Neveu, I., Jehan, F., Jandot-Perrus, M., Wion, D., and Brachet, P. (1993) Enhancement of the synthesis and secretion of nerve growth factor in primary cultures of glial cells by proteases: a possiblc involvement of thrombin. *J. Neurochem.* **60,** 858–867.

Niclou, S., Suidan, H. S., Brown-Luedi, M., and Monard, D. (1994) Expression of the thrombin receptor mRNA in rat brain. *Cell. Mol. Biol.* **40,** 421–428.

Nishino, A., Suzuki, M., Motohashi, O., Umezawa, K., Nagura, H., and Yoshimoto, T. (1993) Thrombin may contribute to the pathophysiology of central nervous system injury. *J. Neurotrauma* **10,** 167–179.

Nishino, A., Suzuki, M., Yoshimoto, T., Otani, H., and Nagura, H. (1994) A novel aspect of thrombin in the tissue reaction following central nervous system injury. *Acta Neurochir.* **60,** 86–88.

Nitsch, C., Scotti, A. L., Monard, D., Heim, C., and Sontag, K.-H. (1993) The glia-derived protease nexin 1 persists for over 1 year in rat brain areas selectively lesioned by transient global ischemia. *Eur. J. Neurosci.* **5,** 292–297.

Nystedt, S., Emilsson, K., Larsson, A.-K., Strombeck, B., and Sundelin, J. (1995) Molecular cloning and functional expression of the gene encoding the human proteinase-activated receptor 2. *Eur. J. Biochem.* **232,** 84–89.

Olesen, O. F. (1994) Proteolytic degradation of microtubule associated protein tau by thrombin. *Biochem. Biophys. Res. Comm.* **201,** 716–721.

Perraud, F., Besnard, F., Sensenbrenner, M., and Labourdette, G. (1987) Thrombin is a potent mitogen for rat astroblasts but not for oligodendroblasts and neuroblasts in primary culture. *Int. J. Dev. Neurosci.* **5,** 181–188.

Pike, C. J., Vaughan, P. J., Cunningham, D. D., and Cotman, C. W. (1996) Thrombin attenuates neuronal cell death and modulates astrocyte reactivity induced by β-amyloid in vitro. *J. Neurochem.* **66,** 1374–1382.

Prescott, S. M., Zimmerman, G. A., and McIntyre, T. M. (1984) Human endothelial cells in culture produce platelet activating factor (1-akyl-2-acetyl-sn-glycero-3-phosphocholine) when stimulated with thrombin. *Proc. Natl. Acad. Sci. USA* **81,** 3534–3538.

Rao, G. N., Delafontaine, P., and Runde, M. S. (1995) Thrombin stimulates phosphorylation of insulin-like growth factor-1 receptor, insulin receptor substrate-1, and phospholipase c-γ1 in rat aortic smooth muscle cells. *J. Biol. Chem.* **270,** 27,871–27,875.

Reddan, J., Dziedzic, D., and McGee, S. (1982) Thrombin induced cell division in rabbit lenses cultured in a completely defined serum-free medium. *Invest. Opthal. and Vis. Sci.* **22,** 486–493.

Santulli, R. J., Derian, C. K., Darrow, A. L., Tomko, K. A., Eckardt, A. J., Seiberg, M., Scarborough, R. M., and Andrade-Gordon, P. (1995) Evidence for the presence of a protease-activated receptor distinct from the thrombin receptor in human keratinocytes. *Proc. Natl. Acad. Sci. USA.* **92,** 9151–9155.

Seiler, S. M., Michel, I. M., and Fenton, J. W. (1992) Involvement of the "tethered ligand" receptor in thrombin inhibition of platelet adenylate cyclase. *Biochem. Biophys. Res. Comm.* **182,** 1296–1302.

Shin, H., Nakajima, T., Kitajima, I., Shigeta, K., Abeyama, K., Imamura, T., Okano, T., Kawahara, K., Nakamura, T., and Maruyama, I. (1995) Thrombin receptor-mediated synovial proliferation in patients with rhematoid arthitis. *Clin. Immunol. Immunopathol.* **76,** 225–233.

Siesjo, B. K., Bengtsson, F., Grampp, W., and Theander, S. (1989) Calcium excitotoxins and neuronal death in the brain. *Ann. NY Acad. Sci.* **568,** 234–251.

Smith, C. D., Carney, J. M., Starko-Reed, P. E., Oliver, C. N., Stadtman, E. R., Floyd, R. A., and Markesbery, W. R. (1991) Excess brain protein oxidation and enzyme dysfunction in normal aging and in Alzheimer's disease. *Proc. Natl. Acad. Sci. USA* **88,** 10,540–10,543.

Smith-Swintosky, V. L., Pettigrew, L. C., Craddock, S. D., Rydel, R. E., and Mattson, M. P. (1994) Secreted forms of β-amyloid precursor protein protect against ischemic brain injury. *J. Neurochem.* **63,** 5191–5193.

Smith-Swintosky, V. L., Zimmer, S., Fenton, J. W., and Mattson, M. P. (1995a) Protease nexin-1 and thrombin modulate neuronal Ca^{2+} homeostasis and sensitivity to glucose deprivation-induced injury. *J. Neurosci.* **15,** 5840–5850.

Smith-Swintosky, V. L., Zimmer, S., Fenton, J. W., and Mattson, M. P. (1995b) Opposing actions of thrombin and protease nexin-1 on amyloid beta-peptide toxicity and on accumulation of peroxides and calcium in hippocampal neurons. *J. Neurochem.* **65,** 1415–1418.

Snider, R. M., and Richelson, E. (1983) Thrombin stimulation of guanosine 3′ 5′-monophosphate formation in murine neuroblastoma cells (clone N1E-115). *Science* **221,** 566–568.

Suidan, H. S., Stone, S. R., Hemmings, B. A., and Monard, D. (1992) Thrombin causes neurite retraction in neuronal cells through activation of cell surface receptors. *Neuron* **8,** 363–375.

Suzuki, M., Motohashi, O., Nishino, A., Shiina, V., Mizoi, K., Oshimoto, T., Kameyama, M., and Onuma, T. (1994) Diphasic increase in thrombin-antithrombin III complex in blood from the internal jugular vein following severe head injury. *Thromb. Haemost.* **71,** 155–157.

Takiguchi, Y., Asai, F., Wada, K., Hayashi, H., and Nakashima, M. (1995) Antithrombotic effects of a novel recombinant hirudin analog CX-397, in a rat arterial thrombosis model. *Br. J. Pharmacol.* **116,** 3056–3060.

Tiruppathi, C., Lum, H., Andersen, T. T., Fenton II, J. W., and Malik, A. B. (1992) Thrombin receptor 14-amino acid peptide binds to endothelial cells and stimulates calcium transients. *Am. J. Physiol.* **263,** L595–L601.

Tordai, A., Fenton II, J. W., Andersen, T., and Gelfand, E. W. (1993) Functional thrombin receptors on human lymphoblastoid cells. *J. Immunol.* **150,** 4876–4886.

Tracy, P. B., Nesheim, M. E., and Mann, K. G. (1981) Coordinate binding of factor Va and factor Xa to the unstimulated platelet. *J. Biol. Chem.* **256,** 743–751.

Turnell, A. S., Brant, D. P., Brown, G. R., Finney, M., Gallimore, P. H., Kirk, C. J., et al. (1995) Regulation of neurite outgrowth from differentiated human neuroepithelial cells: a comparison of the activities of prothrombin and thrombin. *Biochem. J.* **308,** 965–973.

Umemura, K., Kawai, H., Ishihara, H., and Nakashima, M. (1995) Inhibitory effect of clopidogrel, vapiprost and argatroban on the middle cerebral artery thrombosis in the rat. *Jpn. J. Pharmacol.* **67,** 253–258.

van Obberghen-Schilling, E. and Pouyssegur, J. (1993) Signaling pathways of the thrombin receptor. *Thromb. Haemost.* **70,** 163–167.

Vaughan, P. J. and Cunningham, D. D. (1993) Regulation of protease-nexin-1 synthesis and secretion in cultured brain cells by injury-related factors. *J. Biol. Chem.* **268,** 3720–3727.

Vaughan, P. J., Pike, C. J., Cotman, C. W., and Cunningham, D. D. (1995) Thrombin receptor activation protects neurons and astrocytes from cell death produced by environmental insults. *J. Neurosci.* **15,** 5389–5401.

Vu, T.-K. H., Hung, D. T., Wheaton, V. I., and Coughlin, S. R. (1991) Molecular cloning of a functional thrombin receptor reveals a novel proteolytic mechanism of receptor activation. *Cell* **64,** 1057–1068.

Wagner, S. L., Geddes, J. W., Cotman, C. W., Lau, A. L., Isackson, P. J., and Cunningham, D. D. (1989) Protease nexin-1, a protease inhibitor with neurite promoting activity, is reduced in Alzheimer's disease. *Proc. Natl. Acad. Sci. USA.* **86,** 8284–8288.

Wagner, S. L., Van Nostrand, W. E., Lau, A., Farrow, J. S., Suzuki, M., Bartus, R. T., et al. (1993) Codistribution of protease nexin-1 and protease nexin-2 in brains of non-human primates. *Brain Res.* **626,** 90–98.

Weinstein, J. R., Gold, S. J., Cunningham, D. D., and Gall, C. M. (1995) Cellular localization of thrombin receptor mRNA in rat brain: expression by mesencephalic dopaminergic neurons and codistribution with prothrombin mRNA. *J. Neurosci.* **15,** 2906–2919.

Weksler, B. B., Ley, C. W., and Jaffe, E. A. (1978) Stimulation of endothelial cell prostacyclin production by thrombin, trypsin and the ionophore A23187. *J. Clin. Invest.* **62,** 923–930.

Wojtukiewicz, M. Z., Tang, D. G., Ciarelli, J. J., Nelson, K. K., Walz, D. A., Diglio, C. A., Mammen, E. F., and Honn, K. V. (1993) Thrombin increases the metastatic potential of tumor cells. *Int. J. Cancer* **54,** 793–806.

Zhong, C., Hayzer, D. J., Corson, M. A., and Runge, M. S. (1992) Molecular cloning of the rat vascular smooth muscle thrombin receptor. *J. Bio. Chem.* **267,** 16,975–16,979.

Zurn, A. D., Nick, H., and Monard, D. (1988) A glia-derived nexin promotes neurite outgrowth in cultured chick sympathetic neurons. *Dev. Neurosci.* **10,** 17–24.

14

Deleterious and Salutary Effects of Steroid Hormones in the Nervous System

Possible Mediating Cellular Mechanisms

Robert M. Sapolsky

1. INTRODUCTION

For the endocrinologist, one of the more subtle concepts to get across in one's introductory course is the extent to which hormones act by modulating the effects of other factors. This is particularly the case when teaching neuroendocrinology, as one has the hard task of unteaching the commonly held notion that a hormone such as testosterone directly causes aggression, or that glucocorticoids cause depression. Instead, the more accurate picture is one of modulation, of hormones changing the sensitivity of an organism, a neuron, or an ion channel to some other regulatory factor.

In recent years, it has come to be recognized that a number of steroid hormones can modulate the toxicity of necrotic or neurodegenerative insults to the brain. Specifically, it appears that glucocorticoids can endanger neurons, making them less likely to survive various insults. Conversely, estrogenic hormones appear to decrease the vulnerability of various hormones to neurological insults. These modulatory effects are potentially of vast importance in designing future therapeutic interventions to decrease neurological damage, yet also caution us that disturbing consequences could attend the current use of synthetic versions of these hormones.

This chapter will initially review the basic findings regarding glucocorticoid endangerment (as well as one instance of glucocorticoid protection) and estrogenic protection within the central nervous system (CNS), provide an overview of the traditional and nontraditional mechanisms of action of these steroid hormones, and conclude by considering the likely mechanisms underlying these particular instances of hormone modulation.

2. GLUCOCORTICOIDS AS MODULATORS OF INJURY TO THE CENTRAL NERVOUS SYSTEM

The release of glucocorticoids (GCs) is critical for the successful adaptation to physical stressors, because they help to mobilize energy, suppress unessential anabolism, and sharpen cognition. However, when secreted in excess, GCs have dcleterious effects, including exacerbation of adult-onset diabetes, hypertension, and reproductive and

From: Neuroprotective Signal Transduction *edited by M. P. Mattson*
Humana Press Inc., Totowa, NJ

immune suppression (Munck et al., 1984). A literature stretching back to the 1960s indicates that prolonged exposure to GCs can also lead to neuronal damage. Such GC neurotoxicity occurs in the hippocampus, a limbic structure with high concentrations of corticosteroid receptors and marked sensitivity to GC actions. The ability of GCs to directly damage hippocampal neurons is now thought to be relevant to understanding how stress can play a role in hippocampal aging (Meaney et al., 1988; McEwen, 1992), as well as how individual differences in lifelong exposure to stress can give rise to differing degrees of successful aging (Issa et al., 1990). Moreover, indirect evidence now suggests that GC neurotoxicity can occur in the human hippocampus, as well as that of rodents and nonhuman primates (reviewed in Sapolsky, 1996c).

The emergence of fairly convincing evidence for GC neurotoxicity by the mid-1980s prompted the initial work regarding GC neuroendangerment. The latter term refers to circumstances in which GC exposure is of insufficient magnitude or duration to overtly damage neurons, but is sufficient to leave neurons vulnerable and compromised. If such a period of endangerment does not coincide with any additional insult, there are no deleterious consequences. However, if, during such times, there is an additional challenge, neurons are then less likely to survive the insult. This illustrates the modulatory idea outlined at the beginning of this chapter: GCs compromise the ability of neurons to survive coincident challenges.

There is now considerable evidence suggesting such GC neuroendangerment. A first model provides insight into excitotoxic seizures. Both stress and physiological elevation of GC exposure impair the ability of hippocampal neurons to survive seizures (Sapolsky, 1985a,b; Theoret et al., (1985) *Toxicol. Abs.* **6,** 491; Stein and Sapolsky, 1988; Stein-Behrens et al., 1994a,b; Smith-Swintosky et al., 1996), and decrease the threshold for excitotoxin-induced seizures (Talmi et al., 1995). Such endangerment is not limited to the hippocampus, because GCs increase the striatal damage caused by excitotoxins as well (Uhler et al., 1994; Supko and Johnston, 1994). The transition from low basal to high stress levels of GCs augments seizure damage by as much as an order of magnitude, testifying to the potency of this effect. GCs also worsen excitotoxic damage in vitro, in primary hippocampal cultures (Sapolsky et al., 1988; Packan and Sapolsky, 1990; Behl et al. (1995) *Soc. Neurosci. Abst.* **21,** 401.3; Goodman et al., 1996; Rajan et al., 1996).

In a second model, GCs were then shown to augment global ischemic damage to the hippocampus (Sapolsky and Pulsinelli, 1985; Koide et al., 1985; Hall, 1990; Morse and Davis, 1990), neocortex and striatum (Sapolsky and Pulsinelli, 1985; Koide et al., 1986), and also to augment the damage caused in in vitro models of global ischemia (Tombaugh et al., 1992; Tombaugh and Sapolsky, 1993). The issue with infarct damage is less clear, with one report indicating GCs worsening damage (Smith-Swintosky et al., 1996), and one indicating GC protection (de Courten-Myers et al., 1994); the possible resolution to this conflict is discussed below.

Other insults worsened by GCs in vivo include hypoglycemia and antimetabolite toxins (Sapolsky 1985a,b; 1986a,b; Tombaugh et al., 1992), oxygen radical generators (Sapolsky et al., 1988; McIntosh and Sapolsky, 1996; Goodman et al., 1996), and cholinergic or serotonergic toxins (Amoroso et al., 1994; Hortnagl et al., 1993; Johnson et al., 1989). Finally, GCs appear to worsen the toxicity of the β-amyloid protein in vitro (Behl et al. (1995) *Soc. Neurosci. Abst.* **21,** 401.3; Goodman et al., 1996) and the gp120 glycoprotein of the AIDS virus (Brooke et al., 1997). Collectively, these represent an extremely varied array of insults whose toxicities are worsened by GCs, and whatever mechanisms of endangerment are activated by these hormones (to be discussed below), they must be rather broad.

The mechanisms of GC actions in the CNS must also encompass a few instances in which the steroids protect neurons from coincident insults. One example of this was already noted, i.e., the single report of GCs decreasing infarct damage (de Courten-Myers et al., 1994). As a hint of some of the mechanistic issues to be discussed, that study, demonstrating GC protection, utilized supraphysiologic megadoses of the hormone; the study showing GCs increasing infarct damage (Smith-Swintosky et al., 1996) used the high physiological levels that would be generated during stress.

Another example is the observation that GCs decrease the damage caused by traumatic injury to the spinal cord, documented in both preclinical animal studies and in some highly publicized human trials (Young and Flamm, 1982; Braughler and Hall, 1985; Bracken et al., 1990). It forms the basis of one of the few effective interventions that can be done in the immediate aftermath of spinal cord injury. As will be seen, the mechanisms underlying GC neuroendangerment vs neuroprotection are quite different and fairly readily reconciled.

3. ESTROGENS AS MODULATORS OF INJURY TO THE CENTRAL NERVOUS SYSTEM

One of the more striking manifestations of aging in a mammal is the decline of function of the ovarian axis in most females, and a complex literature has shown that part of such reproductive senescence arises from deleterious effects of estrogen in the hypothalamus (reviewed in Finch et al., 1984). More recent work has shown that this involves direct estrogen-induced damage to β-endorphin neurons in the arcuate nucleus (Desjardins et al., 1995).

This instance of estrogen-induced neurotoxicity in the aging hypothalamus bears some resemblance to the neurotoxicity of GCs in the aging hippocampus. Furthermore, like GCs, estrogens can modulate the toxicity of coincident neurological insults. However, the parallelism breaks down, because the estrogenic modulation is protective, rather than endangering.

Some in vitro studies have shown this in recent years. For example, estrogen protects cultured hippocampal neurons against glutamate, hypoglycemia, and a potent pro-oxidant (Goodman et al., 1996). Significant, although lesser, protection was afforded by progesterone, but other steroids, such as testosterone, aldosterone, and vitamin D had no effect; in support of the GC endangerment story, the authors observed that corticosterone significantly worsened the toxicity of these insults. The estrogenic protection was quite potent, causing about a five-fold increase in the LD_{50} for glutamate; moreover, this protection was afforded with relatively physiological concentrations in the high n*M* range.

Virtually identical findings emerged from another study utilizing a less physiological hippocampal cell line, and estrogen concentrations in the mid-μ*M* range (which is likely to be supraphysiological) (Behl et al., 1995). Once again, the steroid decreased the toxicity caused by glutamate and a pro-oxidant, as well as by the amyloid β protein. Again, this effect was hormone-specific, in that other steroids were not protective.

Finally, we have recently uncovered an additional neuroprotective property of estrogenic steroids, namely, protection from gp120, the glycoprotein of HIV, which is thought by many to play a critical role in the neurological damage and dysfunction of

AIDS-related dementia. The glycoprotein both elevates cytosolic calcium (Ca^{2+}) and causes overt toxicity in cultured hippocampal and cortical neurons, and both endpoints are lessened by high n*M* concentrations of estradiol (Brooke et al., 1997).

These findings share the strengths and weaknesses of all in vitro neurobiological studies, in that one can be reasonably certain of the reductive phenomenon that has been uncovered, and be reasonably uncertain as to its relevance to intact nervous systems. The phenomenon of estrogenic neuroprotection has already been shown to be very relevant to the intact nervous system. Specifically, an immensely exciting literature has emerged suggesting that estrogen protects against Alzheimer's disease. These findings have taken three forms:

1. First, a history of postmenopausal estrogen-replacement therapy appears to decrease the likelihood of dementia during aging (Henderson et al., 1994; Paganini-Hill and Henderson, 1994; Simpkins et al., 1994; Mortel and Meyer, 1995; *but see* Brenner et al., 1994, for a failure to observe this relationship). Although the issue is not yet clarified, the protective effects appear to be more relevant to late-onset Alzheimer's, rather than the early-onset version, which is heavily familial. Increasing exposure to estrogen (either in the form of greater or more persistent estrogen treatment) is incrementally associated with an incremental decline in dementia risk in these studies, and the magnitude of the effect is quite large.
2. Among women with Alzheimer's disease, a history of estrogen replacement therapy is associated with less severe symptoms of dementia (i.e., higher performance scores on neuropsychological exams) (Henderson et al., 1994).
3. In perhaps the most surprising facet of this story, the therapeutic administration of estrogen to women already diagnosed with Alzheimer's disease has been reported to cause a significant improvement in neuropsychological performance, cerebral blood flow, and EEG activity (Ohkura et al., 1994, 1995).

Thus, an array of studies, ranging from highly reductive in vitro ones to a literature of proven clinical relevance, documents the neuroprotective potential of estrogen (findings sure to fuel the ongoing debate as to the wisdom of estrogen-replacement therapy, with its assorted benefits and risks). The following section reviews classical and untraditional mechanisms of steroid hormone action, in order to appreciate the means by which GCs and estrogens exert these effects.

4. AN OVERVIEW OF STEROID HORMONE ACTION

4.1. *Classical Actions of Steroid Hormones*

Most steroid hormones act via the long-recognized traditional route: They pass through the lipid membrane and bind to intracellular receptors, which are relatively specific for each class of steroid hormone. Binding to such receptors activates them, allowing the hormone–receptor complex to bind to specific sequences of DNA and regulate genomic events. The elucidation of this pathway constitutes one of the most impressive triumphs of contemporary endocrinology and molecular biology.

Unoccupied steroid receptors can be found in either the cytosol or nucleus (and in the case of the corticosteroid receptors that bind GCs, it remains controversial as to where precisely they are located [reviewed in Howell et al., 1990; McGimsey et al., 1991; Brink et al., 1992]). As originally conceptualized, the unoccupied receptor floated freely, or was perhaps anchored to some cytoskeletal element, awaiting the arrival of the steroidal ligand. Instead, it is now recognized that receptors are part of a complex of ele-

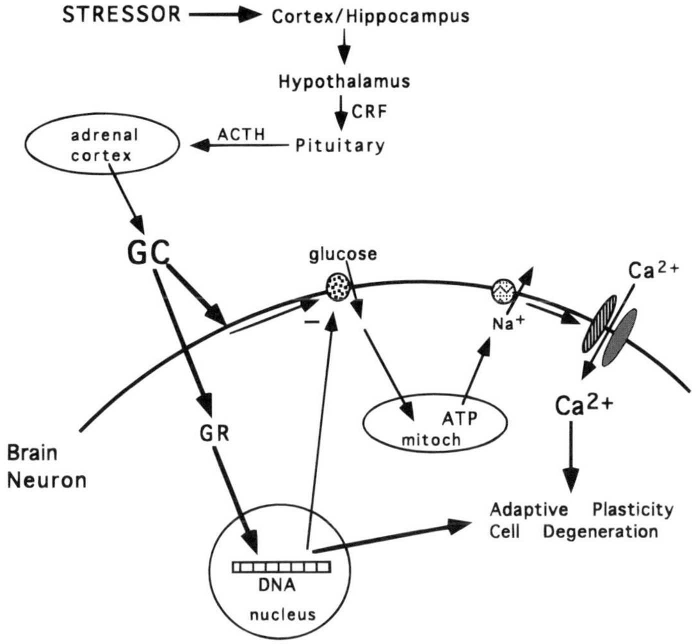

Fig. 1. Mechanisms whereby glucocorticoids affect neuron plasticity and survival. Physiological stressors activate neural pathways in cerebral cortex and hippocampus. The hippocampus provides inputs to the hypothalamus that induce release of corticotrophin-releasing hormone (CRH), which, in turn, stimulates release of adrenocorticotrophin (ACTH) from the anterior pituitary gland. ACTH circulates in the blood and induces production of glucocorticoids (GC) from adrenal cortical cells. GCs circulate in the blood, from which they enter the brain parenchyma. GCs may affect neurons in several ways (lower diagram). GCs may diffuse into the cell cytoplasm, where they bind specific GC receptors (GRs), which then translocate to the nucleus. GRs are transcription factors that bind enhancer DNA sequences in responsive genes, resulting in either activation or suppression of gene expression. Genes regulated by GC may include those involved in regulating neuronal plasticity and survival. For example, GCs may suppress expression of glucose-transporter proteins. This could result in ATP depletion, compromise of ATP-dependent ion transporters (e.g., Na^+/K^+-ATPase), membrane depolarization, and Ca^{2+} influx through voltage-dependent calcium channels. GCs may also act independently of GRs. For example, GCs may bind plasma membrane receptors, or may intercalate into membranes and alter membrane properties in ways that affect function of membrane proteins involved in signal transduction or ion transport. *See text* for further discussion.

ments that regulate the receptor's assembly, stability, affinity for ligand, and ability to become activated and to translocate to the DNA. This has been studied in the most detail for GC receptors (GRs). Such receptors usually form homodimers. In addition, they must be bound to the heat-shock protein hsp90 in order to successfully bind GCs (Picard et al., 1990; Pratt, 1990). The complex can also include a number of other heat-shock proteins, members of the immunophilin protein family, and unsaturated fatty acids (Valette et al., 1991; Tai et al., 1992; Czar et al., 1995). Levels of these cofactors can change as dramatically as those of receptor itself (Patchev et al., 1994), and such regulation appears to be important. For example, when considering a wide variety of tissue

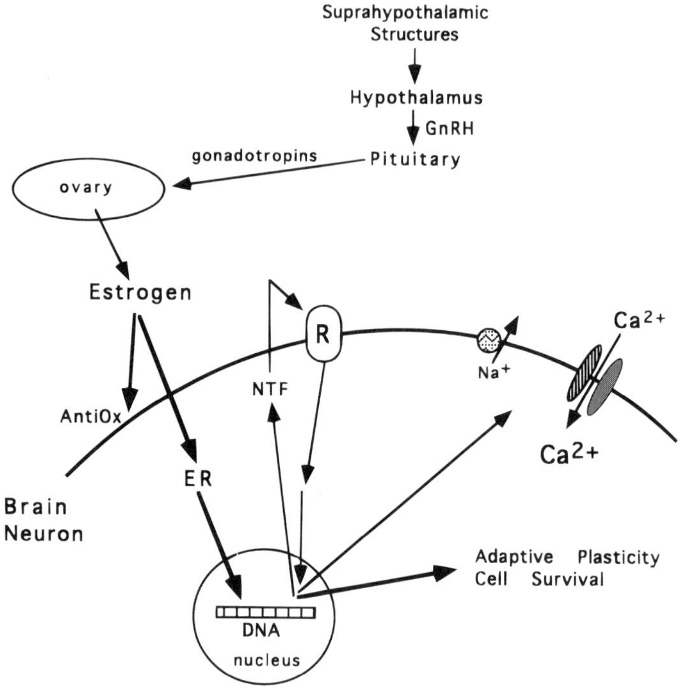

Fig. 2. Mechanisms whereby estrogens affect neuron plasticity and survival. Neural circuits in the brain regulate production of gonadotropin-releasing hormone (GnRH) in the hypothalamus. GnRH in turn stimulates release of gonadotropins into the blood. Gonadotropins stimulate estrogen production by cells in the ovary. Estrogens travel via the blood to the brain, where they can act on neurons in several ways. Estrogens bind specific receptors (ER) in the cell cytoplasm; the activated receptors translocate to the nucleus, where they activate or suppress gene expression. Genes regulated by estrogens include those encoding neurotrophic factors (NTF), such as nerve growth factor. The NTFs promote neuron survival and adaptive plasticity by binding receptors (R) linked to kinase cascades and activation of transcription factors. Estrogens also probably regulate expression of genes involved in regulation of ion homeostasis and free radical metabolism. Estrogens may also act on membrane receptors. In addition, estrogens have inherent antioxidant activity, which may protect neurons against free radical damage, such as that believed to occur in neurodegenerative disorders such as Alzheimer's disease. *See text* for further discussion.

types throughout the body, the degree of sensitivity to a GC signal could better be explained by variability in the amount of a cofactor, such as hsp90, than by variability in the amount of the receptor itself (reviewed in Vamvakopoulous and Chrousos, 1994). Steroid receptors typically contain a number of phosphorylation sites, and changes in the degree of phosphorylation alter the affinity for the ligand, the transcriptional efficacy at the DNA, and the ease of receptor recycling back to the unbound state. With multiple phosphorylation sites, different molecules of receptors might be phosphorylated to different extents, or with different patterns, and this could be a route of microheterogeneity of receptor function within different domains of a cell (reviewed in Orti et al., 1992; Bodwell et al., 1993; Kuiper and Brinkmann, 1994).

Once a hormone–receptor complex forms, the receptor is activated and moves toward the DNA. (Throughout the rest of this section, "activated receptor" and "hormone–

receptor complex" are used interchangeably). For steroid receptors that are cytosolic, the complex must first pass through the nuclear membrane, and this is made possible by a nuclear localization signal sequence on the receptor. A number of shuttle proteins can bind to that sequence and chaperone the hormone–receptor complex to the DNA (LaCasse et al., 1993).

One way in which a hormone–receptor complex can then induce the transcription of a gene is a fairly mechanical one. Activated receptor can disrupt the structure of chromatin, exposing the DNA beneath to various transcription factors (Zaret and Yamamoto, 1984; Santoro et al., 1988). Although the evidence for such a mechanism is quite solid, the bulk of steroid hormone actions in the nucleus appear to involve more proximal interaction with DNA than merely exposing a stretch of DNA. These more proximal forms of interaction require a critical conformational change in the receptor during translocation. The structure of steroid hormone receptors has been elucidated over the last decade, showing them to be part of a receptor superfamily. They all have a similar modular structure, with separate domains relevant to DNA binding, to steroid binding, to transactivation, and to binding to regulatory factors such as the heat-shock proteins (reviewed in Evans, 1988). When receptors are free of ligand, the DNA-binding domain is typically enveloped by the rest of the protein, making it inaccessible to DNA, and part of the activation that occurs on binding of steroidal ligand is conformational unmasking of the DNA-binding domain.

The activated hormone–receptor complex then binds to specific sites on the DNA known as steroid responsive elements (SREs), which are typically up to a few thousand base pairs upstream from a gene whose transcription is regulated by the hormone (Yamamoto, 1985; Beato, 1989; Beato et al., 1989). A number of regulatory events must be explained by the interaction of hormone–receptor complexes with SREs. Different steroid hormones obviously have different effects on target cells, and this must be explained at this molecular level. Furthermore, steroid hormones ultimately lead to increased transcription of a gene in some cases, yet in others, decrease transcription, and the direction of regulation must also be explicable; moreover, in some cases, the same steroid hormone has diametrically opposite effects on the regulation of the same gene in different tissues. Finally, steroid hormones regulate the transcription of different genes (whether positive or negative) with differing affinities. Thus, increasing levels of a steroid hormone can result in enhanced transcription (or repression) of certain genes, as well as producing a hierarchy of hormone concentrations at which transcription is first influenced. On a certain level, this is obvious. Why have incrementally differing levels of circulating hormones if they are not, in some cases, going to have incrementally differing effects on genomic events in target cells? Clearly basal levels of GCs need to have different effects than stress levels on a target cell; similar distinctions must occur between luteul and follicular stage levels of estrogen.

The regulation of transcriptional affinity and specificity, as well as the direction of regulation (i.e., stimulation or repression) by hormone–receptor complexes is anything but simple or intuitive. This is best illustrated with the GC responsive element (GRE). In its canonical form, the GRE is 15 base pairs long, and has two palindromic recognition sites, reflecting the fact that the complex typically contains two homodimerized receptors; each palindromic recognition site is 6 base pairs long, with 3 base pairs as a central linker (Tsai et al., 1988). In theory, one would speculate that the canonical GRE sequence will be quite different from that for the other SREs, and that this would confer specificity of hormone action. Obviously, one wants a fairly reliable molecular mechanism to differentiate a stress hormone from a reproductive hormone, or female- from male-typical hormones. Furthermore, one could imagine a scenario in which the efficacy with which GC–GR complexes regulate genomic events would be a function of approximation of fit to the canonic GRE; that is, genes that have upstream GREs that are closest to the canonical sequence are most potently regulated. Additionally, or alternatively, one might imagine that GREs could occur in series, with increasing numbers of them in the upstream region conferring greater regulatory potency to hormone–receptor complexes.

There are indeed varying numbers of repeating SREs upstream of genes, and individual SREs typically do show varying degrees of degeneracy from the canonical sequence. However, here is where some daunting complexities arise (reviewed in Zeelan, 1990; Funder, 1993; Pearce and Yamamoto, 1993):

1. Neither "goodness of fit" to the canonical form nor numbers of SREs in sequence are particularly good predictors of potency of transcriptional regulation, let alone whether such regulation is positive or negative.
2. The DNA-binding domain of different steroid hormone receptors (the GR, the mineralocorticoid receptor, the progestin receptor, and the androgen receptor) differ in only 4–6 of 66 amino acids, seriously limiting the extent of specificity that can be coded for on that level.
3. There is extensive overlap among different SREs in the canonical sequence, further limiting the possibility of hormonal specificity at that level (Arriza et al., 1987; Cato et al., 1986; Strahle et al., 1987).

How then is there coding for hormone-specific genomic effects, differing potencies of hormonal effects, or the direction of regulation? The answer appears to reside in other factors that the hormone–receptor complex encounters besides the DNA. The neighborhood of an SRE is dense with other regulatory sequences and transcriptional factors, and there are complex interactions among them. In some cases, the activated GR, by interacting with its GRE, can synergize with another transcriptional regulator and its responsive element. Synergizing can readily involve protein–protein interactions between the GR and the other transcriptional factor, which will stabilize them and prolong the duration of their actions at the DNA. This appears to happen, and probably explains the finding that distance between different responsive elements will influence the potency of their synergies. In other circumstances, the spacing of responsive elements can explain some of the inhibitory actions of steroid hormones. For example, GCs block transcriptional cAMP responses in some tissues by binding to a GRE that partially overlaps a cAMP-responsive enhancer. Thus, sheer steric hindrance accounts for that regulation (Akerblom et al., 1988).

Steroid receptors can bind directly to other proteins; activated GRs, for example, bind to nuclear factor I and octamer transcription factors (Funder, 1993). In some cases, such protein–protein interactions entirely explain the hormone action. For example, activated GR blocks AP-1 from inducing expression of collagenase. This effect is independent of GR binding to DNA and of protein synthesis, and occurs at a far lower concentration than needed for genomically mediated GR effects. Moreover, coprecipitation indicates direct GR–AP-1 binding (Jonat et al., 1990).

The combination of all of these potential regulatory mechanisms goes a long way toward explaining some of the informational tasks expected of a steroid hormone–receptor complex. A particularly interesting example is seen with GRs and mineralocorticoid receptors. The two are equivalently active at a GRE. However, the GRE can be imbedded in a 25-nucleotide-long composite response element, which also contains an AP1-binding site. This site binds either heterodimers of c-Fos and c-Jun, or c-Jun homodimers. Activated GRs block AP1-mediated transcription, but activated mineralocorticoid receptors do not (Pearce and Yamamoto, 1993). This implies that specificity is not at the level of the receptor's DNA-binding domain and its interaction with the SRE, but in the interaction between some other domain of the receptor and these other transcriptional factors. Another example of such interactions is a protein called coactivator-1, which stimulates transactivation by a variety of steroid receptors at a particular promoter, without itself altering basal activity of the promoter (Onate et al., 1995). This represents the introduction of conditional regulation: an effect by coactivator-1, if, and only if, another transcriptional regulatory is present.

As noted, steroid hormone action represents one of the most exciting and rapidly changing topics in endocrinology, and no doubt many details of this review will be obsolete by the time this is published. However, details of work in progress serve to give a good sense of the nature and complexity of traditional steroid hormone mechanisms of action.

4.2. *Nontraditional Mechanisms of Steroid Hormone Action*

The preceding section describes what is generally viewed as the classical route of steroid hormone action. Naturally, steroids turn out to have a variety of additional and nontraditional mechanisms of action.

One such mode would initially seem to qualify as such a nontraditional route. As outlined, steroids can increase or decrease the levels of particular proteins as an outcome of the hormones' roles as both positive and negative transcriptional regulators. However, steroids can also change the levels of some proteins by acting posttranscriptionally. For example, GCs inhibit type I and IV collagen synthesis in primary liver cells through both the expected transcriptional effects, and a posttranscriptional shortening of mRNA half-life (Weiner et al., 1987). Similarly, GCs increase the levels of cytochrome P450d in hepatocytes through a posttranscriptional effect (Silver et al., 1990). Insofar as these steroid actions are not at the transcriptional level, they would appear to be nontraditional. However, the appearances are probably deceptive, in that the posttranscriptional regulation is probably secondary to a transcriptional change in the levels of posttranscriptional enzymes (Weiner et al., 1987).

One mode of steroid hormone action already discussed would qualify as nontraditional, if one were a purist and defined traditional actions as exclusively involving interaction of activated hormone–receptor complex with DNA. Within that narrow definition, the cases in which transcriptional regulation arises purely from protein– protein interactions between receptor and transcriptional factors (Jonat et al., 1990), or with chromatin (Santoro et al., 1988) would count as nontraditional.

For most in the field, though, the relatively few instances of nontraditional steroid hormone effects occur far from the nucleus. Some of these actions reflect the molecular structure of steroid hormones, rather than the hormones' abstract roles as carrying information. Because of their steroidal ring structure and lipophilic nature, steroids readily penetrate cell membranes and can fluidize them in the process, changing the affinity of membrane receptors, or the coupling between receptors and ion channels. Nonspecific fluidization is usually inferred when the same effect is provoked by a broad range of steroids, including those not thought to be traditional hormones (such as cholesterol). In addition, because of their complex ring structure, steroid molecules can act as antioxidants through a variety of mechanisms (Lacort et al., 1995; Subbiah et al., 1993; Parthasarathy et al., 1994). Estrogenic compounds, for example, appear to be particularly effective at antagonizing iron-catalyzed oxidant reactions (Lacort et al., 1995).

Some evidence has accumulated that steroid hormones can also interact with membrane-bound receptors, as evidenced by specific, saturable, competition-sensitive binding to membrane fractions by the steroids (Towle and Sze, 1983; Gametchu, 1987; Quelle et al., 1988; Orchinik et al. (1991) *Soc. Neurosci. Abst.* **17,** 561.2). Such receptors could be specific for steroids (i.e., principally or exclusively binding some steroid hormone), or may primarily bind some other ligand, with that binding modulated by steroids. As an example of the latter, progesterone metabolites bind to the GABA receptor, interacting with the site that also binds barbiturates. Much like barbiturates, these steroids increase chloride influx into neurons, a discovery which clarified the long-recognized ability of these particular steroids to act as anesthetics (Majewska et al., 1986; Lambert et al., 1987). In addition, glucocorticoids may act by intercalating into the plasma membrane phospholipid bilayer and altering its chemistry in ways that affect the function of signal transduction pathways. For example, glucocorticoids may inhibit coupling of certain receptors to their GTP-binding effector proteins (Mattson and Mrotek, 1985).

This brief review covers the current state of knowledge concerning both the traditional (i.e., genomic) and nontraditional mechanisms of steroid hormone action. As will now be seen, this full array of mechanisms is relevant for understanding steroidal neuroprotection and neuroendangerment.

5. LIKELY MECHANISMS UNDERLYING THE PROTECTIVE AND ENDANGERING EFFECTS OF STEROIDS IN THE BRAIN

The ability of GCs and estrogens to modulate the toxicity of neurological insults has, for the most part, been recognized only recently. One disadvantage of dealing with such current findings is that there is relatively little known about the underlying mechanisms. Consequently, much of the current state of knowledge reviewed here is inferential.

Most broadly, modulatory effects are considered to be traditional or nontraditional, based on the following criteria (Maclusky, 1987; Hua and Chen, 1989; Rose et al., 1993; Sze and Iqbal, 1994):

1. Time-course: Steroid effects taking more than 30–60 min to emerge are assumed to be protein synthesis-dependent and likely to involve classic genomic mechanisms. Until recently, steroid effects that were demonstrable in a shorter time were thought to be nongenomic. However, there is now less certainty about such rapidity precluding a genomic effect (reviewed in McEwen, 1994).
2. Concentration requirement: Effects whose most minimal manifestation requires supraphysiological steroid concentrations, or at least high physiological ones far in excess of the receptor kDa, are immediate suspects for nontraditional routes of actions.
3. Receptor dependency: Effects (related to high steroid–low receptor concentrations) that occur independent of steroid receptor binding (i.e., which persist despite the use of specific receptor antagonists) are candidates for nontraditional mechanisms.

4. Independence from the nucleus: Effects that can be shown in synaptosomes or membrane preparations to have bypassed the requirement for genomic interactions count as nontraditional.
5. Requirement of protein synthesis: This is the most time-honored test for whether a steroid effect is genomically mediated in the traditional manner. However, for the purist with contemporary molecular knowledge of steroid action, the requirement of protein synthesis is no longer sufficient. As discussed above, activated steroid receptors can regulate gene transcription without actually interacting with the DNA, instead having protein–protein interactions with other transcriptional factors or with chromatin, and such interactions technically should count as nontraditional. By the more stringent criterion of requiring direct DNA interaction to consider an action traditional, virtually no steroid effect relevant to this chapter has yet been demonstrated to be traditional via that route. However, as will be seen, in a few cases, the gene whose regulation seems to be at the center of the endangerment or protection has been shown to have an upstream responsive element for the relevant hormone, suggesting a true transcriptional regulatory effect.
6. Change in mRNA level for a regulated protein: Steroid effects that produce changes in mRNA levels are generally thought to indicate traditional hormone actions. However, this interpretation has obvious problems: If a nuclear runoff assay indicates that the effect is actually post-transcriptional, involving a change in message stability, this may or may not imply an indirect genomic effect. Morever, as just discussed, if a runoff assay indicates a direct transcriptional effect, it might still be secondary to a nontraditional protein–protein interaction.

Thus, not much is understood at this point, and a traditional mechanism (or the closely related protein–protein interaction in the neighborhood of DNA) is inferred in cases in which the steroid mechanism is relatively slow, receptor-mediated, occurs at physiological concentrations, requires an intact neuron, and involves changes in mRNA levels of a particular protein and protein synthesis. As will be seen, most of the endangerment and protection afforded by the steroids under discussion fall into this category, with little mechanistic understanding beyond that.

5.1. Glucocorticoids

A number of mechanisms are thought to underlie GC neuroendangerment, and the bulk of them are probably genomically mediated. This conclusion is based in part on the nature of the individual effects about to be reviewed. In addition, endangerment is not typically triggered by other, non-GC steroids, and the endangerment is GR-receptor-mediated (Packan and Sapolsky, 1990; Mizoguchi et al., 1992; Talmi et al., 1995, 1996). As outlined, this does not guarantee a classical, genomic mechanism of action, but is taken by most as a strong indicator of a genomic effect.

5.1.1. DNA Fragmentation

It is important, perhaps, to begin with a mechanism that does not appear to be relevant to GC neuroendangerment, simply because it is central to some well-understood catabolic GC actions in other tissues. The immunosuppressive actions of GCs have been known for decades and involve dramatic lymphocytolysis. Such cell killing has long been one of the most-studied of models of programmed cell death, and involves the apoptotic fragmentation of DNA at internucleosomal sites. This has been shown to depend on protein synthesis, and appears to involve GCs either inducing an endonuclease, or inducing a Ca^{2+} channel whose subsequent localization in the nuclear membrane allows Ca^{2+} influx to activate a Ca^{2+}-dependent endonuclease (Evans-Storms and Cidlowski, 1995). Parsimony would initially suggest that GCs wreak havoc in the

CNS through a similar route. Moreover, a veritable deluge of papers in recent years indicates that necrotic insults, such as seizure or ischemia, trigger apoptotic DNA cleavage in neurons, counter to prior dogma. Despite this, there is little evidence favoring GC neuroendangerment involving DNA fragmentation, and one paper suggests that it does not occur (Masters et al., 1989).

5.1.2. *Inhibition of Glucose Transport*

This route of endangerment is also anchored in the ancient history of GC actions in peripheral tissue; however, this route does appear to apply to neurons. As part of the teleology of GC actions during physical stressors, energy is mobilized from storage sites throughout the body and diverted to exercising muscle (often, after an initial detour to the liver for gluconeogenesis). One feature of such adaptive energy mobilization is the inhibition of glucose transport into unessential tissues: In effect, it would be grossly illogical, for example, to be transporting glucose into fat cells while one is attempting to sprint away from a predator. Glucocorticoids inhibit as much as 75% of glucose transport in adipocytes, fibroblasts, and immune cells (Munck, 1971). The effect is protein synthesis-dependent and is generally conceded to arise from two sequential genomic effects. First, over the course of minutes to hours, GCs are thought to increase synthesis of a protein which sequesters glucose tranporters from the cell membrane to intracellular storage sites. The net result is no change in total cellular concentrations of transporter, but a redistribution of them away from active transport sites (Carter-Su and Okamoto, 1985; Horner et al., 1987); insulin mobilizes glucose transport, in part, via redistribution of transporters from such storage sites back to the cell membrane. Second, over the course of hours to days, GCs either inhibit transcription of the gene for the glucose transporter (through a direct genomic or protein–protein mechanism), or destabilize levels of mRNA for the glucose transporter; the net result is a decrease in the total amount of glucose transporter in a cell (Garvey et al., 1989). Collectively, GCs bring about a decrease in the rate of glucose entry into a cell.

A number of groups have now shown that GCs cause a similar decrease in glucose transport in the brain. In vivo studies with the 2-deoxyglucose method indicate inhibition throughout the brain, with the effect being particularly pronounced in the hippocampus (Kadekaro et al., 1988; Bryan and King (1988) *Soc. Neurosci. Abst.* 399.11; Freo et al., 1992; Doyle et al., 1993). As a demonstration of the likely dependency of this effect on relatively slow genomic mechanisms, GC exposure for only 15 min is insufficient to inhibit transport (Seckl et al., 1991). In vitro studies with primary cultures indicate that the GC inhibition of transport is exclusive to neurons and glia of the hippocampus (Horner et al., 1990; Virgin et al., 1991). This seeming paradox is probably resolved with the explanation that GCs inhibit transport across the endothelial border of the blood–brain barrier throughout the brain (Olgemoller et al., 1985), coupled with additional inhibition directly at hippocampal neurons and glia.

When compared to peripheral tissues, the effect is quite mild (on thc order of 25%), too little to cause overt death, even in a cell type as demanding of energy and dependent on glucose as a neuron. Instead, this appears to be the centerpiece of the modulatory endangerment. For example, GCs do not cause a decline in basal ATP concentrations or mitochondrial enzyme activity in hippocampal neurons and glia, but exacerbate the decline induced by a coincident insult (Tombaugh and Sapolsky, 1992; Lawrence and

Sapolsky, 1994; Bennett et al. (1993) *Soc. Neurosci. Abst.* **19,** 78.12). Although this GC effect is not associated with a change in basal hippocampal metabolic rate (as assessed by lactrographic or microphysiometric techniques), again, it leads to an exacerbation of the changes induced by a coincident insult (Krugers et al., 1992; Ajilore and Sapolsky, 1997). Many of the critical steps involved in necrotic neuron death are highly energy-dependent, and the catabolic GC effects on neuronal and glial energetics seem to worsen these steps. Thus, during a necrotic insult such as seizure, the typical synaptic accumulation of glutamate or aspartate is worsened by GCs or by stress (Stein-Behrens et al., 1992, 1994a); an elaboration on this finding is that seizure induced glutamate accumulation is exaggerated in aged rats, who have elevated GC levels (Lowy et al., 1995). The costly process of glutamate and aspartate reuptake from the synapse appears to be particularly energy-sensitive during an insult, and this is the step most likely exacerbated by the steroid (Chou et al., 1994). As evidence that this GC effect is derived from the energetic actions of the steroid, it is prevented by energy supplementation (Stein-Behrens et al., 1992). Glucocorticoids also exacerbate the accumulation of free cytosolic calcium in the postsynaptic neuron (Elliott and Sapolsky, 1992, 1993). This would be expected as a simple outcome of the worsening of the extracellular glutamatergic milieu. In addition, the energetic consequences of GC actions should directly impair the ability of neurons to carry out costly calcium sequestering and efflux mechanisms, and this appears to occur (Elliott and Sapolsky, 1993); again, the effect is reversible with energy supplementation. Finally, as one would expect, GC induced worsening of the calcium tone in a neuron undergoing a necrotic insult is associated with exacerbation of calcium-dependent degenerative events, such as cytoskeletal and microtubule pathology, and oxygen radical accumulation (Elliott et al., 1993; Stein-Behrens et al., 1994b; McIntosh and Sapolsky, 1996); again, these GC effects are generally reversible with energy supplementation.

Thus, the mild inhibition of glucose transport in the brain, a seemingly classic steroid effect, has a wide variety of endangering consequences. It should be noted here, however, that these findings do not preclude additional routes by which GCs cause an energetic vulnerability in neurons.

5.1.3. Disruption of Neuronal Defenses

Amid the torrents of glutamate and calcium loosed by necrotic insults, neurons possess a number of protective mechanisms. One example would be the Ca^{2+}-, nitric oxide- or proton-mediated feedback loops to inhibit Ca^{2+} currents at the NMDA receptor and the voltage-gated calcium channel (Lieberman and Mody, 1994; de Leon et al., 1995; Traynelis and Cull-Candy, 1990; Lipton et al., 1993). Another is the induction of stress proteins such as hsp72 (Welch, 1992), or the induction of antioxidant enzymes. Yet another potent route of neuronal defensive mobilization involves the release of inhibitory neurotransmitters, such as adenosine, GABA, and taurine, in response to glutamate exposure. These retaliatory neurotransmitters inhibit subsequent glutamate release (Huxtable, 1989; Dragunow et al., 1985).

Some recent studies indicate that GCs not only exacerbate the offensive components of the damaging glutamate–Ca^{2+} cascade, but also impair the defensive components, and an extensive literature now demonstrates the damaging potential of impairing such defenses during an insult. These novel GC actions are clearest for the retaliatory neurotransmitters and their protective actions. Microdialysis studies indicate that GCs decrease the extracellular mobilization of GABA and adenosine (but not taurine) during insults (Ravindran et al., 1994; Dash et al. (1995) *Soc. Neurosci. Abst.* **21,** 347.9). Moreover, GCs blunt the GABA ergic inhibitory postsynaptic potentials that typically emerge as a protective response to sustained glutamate exposure (Joels and de Kloet, 1992, 1993). Dosage and time-course requirements suggest that the electrophysiological effects are classically genomic, but no information is available regarding the mechanisms for the microdialysis findings.

A second example of a neuronal defense impaired by GCs is that the steroids decrease the activity of Cu–Zn-superoxide dismutase and glutathione peroxidase in the hippocampus and cortex (McIntosh and Sapolsky, submitted); the mechanism underlying that is completely unknown. Finally, there are some hints that GCs might disrupt some aspects of stress-protein induction. Although the steroids augment excitotoxin-triggered induction of hsp70 in the hippocampus and cortex (Lowy et al., 1994) and heat-shock-triggered induction of mRNA for hsp32 (Maines et al., 1995), they decrease hsp32 protein levels (Weber et al., 1994). Thus, these recently uncovered GC actions are not at all understood on the mechanistic level.

5.1.4. *Inhibition of Neurotrophin Synthesis*

In another possible route of GC-induced endangerment in the hippocampus, the steroids both decrease basal levels of mRNA for the neurotophin BDNF and blunt the induction of BDNF mRNA following an excitotoxic insult (Smith et al., 1995; Cosi et al., 1993; but *see* Barbany and Persson, 1993). This is presumed to reflect a classic genomic effect, but has not been tested rigorously. There have also been reports of GC effects on NGF mRNA levels, but there has been no consistency to the direction of these effects (Foreman et al., 1993; Scully and Otten, 1993; Ueyama et al. (1995) *Soc. Neurosci. Abst.* **21,** 126.2; Niu et al. (1995) *Soc. Neurosci. Abst.* **21,** 417.4). The implications of the inhibition of BDNF message levels are not clear for a number of reasons: It is not yet known if this inhibition translates into decreased levels of the protein itself; GCs have also been reported to increase mRNA levels for the Trk B receptor for BDNF, potentially offsetting the effect on BDNF; it is not clear if a decrease in levels of a neurotrophin will be relevant to the relatively rapid phenomenon of GC neuroendangerment, or, more plausibly, to overt killing of hippocampal neurons by sustained GC overexposure over far longer periods.

5.1.5. *GC Endangerment vs GC Protection*

Thus, there are a variety of mechanisms that probably give rise to the neuroendangering effects of GCs, and the bulk of these are likely to involve the classic genomic pathway. As noted, GCs decrease spinal cord damage in response to traumatic injury (Young and Flamm, 1982; Braughler, 1984; Bracken et al., 1990) and decrease glutamatergic toxicity in spinal cord neurons in culture (Ogata et al., 1993). In contrast to GC-mediated neuroendangerment, this case of neuroprotection is likely to involve a nontraditional mechanism of action: The hormones are only effective when administered in megadose quantities that are orders of magnitude above the kDa of the GR (Bracken et al., 1984, 1985, 1990; Braughler et al., 1987); the protective effects are extremely rapid (Hall et al., 1987); the spinal cord does not have particularly high concentrations of GRs; GCs protect against trauma only so long as the actual steroid molecules are physically present (Braughler and Hall, 1984); synthetic 21-aminosteroids (the so-called "lazaroid" compounds), which have no activity at any known steroid receptor, are even better than GCs at protecting against spinal cord injury (Hall et al., 1987); among various GCs and lazaroid derivatives, the extent of spinal cord protection is best predicted by the extent of the steroid's lipid solubility and ability to inhibit lipid peroxidation (Hall et al., 1987). Collectively, these observations argue strongly against GCs being protective through a classic genomic mechanism. Instead, it seems most likely that they intercalate into membranes and stabilize them against peroxidative attack after spinal cord injury. This nontraditional route of protection also probably explains the contradictory finding regarding the effects of GCs on infarct damage in the brain. The study showing GC exacerbation of damage utilized high physiological concentrations of GCs (Smith-Swintosky et al., 1996), but the demonstration of protection utilized supraphysiological megadose concentrations (de Courten-Meyers et al., 1994).

5.2. *Estrogens*

As discussed above, estrogenic steroids decrease excitotoxic, metabolic, and oxidative damage to both primary hippocampal neurons in culture and a hippocampal cell line (Goodman et al., 1996; Behl et al., 1995). These instances of neuroprotection appear to involve nontraditional mechanisms, since the protection is independent of the presence or availability of estrogen receptors and occurs rapidly and in the absence of protein synthesis; moreover, incremental increases in protection are seen with increments of estrogen exposure well above the kDa of the estrogen receptor.

Instead, estrogenic protection is most plausible through an antioxidant mechanism. The most obvious support for this hypothesis is that in both of these studies the researchers decreased the toxicity of oxygen radical generators; however, this could arise from actions distal to the biochemistry of oxygen radicals. As more direct evidence, estrogen concentrations in the same range as utilized in these studies act as antioxidants and decrease oxidative damage in cardiac tissue (Subbiah et al., 1993), liver microsomes (Lacort et al., 1995; Ruiz-Larrea et al., 1994) and vascular endothelium (Keaney et al., 1994). The notion that these antioxidant effects arise from direct intercalation into the membrane is based on a number of findings. First, the antioxidant activities of estrogen can be receptor-independent (Lacort et al., 1995). Next, the efficacious estrogen concentrations required are typically in the supraphysiological range, much as in the GC megadose studies with spinal cord injury. Finally, to my knowledge, there is no identified pathway by which estrogens exert classic genomic effects on a gene relevant to oxygen radical biochemistry.

A second mechanism underlying the neuroprotection is that estrogen blunts glutamate-induced Ca^{2+} mobilization in postsynaptic neurons (Goodman et al., 1996). This could be caused by a direct, receptor-independent effect of estrogen on Ca^{2+} dynamics; there is little evidence to suggest this, however. Alternatively, it may be secondary to estrogen's antioxidant effects, as speculated by the authors. Oxygen radicals can increase cytosolic Ca^{2+} concentrations in neurons through a number of mechanisms (Araki et al., 1992). However, Goodman and coworkers did not show that estrogen reduces oxygen radical levels, but rather that it reduces peroxidative lipid damage. There has not yet been a direct demonstration that such lipid damage elevates cytosolic calcium concentrations, but the authors emphasize that an intact and functional membrane would be an important prerequisite for calcium homeostasis.

Thus, estrogenic protection against acute necrotic insults appears to result from nontraditional protection against oxidative attack; it will be interesting to see if estrogen can also be used clinically to decrease spinal cord damage. Ironically, it appears that in the sole example in which estrogen exposure is neurotoxic, the damage is oxidative in nature. As reviewed above, prolonged estrogen exposure damages β-endorphin neurons in the arcuate nucleus of the hypothalamus, a critical step in a cascade of degeneration that gives rise to reproductive senescence in the female rodent. Such damage can be prevented with antioxidants (Desjardins et al., 1995); the mechanism for such estrogen-mediated oxidative damage is unknown. The reader should recall the GC story, in which high physiological concentrations of that steroid increase infarct damage, but megadoses decrease; in the present case, it is high physiological levels of estrogen that are associated with hypothalamic damage, but again, it is megadose steroid regimes that protect against necrotic insults. What is striking in this case is that estrogenic protection and endangerment both revolve around oxidative biochemistry.

The other instance of estrogenic protection, namely, the dramatic association of estrogen replacement therapy with protection from Alzheimer's disease, is a complete mystery. One can certainly speculate that this could arise solely from the antioxidant properties of the hormone, and there is a growing feeling that neuron loss in Alzheimer's disease is oxidative in flavor (reviewed in Coyle and Puttfarcken, 1993). However, two additional and interrelated genomic routes of estrogen action are probably relevant to this exciting example of neuroprotection.

The first concerns the ability of estrogen to increase the density of dendritic spines in the hippocampus (McEwen and Woolley, 1994; Woolley and McEwen, 1993, 1994). This appears to be a classic genomic action of the steroid (McEwen, 1994), and has a number of surprising features to it:

1. Although representing a classic mechanism of steroid action, the effect appears to be unorthodox insofar as it is transsynaptic. As understood at this point, the increase in spine density, specific to pyramidal neurons of the CA1 region of the hippocampus, is secondary to direct estrogenic effects on neighboring GABAergic interneurons, which, in some manner, induce more pyramidal cell spines through an NMDA-receptor-dependent mechanism (Woolley and McEwen, 1994).
2. One is certainly accustomed to hormonal regulation of neuronal morphology in the developing brain, but this effect occurs in the adult. Moreover, it is highly relevant to the quotidian physiology of the hippocampus: Hippocampal spine density fluctuates over the course of the 4-d estrous cycle in the female rat as a result of this effect (Woolley and McEwen, 1993).
3. Finally, the magnitude of the spine density changes is probably enough to explain some of the documented effects of estrogen and reproductive cyclicity on cognition in females in a wide variety of species, including humans (Hampson and Kimura, 1992).

How estrogens induce dendritic spines is not known on a more reductive level. However, a perfectly plausible mechanism is the second genomic effect of estrogen, namely, the ability of the steroid to induce neurotrophins in the hippocampus and cortex. Estrogen increases levels of nerve growth factor, as well as levels of mRNAs for BDNF (Singh et al. (1993) *Soc. Neurosci. Abst.* **23,** 514.11; Singh et al., 1995; Sohrabji et al., 1995). This effect is likely to be genomic, based on dosage, time-course, and the demonstration of an estrogen response element in the gene encoding BDNF (Sohrabji et al., 1995).

There is currently intense interest in neurotrophins, not only in terms of understanding neuronal development and in potentially exploiting neurotrophins for repairing damaged brain tissue, but also for understanding the dynamic process of expanding and contracting dendritic processes and spines that seems to underlie normal brain function. This mechanism is an ideal candidate for explaining the action of estrogen on dendritic spine density. To my knowledge, it is not known whether estrogen regulates dendritic spine density in the cortex; thus, the demonstration of this neurotrophin effect in both hippocampus and cortex does not disqualify it as an explanation for the dendritic spine effect that is at present exclusive to the hippocampus.

The estrogenic effects on dendritic spine density and neurotrophin levels could help explain the protective effects of estrogen against Alzheimer's disease. Damage in that disease is centered in the hippocampus and cortex, and the protective effects of neurotrophins against insults to neurons is now well documented (*see*, for example, Mattson et al., 1995). Obviously, understanding more fully the route of protection in this case is of enormous importance.

6. CONCLUSIONS

The first hints as to how steroid hormones may modulate the likelihood of a neuron surviving a necrotic or neurodegenerative insult are being elucidated. It is a testimony to the versatility of steroid hormone action that, though GCs are predominantly neuroendangering, there are nevertheless circumstances in which they protect against an insult, and that though estrogens are predominantly neuroprotective, there is still at least one example of those hormones being damaging to neurons. On a more mechanistic level, it is also striking how heterogeneous the paths of endangerment and of protection appear to be. In some cases, there is a pleasing symmetry to what is understood; for example, estrogenic protection and GC endangerment probably arise, in part, from their respective abilities to promote and to inhibit synthesis of neurotrophins. But there is little else about this literature that is that cleanly symmetrical. In at least one instance, both estrogenic and GC neuroprotection appear most readily explained by their similar abilities to protect from peroxidative attack; conversely, both can endanger in at least one circumstance through what appear to be pro-oxidant mechanisms. The protective, antioxidant roles of both hormones probably centers around their ability to intercalate into membrane; the opposite, pro-oxidant activities of GCs probably involves augmentation of Ca^{2+}-dependent oxygen radical generation and inhibition of the activity of antioxidant enzymes. Furthermore, although there is a superficial symmetry seen when estrogen decreases glutamatergic tone and GCs enhance it, the former effect is probably secondary to the antioxidant actions, but the latter effect appears to result from the disruptive effects of GCs on neuronal energetics.

Thus, there is little reason to think that estrogen is predominantly protective and GCs predominantly endangering through their simply having opposite effects on neurons. Another tentative theme that emerges from examining the broad mechanisms of actions of these two hormones concerns the dichotomy between physiology and pharmacology. In general, the endangering and damaging effects of these hormones appear to happen in the physiological (albeit the high physiological) range. Considerable work has gone into showing that it is the physiological exposure to endogenous GC and estrogen levels over the lifetime that acts as a pacemaker of hippocampal and hypothalamic aging, respectively; furthermore, the endogenous secretion of GCs at the time of a necrotic insult is sufficient to activate the cascade of endangerment. In contrast, at least some of the estrogenic protection in in vitro systems, and the GC protection against spinal cord injury, both require markedly supraphysiological concentrations. Whether it is the physiological features of estrogen replacement therapy (i.e., the fact that levels are typically maintained in the roughly physiological range), or its pharmacological properties (that the circadian pattern of levels generated is not physiological) that protects against Alzheimer's disease must be determined.

It is not a mere academic issue to understand the mechanisms by which these hormones protect on one hand and endanger on the other, and when those transitions occur. The potential of estrogen to protect against something as nightmarish as Alzheimer's, as well as against necrotic insults, could be extraordinarily beneficial; in contrast, the carcinogenic potential of estrogen therapy is not trivial. Similarly, it is vital to understand how endogenous stress-induced secretion of GCs over a lifetime plays a role in brain aging; this is particularly important in light of recent findings that suggest that the GC hypersecretion associated with Cushing's syndrome, major depression, and trauma can all result in selective hippocampal atrophy in the human (reviewed in Sapolsky, 1996c). Morever, the fact that approx 16 million prescriptions are written annually in the United States for synthetic GCs (Eufermiond, 1990) is disquieting in terms of possible neurologic complications, but should also be a reminder that they are prescribed for good reasons; thus, there may be some difficult decisions to be made as to when the administration of GCs is advisable. Collectively, the current ignorance about the mechanisms of these steroid actions, coupled with their potential clinical relevance, should be a strong impetus to further research in these areas.

REFERENCES

Ajilore, O. and Sapolsky, R. (1997) Application of silicon microphysiometry to tissue slices: detection of metabolic correlations of selective vulnerability. *Brain Res.* **752,** 99–106.

Akerblom, I., Slater, E., Beato, M., Baxter, J., and Mellon, P. (1988) Negative regulation by glucocorticoids through interference with the cAMP responsive enhancer. *Science* **241,** 350–353.

Amoroso, D., Kindel, G., Wulfert, E., and Hanin, I. (1994) Long-term exposure to high levels of corticosterone aggravates AF64A-induced cholinergic hypofunction in rat hippocampus in vivo. *Brain Res.* **661,** 9–16.

Araki, N., Greenberg, J., Uematsu, D., Sladky, J., and Reivich, M. (1992) Effect of superoxide dismutase on intracellular calcium in stroke. *J. Cereb. Blood Flow Metab.* **12,** 43.

Arriza, J., Winberger, C., Cerelli, G., Glaser, T., Handelin, B., Housman, D., and Evans, R. (1987) Cloning of human mineralocorticoid receptor cDNA: structure and functional kinship with the glucocortiocid receptor. *Science* **237,** 268–275.

Barbany, G. and Persson, H. (1993) Adrenalectomy attenuates kainic acid-elicited increases of mRNAs for neurotrophins and their receptors in the rat brain. *Neuroscience* **54,** 909–918.

Beato, M. (1989) Gene regulation by steroid hormones. *Cell* **56,** 335–344.

Beato, M., Chalepakis, G., Schauer, M., and Slater, E. (1989) DNA regulatory elements for steroid hormones. *J. Steroid Biochem.* **32,** 737–747.

Behl, C., Widmann, M., Trapp, T., and Holsboer, F. (1995) 17-Beta estradiol protects neurons from oxidative stress-induced cell death in vitro. *Biochem. Biophys. Res. Comm.* **216,** 473–482.

Bodwell, J., Hu, L., Hu, J., Orti, E., and Munck, A. (1993) Glucocorticoid receptors: ATP-dependent cycling and hormone-dependent hyperphosphorylation. *J. Steroid. Biochem. Molec. Biol.* **47,** 31–38.

Bracken, M., Collins, W., and Freeman, D. (1984) Efficacy of methylprednisolone in acute spinal cord injury. *JAMA* **251,** 45–52.

Bracken, M., Shepard, M., Collins, W., Holford, T., Young, W., Baskin, D., et al. (1990) A randomized controlled trial of methylprednisolone or naloxone in the treatment of acute spinal cord injury. *New Engl. J. Med.* **322,** 1405–1410.

Bracken, M., Shepard, M., and Hellenbrand, K. (1985) Methylprednisolone and neurological function one year after spinal cord injury. *J. Neurosurg.* **63,** 704–713.

Braughler, J. and Hall, E. (1984) Effects of multi-dose methylprednisolone sodium succinate administration on injured cat spinal cord neurofilament degradation and energy metabolism. *J. Neurosurg.* **61,** 290–295.

Braughler, J. and Hall, E. (1985) Current application of high dose steroid therapy for CNS trauma. *J. Neurosurg.* **62,** 806–812.

Braughler, J., Hall, E., Means, E., Waters, T., and Anderson, D. (1987) Evaluation of an intensive methylprednisolone sodium succinate dosing regimen in experimental spinal cord injury. *J. Neurosurg.* **67,** 102–105.

Brenner, D., Kukull, W., Stergachis, A., van Belle, G., Bowen, J., McCormick, W., Teri, L., and Larson, E. (1994) Postmenopausal estrogen replacement therapy and the risk of Alzheimer's disease: a population-based case-control study. *Am. J. Epidemiol.* **140,** 262–267.

Brink, M., Humbel, B., de kloet, E., and van Driel, R. (1992) The unliganded glucocorticoid receptor is localized in the nucleus, not in the cytoplasm. *Endocrinology* **130,** 3575–3581.

Brooke, S., Chan, R., Howard, S., and Sapolsky, R. (1997) Endocrine modulation of the neurotoxicity of gp120: implications for AIDS-related dementia complex. *Proc Natl. Acad. Sci. USA,* in press.

Carter-Su, C. and Okamoto, K. (1985) Effects of glucocorticoids on hexose transport in rat adipocytes: evidence for decreased transporters in the plasma membrane. *J. Biol. Chem.* **260,** 11,091–11,098.

Cato, A., Miksicek, R., Schultz, G., Arnemann, J., and Beato, M. (1986) The hormone regulatory element of mouse mammary tumor virus mediates progesterone induction. *EMBO J.* **5,** 2237–2240.

Chou, Y., Lin, W., and Sapolsky, R. (1994) Glucocorticoids increase extracellular [3H]D-aspartate overflow in hippocampal cultures during cyanide-induced ischemia. *Brain Res.* **654,** 8–15.

Cosi, C., Spoerri, P., Comelli, M., Guidolin, D., and Skaper, S. (1993) Glucocorticoids depress activity-dependent expression of BDNF mRNA in hippocampal neurones. *Neuroreport* **4,** 527–531.

Coyle, J. and Puttfarcken, P. (1993) Oxidative stress, glutamate, and neurodegenerative disorders. *Science* **262,** 689–695.

Czar, M., Lyons, R., Welsh, M., Renoir, J., and Pratt, W. (1995) Evidence that the FK506-binding immunophilin heat shock protein 56 is required for trafficking of the glucocorticoid receptor from the cytoplasm to the nucleus. *Mol. Endocrinol.* **9,** 1549–1560.

de Courten-Myers, G., Kleinholz, M., Wagner, K., Xi, G., and Myers, R. (1994) Efficacious experimental stroke treatment with high-dose methylprednisolone. *Stroke* **25,** 487–493.

de Leon, M., Wang, Y., Jones, L., Perez-Reyes, E., Wei, X., Soong, T., Snutch, T., and Yue, D. (1995) Essential calcium-binding motif for calcium-sensitive inactivation of L-type calcium channels. *Science* **270,** 1502–1505.

Desjardins, G., Beaudet, A., Meaney, M., and Brawer, J. (1995) Estrogen-induced hypothalamic beta-endorphin neuron loss: a possible model of hypothalamic aging. *Exp. Gerontol.* **30,** 253–267.

Doyle, P., Rohner-Jeanrenaud, F., and Jeanrenaud, B. (1993) Local cerebral glucose utilization in brains of lean and genetically obese (fa/fa) rats. *Am. J. Physiol.* **264,** E29–E36.

Dragunow, M., Goddard, G., and Laverty, A. (1985) Is adenosine an endogenous anticonvulsant? *Epilepsia* **26,** 480–490.

Elliott, E., Mattson, M., Vanderklish, P., Lynch, G., Chang, I., and Sapolsky, R. (1993) Corticosterone exacerbates kainate-induced alterations in hippocampal tau immunoreactivity and spectrin proteolysis in vivo. *J. Neurochem.* **61,** 57–64.

Elliott, E. and Sapolsky, R. (1992) Corticosterone enhances kainic acid-induced calcium mobilization in cultured hippocampal neurons. *J. Neurochem.* **59,** 1033–1039.

Elliott, E. and Sapolsky, R. (1993) Corticosterone impairs hippocampal neuronal calcium regulation: possible mediating mechanisms. *Brain Res.* **602,** 84–89.

Eufermiond, M. (1990) Advances in the therapy of steroid-induced osteoporosis. *Geriatr. Med. Today* **9,** 41–54.

Evans, R. (1988) The steroid and thyroid hormone receptor superfamily. *Science* **240,** 889–895.

Evans-Storms, R. and Cidlowski, J. (1995) Regulation of apoptosis by steroid hormones. *J. Steroid Biochem. Mol. Biol.* **53,** 1–12.

Finch, C., Felicio, L., Mobbs, C., and Nelson, J. (1984) Ovarian and steroidal influences on neuroendocrine aging processes in female rodents. *Endocr. Rev.* **5,** 467–497.

Foreman, P., Taglialatela, G., Angelucci, L., Turner, C., and Perez-Polo, J. (1993) Nerve growth factor and p75NGFR factor receptor mRNA change in rodent CNS following stress activation of the HPA axis. *J. Neurosci. Res.* **36,** 10–18.

Freo, U., Holloway, H., Kalogeras, K., Rapoport, S., and Soncrant, T. (1992) Adrenalectomy or metyrapone-pretreatment abolishes cerebral metabolic responses to the serotonin agonist DOI in the hippocampus. *Brain Res.* **586,** 256–261.

Funder, J. (1993) Mineralocorticoids, glucocorticoids, receptors and response elements. *Science* **259,** 1132–1133.

Gametchu, B. (1987) Glucocorticoid receptor-like antigen in lymphoma cell membranes: Correlation to cell lysis. *Science* **236,** 456–461.

Garvey, W., Huecksteadt, T., Lima, F., and Birnbaum, M. (1989) Expression of a glucose transporter gene cloned from brain in cellular models of insulin resistance: dexamethasone decreases transporter mRNA in primary cultured adipocytes. *Mol. Endocrinol.* **3,** 1132–1137.

Goodman, Y., Bruce, A., Cheng, B., and Mattson, M. (1996) Estrogens attenuate excitotoxicity, oxidative injury, and amyloid b-peptide toxicity in hippocampal neurons. *J. Neurochem.* **66,** 1836–1844.

Hall, E. (1990) Steroids and neuronal destruction or stabilization, in *Steroids and Neuronal Activity,* Ciba Foundation Symposium 153. Wiley, Chichester, pp. 206–215.

Hall, E., McCall, J., Chase, R., Yonkers, P., and Braughler, J. (1987) A nonglucocorticoid steroid analog of methylprednisolone duplicates its high-dose pharmacology in models of central nervous system trauma and neuronal membrane damage. *J. Pharm. Exp. Ther.* **242,** 137–145.

Hampson, E. and Kimura, D. (1992) Sex differences and hormonal influences on cognitive function in humans, in *Behavioral Endocrinology* (Becker, J., Breedlove, S., and Crews, D., eds.). MIT, Cambridge, MA, pp. 357–373.

Henderson, V., Paganini-Hill, A., Emanuel, C. Dunn, M., and Buckwalter, J. (1994) Estrogen replacement therapy in older women. Comparisons between Alzheimer's disease cases and nondemented control subjects. *Arch. Neurol.* **51,** 896–900.

Horner, H., Packan, D., and Sapolsky, R. (1990) Glucocorticoids inhibit glucose transport in cultured hippocampal neurons and glia. *Neuroendocrinology* **52,** 57–62.

Horner, H., Munck, A., and Lienhard, G. (1987) Dexamethasone causes translocation of glucose transporters from the plasma membrane to an intracellular site in human fibroblasts. *J. Biol. Chem.* **262,** 17,696–17,703.

Hortnagl, H., Berger, M., Havelec, L., and Hornykiewicz, O. (1993) Role of glucocorticoids in the cholinergic degeneration in rat hippocampus induced by ethylcholine aziridinium (AF64A). *J. Neurosci.* **13,** 2939–2946.

Howell, G., Gustafsson, J., and Lefebvre, Y. (1990) Glucocorticoid receptor identified on nuclear envelopes of male rat livers by affinity labelling and immunochemistry. *Endocrinology* **127,** 1087–1096.

Hua, S. and Chen, Y. (1989) Membrane receptor-mediated electrophysiological effects of glucocorticoids on mammalian neurons. *Endocrinology* **124,** 687–691.

Huxtable, R. (1989) Taurine in the CNS and the mammalian actions of taurine. *Prog. Neurobiol.* **32,** 471–503.

Issa, A., Rowe, W., Gauthier, S., and Meaney, M. (1990) Hypothalamic-pituitary-adrenal activity in aged, cognitively impaired and cognitively unimpaired rats. *J. Neurosci.* **10,** 3247–3253.

Joels, M. and de Kloet, E. (1992) Coordinative mineralocorticoid and glucocorticoid receptor-mediated control of response to serotonin in rat hippocampus. *Neuroendocrinology* **55,** 344–350.

Joels, M. and de Kloet, E. (1993) Corticosteroid actions on amino acid-mediated transmission in rat CA1 hippocampal cells. *J. Neurosci.* **13,** 4082 4090.

Johnson, M., Stone, D., Bush, L., Hanson, G., and Gibb, J. (1989) Glucocorticoids and 3,4-methylenedioxymethamphetamine (MDMA)-induced neurotoxicity. *Eur. J. Pharmacol.* **161,** 181.

Jonat, C., Rahmsdorf, H., Park, K., Cato, A., Gebel, S., Ponta. H., and Herrlich, P. (1990) Antitumor promotion and antiinflammation down modulation of AP-1 (Fos Jun) activity by glucocorticoid hormone. *Cell* **62,** 1189.

Kadekaro, M., Masanori, I., and Gross, P. (1988) Local cerebral glucose utilization is increased in acutely adrenalectomized rats. *Neuroendocrinology* **47,** 329–334.

Keaney, J., Shwaery, G., Xu, Z., Nicolosi, R., Loscalzo, J., Foxall, T., and Vita, J. (1994) 17-Beta-estradiol preserves endothelial vasodilator function and limits low-density lipoprotein oxidation in hypercholesterolemic swine. *Circulation* **89,** 2251–2259.

Koide, T., Wieloch, T., and Siesjo, B. (1986) Chronic dexamethasone pretreatment aggravates ischemic neuronal necrosis. *J. Cereb. Blood Flow Metab.* **6,** 395–403.

Krugers, H., Jaarsma, D., and Korf, J. (1992) Rat hippocampal lactate efflux during electroconvulsive shock or stress is differently dependent on entorhinal cortex and adrenal integrity. *J. Neurochem.* **58,** 826–831.

Kuiper, G. and Brinkmann, A. (1994) Steroid hormone receptor phosphorylation: Is there a physiological role? *Mol. Cell Endocrinol.* **100,** 103–107.

LaCasse, E., Lochnan, H., Walker, P., and Lefebvre, Y. (1993) Identification of binding proteins for nuclear localization signals of the glucocorticoid and thyroid hormone receptors. *Endocrinology* **132,** 1017–1025.

Lacort, M., Leal, A., Liza, M., Martin, C., Martinez, R., and Ruiz-Larrea, M. (1995) Protective effect of estrogens and catecholestrogens against peroxidative membrane damage in vitro. *Lipids* **30,** 141–146.

Lambert, J., Peters, J., and Cottrell, G. (1987) Actions of synthetic and endogenous steroids on the GABA A receptor. *Trends Pharm. Sci.* **8,** 224.

Lawrence, M. and Sapolsky, R. (1994) Glucocorticoids accelerate ATP loss following metabolic insults in cultured hippocampal neurons. *Brain Res.* **646,** 303–309.

Lieberman, D. and Mody, I. (1994) Regulation of NMDA channel function by endogenous calcium-dependent phosphatase. *Nature* **369,** 235–239.

Lipton, S., Choi, Y., Pan, Z., Lei, S., Chen, H., Sucher, N., et al. (1993) A redox-based mechanism for the neuroprotective and neurodestructive effects of nitric oxide and related nitroso-compounds. *Nature* **364,** 626–629.

Lowy, M., Wittenberg, L., and Yamamoto, B. (1995) Effect of acute stress on hippocampal glutamate levels and spectrin proteolysis in young and aged rats. *J. Neurochem.* **65,** 268–274.

Lowy, M., Wittenberg, L., and Novotncy, S. (1994) Adrenalectomy attenuates kainic acid-induced spectrin proteolysis and heat shock protein 70 induction in hippocampus and cortex. *J. Neurochem.* **63,** 886–893.

Maclusky, N. (1987) Relationship between nuclear steroid binding and steroid-induced responses, in *Steroid Hormone Receptors* (Clark, C., ed.). Horwood, Chichester, UK.

Maines, M., Eke, B., Weber, C., and Ewing, J. (1995) Corticosterone has a permissive effect on expression of heme oxygenase-1 in CA1–CA3 neurons of hippocampus in thermal stressed rats. *J. Neurochem.* **64,** 1769–1774.

Majewska, M., Harrison, N., Schwartz, R., Barker, T., and Paul, S. (1986) Steroid hormone metabolites are barbiturate-like modulators of the GABA receptor. *Science* **232,** 1004–1007.
Masters, J., Finch, C., and Sapolsky, R. (1989) Glucocorticoid endangerment of hippocampal neurons does not involve DNA cleavage. *Endocrinology* **124,** 3083–3089.
Mattson, M. P. and Mrotek, J. (1985) Exogenous steroids alter steroidogenesis in cultured Y-1 adrenal tumor cells by actions preceding cyclic AMP. *Steroids* **46,** 619–637.
Mattson, M., Lovell, M., Furukawa, K., and Markesbery, W. (1995) Neurotrophic factors attenuate glutamate-induced accumulation of peroxides, elevation of intracellular calcium concentration, and neurotoxicity and increase antioxidant enzyme activities in hippocampal neurons. *J. Neurochem.* **65,** 1740–1751.
McEwen, B. (1992) Reexamination of the glucocorticoid hypothesis of stress and aging. *Prog. Brain Res.* **93,** 365–393.
McEwen, B. (1994) Steroid hormone actions on the brain: When is the genome involved? *Hormones Behav.* **28,** 396–405.
McEwen, B. and Woolley, C. (1994) Estradiol and progesterone regulate neuronal structure and synaptic connectivity in adult as well as develping brain. *Exp. Gerontol.* **29,** 431–436.
McGimsey, W., Cidlowski, J., Stumpf, W., and Sar, M. (1991) Immunocytochemical localization of the glucocorticioid receptor in rat brain, pituitary, liver, and thymus with two new polycolonal antipeptide antibodies. *Endocrinology* **129,** 3064–3072.
McIntosh, L. and Sapolsky, R. (1996) Glucocorticoids increase oxidative parameters in hippocampal and cortical neuronal cultures exposed to adriamycin. *Exp. Neurol.* **141,** 201–208.
McIntosh, L. and Sapolsky, R. (1997) Glucocorticoids may enhance oxygen radical-mediated neurotoxicity. *Neurotoxicology* **17,** 873–881.
Meaney, M., Aitken, D., Bhatnager, S., Van Berkel, C., and Sapolsky, R. (1988) Postnatal handling attenuates neuroendocrine, anatomical, and cognitive impairments related to the aged hippocampus. *Science* **238,** 766–771.
Mizoguchi, K., Kunishita, T., Chui, D., and Tabira, T. (1992) Stress induces neuronal death in the hippocampus of castrated rats. *Neurosci. Lett.* **138,** 157–160.
Morse, J. and Davis, J. (1990) Regulation of ischemic hippocampal damage in the gerbil: adrenalectomy alters the rate of CA1 cell disappearance. *Exp. Neurol.* **110,** 86–92.
Mortel, K. and Meyer, J. (1995) Lack of postmenopausal estrogen replacement therapy and the risk of dementia. *J. Neuropsych. Clin. Neurosci.* **7,** 334–337.
Munck, A. (1971) Glucocorticoid inhibition of glucose uptake by peripheral tissues: old and new evidence, molecular mechanisms, and physiological significance. *Perspect. Biol. Med.* **14,** 265–286.
Munck, A., Guyre, P., and Holbrook, N. (1984) Physiological actions of glucocorticoids in stress and their relation to pharmacological actions. *Endocr. Rev.* **5,** 25–44.
Ogata, T., Nakamura, Y., Tsuji, K., Shibata, T., and Kataoka, K. (1993) Steroid hormones protect spinal cord neurons from glutamate toxicity. *Neuroscience* **55,** 445–454.
Ohkura, T., Isse, K., Akazawa, K., Hamamoto, M., Yaoi, Y., and Hagino, N. (1994) Evaluation of estrogen treatment in female patients with dementia of the Alzheimer type. *Endocr. J.* **41,** 361–371.
Ohkura, T., Isse, K., Akazawa, K., Hamamoto, M., Yaoi, Y., and Hagino, N. (1995) Long-term estrogen replacement therapy in female patients with dementia of the Alzheimers type: 7 case reports. *Dementia* **6,** 99.
Olgemoller, B., Schon, J., and Wieland, O. (1985) Endothelial plasma membrane is a glucocorticoid-regulated barrier for the uptake of glucose into the cell. *Mol. Cell Endocrinol.* **43,** 165–173.
Onate, S., Tsai, S., Tsai, M., and O'Malley, B. (1995) Sequence and characterization of a coactivator for the steroid hormone receptor superfamily. *Science* **270,** 1354–1357.
Orchinik, M., Murray, T., and Moore, F. (1991) A corticosteroid receptor in neuronal membranes. *Science* **252,** 1848–1851.

Orti, E., Bodwell, J., and Munck, A. (1992) Phosphorylation of steroid hormone receptors. *Endocr. Rev.* **13,** 105–128.

Packan, D. and Sapolsky, R. (1990) Glucocorticoid endangerment of the hippocampus: tissue, steroid and receptor specificity. *Neuroendocrinology* **51,** 613–620.

Paganini-Hill, A. and Henderson, V. (1994) Estrogen deficiency and risk of Alzheimer's disease in women. *Am. J. Epidemiol.* **140,** 256–261.

Parthasarathy, S., Morales, A., and Murphy, A. (1994) Antioxidant: a new role for RU-486 and related compounds. *J. Clin. Invest.* **94,** 1990–1995.

Patchev, V., Brady, L., Karl, M., and Chrousos, G. (1994) Regulation of HSP90 and corticoseroid receptor mRNA by corticosterone in vivo. *Mol. Cell Endocrinol.* **103,** 57–64.

Pearce, D. and Yamamoto, K. (1993) Mineralocorticoid and glucocorticoid receptor activities distinguished by nonreceptor factors at a composite response element. *Science* **259,** 1161–1165.

Picard, D., Khursheed, B., Garabedian, M., Fortin, M., Lindquist, S., and Yamamoto, K. (1990) Reduced levels of hsp90 compromise steroid receptor action in vivo. *Nature* **348,** 166–168.

Pratt, W. (1990) Interaction of hsp90 with steroid receptors: organizing some diverse observations and presenting the newest concepts. *Mol. Cell Endocrinol.* **74,** C69–C76.

Quelle, F., Smith, R., Hrycyna, C., Kaliban, T., Crooks, J., and O'Brien, J. (1988) 3H-Dexamethasone binding to plasma membrane-enriched fractions from liver of nonadrenalectomized rats. *Endocrinology* **123,** 1642.

Rajan, V., Edwards, C., and Seckl, J. (1996) 11β-hydroxysteroid dehydrogenase in cultured hippocampal cells reactivates inert 11-dehydrocorticosterone, potentiating neurotoxicity. *J. Neurosci.* **16,** 65–70.

Ravindran, J., Shuaib, A., Ijaz, S., Galazka, P., Waqar, T., Ishaqzay, R., Miyashita, H., and Liu, L. (1994) High extracellular GABA levels in hippocampus as a mechanism of neuronal protection in cerebral ischemia in adrenalectomized gerbils. *Neurosci Lett.* **176,** 209–215.

Rose, J., Moore, F., and Orchinik, M. (1993) Rapid neurophysiological effects of corticosterone on medullary neurons: relationship to stress-induced suppression of courtship clasping in an amphibian. *Neuroendocrinology* **57,** 815–821.

Ruiz-Larrea, M., Leal, A., Liza, M., Lacort, M., and de Groot, H. (1994) Antioxidant effects of estradiol and 2-hydroxyestradiol on iron-induced lipid peroxidation of rat liver microsomes. *Steroids* **59,** 383–388.

Santoro, C., Mermod, N., Andrews, P., and Tijan, R. (1988) A family of human CCAAT-bos-binding proteins active in transcription and DNA replication: cloning and expression of multiple cDNAs. *Nature* **334,** 218–224.

Sapolsky, R. (1985a) A mechanism for glucocorticoid toxicity in the hippocampus: increased neuronal vulnerability to metabolic insults. *J. Neurosci.* **5,** 1227–1234.

Sapolsky, R. (1985b) Glucocorticoid toxicity in the hippocampus: temporal aspects of neuronal vulnerability. *Brain Res.* **339,** 300–307.

Sapolsky, R. (1986a) Glucocorticoid toxicity in the hippocampus: temporal aspects of synergy with kainic acid. *Neuroendocrinology* **43,** 386–394.

Sapolsky, R. (1986b) Glucocorticoid toxicity in the hippocampus: reversal by supplementation with brain fuels. *J. Neurosci.* **6,** 2240–2246.

Sapolsky, R. (1996c) Why stress is bad for your brain. *Science* **273,** 749, 750.

Sapolsky, R., Packan, D., and Vale, W. (1988) Glucocorticoid toxicity in the hippocampus: in vitro demonstration. *Brain Res.* **453,** 367–375.

Sapolsky, R. and Pulsinelli, W. (1985) Glucocorticoids potentiate ischemic injury to neurons: Therapeutic implications. *Science* **229,** 1397–1400.

Scully, J. and Otten, U. (1993) Glucocorticoid modulation of neurotrophin expression in immortalized mouse hippocampal neurons. *Neurosci Lett.* **155,** 11–16.

Seckl, J., Kelly, P., and Sharkey, J. (1991) Glycyrrhetinic acid, an inhibitor of 11beta-hydroxysteroid dehydrogenase, alters local cerebral glucose utilization in vivo. *J. Steriod Biochem. Mol. Biol.* **39,** 777–785.

Silver, G., Reid, L., and Krauter, K. (1990) Dexamethasone mediated regulation of 3-methylcholanthrene induced cytochrome P450D mRNA accumulation in primary rat hepatocyte cultures. *J. Biol. Chem.* **265,** 3134–3138.

Simpkins, J., Singh, M., and Bishop, J. (1994) The potential role for estrogen replacement therapy in the treatment of the cognitive decline and neurodegeneration associated with Alzheimer's disease. *Neurobiol. Aging* **15,** S195–S197.

Singh, M., Meyer, E., and Simpkins, J. (1995) The effect of ovariectomy and estradiol replacement on BDNF mRNA expression in cortical and hippocampal brain regions of female Sprague-Dawley rats. *Endocrinology* **136,** 2320–2324.

Smith, M., Makino, S., Kvetnansky, R., and Post, R. (1995) Stress and glucocorticoids affect the expression of BDNF and NT-3 mRNAs in the hippocampus. *J. Neurosci.* **15,** 1768–1777.

Smith-Swintosky, V., Pettigrew, L., Sapolsky, R., Phares, C., Craddock, S., Brooke, S., and Mattson, M. P. (1996) Metyrapone, an inhibitor of glucocorticoid production, reduces brain injury induced by focal and global ischemia and seizures. *J. Cereb. Blood Flow Metab.* **16,** 585–598.

Sohrabji, F., Miranda, R., and Toran-Allerand, C. (1995) Identification of a putative estrogen response element in the gene encoding BDNF. *Proc. Natl. Acad. Sci. USA* **92,** 11,110–11,114.

Stein, B. and Sapolsky, R. (1988) Chemical adrenalectomy reduces hippocampal damage induced by kainic acid. *Brain Res.* **473,** 175–181.

Stein-Behrens, B., Elliott, E., Miller, C., Schilling, J., Newcombe, R., and Sapolsky, R. (1992) Glucocorticoids exacerbate kainic acid-induced extracellular accumulation of excitatory amino acids in the rat hippocampus. *J. Neurochem.* **58,** 1730–1738.

Stein-Behrens, B., Lin, W., and Sapolsky, R. (1994a) Physiological elevations of glucocorticoids potentiate glutamate accumulation in the hippocampus. *J. Neurochem.* **63,** 596–603.

Stein-Behrens, B., Mattson, M. P., Chang, I., Yeh, M., and Sapolsky, R. (1994b) Stress exacerbates neuron loss and cytoskeletal pathology in the hippocampus. *J. Neurosci.* **14,** 5373–5380.

Strahle, U., Klock, G., and Schutz, G. (1987) A DNA sequence of l5 base pairs is sufficient to mediate both glucocorticoid and progesterone induction of gene expression. *Proc. Natl. Acad. Sci. USA* **84,** 7871–7875.

Subbiah, M., Kessel, B., Agrawal, M., Rajan, R., Abplanalp, W., and Rymaszewski, Z. (1993) Antioxidant potential of specific estrogens on lipid peroxidation. *J. Clin. Endocrinol. Metab.* **77,** 1095–1097.

Supko, D. and Johnston, M. (1994) Dexamethasone potentiates NMDA receptor-mediated neuronal injury in the postnatal rat. *Eur. J. Pharmacol.* **270,** 105–109.

Sze, P. and Iqbal, Z. (1994) Glucocorticoid actions on synaptic plasma membranes: modulation of [125I]Calmodulin binding. *J. Steroid Biochem. Mol. Biol.* **48,** 179–186.

Tai, P., Albers, M., Chang, H., Faber, L., and Schreiber, S. (1992) Association of a 59-Kilodalton immunophilin with the glucocorticoid receptor complex. *Science* **256,** 1315–1318.

Talmi, M., Carlier, E., Benbgelloun, W., and Soumireu-Mourat, B. (1996) Chronic RU 486 treatment reduces age-realted alterations of mouse hippocampal function. *Neurobiol. Aging* **17,** 9–17.

Talmi, M., Carlier, E., Bengelloun, W., and Soumireu-Mourat, B. (1995) Synergetic action of corticosterone on kainic acid-induced electrophysiological alterations in the hippocampus. *Brain Res.* **704,** 97–104.

Tombaugh, G. and Sapolsky, R. (1992) Corticosterone accelerates hypoxia-induced ATP loss in cultured hippocampal astrocytes. *Brain Res.* **588,** 154–158.

Tombaugh, G. and Sapolsky, R. (1993) Endocrine features of glucocorticoid endangerment in hippocampal astrocytes. *Neuroendocrinology* **57,** 7–15.

Tombaugh, G., Yang, S., Swanson, R., and Sapolsky, R. (1992) Glucocorticoids exacerbate hypoxic and hypoglycemic hippocampal injury in vitro: biochemical correlates and a role for astrocytes. *J. Neurochem.* **59,** 137–145.

Towle, A. and Sze, P. (1983) Steroid binding to synaptic plasma membrane: differential binding of glucocorticoids and gonadal steroids. *J. Steroid Biochem.* **18,** 135–143.

Traynelis, S. and Cull-Candy, S. (1990) Proton inhibition of NMDA receptors in cerebellar neurons. *Nature* **345,** 347–350.

Tsai, S., Carlstedt-Duke, J., Weigel, N., Dahlman, K., Gustafsson, J., Tsai, M., and O'Malley, B. (1988) Molecular interactions of steroid hormone receptor with its enhancer element: evidence for receptor dimer formation. *Cell* **55,** 361–370.

Uhler, T., Frim, D., Pakzaban, P., and Isacson, O. (1994) The effects of mega-dose methylprednisolone and U-78517F on glutamate receptor mediated toxicity in the rat neostriatum. *Neurosurgery* **34,** 122–127.

Vallette, G., Vanet, A., Sumida, C., and Nunez, E. (1991) Modulatory effects of unsaturated fatty acids on the binding of glucocorticoids to rat liver glucocorticoid receptors. *Endocrinology* **129,** 1363–1369.

Vamvakopoulos, N. and Chrousos, G. (1994) Hormonal regulation of human CRH gene expression: implications for the stress response and immune/inflammatory reaction. *Endocr. Rev.* **15,** 409–420.

Virgin, C., Ha, T., Packan, D., Tombaugh, G., Yang, S., Horner, H., and Sapolsky, R. (1991) Glucocorticoids inhibit glucose transport and glutamate uptake in hippocampal astrocytes: implications for glucocortiocid neurotoxicity. *J. Neurochem.* **57,** 1422–1428.

Weber, C., Eke, B., and Maines, M. (1994) Corticosterone regulates heme oxygenase-2 and NO synthase transcription and protein expression in rat brain. *J. Neurochem.* **63,** 953–960.

Weiner, F., Czaja, M., Jefferson, D., Giambrone, M., Turkaspa, R., Reid, L., and Zern, M. (1987) The effects of dexamethasone on in vitro collagen gene expression. *J. Biol. Chem.* **262,** 6955–6958.

Welch, W. (1992) Mammalian stress response: cell physiology, structure/function of stress proteins, and implications for medicine and disease. *Physiol. Rev.* **72,** 1063–1081.

Woolley, C. and McEwen, B. (1993) Roles of estradiol and progesterone in regulation of hippocampal dendritic spine density during the estrous cycle in the rat. *J. Comp. Neurol.* **336,** 293–306.

Woolley, C. and McEwen, B. (1994) Estradiol regulates hippocampal dendritic spine density via an NMDA receptor-dependent mechanism. *J. Neurosci.* **14,** 7680–7687.

Yamamoto, K. (1985) Steroid receptor regulated transcription of specific genes and gene networks. *Ann. Rev. Genet.* **19,** 209–252.

Young, W. and Flamm, E. (1982) Effect of high-dose corticosteroid therapy on blood flow, evoked potentials, and extracellular calcium in experimental spinal injury. *J. Neurosurg.* **57,** 667–673.

Zaret, K. and Yamamoto, K. (1984) Reversible and persistent changes in chromatin structure accompany activation of a glucocorticoid dependent enhancer element. *Cell* **38,** 29–38.

Zeelan, F. (1990) Determination of the affinity of a steroid for its receptor is not sufficient to measure its intrinsic hormonal activity. *Steroids* **55,** 325–326.

15

Therapeutic Potential of Leukocyte Antiadhesion Strategies in CNS Ischemia

Wayne M. Clark and Justin A. Zivin

1. LEUKOCYTE POTENTIATION OF CNS ISCHEMIC INJURY

Leukocyte appearance in central nervous system (CNS) ischemic tissue has previously been considered to represent a pathophysiological response to existing injury. Recent evidence suggests that leukocytes may also be directly involved in the pathogenesis and extension of CNS ischemia/reperfusion (I/R) injury (Engler et al., 1986; Schmid-Schonbein and Engler, 1986). Two proposed mechanisms of leukocyte involvement in I/R are: direct microvascular occlusion after endothelial and basement membrane adhesion (Schmid-Schonbein, 1987; Mori et al., 1992); and transendothelial migration, with secondary CNS tissue infiltration and neuronal cytotoxic injury (Fassbender et al., 1995). Initial adhesion of leukocytes to microvascular endothelium is essential for initiation of either of these mechanisms (*see* Fig. 1A).

Leukocyte infiltration into the brain has been shown to occur rapidly following a stroke. In focal experimental CNS ischemia, granulocyte infiltration occurs as early as 4 h and peaks at 24 h, with marked infiltration seen in reperfusion stroke models (Barone et al., 1992; Dereski et al., 1992; Hallenbeck et al., 1986; del Zoppo et al., 1991). Recent studies using filament middle cerebral artery occlusion models have found neutrophil transmigration across postcapillary venules beginning at 4–6 h after MCA occlusion. The peak number of intravascular neutrophils was detected at 12 h; maximum invasion into the brain parenchyma was seen at 24 h after reversible MCA occlusion. Monocytes were also observed appearing somewhat later (Zea-Longa et al., 1989; Garcia et al., 1993, 1994). Early leukocyte infiltration has also been observed in clinical stroke (Wang et al., 1993).

Leukocyte capillary plugging may also be the major mechanism of the no-reflow phenomenon. This phenomenon, first described in the CNS by Ames and colleagues (1968), is defined as the incomplete restoration of normal blood flow following a period of ischemia. Areas of parenchyma that might be viable when blood flow returns to the large vessels are not adequately reperfused and ultimately die. The no-reflow phenomenon has been observed under conditions of focal cerebral I/R (Cromwell and Olsson, 1972; Cromwell et al., 1970; del Zoppo et al., 1991; Mori et al., 1992) and after global cerebral ischemia (Ames et al., 1968; Ginsberg and Myers, 1972; Harrison et al., 1975; Hossman

From: Neuroprotective Signal Transduction *edited by M. P. Mattson*
Humana Press Inc., Totowa, NJ

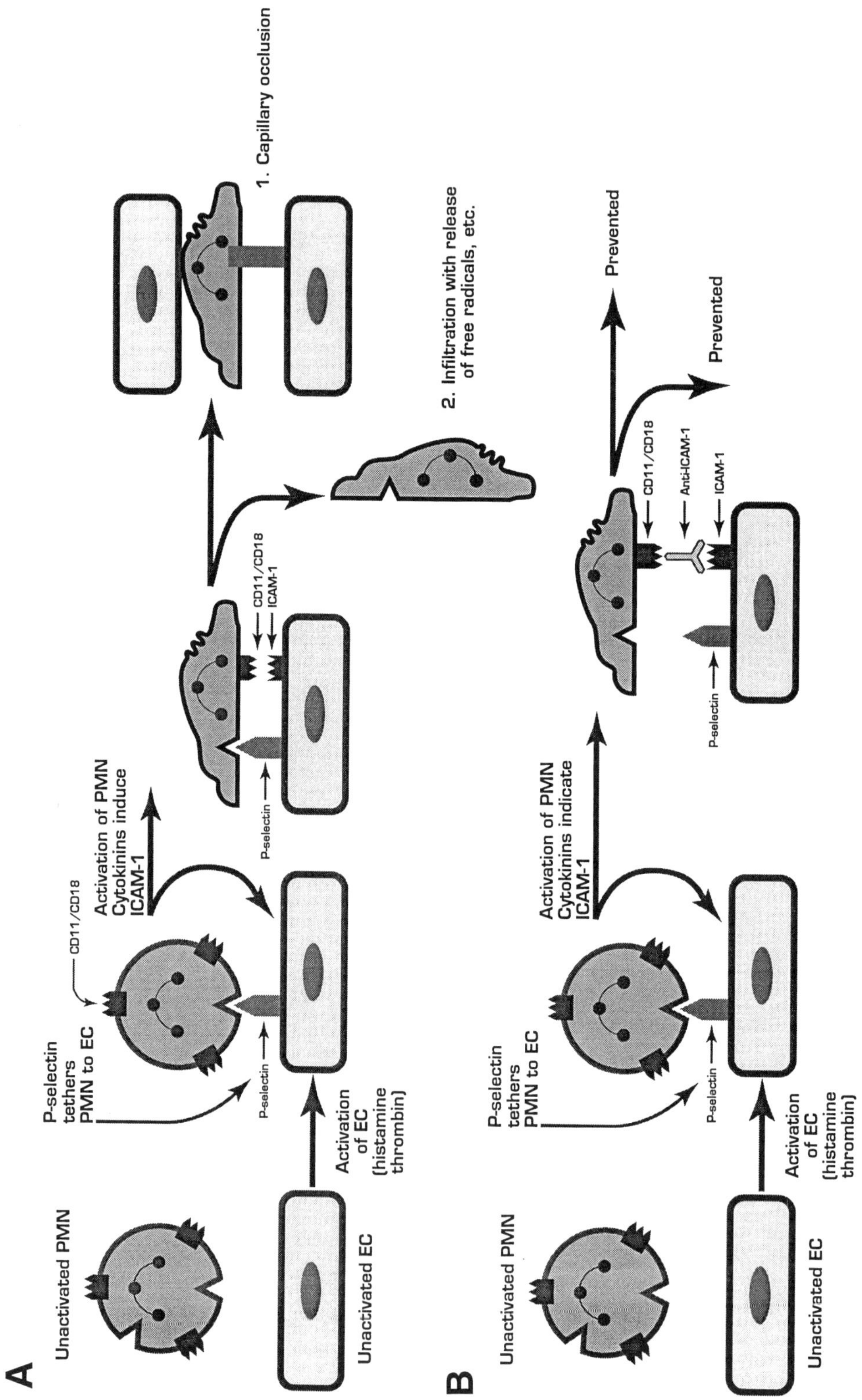
A
Unactivated PMN
Unactivated EC
Activation of EC (histamine thrombin)
P-selectin tethers PMN to EC
P-selectin
CD11/CD18
Activation of PMN Cytokinins induce ICAM-1
P-selectin
CD11/CD18
ICAM-1
1. Capillary occlusion
2. Infiltration with release of free radicals, etc.
B
Unactivated PMN
Unactivated EC
Activation of EC (histamine thrombin)
P-selectin tethers PMN to EC
P-selectin
Activation of PMN Cytokinins indicate ICAM-1
P-selectin
CD11/CD18
Anti-ICAM-1
ICAM-1
Prevented
Prevented

et al., 1973). In addition to direct capillary occlusion by leukocytes, other factors that may contribute to the no-reflow phenomenon include endothelial cell swelling and platelet aggregation (Mori et al., 1992).

Leukocytes are also felt to potentiate I/R injury by direct infiltration into brain parenchyma. Once there, they appear to be one of the major producers of oxygen free radicals and proinflammatory mediators termed cytokines (Hallenbeck and Dutka, 1990). The released proinflammatory cytokines, particularly IL-1β and IL-6, can further potentiate ischemic injury by aiding the recruitment of more leukocytes, and can also produce direct endothelial cell injury. The role of leukocyte adhesion and infiltration in potentiating CNS I/R injury is supported by older experimental studies in which systemic leukocyte depletion was produced using either antiserum, cyclophosphamide, or mechlorethamine, and then subjecting the animals to CNS ischemia. These studies showed that leukocyte depletion improved blood flow in an embolic stroke model (Dutka et al., 1989), reduced infarct size in a model of thromboembolism (Bednar et al., 1991), and improved blood flow in a forebrain ischemia model (Grøgaard et al., 1989). Although these studies support leukocyte involvement in stroke, they use toxic leukocyte depletion treatments that cannot be used in clinical stroke therapy. Instead of depleting leukocytes, an alternate approach that may be therapeutically feasible is to block leukocyte adhesion to the endothelium through the use of specific antiadhesion monoclonal antibodies (*see* Fig. 1B). By blocking leukocyte adhesion to endothelial cells, both capillary plugging and tissue infiltration may be prevented.

2. INTEGRIN SIGNALING: MECHANISM OF LEUKOCYTE ADHESION TO ENDOTHELIAL CELLS

The migration of leukocytes from the blood into the brain is a multistep process. First, primary leukocyte–endothelial cell interaction (rolling) is mediated by selectins involving P-selectin and E-selectin on the surface of the endothelial cells, and L-selectin on leukocytes (Fassbender et al., 1995; *see* Fig. 2). On activation, firm adherence of leukocytes to the endothelial lining (sticking) is mediated by a leukocyte membrane glycoprotein receptor complex termed CD-18, or β_2-integrin, and its endothelial ligand, the intracellular adhesion molecule (ICAM-1) (Smith et al., 1988; Argenbright et al., 1991).

Fig. 1. (A) Schematic representation of the process of endothelial-dependent leukocyte adhesion. Endothelial cells activated by histamine or thrombin rapidly translocate P-selectin to their surfaces (also E-selectin, not shown), which tethers leukocytes to the endothelial cell. This tethering does not require an active response from the leukocyte. Once tethered, other factors, including platelet-activating factor and cytokines, are released, which stimulates a leukocyte activation response. This response includes shape changing and increased surface expression of CD-11/CD-18. The CD-11/CD-18 then binds to the corresponding ICAM-1 receptor on the endothelial cell, leading to firm endothelial attachment. This attachment may produce either direct obstruction of the microcirculation or lead to infiltration into the surrounding brain parenchyma. **(B)** Schematic representation showing that, through the use of monoclonal antibodies directed against the anti-ICAM-1 receptor, the CD-11/CD-18 to ICAM-1 attachment is prevented. This, in turn, prevents subsequent microvessel obstruction and leukocyte infiltration.

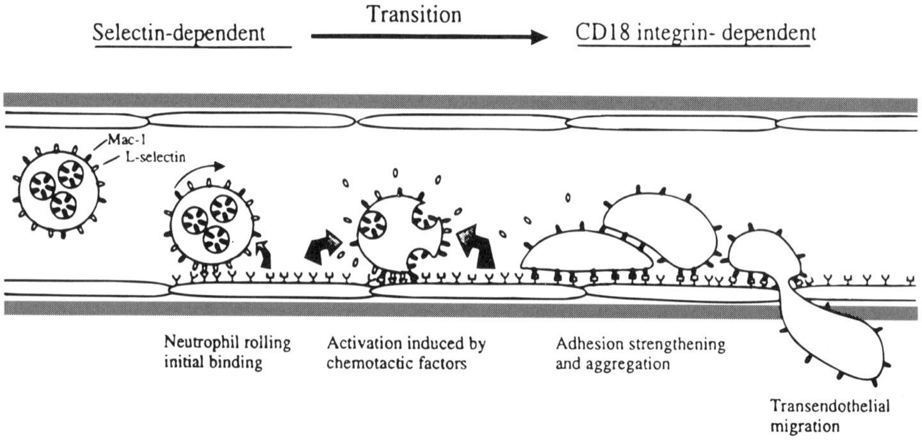

Fig. 2. A model for neutrophil interaction with inflamed endothelium. Neutrophil localization and activation are multistep processes, which must be tightly regulated to minimize injury to healthy tissue. Endothelial cells adjacent to a site of inflammation are stimulated by cytokines, such as tumor necrosis factor and interleukin-1, to express the adhesion molecules E- and P-selectins and intercellular adhesion molecule (ICAM-1) (the two Y-shaped symbols), which guide localization of leukocytes. The adhesion molecule L-selectin (open symbols), present on the neutrophil surface, mediates the initial interaction between the unactivated neutrophil and the endothelium, which can be observed as neutrophil rolling and reversible sticking. (Both the selectins and ICAM-1 appear to have multiple ligands. This illustration is not intended to show all possible molecular pairings.) Only these transiently bound neutrophils are exposed to low concentrations of chemoattractants (large arrows) for a sufficient time to be programmed for activation. The activation triggers a rapid transition in neutrophil adhesiveness: There is a transient increase in L-selectin-mediated adhesion, followed by a rapid shedding of the cell-surface L-selectin, and, concomitantly, activation of Mac-1 mediated adhesiveness (closed symbols). This is accomplished by a rapid increase in cell-surface expression of Mac-1 from intracellular stores. Thus, activation rapidly shuts down one adhesion pathway L-selectin and turns on another (Mac-1). Engagement of Mac-1 results in adhesion strengthening, neutrophil aggregation, and transendothelial migration, allowing neutrophils to enter the inflamed tissue. (Reproduced with permission from Kishimoto, 1991.)

The leukocyte membrane glycoprotein complex predominantly responsible for final endothelial adherence is termed CD-18, or alternately, β_2-integrin. This complex consists of three heterodimers. All three share an identical β-unit (also frequently called CD-18) and are distinguished from each other by distinct α-subunits. The three α-subunits are termed Leukocyte Function Antigen (LFA-1 or CD-11a, present on all leukocytes), MAC-1 (CD-11b, present mostly on PMNs and monocytes), and P150 (CD-11c, present on neutrophils and monocytes).

Corresponding counterreceptors for the CD-18 integrin complex are the intracellular adhesion molecule (ICAM) family of adhesion molecules (Smith et al., 1989). The terminology used to describe the ICAM receptor is variable in the literature. At times, ICAM-1 is used as the designation to describe the molecule that binds with the CD-18 complex, with various subunits of the ICAM receptor binding to either the MAC-1 α-subunit, the LFA-1 α-subunit, or the common β-subunit (Smith et al., 1989). At other times, the designation ICAM-1 refers only to the component that binds to the MAC-1 subunit, and ICAM-2 corresponds to the part of the receptor that binds to the LFA-1 α-subunit (Hossman et al., 1973). In this chapter, we will use the first designation, in which ICAM-1 corresponds to the endothelial receptor that binds to the entire CD-18 complex. ICAM-1 is inducible with inflammatory cytokines such as IL-1 and TNF-α (Kochanek and Hallenbeck, 1992).

We have previously demonstrated that CD-18–ICAM-1-mediated neutrophil adhesion increases following clinical and experimental stroke (Clark et al., 1992, 1993). One of the strongest areas of support for the involvement of leukocytes in CNS I/R injury comes from therapeutic preclinical studies that reduced CNS injury by preventing leukocyte adhesion and infiltration, using specific monoclonal antibodies (MAbs).

3. ANTIADHESION MONOCLONAL ANTIBODIES

3.1. Monoclonal Antibodies Directed Against the β_2-Integrin (CD-18)

Using specific mouse MAbs directed at various components of the CD-18 complex, previous studies had shown a reduction of ischemic injury and decreased leukocyte tissue infiltration in heart, lung, and intestinal ischemia models (Arfors et al., 1987; Hernandez et al., 1987; Price et al., 1987; Simpson et al., 1988; Vedder et al., 1988). We were the first to test the therapeutic potential of CD-18 antibodies in two CNS ischemia models in rabbits (Clark et al., 1991a). We used a mouse MAb directed against the CD-18 β-complex in two animal models. In the first model, the treatment was given just prior to injecting multiple nondissolvable microspheres into the cerebral circulation. The animals underwent a neurologic evaluation at 18 h after embolization. In this experiment, no beneficial effects were observed with the use of the anti-CD-18 antibody treatment. In the second experiment, the anti-CD-18 was given just prior to the onset of spinal cord ischemia. After varying periods of ischemia, reperfusion was allowed to occur. The animals were then evaluated 18 h later to determine neurologic function. In this reperfusion model, animals treated with anti-CD-18 had an approx 50% increase in the duration of ischemia they could tolerate (*see* Fig. 3). These results supported the role of leukocytes as active participants in the potentiation of CNS ischemic injury. We further suggested that the greatest benefit with the use of antiadhesion therapies will be in ischemic conditions in which reperfusion occurs.

Treatment with MAbs against the CD-18 complex has also been found to be beneficial in a rat middle cerebral artery (MCA) occlusion model. Chen et al. (1994) administered an antibody directed against the MAC-1 component (anti-CD-11b); on reperfusion, the middle cerebral artery had been occluded with a filament. After a 2-h period of occlusion, the filament was removed to permit reperfusion. After 46 h of reperfusion, the animals were rated for neurologic function and then sacrificed for infarct size determination. The animals who received 2 mg/kg of the anti-CD11b antibody had improved neurologic scores and a 43% reduction in their infarct size. The study also demonstrated that the infiltration of neutrophils was significantly reduced in the anti-CD11b treated animals. Using the same animal model, this group of investigators (Chopp et al., 1994) found that the administration of an anti-CD11b antibody after the onset of CNS ischemia was also effective. In the latter study, animals underwent 2 h of middle cerebral artery occlusion and the antibody was given at 1 h of reperfusion (3 h

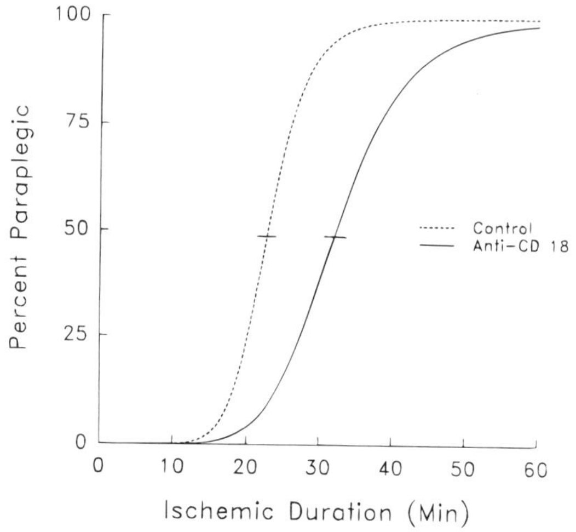

Fig. 3. Quantal dose-response curves for anti-CD-18 in rabbit spinal cord ischemia model; percentage of paraplegic animals vs duration of spinal cord ischemia. Midpoint of curve, E_{50}, is ischemic duration that causes 50% of rabbits to be paraplegic. Error bar shows SEM or the E_{50}.

after the onset of ischemia). Animals were again rated at 46 h and infarct volume was determined. The infarct volume was 28% smaller in the anti-CD11b group, compared to the control group. Again, the number of infiltrating leukocytes was also significantly reduced. These data not only confirmed the earlier therapeutic potential of antiadhesion antibodies, but broadened the therapeutic time by demonstrating a beneficial effect at 3 h after the onset of ischemia.

Matsuo et al. (1994) also found that MAbs against the CD-18 complex were effective in the rat middle cerebral artery occlusion model. In this study, the investigators actually used three separate antibodies. The first was directed against the CD-11a component (LFA-1), the second directed against the β-unit (CD-18), and the third directed against ICAM-1. Compared to sham-operated animals, each of these antibodies produces significant reduction in infarct size and leukocyte infiltration at 24 h. Although all three antibodies produced a significant improvement, the greatest effect was seen with the anti-ICAM-1 antibody (~80% infarct size reduction). The next most effective was the antibody targeting the common β-unit (~60% infarct size reduction), and the least effective was the one solely directed against LFA-1 component (~50% infarct size reduction). The results of the study suggest that all components of the CD-18 receptor complex appear to be involved in leukocyte adhesion and infiltration. However, a greater benefit does appear to occur with blockage of either the common β-unit or the corresponding ICAM-1 receptor.

Lindsberg et al. (1995) also found that antiadhesion therapy was useful postreperfusion in a rabbit spinal cord injury model. One mg/kg of a MAb directed against the β-unit (CD-18) was given after 30 min of reperfusion (1 h after onset of ischemia). The animals that received the anti-CD-18 therapy showed significant improvements in spinal cord blood flow, neurologic function, preservation of the blood–brain barrier (as measured by extravasation of Evans blue albumin), and an improvement in the number of surviving motor neurons at 6 h after the insult.

The one negative study that has been published using antiadhesion agents is that of Takeshima et al. (1992) in a middle cerebral artery reperfusion model in cats. In this study, 2 mg/kg of an anti-CD-18 (common β-unit) was given to cats 45 min after the onset of middle cerebral artery ischemia. The animals underwent a total of 90 min of ischemia, followed by 180 min of reperfusion, and were then sacrificed. In this study, no differences were seen at 180 min, in terms of blood flow to the two hemispheres and in infarct volume at 180 min, as measured by triphenyltetrazolium chloride staining. One of the major differences in this study, compared to the prior work, is in the timing of the evaluation. These animals were evaluated at 3 h of reperfusion. It is likely that much of the histologically detectable injury that occurs from white cell influx happens in a more delayed fashion. This may explain the failure to find a benefit at 3 h. Also, the triphenyltetrazolium chloride staining may be an inexact way of determining infarct size, particularly when this is attempted at 3 h postischemia (Cole et al., 1990).

Overall, the above studies support the beneficial effects of using antibodies directed against the CD-18 complex in reducing CNS reperfusion injury. In clinical stroke treatment, reperfusion injury may be of particular importance following the use of thrombolytic agents. Consequently, an ideal combined therapeutic approach would utilize both a thrombolytic agent and an agent that reduces reperfusion injury. To test this combined approach, we studied the use of tissue plasminogen activator (rtPA) and anti-CD-18 antibody in a rabbit model of multiple cerebral emboli (Bowes et al., 1995). In this study, animals were injected with multiple autologous small blood clots. Five minutes postischemia onset, animals received either saline or 1 mg/kg of anti-CD-18 (that binds to the β-unit). At 30 min postembolization, the animals then received either saline or 3 mg/kg of rtPA. The neurologic impairment of the animals was rated 18 h later. Both rtPA and anti-CD-18 treatment significantly improved the neurologic function at 18 h. However, the combination of the two was no more effective than either substance alone. A possible explanation of this failure to detect a synergistic effect of the two agents is that the blood clot model may have an efficacy ceiling effect. As such, monotherapy with either agent produced the maximal possible benefit that could be seen with this model and consequently, further improvement with combined therapy could not be accurately measured. This study did not evaluate whether combined therapy with rtPA and anti-CD-18 could prolong the therapeutic window for the rtPA treatment.

The above studies suggest that antibodies directed against the common β-unit (anti-CD-18 antibodies) and those directed against the Mac-1 component (CD-11b) are effective at reducing CNS reperfusion injury. To our knowlege, antibodies directed against P150 have not been tested in CNS ischemia. Currently, there are no clinical trials using antibodies directed against the CD-18 complex in stroke. This is a result, in part of concerns regarding blocking other leukocyte-related functions with the CD-18 antibody, i.e., such a treatment could potentially increase the risk of systemic infection.

3.2. *Monoclonal Antibodies Directed Against Anti-ICAM-1*

Previous studies using anti-ICAM-1 MAb treatments in animal models have shown reductions in inflammation and asthma (Wegner et al., 1990), in kidney rejection (Cosimi et al., 1990), and in lung injury (Barton et al., 1989). Based on the encouraging results from these studies, as well as the apparent success of anti-CD-18 therapy in reducing CNS ischemia, we tested the ability of the monoclonal antibody directed against ICAM-1 to reduce experimental CNS ischemic injury (Clark et al., 1991b). In this study, 1 mg/kg of anti-ICAM-1 was given to rabbits 30 min before producing either reversible spinal cord ischemia or irreversible multiple brain emboli. Similar to our anti-CD-18 results, we found that pretreatment with anti-ICAM-1 produced a marked improvement in neurologic function in the spinal cord reperfusion CNS-injury model, but no benefit was seen with anti-ICAM-1 treatment in the nonreperfusion cerebral microsphere emboli model. We also found qualitative reductions in spinal cord infiltration in animals treated with anti-ICAM-1. We then tested these results in a multiple cerebral embolism model with reperfusion (Bowes et al., 1993). In this study, animals were treated with either saline or 1 mg/kg of anti-ICAM-1 5 min after being injected with multiple blood clot cerebral emboli. The difference in this model compared to the nondissolvable microspheres is that autologous blood clots may undergo spontaneous lysis, which produces reperfusion. In the multiple-clot model with reperfusion, treatment with anti-ICAM-1 significantly increased the number of emboli required to produce neurologic impairment at 18 h. Taken together, our studies indicate that treatment with anti-ICAM-1 is effective in CNS ischemia, but the beneficial action appears only when there is significant reperfusion involved.

The finding that anti-ICAM-1 treatment is beneficial in a reperfusion model, but not in a nonreperfusion model, is supported by work done by Zhang et al. (1995), in which the filament model of middle cerebral artery occlusion was used. Half the animals underwent transient MCA occlusion (2 h of ischemia followed by reperfusion), and the other half underwent permanent occlusion of the middle cerebral artery. At 1 h of reperfusion (3 h postischemia), animals received 1 mg/kg of an MAb directed against ICAM-1 (1A29), or saline. Animals were sacrificed at 1 wk and infarct size was determined. In animals that underwent ischemia with reperfusion, there was a significant reduction (44%) in the infarct size seen in animals treated with anti-ICAM-1. In contrast, there was no benefit seen with anti-ICAM-1 treatment in the animals subjected to permanent middle cerebral artery occlusion (*see* Fig. 4). By finding that anti-ICAM-1 treatment is still beneficial when given up to 3 h after the onset of ischemia, this study gives hope that these agents may prove to be clinically useful. An additional important point of this study is that a benefit in infarct size was seen when the animals were evaluated at 1 wk. By this time, many of the effects related to transient cerebral edema would have disappeared and this confounding variable may not have biased the results.

In addition to reducing neurologic impairment, treatment with anti-ICAM-1 also appears to significantly reduce leukocyte infiltration. Zhang et al. (1994) used 1 mg/kg of anti-ICAM-1 in the rat MCA occlusion model. Animals were treated with 1A29 MAb administered at the time of initiation of reperfusion (i.e., after 2 h of ischemia). At 46 h postischemia, the infarct volume was determined histologically, and animals treated with anti-ICAM-1 had a 40% reduction in their infarct size. The authors then examined whether anti-ICAM-1 treatment significantly reduced neutrophil infiltration by using a myloperoxidase activity assay. Animals treated with anti-ICAM-1 had a 50% reduction in the number of neutrophils that infiltrated the ischemic area. We recently have confirmed the ability of anti-ICAM-1 treatment to reduce leukocyte infiltration in CNS ischemia. Using specific immunohistochemistry, we found that treatment with anti-ICAM-1 significantly inhibited both mononuclear cell and neutrophil infiltration into the brain at 24 h in a rat model of transient forebrain ischemia with reperfusion (Clark

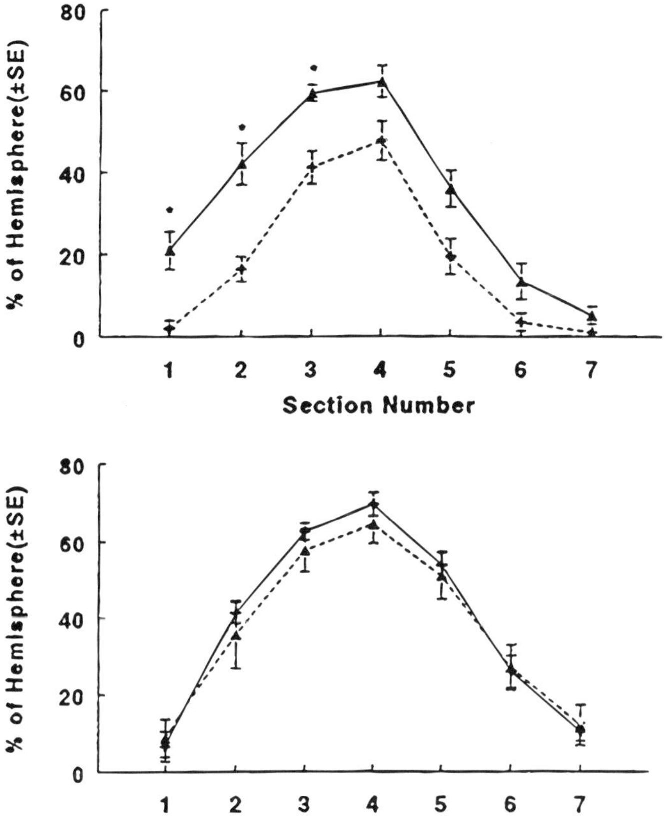

Fig. 4. Line graphs show percentage of infarct area in each coronal section in antibody-treated (+) ($n = 11$) and vehicle-treated (s) ($n = 9$) rats in the transient middle cerebral artery (MCA) occlusion group (top) and in antibody-treated (+) ($n = 9$) and vehicle-treated (s) ($n = 7$) rats in the permanent MCA occlusion group (bottom). Infarct areas were reduced in rats subjected to transient MCA occlusion and treated with anti-intercellular adhesion molecule-1 antibody compound compared with vehicle-treated animals 1 wk after onset of ischemia. Significant differences in infarct areas between the two groups were detected in coronal sections 1, 2, and 3. $*p > 05$ vs vehicle-treated animals (Zhang et al., 1995). (Reproduced with permission, from Zhang et al., 1995.)

et al., 1995a). These studies support the theory that at least some of the beneficial effects seen with anti-ICAM-1 treatment relate to its ability to keep leukocytes out of the brain ischemic tissue.

The above studies suggest that anti-ICAM-1 treatment may be beneficial in reducing CNS reperfusion injury. Again, looking toward possible clinical application, we studied whether combined rtPA and anti-ICAM-1 therapy would have a synergistic effect (Bednar et al., 1991). Using the autologous multiple cerebral clot model in rabbits, we found that rtPA therapy alone was not effective when given 2 h after the onset of ischemia. However, when rtPA therapy was combined with anti-ICAM therapy (anti-ICAM initiated 15 min after embolization), treatment at 2 h with rtPA did produce a significant reduction in neurologic injury. This finding supports the potential use of anti-ICAM-1 treatment along with rtPA in future clinical studies.

Histologic studies have shown that ICAM-1 expression on endothelial cells is minimally present in normal brain. However, it is strongly expressed in areas of focal brain ischemia, beginning as early as 3 h after vessel occlusion (Clark et al., 1995b). A potential advantage of this finding is that treatment with mAbs directed against ICAM-1 will only interfere with leukocyte adhesion at selective areas in which brain ischemia are present. They may therefore be less likely to interfere with other leukocyte functions, particularly the ability to fight infections.

An increased risk of infection is one potential side effect of the antiadhesion therapies that is of concern to clinicians. By interfering with ICAM-1 interactions with CD-18, it is possible that a condition similar to CD-18-deficient (LAD) patients will be caused. These patients exhibit multiple frequent infections with little ability to produce an appropriate leukocyte infiltration (Anderson and Springer, 1987; Todd and Freyer, 1988). To test whether antiadhesion therapies would increase the risk of infection, a recent study injected animals with subcutaneous doses of *staph aureus* in conjunction with anti-CD-18 therapy. These animals were found to have much larger abcesses than control animals (Sharar et al., 1991). In contrast to this, animals treated with anti-ICAM-1 antibody in conjunction with *staph aureus* did not have an increased size of abscesses seen (Mileski et al., 1993). A possible explanation as to why CD-18 therapies promoted infection but anti-ICAM-1 did not is because, as well as being an adhesion molecule, CD-18 is also a complement receptor. Since complement is one of the primary means by which leukocytes internalize and destroy bacteria, an anti-CD-18 treatment would not only prevent leukocyte adhesion, it would also paralyze the neutrophils' ability to destroy bacteria. It appears, therefore, that anti-ICAM-1 therapies will have greater clinical usefulness.

3.3. *Monoclonal Antibodies Directed Against Selectins*

Although the majority of leukocyte to endothelial cell adhesion is mediated by the CD-18 to anti-ICAM-1 receptor interaction, there is additional leukocyte adhesion that is mediated by a family of adhesion receptors termed the selectins (Williams and Hellewell, 1992) (*see* Fig. 1). Although circulating forms of selectins (sL-selectin, sE-selectin) have been observed after stroke in patients, to our knowledge, anti adhesion MAbs directed against selectins have not been evaluated in their ability to reduce CNS ischemic injury (Kochanek and Hallenbeck, 1992).

4. ADDITIONAL ANTIADHESION THERAPIES

Although the above studies suggest that antiadhesion MAb therapy may play a role in future stroke treatment, these agents are currently not available outside of ongoing clinical trials. Leukocyte-depleting strategies are not appropriate to the clinical population, and general strategies aimed at suppressing immune function either have not shown any benefit to stroke patients (steroid treatment) or could increase the risk of secondary infection (Price et al., 1987). One currently available treatment that appears to reduce leukocyte adhesion is the antibiotic doxycycline. Doxycycline, a member of the tetracycline family of antibiotics, interferes with leukocyte function by binding divalent cations, which are required for the CD-18 receptor complex to function (Gabler and Tsukuda, 1991). In vitro studies have shown that doxycycline inhibits leukocyte adhesion (Gabler and Creamer, 1991). We recently found doxycycline to be beneficial in a

CNS reperfusion injury model (Clark et al., 1994). Using a dose of 10 mg/kg of doxycycline (equivalent to 100 mg bid clinical dose), we found that treatment 30 min before reversible spinal cord ischemia produced a significant reduction in neurologic impairment at 18 h. Although doxycycline may also have clinical efficacy in stroke, it has not been studied in a randomized clinical trial.

5. ANTIADHESION STRATEGIES: CURRENT CLINICAL TRIALS

Based on the encouraging preclinical results outlined above, a multicenter, international study investigated the clinical therapeutic potential of anti-ICAM-1 therapy. The study, "A Double-blind, randomized, placebo-controlled, parallel group trial of the efficacy and safety of anti-ICAM-1 (Enlimomab) compared to placebo administered within 6 h of the onset of stroke symptoms, for treatment of acute ischemic stroke," is sponsored by Boehringer-Ingelheim Pharmaceuticals (Ridgefield, CT). We recently presented results of this 600 patient study, conducted in the United States and Europe, at the American Academy of Neurology Meeting (The Enlimomab Acute Stroke Trial Investigators, 1997). Six hundred twenty-five patients were enrolled with primary endpoints consisting of functional outcome at 3 mo. Unfortunately, patients treated with Enlimomab were significantly less likely to have a good functional outcome, compared to placebo patients. Patients treated with Enlimobab also had a higher mortality rate (22 vs 16%) compared to placebo and had a higher incidence of fever (51 vs 27%) and total infections (55 vs 42%). The reason why this clinical trial produced an adverse treatment effect with anti-ICAM therapy in contrast to the multiple animal studies, which produce a beneficial treatment effect, is still unknown. Possible reasons being investigated include that this trial used a mouse MAb treatment for 5 d. This foreign antibody may actually have increased the inflammatory response, which counteracted the desired leukocyte-blocking effects. Part of the adverse treatment effect may have also been explained by the increased incidence of fever in patients treated with Enlimomab. In this trial, fever was significantly associated with a poor outcome, and since a higher percentage of Enlimomab-treated patients had fevers, this could have been a mechanism for some, although not all of the adverse effects seen. The results of this clinical trial suggest that the leukocyte adhesion system is not as simple as we first had hoped. Further work using nonimmunogenic molecules or smaller site-specific antiadhesion therapies will be needed to fully investigate the potential role of anti-adhesion strategies in the CNS.

6. SUMMARY

The above studies support the deleterious role of leukocytes in CNS ischemia and suggest that mAbs directed against CD-18 or ICAM-1 may both be able to reduce CNS ischemic injury. ICAM-1 may have the most clinical potential, given that it appears less likely to promote concurrent infections. These agents appear to be most effective in CNS reperfusion models. Consequently, these agents may be of particular benefit in clinical situations associated with reperfusion, i.e., following thrombolysis, especially when rtPA is induced, and some experimental data exists to support this synergistic effect. Whether these agents will ultimately prove to be clinically useful awaits the results of the anti-ICAM clinical stroke trial and future anti-ICAM/tPA trials. However, given the strong preclinical support for their usefulness, it is likely that they will be part of a stroke cocktail that will be employed in future stroke therapeutic efforts.

ACKNOWLEDGMENTS

The authors wish to thank Sandi Hungerford for her assistance in the preparation of this manuscript.

REFERENCES

Ames, A., Wright, R. L., Kowada, M., Thurston, J. M., and Majno G. (1968) Cerebral ischemia: the no-reflow phenomenon. *Am. J. Pathol.* **52,** 437–453.

Anderson, D. C. and Springer, T. A. (1987) Leukocyte adhesion deficiency: an inherited defect in the Mac-1, LFA-1, and p150,95 glycoproteins. *Ann. Rev. Med.* **38,** 175.

Arfors, K., Lundberg, C., Lindbom, L., Lundberg, K., Beatty, P. G., and Harlan, J. M. (1987) A monoclonal antibody to the membrane glycoprotein complex CD-18 inhibits PMN accumulation and plasma leakage in vivo. *Blood* **69,** 338–340.

Argenbright, L., Letts, L., and Rothlein, R. (1991) Monoclonal antibodies to the leukocyte membrane CD18 glycoprotein complex and to intercellular adhesion molecule-1 inhibit leukocyte-endothelial adhesion in rabbits. *J. Leukocyte Biol.* **49,** 253–257.

Barone, F. C., Schmidt, D. B., Hillegass, L. M., Price, W. J., White, R. F., Feuerstein, G. Z., et al. (1992) Reperfusion increases neutrophils and leukotriene B_4 Receptor Binding in rat focal ischemia. *Stroke* **23,** 1337–1348.

Barton, R. W., Rothlein, R., Ksiazek, J., and Kennedy, C. (1989) The effect of anti-intercellular adhesion molecule-1 on phorbol ester induced rabbit lung inflammation. *J. Immunol.* **142,** 1278–1282.

Bednar, M. M., Raymond, S., McAuliffe, T., Lodge, P. A., and Gross, C. E. (1991) The role of neutrophils and platelets in a rabbit model of thromboembolic stroke. *Stroke* **22,** 44–50.

Bowes, M. P., Rothlein, R., Fagan, S. C., and Zivin, J. A. (1995) Monoclonal antibodies preventing leukocyte activation reduce experimental neurologic injury and enhance efficacy of thrombolytic therapy. *Neurology* **45,** 815–819.

Bowes, M. P., Zivin, J. A., and Rothlein, R. (1993) Monoclonal antibody to the ICAM-1 adhesion site reduces neurological damage in a rabbit cerebral embolism stroke model. *Exp. Neurol.* **119,** 215–219.

Chen, H., Chopp, M., Zhang, R. L., Bodzin, G., Chen, Q., Rusche, J. R., and Todd 3rd., R. F. (1994) Anti-CD11b monoclonal antibody reduces ischemic cell damage after transient focal cerebral ischemia in rat. *Ann. Neurol.* **35,** 458–463.

Chopp, M., Zhang, R. L., Chen, H., Li, Y., Jiang, N., and Rusche, J. R. (1994) Postischemic administration of an anti-Mac-1 antibody reduces ischemic cell damage after transient middle cerebral artery occlusion in rats. *Stroke* **25,** 869–876.

Clark, W. M., Calcagno, F. A., and Gabler, W. L. (1994) Reduction of central nervous system reperfusion injury in rabbits using doxycycline treatment. *Stroke* **25,** 1411–1416.

Clark, W., Coull, B., Corliss, L., Beamer, N., Austin, T., and de Garmo, P. (1992) Role of leukocyte adhesion in clinical stroke. *J. Stroke Cerebrovasc. Dis.* **2,** 80–84.

Clark, W. M., Lauten, J. D., Lessov, N., Woodward, W., and Coull, B. (1995a). The influence of anti-adhesion therapies on leukocyte subset accumulation in central nervous system ischemia in rats. *J. Mol. Neurosci.* **6,** 43–50.

Clark, W. M., Lauten, J. D., Lessov, N., Woodward, W., and Coull, B. (1995b) Time course of ICAM-1 expression and leukocyte subset infiltration in rat forebrain ischemia. *Mol. Chem. Neuropath.* **26,** 213–230.

Clark, W., Madden, K., Rothlein, R., and Zivin, J. A. (1991a) Reduction of CNS ischemic injury using leukocyte adhesion antibody treatment. *Stroke* **22,** 877–883.

Clark, W., Madden, K., Rothlein, R., and Zivin, J. A. (1991b) Reduction of central nervous system ischemic injury by monoclonal antibody to intercellular adhesion molecule. *J. Neurosurg.* **75,** 623–627.

Clark, W., Walsh, C., Briley, D., and Coull, B. (1993) Neutrophil adhesion in central nervous system ischemia in rabbits. *Brain Behav. Immun.* **7,** 63–69.

Cole, D. J., Drummond, J. C., Ghazal, E. A., and Shapiro, H. M. (1990) A reversible component of cerebral injury as identified by the histochemical stain 2,3,5-triphenyltetrazolium chloride (TTC). *Acta Neuropathol.* **80,** 152–155.

Cosimi, A. B., Conti, D., Delmonico, F., Preffer, F. I., Wee, S. L., Rothlein, R., Faanes, R., and Colvin, R. B. (1990) In vivo effects of monoclonal antibody to ICAM in nonhuman primates with renal allografts. *J. Immunol.* **144,** 4604–4612.

Cromwell, R. M. and Olsson, Y. (1972) Impaired microvascular filling after focal cerebral ischemia in the monkey. *Neurology* **22,** 500–504.

Cromwell, R. M., Olsson, Y., Klatzo, I., and Ommaya, A. (1970) Temporary occlusion of the middle cerebral artery in monkeys. Clinical and pathological observations. *Stroke* **1,** 439–448.

del Zoppo, G. J., Schmid-Schonbein, G., Mori, E., Copeland, B. R., and Chang, C. M. (1991) Polymorphonuclear leukocytes occlude capillaries following middle cerebral artery occlusion and reperfusion in baboons. *Stroke* **22,** 1276–1283.

Dereski, M. O., Chopp, M., Knight, R. A., and Garcia, J. H. (1992) Focal cerebral ischemia in the rat: temporal profile of neutrophil responses. *Neurosci. Res. Comm.* **11,** 179–185.

Dutka, A. J., Kochanek, P. M., and Hallenbeck, J. M. (1989) Influence of granulocytopenia on canine cerebral ischemia induced by air embolism. *Stroke* **20,** 390–395.

Enlimomab Acute Stroke Trial Investigators (1997) The Enlimomab Acute Stroke Trial results: final results. *Neurology* **48,** 270.

Engler, R., Dahlgren, M., Morris, D., Peterson, M. A., and Schmid-Schonbein, G. W. (1986) Role of leukocytes in response to acute myocardial ischemia and reflow in dogs. *Am. J. Physiol.* **251,** 314–322.

Fassbender, K., Mössner, R., Motsch, L., Kischka, U., Grau, A., and Hennerici, M. (1995) Circulating selectin-and immunoglobulin-type adhesion molecules in acute ischemic stroke. *Stroke* **26,** 1361–1364.

Gabler, W. and Creamer, H. (1991) Suppression of human neutrophil function by tetracyclines. *J. Periodont. Res.* **26,** 52–58.

Gabler, W. and Tsukuda, N. (1991) The influence of divalent cations and doxycycline on iodoacetamide-inhibitable leukocyte adherence. *Res. Commun. Chem. Pathol. Pharmacol.* **74,** 131–140.

Garcia, J. H., Liu, K. F., Yoshida, Y., Lian, J., Chen, S., and del Zoppo, G. J. (1994) The influx of leukocytes and platelets in an evolving brain infarct (Wistar rat). *Am. J. Pathol.* **144,** 189–199.

Garcia, J. H., Yoshida, Y., Lian, J., et al. (1993) Experimental occlusion of a middle cerebral artery (MCA) is accompanied by early polymorphonuclear leukocyte infiltration. *Stroke* **24,** 166.

Ginsberg, M. D. and Myers, R. E. (1972) The topography of impaired microvascular perfusion in the primate brain following total circulatory arrest. *Neurology* **22,** 998–1011.

Grøgaard, B., Schürer, L., and Gerdin, B. (1989) Delayed hypoperfusion after incomplete fore-brain ischemia in the rat: the role of polymorphonuclear leukocytes. *J. Cereb. Blood Flow Metab.* **9,** 500–505.

Hallenbeck, J. M. and Dutka, A. J. (1990) Background review and current concepts of reper-fusion injury. *Arch. Neurol.* **47,** 1245–1254.

Hallenbeck, J. M., Dutka, A. J., Tanishima, T., Kochanek, P. M., Kumaroo, K. K., Thompson, C. B., Obrenovitch, T. P., and Contreras, T. J. (1986) Polymorphonuclear leukocyte accumula-tion in brain regions with low blood flow during the early postischemic period. *Stroke* **17,** 246–253.

Harrison, M. J. G., Sedal, L., Arnold, J., and Russell, R. W. (1975) No-reflow phenomenon in the cerebral circulation of the gerbil. *J. Neurol. Neurosurg. Psychiatry* **38,** 1190–1193.
Hernandez, L. A., Grisham, M. B., Twohig, B., Arfors, K. E., Harlan, J. M., and Granger, D. N. (1987) Role of neutrophils in ischemia reperfusion induced microvascular injury. *Am. J. Physiol.* **253,** H699–H703.
Hossmann, K. A., Lechape-Gruter, H., and Hossmann, V. (1973) The role of cerebral blood flow for the recovery of the brain after prolonged ischemia. *Z. Neurol.* **204,** 281–299.
Kishimoto, T. K. (1991) A dynamnic model for neutrophil localization to inflammatory sites. *J. NIH Res.* **3,** 75–77.
Kochanek, P. M. and Hallenbeck, J. M. (1992) Polymorphonuclear leukocytes and monocytes/macrophages in the pathogenesis of cerebral ischemia and stroke. *Stroke* **23,** 1367–1379.
Lindsberg, P. J., Siren, A. L., Feuerstein, G. Z., and Hallenbeck, J. M. (1995) Antagonism of neutrophil adherence in the deteriorating stroke model in rabbits. *J. Neurosurg.* **82,** 269–277.
Matsuo, Y., Onodera, H., Shiga, Y., Shozuhara, H., Ninomiya, M., Kihara, T., et al. (1994) Role of cell adhesion molecules in brain injury after transient middle cerebral artery occlusion in the rat. *Brain Res.* **656,** 344–352.
Mileski, W. J., Sikes, P., and Atiles, L. (1993). Inhibition of leukocyte adherence and susceptibility to infection. *J. Surg. Res.* **54,** 349–354.
Mori, E., del Zoppo, G. J., Chambers, J. D., Copeland, B. R., and Arfors, K. E. (1992) Inhibition of polymorphonuclear leukocyte adherence suppresses no-reflow after focal cerebral ischemia in baboon. *Stroke* **23,** 712–718.
Price, T., Beaty, B., and Corpuz, S. (1987) In vivo inhibition of neutrophil function in the rabbit using monoclonal antibody to CD-18. *J. Immunol.* **139,** 4174–4177.
Schmid-Schonbein, G. (1987) Capillary plugging by granulocytes and the no-reflow phenomenon in the microcirculation. *FASEB J.* **46,** 2397–2401.
Schmid-Schonbein, G. and Engler, R. (1986) Granulocytes as active participants in acute myocardial ischemia and infarction. *Am. J. Cardiovasc. Pathol.* **1,** 15–29.
Sharar, S. R., Winn, R. K., Murray, C. E., Harlan, J. M., and Rice, C. L. (1991) A CD18 monoclonal antibody increases the incidence and severity of subcutaneous abscess formation after high-does staphylococcus aureus injection in rabbits. *Surgery* **110,** 213–220.
Simpson, P. J., Todd 3rd, R. F. Fantone, J. C., Mickelson, J. K., Griffin, J. D., and Lucchesi, B. R. (1988) Reduction of experimental canine myocardial reperfusion injury by a monoclonal antibody that inhibits leukocyte adhesion. *J. Clin. Invest.* **81,** 624–629.
Smith, C. W., Martin, S., and Rothlein, R. (1989) Cooperative interactions of LFA1 and MAC1 with intracellular adhesion molecule in facilitating adherence and transendothelial migration of human neutrophils in vitro. *J. Clin. Invest.* **83,** 2008–2017.
Smith, C. W., Rothlein, R., Hughes, B., Mariscalco, M. M., Rudloff, H. E., Schmalstieg, F. C., and Anderson, D. C. (1988) Recognition of an endothelial determinant for CD 18-dependent human neutrophil adherence and transendothelial migration. *J. Clin. Invest.* **82,** 1746–1756.
Takeshima, R., Kirsch, J. R., Koehler, R. C., Gomoll, A. W., and Traystman, R. J. (1992) Monoclonal leukocyte antibody does not decrease the injury of transient focal cerebral ischemia in cats. *Stroke* **23,** 247–252.
Todd, R. F. III, and Freyer, D. R. (1988) The CD11/CD18 leukocyte glycoprotein deficiency. *Hematol. Oncol. Clin.* **2,** 13.
Vedder, N., Winn, R. K., Rice, C. L., Chi, E. Y., Arfors, K. E., and Harlan, J. M. (1988) A monoclonal antibody to the adherence promoting leukocyte glycoprotein, CD-18, reduces organ injury and improves survival from hemorrhagic shock and resuscitation in rabbits. *J. Clin. Invest.* **81,** 939–944.
Wang, P., Kao, M., and Wand, S. (1993) Leukocyte infiltration in acute hemispheric stroke. *Stroke* **24,** 236–240.

Wegner, C. D., Gundel, R. H., Reilly, P., Haynes, N., Letts, L. G., and Rothlein, R. (1990) Intercellular adhesion molecule-1(ICAM-1) in the pathogenesis of asthma. *Science* **247,** 456–459.

Williams, T. J., and Hellewell, P. G. (1992) Endothelial cell biology: adhesion molecules involved in the microvascular inflammatory response. *Am. Rev. Respir. Dis.* **146,** 45–50.

Zea-Longa, E., Weinstein, P. R., Carlson, S., and Cummins, R. (1989) Reversible middle cerebral artery occlusion without cranectomy in rats. *Stroke* **20,** 84–91.

Zhang, R. L., Chopp, M., Jiang, N., Tang, W. X., Prostak, J., Manning, A. M., and Anderson, D. C. (1995) Anti-intercellular adhesion molecule-1 antibody reduces ischemic cell damage after transient but no permanent middle cerebral artery occlusion in the wistar rat. *Stroke* **26,** 1438–1443.

Zhang, R. L., Chopp, M., Li, Y., Zaloga, C., Jiang, N., Jones, M. L., Miyasaka, M., and Ward, P. A. (1994) Anti-ICAM-1 antibody reduces ischemic cell damage after transient middle cerebral artery occlusion in the rat. *Neurology* **44,** 1747–1751.

16
Neuroprotective Strategies Based on Targeting of Postreceptor Signaling Events

Mark P. Mattson

1. INTRODUCTION

Other chapters in this book have considered how some of the major signaling systems keep neurons alive and functioning properly, and how these systems may be involved in various neurodegenerative conditions. This last chapter is a signal transduction hodge-podge, reviewing some of the same neuroprotective signaling pathways discussed in other chapters, as well as presenting novel neuroprotective signaling mechanisms not considered elsewhere. It emphasizes discussion of postreceptor molecular targets for possible therapeutic intervention. The molecular complexity of signaling pathways in the nervous system is greater than in any other organ system, and it is therefore predictable that there should be a relatively large number of signaling mechanisms that impact on neuronal survival. The development of pharmacological interventions for diseases of the nervous system, as in other organ systems, has relied heavily on natural compounds produced by organisms ranging from plants to bacteria to mammals. As will become evident in this chapter, many such natural compounds can be put to therapeutic use in neurodegenerative conditions ranging from stroke to epilepsy to Alzheimer's disease (AD).

Neuroprotective agents can be broadly classified into two categories: those that block or counteract cytotoxic cascades and those that activate neuroprotective pathways. Current efforts in drug discovery for neurodegenerative conditions are focused largely on the former category of neuroprotective agents. Examples include: Na^+-channel blockers for the treatment of severe epileptic seizures (Macdonald and Kelly, 1993); NMDA and AMPA receptor antagonists for ischemic stroke (Kochhar et al., 1988; Ohno et al., 1992; Li and Buchan, 1993); and antioxidants for treating AD, Parkinsonism, and amyotrophic lateral sclerosis (Bowling et al., 1993; Behl et al., 1994; Berry et al., 1994; Goodman and Mattson, 1994a; Goodman et al., 1994). One purpose of bringing together the present collection of articles in one book is to emphasize the second category of neuroprotective agents, those that activate neuroprotective signal transduction pathways, with the hope of fostering new avenues of drug discovery. This last chapter presents some recent data suggesting novel therapeutic targets based on signaling cascades activated by neuroprotective factors, as well as those activated by potentially toxic signals such as reactive oxygen species (ROS) and Ca^{2+}.

From: Neuroprotective Signal Transduction *edited by M. P. Mattson*
Humana Press Inc., Totowa, NJ

The preface to this book attempted to place the broad view of its subject matter in an evolutionary context. When one takes such a view, it becomes clear that many of the same mechanisms of cell injury and repair that occur in the nervous system are also operative in other types of tissues. In this regard, accumulation of free radicals and disruption of ion homeostasis are convergence points for cell death processes in general. Accordingly, many compounds that have proven effective in paradigms of CNS injury are also known to be effective in nonneural systems. On the other hand, the death of nerve cells does not carry the same implications for survival of the organism as does the death of skeletal muscle cells, liver cells, or fibroblasts. Thus, whereas skeletal muscle cells, hepatocytes, and fibroblasts can be replaced through mitosis of surviving neighbors, nerve cells are not replaced. It is my view that (for the latter evolutionary reason), the nervous system has adapted signal transduction pathways that do not promote survival (or that even induce cell death) in mitotic cells to function in survival-promoting capacity in neurons. An emerging example is tumor necrosis factor-α (TNFα), which can damage and kill many types of mitotic cells, but protects CNS neurons against excitotoxic, metabolic, and oxidative insults (*see* Chapter 9). Mirror-image parallels between mechanisms regulating survival of mitotic cells and neurons have also become apparent from studies of cytoskeletal function. For example, anticancer agents that prevent cell division (e.g., the microtubule-stabilizing agent taxol and the microfilament-disrupting agent cytochalasin D) can protect neurons against excitotoxic and oxidative injury (Furukawa and Mattson, 1995a; Furukawa et al., 1995; and *see below*).

2. CONVENTIONAL SIGNALING PATHWAYS

2.1. *Receptor Tyrosine Kinase Pathways*

Many different growth factors that can protect neurons against excitotoxic, metabolic, and oxidative insults engage receptors that are transmembrane proteins with intrinsic tyrosine kinase activity (*see* Lindvall et al., 1994; Mattson and Scheff, 1994; Mattson and Furukawa, 1996, for review). Examples include basic fibroblast growth factor (bFGF), nerve growth factor (NGF), brain-derived neurotrophic factor (BDNF), neurotrophin-3 (NT-3), neurotrophin-4/5 (NT4/5), insulin-like growth factor-1 (IGF-1), and platelet-derived growth factors (PDGFs) (Table 1, Fig. 1). In most cases, ligand binding induces receptor dimerization, which, in turn, stimulates transautophosphorylation of the receptors. Receptor tyrosine phosphorylation causes activation of guanine nucleotide-binding proteins (e.g., members of the Ras family), which are localized at the membrane–cytosol interface. Growth factor receptors activate Ras by bringing together a protein complex consisting of SH2 proteins, a guanine nucleotide exchange factor, and Ras. Activated Ras then transmits its signal through a series of kinases, including Raf, MEK, and mitogen-activated protein (MAP) kinase (Keegan and Halegoua, 1993; Seger and Krebs, 1995). Phosphorylated MAP kinase then translocates to the nucleus, where it activates transcription factors either indirectly (e.g., cyclic AMP [cAMP] response-element binding protein) or directly (e.g., Elk-1). A similar set of cytoplasmic reactor proteins appears to mediate signal transduction of several different receptor tyrosine kinases. These include: adaptor proteins containing SH2 (src homology-2) domains (e.g., Grb2 and Shc); guanine nucleotide exchange factors such as Sos (son of sevenless); and GTP-binding proteins (e.g., Ras). The GTP-binding protein Ras is required for

Table 1
Examples of Compounds That Activate Neuroprotective Signaling Pathways with Potential Clinical Utility in Ischemic and Traumatic Brain Injury, and in Neurodegenerative Disorders

Endogenous ligands	Agents that activate downstream events
Basic fibroblast growth factor	K-252 compounds and staurosporine (stimulates tyrK)
Insulin-like growth factors	Ceramide (activates NFκB)
Nerve growth factor	cGMP and cAMP analogs
Brain-derived neurotrophic factor	Phosphodiesterase inhibitors
Neurotrophin-3	Potassium channel openers
Neurotrophin-4/5	Cytochalasins
Transforming growth factor-β	Taxol
Platelet-derived growth factors	Antioxidants
Secreted forms of βAPP	Vitamin E (α-tocopherol)
Glia-derived neurotrophic factor	Tert-phenyl-butyl-nitrone
Tumor necrosis factors	Nordihydroguaiaretic acid
Ciliary neurotrophic factor	Glutathione, *N*-acetyl-cysteine
Protease nexin-1	Propyl gallate
Trophic factor inducers	Estrogens
Nabeshima	
Selegiline	

See text and references cited therein for primary data on which this table was based.

signaling downstream events in response to activation of several different neurotrophic factor receptors, because Ras-neutralizing Fab fragments blocked the abilities of NGF, LIF, and CNTF to promote neuronal survival in sympathetic ganglion cell cultures (Nobes and Tolkovsky, 1995). In the latter study, the Ras-neutralizing antibody fragments also blocked cAMP-induced cell survival, suggesting that cAMP may somehow signal via the receptor tyrosine kinase cascades. In a more recent related study, Rosen and Greenberg (1996) showed that Ca^{2+} influx induced by KCl induces tyrosine phosphorylation of the EGF receptor and association of protein complexes containing Grb2, Shc, and Sos. These findings may explain the well-known abilities of KCl and cAMP to promote survival of many different types of neurons in cell culture, and suggest that receptor tyrosine kinase cascades may be convergence points for the neurotrophic activities of a variety of activity-dependent second messenger pathways (Fig. 1).

In addition to transcription factors, MAP kinases phosphorylate a variety of substrates that likely mediate rapid responses to growth factors (e.g., microtubule-associated proteins). The events resulting from activation of receptor tyrosine kinase cascades that confer cellular resistance to various insults are beginning to be elucidated; they include stimulation of the expression of antioxidant enzymes and modulation of proteins involved in regulation of ion homeostasis (Fig. 1). For example, bFGF induced an increase in Cu–Zn-SOD and glutathione reductase activities, BDNF induced glutathione peroxidase activity, and NGF increased catalase activity in rat hippocampal cell cultures (Mattson et al., 1995a). PDGFs induced increases in superoxide dismutases, catalase, and glutathione peroxidase in embryonic rat hippocampal cell cultures (Cheng and

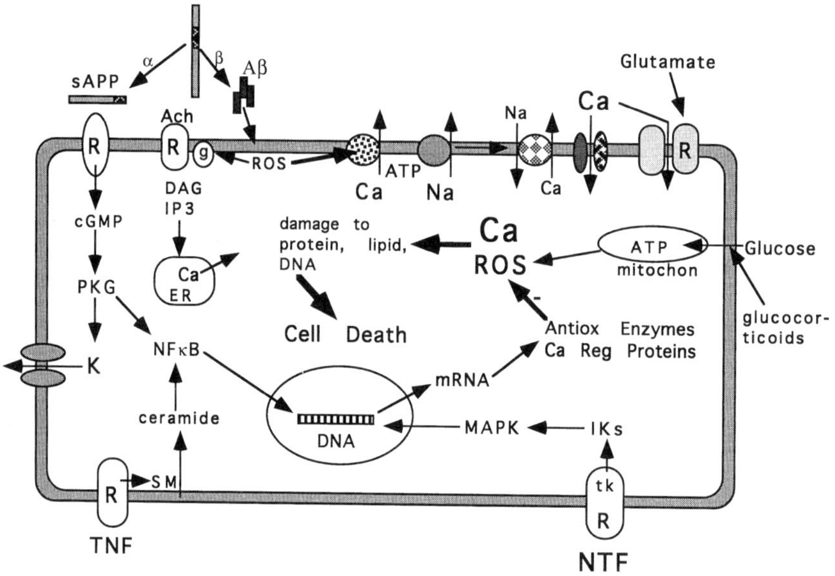

Fig. 1. Overview of mechanisms of neuronal cell injury and some cellular signaling pathways that protect therefrom. Two major convergence points for induction of cell injury are excessive accumulation of calcium and ROS. Several conditions can promote disruption of ion homeostasis and elevation of $[Ca^{2+}]_i$, including: the neurotransmitter glutamate, which induces calcium influx through receptor channels and voltage-dependent calcium channels (VDCC); reduction in energy availability (e.g., hypoglycemia or ischemia), which depletes ATP, resulting in failure of ATP-dependent ion extrusion mechanisms, including the plasma membrane Na^+/K^+- and Ca^{2+}-ATPases; accumulation of the amyloid β-peptide (Aβ), which can induce lipid peroxidation, impair ion-motive ATPases, and increase neuronal vulnerability to excitotoxicity. Elevation of $[Ca^{2+}]_i$ and ROS both contribute to damage to proteins, lipids, and DNA, and ultimately to cell death. Membrane oxidation induced by Aβ, for example, can impair transduction pathways for receptors linked to GTP-binding proteins (g) such as muscarinic acetylcholine receptors (Ach R). Neurotrophic factors (NTF) that activate receptor tyrosine kinases (Trk) induce a cascade of phosphorylation reactions involving intermediate kinases (IK), mitogen-activated protein kinases (MAPK), and transcription factors. Genes whose expression is induced by NTFs include proteins involved in regulation of $[Ca^{2+}]_i$ (e.g., calcium-binding proteins) and antioxidant enzymes (e.g., superoxide dismutases and catalase). Binding of tumor necrosis factor (TNF) to its receptor (R) induces hydrolysis of membrane-associated sphingomyelin, release of ceramide, and activation of the transcription factor NFκB. NFκB may regulate expression of neuroprotective gene products. Neuronal activity induces release of secreted forms of βAPP (sAPP), which can activate putative cell surface receptors linked to cGMP production. The cGMP then activates a cGMP-dependent protein kinase (PKG), which can activate potassium (K) channels, perhaps by inducing activation of a protein phosphatase that dephosphorylates the channel protein. By hyperpolarizing the membrane, sAPP can counteract the depolarizing actions of glutamate, energy failure, and Aβ.

Mattson, 1995). Basic FGF induced expression of the 28 kDa Ca^{2+}-binding protein calbindin (Collazo et al., 1992) and suppressed the expression of the glutamate-binding subunit of an NMDA-receptor complex in cultured hippocampal neurons (Mattson et al., 1993a).

Many different growth factors that signal via receptor tyrosine kinases have been reported to protect neurons against excitotoxic, metabolic, and oxidative insults, both in vitro and in vivo (*see* other chapters in this volume, and Mattson and Furukawa, 1996, for review). Continuous intraventricular infusion of BDNF protected hippocampal CA1 neurons against ischemic injury in adult rats (Beck et al., 1994). As with other neurotrophin receptors, the BDNF receptor tyrosine kinase (Trk B) appears to activate a pathway leading to MAP kinase activation. Because levels of several different ligands for receptor tyrosine kinases (e.g., NGF, BDNF, bFGF, IGF-1) are rapidly increased following brain injuries, it is likely that they activate neuroprotective pathways that help to limit the scope of cell injury and death. Indeed, Hu and Wieloch (1994) provided evidence for activation of receptor tyrosine kinases and MAP kinases in rat brain, following transient cerebral ischemia. There is also evidence that brain injury induces upregulation of receptor tyrosine kinases, including the bFGF receptor (Takami et al., 1993) and the high-affinity BDNF receptor (Merlio et al., 1993). Although activation of receptor tyrosine kinases appears to represent a common neuroprotective signaling mechanism in the nervous system, tyrosine phosphorylation *per se* is not necessarily beneficial. For example, Kindy (1993) reported that tyrosine kinase inhibitors protected hippocampal neurons in gerbils against delayed neuronal death following ischemia. Events downstream of tyrosine phosphorylation events may also prove useful targets for neuroprotective therapy. For example, compounds that activate or inhibit specific transcription factors are being identified (Butt and Karathanasis, 1995).

Protein kinases have proven to be targets for many different naturally occurring compounds, and, in that regard, cascades initiated by receptor tyrosine kinases are amenable to pharmacological manipulation. However, as with most drug targets, it is much easier to identify inhibitors/antagonists than activators/agonists. Accordingly, several different tyrosine kinase inhibitors have been identified, including genistein, lavendustin-A, and the tyrphostins (Levitzki and Gazit, 1995). Such tyrosine kinase inhibitors are useful for establishing whether tyrosine phosphorylation is involved in a particular growth factor response, but they suppress activation of neuroprotective signaling pathways. Neuroprotective compounds, on the other hand, should stimulate the neurotrophic factor signaling mechanisms. One class of such tyrosine kinase activators was recently identified and includes the bacterial alkaloids K252a, K252b, and staurosporine (Cheng et al., 1994a; Goodman and Mattson, 1994b). In the latter studies, it was shown that concentrations of the alkaloids in the range of 10–1000 p*M* protected cultured hippocampal neurons against glucose deprivation-induced injury, excitotoxicity, and oxidative insults ($FeSO_4$ and amyloid β-peptide). Although much higher concentrations of K252a and staurosporine can inhibit several different kinases, the lower neuroprotective concentrations induced tyrosine phosphorylation of several different proteins, including those with mol wt essentially identical to MAP kinases and neurotrophic factor receptors (Cheng et al., 1994a). The increases in $[Ca^{2+}]_i$ and free radicals normally induced in neurons by excitotoxic and oxidative insults were significantly attenuated in hippocampal cultures pretreated with K252a, K252b, or staurosporine. One or more of these compounds were reported to protect brain cells in vivo against ischemic injury (Hara et al., 1990; Ohno et al., 1991) and excitotoxic insults (Nabeshima et al., 1991; Roberts-Lewis and Siman, 1993; Smith-Swintosky et al., 1996a).

Finally, neuroprotective compounds are being identified that appear to act by inducing production of endogenous neurotrophic factors. For example, Nabeshima et al. (1994) have shown that idebenone and propentofylline induce NGF production and prevent cognitive dysfunction in basal forebrain-lesioned rats. Semkova et al. (1996) recently reported that selegiline induces NGF production in cultured cortical astrocytes and in intact rat cerebral cortex following intraperitoneal administration in vivo. Moreover, selegiline protected cortical tissue from ischemic injury induced by middle cerebral artery occlusion.

2.2. *Cyclic Nucleotide Pathways*

cAMP is produced upon activation of cell surface receptors linked to the enzyme adenylyl cyclase; cyclic GMP (cGMP) is generated when soluble or membrane-receptor guanylyl cyclases are activated (Fig. 2). The fact that cAMP and cGMP mediate neurotrophic and neuroprotective signal transduction pathways has been known for some time. In sympathetic neurons, membrane-permeant cAMP analogs protect against cell death induced by NGF withdrawal (Rydel and Greene, 1988). In hippocampal cell, cultures, dibutyryl cAMP and forskolin (an activator of adenylate cyclase) protect neurons against glutamate toxicity (Mattson and Kater, 1988). Rolipram, a cAMP-selective phosphodiesterase inhibitor, protects hippocampal CA1 neurons against injury induced by transient forebrain ischemia in adult gerbils (Kato et al., 1995).

Alterations in signaling pathways that employ cAMP have been reported in several neurodegenerative disorders. For example, Gibson and co-workers (Huang and Gibson, 1993) showed that cAMP production in response to β-adrenergic receptor stimulation in cultured fibroblasts from AD patients was reduced to 20% of the levels observed in control fibroblasts. Cerebellar granule cells in slice cultures were protected against excitotoxicity by cGMP (Garthwaite and Garthwaite, 1988). cGMP analogs protected cultured rat hippocampal neurons against glucose deprivation-induced excitotoxic injury (Barger et al., 1995a). cGMP appears to mediate the excitoprotective activity of the secreted form of β-amyloid precursor protein (sAPPα) in hippocampal neurons because sAPPα increased cGMP levels, a cGMP-dependent protein kinase inhibitor blocked the neuroprotective action of sAPPα, and 8-bromo-cGMP mimicked the excitoprotective actions of sAPPα (Barger et al., 1995a). In addition, cGMP appears to mediate the excitoprotective action of somatostatin in cultured rat cortical neurons (Forloni et al., 1996). cGMP may exert its excitoprotective actions by inducing activation of K^+ channels (Furukawa et al., 1996; and *see* Chapter 1). Indeed, K^+ channel activators, such as diazoxide and levocromakalim, are effective in protecting neurons against metabolic, excitotoxic, and oxidative insults (Goodman and Mattson, 1996b).

The production of cGMP in response to sAPPα appears to be mediated by a membrane-associated guanylate cyclase, rather than by nitric oxide (NO)-mediated activation of soluble guanylate cyclase, because sAPPα-induced cGMP production in membranes isolated from early postnatal neocortical cells and inhibitors of nitric oxide synthase (NOS; e.g., *N*-methylarginine) did not block sAPPα-induced cGMP production in intact cultured hippocampal neurons (Barger and Mattson, 1995a). In cultured PC12 cells and sympathetic neurons, cGMP analogs and the cGMP phosphodiesterase inhibitor dipyridamole delayed cell death following withdrawal of trophic factors (Farinelli et al., 1996). In the latter systems, agents that induced NO production

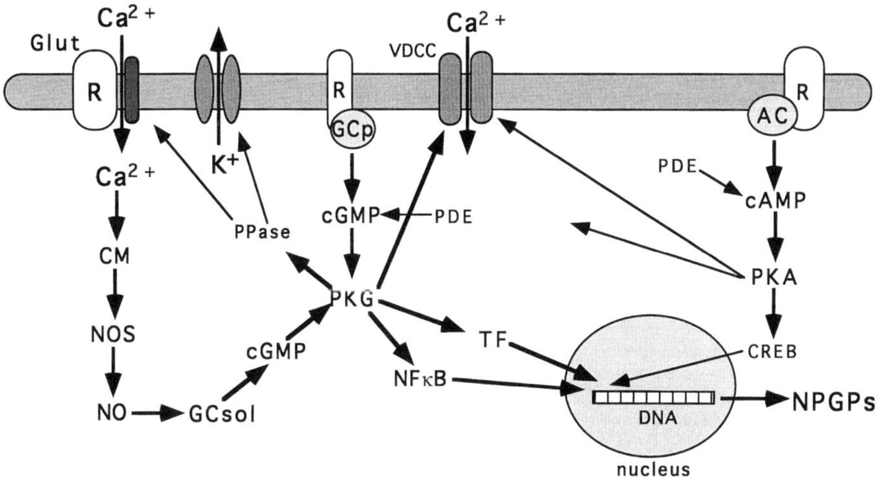

Fig. 2. Cyclic nucleotide neuroprotective signaling mechanisms. cGMP is produced in neurons via two pathways. In one pathway, elevation of $[Ca^{2+}]_i$ results in calmodulin (CM)-mediated activation of the cytosolic enzyme nitric oxide synthase (NOS) and nitric oxide (NO) production. NO then activates a soluble guanylyl cyclase (GCsol), resulting in cGMP production from GTP. In the other pathway, ligands bind to cell surface receptors with intrinsic guanylyl cyclase activity in a cytoplasmic domain (particulate guanylyl cyclase: GCp). cGMP then activates a cGMP-dependent protein kinase (PKG). PKG may mediate neuroprotective actions in several ways, including: PKG activates a protein phosphatase (PPase), which dephosphorylates and thereby activates K^+ channels, resulting in reduced membrane excitability and decreased vulnerability to excitotoxicity; direct or indirect (via an intervening PPase) modulation of glutamate (Glut) receptor channels or voltage-dependent calcium channels (VDCC); activation of the transcription factor NFκB or other transcription factors (TF), which induce the expression of neuroprotective gene products (NPGPs). cAMP is produced as the result of activation of receptors linked to adenylyl cyclase (AC), an enzyme that converts ATP to cAMP. cAMP may rapidly modulate Ca^{2+} influx by acting on glutamate receptor channels or VDCC, or it may activate transcription-dependent neuroprotective pathways involving the cAMP responsive element binding protein (CREB). cGMP and cAMP are inactivated by specific phosphodiesterases (PDE); agents that modulate PDE activity may therefore influence levels of activation of the neuroprotective cyclic nucleotide pathways.

(e.g., sodium nitroprusside and S-nitroso-*N*-acetylpenicillamine) also delayed neuronal death, and an inhibitor of soluble guanylate cyclase LY 83583 blocked promotion of cell survival following trophic factor withdrawal, indicating that cGMP mediated the survival-promoting actions of NO in these systems.

Approaches to treating neurodegenerative disorders based on cyclic nucleotide pathways have, perhaps, been understudied. A variety of types of compounds are available to manipulate both cAMP and cGMP pathways. Several different phosphodiesterase inhibitors have been identified and studied, mainly from the standpoint of their possible uses in immunological disorders (Murray and England, 1992). However, such compounds may have neuroprotective activities, as suggested by data showing that dipyridamole protects sympathetic neurons against NGF withdrawal (Farinelli et al.,

1996). In addition, agents that affect downstream targets of cyclic nucleotides are attractive neuroprotective agents. For example, cGMP induces activation of K^+ channels in cultured hippocampal neurons, and K^+ channel activators, such as diazoxide, are very effective neuroprotective agents (Goodman and Mattson, 1996b).

2.3. *Inositol Phospholipid Pathways*

A variety of intercellular signals activate receptors linked to GTP-binding proteins that are, in turn, coupled to phospholipases that induce hydrolysis of phosphatidylinositol *bis*-phosphate (PIP_2) (*see* Nishizuka, 1995, for review). This class of receptors, which includes muscarinic acetylcholine receptors and a subtype of glutamate receptors, are referred to as metabotropic receptors, to contrast them with ionotropic receptors, which are ligand-activated ion channels (Fig. 1). Two important intracellular messengers cleaved from the PIP_2 are diacylglycerol and inositol trisphosphate (IP_3). Diacylglycerol activates the Ca^{2+}-dependent enzyme protein kinase C (PKC); IP_3 activates a channel in the smooth endoplasmic reticulum that induces Ca^{2+} release from that organelle. Neurons in many different brain regions receive cholinergic input (e.g., hippocampal pyramidal neurons and neocortical neurons throughout the brain), and activation of such cholinergic pathways may modify neuronal vulnerability to excitotoxicity under conditions of metabolic compromise and/or oxidative stress. Initial studies showed that activation of muscarinic cholinergic receptors in rat hippocampal cell cultures enhanced excitotoxicity induced by glutamate (Mattson, 1989). Subsequent cell culture and in vivo studies support the cholinergic excitotoxicity hypothesis. For example, Olney and coworkers provided evidence that acetylcholine makes a major contribution to the excitotoxic death of cingulate gyrus neurons in adult rats exposed to phencyclidine (*see* Olney, 1994, for review). Moreover, muscarinic cholinergic receptor antagonists reduce neuronal damage induced by traumatic brain injury in adult rats (*see* Hayes et al., 1992, for review). Other PIP_2-linked signaling systems of relevance to mechanisms of neuronal degeneration in both acute and chronic neurodegenerative disorders are those activated by glutamate and thrombin (Cunningham et al., 1993; Schoepp and Conn, 1993; O'Neill et al., 1994; Savolainen et al., 1994; Smith-Swintosky et al., 1995). Although signals that activate the PIP_2 pathway can potentiate neurodegeneration, they may in some cases exert neuroprotective actions. Activation of metabotropic glutamate receptors has been reported to reduce neuronal vulnerability to excitotoxicity (Koh et al., 1991). Gereau and Conn (1994) reported that activation of metabotropic glutamate receptors potentiated cAMP responses and suppressed synaptic transmission in hippocampus; they provided evidence that this mechanism was independent of kinase activation. On the other hand, Aniksztejn et al. (1991) and O'Connor et al. (1994) showed that activation of metabotropic glutamate receptors enhanced responses of hippocampal neurons to NMDA, and Gerber et al. (1992) showed that metabotropic receptor activation resulted in suppression of K^+ currents in hippocampal pyramidal neurons; these consequences of metabotropic glutamate receptor activation would be expected to increase neuronal excitability and hence vulnerability to excitotoxicity. As described in Chapters 12 and 13 in this volume, thrombin can increase neuronal vulnerability to excitotoxicity and oxidative insults. These neurotoxic actions of thrombin are mediated by receptors linked to Ca^{2+} release from intracellular stores (Smith-Swintosky et al., 1995; Mattson and Begley, 1996).

Both PKC activation and IP_3-induced Ca^{2+} release may contribute to the neurodegenerative consequences of activation of PIP_2-linked receptors. PKC inhibitors (e.g., H-7) protect cultured human cortical neurons against excitotoxicity and prevent neurofibrillary tangle-like alterations in the cell cytoskeleton (Mattson, 1991). Downregulation of PKC by prior exposure to phorbol ester activators of this enzyme results in increased resistance of cultured cerebellar granule cells to excitotoxicity (Favaron et al., 1990). Activation of the PIP_2 pathway may also promote neuronal degeneration as the result of Ca^{2+} release from intracellular stores. Inhibitors of Ca^{2+} release from IP_3-sensitive pools (e.g., dantrolene) were reported to afford protection against excitotoxicity in cortical cell cultures (Frandsen and Schousboe, 1993). Moreover, Chiou and Hong (1995) reported that several trimethoxybenzoate compounds, which are known to inhibit Ca^{2+} release from endoplasmic reticulum stores, protected cortical neurons against focal ischemic injury in adult rats.

Alterations in PIP_2 and related signaling pathways have been reported in several different neurodegenerative disorders. Huang et al. (1995) found that IP_3 accumulation in response to bradykinin was significantly increased in fibroblasts from Alzheimer's patients, compared to fibroblasts from age-matched controls. Alterations in PKC activities and subcellular localization were reported to occur in association with neurofibrillary tangles and neuritic plaques in AD (*see* Jin and Saitoh, 1995, for review). Levels of PKC and IP_3 were reduced in striatum from patients with Huntington's disease, but were unaltered in basal ganglia tissues from patients with Parkinson's disease (Tanaka et al., 1993).

2.4. *Sphingomyelin–Ceramide–NFκB Pathway*

A very intriguing signal transduction pathway that appears to play important roles in cellular responses to brain injury involves the nuclear transcription factor NFκB (Mattson et al., 1997; and *see* Chapter 11). NFκB (in the inactive form) consists of three subunits localized in the cytosolic compartment, namely, p50, p65, and IκB (*see* Liou and Baltimore, 1993, for review). Activation of NFκB involves dissociation of IκB from the p50–p65 (transcription factor) dimer, and can be induced by several different stimuli, including cytokines, such as tumor necrosis factor-α (TNFα), and exposure to reactive oxygen species (Fig. 3). The signal transduction pathway by which TNFα activates NFκB involves binding to a receptor called p55 (or TNFR-1). The p55 receptor is linked to activation of an acidic sphingomyelinase that cleaves membrane-associated sphingomyelin, resulting in the release of ceramide (Kolesnick and Golde, 1994). Ceramide then activates a kinase that phosphorylates the IκB subunit of NFκB, thereby inducing ubiquitin-dependent degradation of IκB. The p50/p65 dimer then translocates to the nucleus, where it binds specific sites in the enhancer region of NFκB-responsive genes. The manner in which ROS induce NFκB is not completely understood, but it has been proposed that oxidation of IκB renders it vulnerable to proteolytic degradation (Schreck et al., 1991).

NFκB has been implicated as a mediator of cell injury and death in certain types of tumor cells and in nonneural cells (Wong, 1995), but recent studies of the direct actions of TNF on neuronal survival and vulnerability to excitotoxic, metabolic, and oxidative insults suggest that NFκB activates neuroprotective genetic programs (Barger et al., 1995b, 1996; Goodman and Mattson, 1996a). TNF protected cultured rat hippocampal

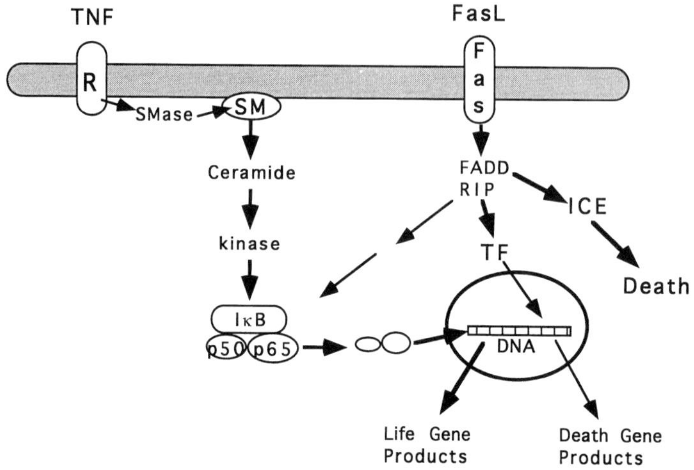

Fig. 3. Signal transduction pathways of tumor necrosis factor (TNF) and Fas ligand. Binding of TNF to the 55 kDa receptor (R) activates a sphingomyelinase, resulting in release of ceramide from membrane-associated sphingomyelin. Ceramide may then activate a kinase that, in turn, phosphorylates IκB, the inhibitory subunit of the NFκB complex. On phosphorylation, IκB dissociates from the p50/p65 protein dimer and the p50/p65 dimer translocates to the nucleus and binds to enhancer DNA in responsive genes. Some genes induced by NFκB encode protein products such as antioxidant enzymes that exert neuroprotective activities (life genes). Binding of Fas ligand to its receptor activates a death domain in the receptor resulting in interaction with other proteins, including Fas-associated protein with death domain (FADD) and receptor-interacting protein (RIP) (*see* Fraser and Evan, 1996, for review). Downstream events in the Fas signaling pathway include activation of interleukin-1β-converting enzyme (ICE) and transcription factors (TF) that may induce expression of death genes or suppress production of life genes.

neurons against glucose deprivation-induced injury, excitotoxicity, and exposure to amyloid β-peptide (Cheng et al., 1994c; Barger et al., 1995b). The latter studies showed that increases of $[Ca^{2+}]_i$ and ROS induced by the different insults were attenuated in neurons pretreated with TNF. The neuroprotective actions of TNF were apparently mediated by the p55 receptor–NFκB pathway because human TNF, which activates only the p55 receptor in rodent cells, was effective (Cheng et al., 1994c; Barger et al., 1995b); TNF-induced NFκB DNA-binding activity in the neurons (Barger et al., 1995b); and the membrane-permeant ceramide analog exhibited concentration-dependent neuroprotective activity (Goodman and Mattson, 1996a). Endogenous TNF, produced in response to brain injury, appears to serve a neuroprotective function in vivo, because neuronal damage induced by ischemia or excitotoxins was exacerbated in mice lacking TNF receptors (Bruce et al., 1996).

Recent findings suggest that the sphingomyelin–ceramide–NFκB pathway may also mediate cell survival responses to NGF. Dobrowsky et al. (1994) reported that NGF induced sphingomyelin hydrolysis in cultured T9 glioma cells, and a membrane-permeant ceramide analog mimicked the effects of NGF on cell differentiation. Moreover, expression of an EGF receptor–p75 NGF receptor chimera resulted in cells in which EGF activated the sphingomyelin pathway.

2.5. *Steroid Signaling Pathways*

In Chapter 14, Sapolsky reviews the considerable evidence that glucocorticoids and estrogens play important roles in modulating neuronal vulnerabilty to excitotoxic, metabolic, and oxidative insults. The latter chapter also described both classic steroid receptor-mediated signaling pathways and signaling pathways that may be initiated at the plasma membrane. This section considers a few additional aspects of steroid actions in the nervous system that may be of relevance to their impact on neuronal survival and function, and on neurodegenerative disorders. One emerging view is that estrogenic steroids can act as antioxidants, and, in this way, protect neurons against the kinds of oxidative insults that are central to many different neurodegenerative conditions. Studies of mechanisms of atherosclerosis have provided strong evidence that estrogens (17β-estradiol and estriol) exhibit antioxidant properties, and that these steroids can suppress oxidation of low-density lipoprotein and protect the vascular endothelium against oxidative damage (Keaney et al., 1994). Cell culture studies showed that estrogens can protect neurons against excitotoxicity and oxidative insults, including exposure to $FeSO_4$ and amyloid β-peptide (Behl et al., 1995; Goodman et al., 1996); (Fig. 4). By directly monitoring lipid peroxidation levels and $[Ca^{2+}]_i$ in the hippocampal neurons, it was demonstrated that 17β-estradiol and estriol suppress lipid peroxidation and disruption of Ca^{2+} homeostasis that otherwise occurs following exposure to glutamate, $FeSO_4$ and amyloid β-peptide (Goodman et al., 1996). The concentrations of estrogens required to acheive these antioxidant-like effects are in the range of 100 n*M* to 10 μ*M*, which are considerably higher than levels normally present in blood. However, because of their hydrophobicity, steroids may accumulate at micromolar levels within cell membranes, which could occur in neurons in the brain.

Glucocorticoids can increase neuronal vulnerability to metabolic, excitotoxic, and oxidative insults (Sapolsky, 1994; Stein-Behrens et al., 1994; Goodman et al., 1996; Smith-Swintosky et al., 1996b). Glucocorticoids may act by suppressing glucose uptake into neurons and thereby promoting ATP depletion (Sapolsky, 1994). Antagonists and agonists of glucocorticoid and estrogen receptors are obvious candidate compounds for therapeutic intervention in both acute and chronic neurodegenerative conditions. In the case of glucocorticoids, several sites of therapeutic intervention are possible. One approach is to prevent glucocorticoid production in adrenocortical cells. Metyrapone, an 11-β-hydroxylase inhibitor, significantly reduces stress-induced elevations of circulating corticosterone levels and protected rat brain cells against injury in two different stroke models (middle cerebral artery occlusion and transient global forebrain ischemia) and a kainic acid seizure model (Smith-Swintosky et al., 1996b).

3. OTHER SIGNALING PATHWAYS

3.1. *Cytoskeleton*

Cytoskeletal alterations in neurons exposed to neurotoxic conditions have been widely reported in cell culture and animal models of both acute and chronic neurodegenerative disorders. For example, ischemic and excitotoxic insults induce microtubule depolymerization, proteolysis of microtubule-associated proteins (e.g., MAP2), and disruption of actin filaments and spectrin (Siman and Noszek, 1988; Stein-Behrens et al., 1994; Geddes et al., 1994). Neurons degenerating in Alzheimer's disease manifest

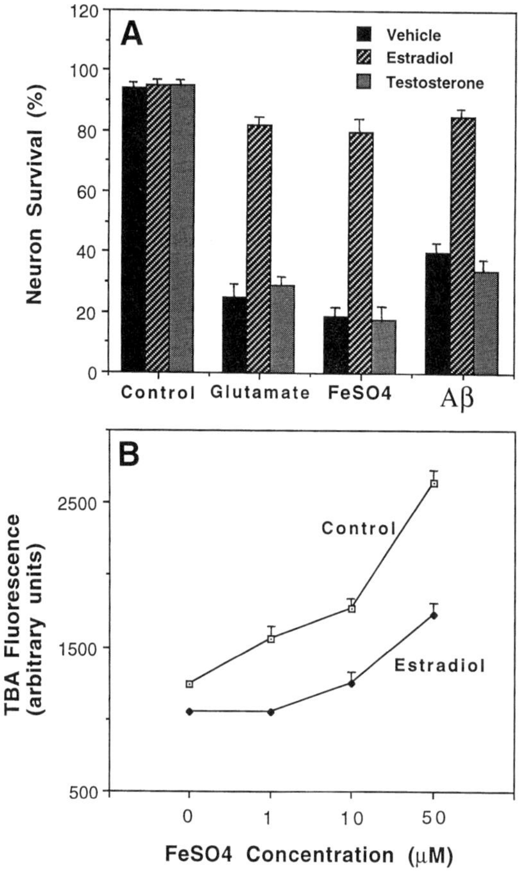

Fig. 4. Estrogens suppress lipid peroxidation and neuronal injury induced by oxidative insults. **(A)** Embryonic rat hippocampal cell cultures were pretreated for 2 h with vehicle, 1 μ*M* 17β-estradiol or 1 μ*M* testosterone. Cells were then exposed to 10 μ*M* glutamate (20 h), 5 μ*M* $FeSO_4$ (6 h), or 5 μ*M* Aβ25–35 (24 h), and neuronal survival was quantified. Values represent the mean and SEM of four separate cultures. **(B)** Membranes prepared from adult rat cortex were preincubated for 30 min with vehicle (Control) or 50 μ*M* 17β-estradiol, and were then exposed for 30 min to the indicated concentrations of $FeSO_4$. Levels of the lipid peroxidation product malondialdehyde were then quantified, using the thiobarbituric acid fluorescence method. Values represent the mean and SEM of 4–6 assays. (Adapted with permission from Goodman et al., 1996.)

microtubule loss and accumulation of abnormal filaments comprised of the microtubule-associated protein tau (Selkoe, 1991). One prevalent interpretation of such cytoskeletal alterations is that they contribute to cell injury and death processes (Trojanowski and Lee, 1995). However, studies that have moved beyond description of the cytoskeletal alterations to address their role in the cell death process have revealed that at least some of the cytoskeletal changes that occur in injured neurons are functionally involved in neuroprotective pathways.

The state of actin polymerization is highly sensitive to $[Ca^{2+}]_i$ so that transient elevation of $[Ca^{2+}]_i$ promotes microfilament formation; prolonged elevation of $[Ca^{2+}]_i$ results

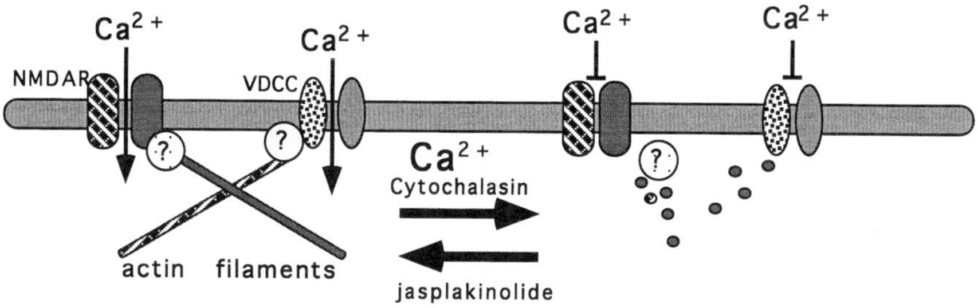

Fig. 5. Actin filaments regulate Ca^{2+} influx and thereby influence neuronal vulnerability to excitotoxicity. Actin filaments are associated with the cytoplasmic face of the plasma membrane, where they may interact, either directly or via intermediary actin binding proteins, with NMDA receptor channels and VDCCs. Depolymerization of actin filaments in response to Ca^{2+} influx, or to actin-depolymerizing agents such as cytochalasins, results in reduced opening of NMDA receptors and VDCC and, accordingly, reduced Ca^{2+} influx. Studies have shown that cytochalasins can protect nurons against excitotoxicity by suppressing Ca^{2+} influx. In contrast, the actin-stabilizing agent jasplakinolide enhances Ca^{2+} influx and increases neuronal vulnerability to excitotoxicity. Data suggest that actin depolymerization in response to Ca^{2+} influx is part of a feedback mechanism to limit further calcium influx and protect neurons from excitotoxicity.

in microfilament depolymerization (Bernstein and Bamburg, 1985; Mattson, 1992). Rapid actin polymerization in response to environmental signals that elevate $[Ca^{2+}]_i$ transiently (e.g., brief exposures to glutamate) mediates responses of growth cones (i.e., filopodial protrusion) and postsynaptic spines to physiologically relevant inputs. On the other hand, sustained elevations of $[Ca^{2+}]_i$ that occur in pathological conditions (e.g., excitotoxicity) result in actin filament depolymerization. We have found that actin depolymerization is mechanistically linked, in a feedback pathway, to a reduction in calcium influx through NMDA receptors and voltage-dependent calcium channels (Furukawa et al., 1995) (Fig. 5). When cultured rat hippocampal neurons are pretreated with the actin-disrupting agent cytochalasin D prior to exposure to glutamate, the $[Ca^{2+}]_i$ responses to glutamate are attenuated and the cells are relatively resistant to excitotoxicity. In addition, cytochalasin D protected cultured hippocampal neurons against the toxicity of amyloid β-peptide (Furukawa and Mattson, 1995a), consistent with a role for calcium influx in the toxicity of that amyloidogenic peptide (Mark et al., 1995). On the other hand, stabilization of actin filaments with jasplakinolide enhances glutamate-induced calcium influx and neurotoxicity (Furukawa et al., 1995). Taken together with patch clamp analyses of the effects of cytochalasins on ion currents through NMDA receptors (Rosenmund and Westbrook, 1993) and voltage-dependent calcium channels (Johnson and Byerly, 1993), these data indicate that actin filaments regulate NMDA receptors and voltage-dependent Ca^{2+} channels. Although thc molecular basis of calcium channel regulation by actin has not been established, it probably involves one or more actin-binding proteins that link actin filaments with membrane components (Sheterline et al., 1995).

Microtubules may also be more than passive bystanders in neurodegenerative processes. It is well known that the microtubule-disrupting agent colchicine can be neurotoxic. However, only recently has the mechanism of colchicine toxicity been elucidated, and, surprisingly, it appears to involve destabilization of Ca^{2+} homeostasis (Furukawa and Mattson, 1995b). Exposure of cultured hippocampal neurons to colchicine resulted in an elevation of $[Ca^{2+}]_i$ that preceded cell degeneration and death. Removal of extracellular Ca^{2+} significantly reduced the neurotoxicity of colchicine. When hippocampal cultures were pretreated with the microtubule-stabilizing agent taxol, they were protected against excitotoxic insults and $[Ca^{2+}]_i$ was stabilized (Furukawa and Mattson, 1995b).

3.2. *Neuropeptides*

Strijbos et al. (1994) reported that a corticotrophin-releasing factor antagonist (α-helical CRF9-41) protected rat brain cells against excitotoxic and ischemic injury in vivo, and provided evidence that these actions were not mediated by glucocorticoids. Baskin et al. (1994) showed that dynorphin and related opioid agonists can reduce ischemic damage to neurons in a cat focal cerebral ischemia model. Dynorphin receptors are coupled to a signal transduction pathway involving activation of a GTP-binding protein linked negatively to adenylate cyclase and to inhibition of voltage-dependent Ca^{2+} channels (Xie et al., 1994). Administration of calcitonin gene-related peptide (CRGP) to adult rats reduced cortical infarct volume following middle cerebral artery occlusion (Holland et al., 1994). CGRP stimulates receptors linked to elevation of cAMP levels and activation of ATP-sensitive K^+ channels in several different cell types, including vascular smooth muscle (Quayle et al., 1994).

4. ROLES FOR CALCIUM IN NEURONAL INJURY AND NEUROPROTECTION

Many different physiological signals and pathological insults cause an elevation of $[Ca^{2+}]_i$ in neurons (*see* Mattson, 1992, for review). Physiological signals that stimulate Ca^{2+} influx through plasma membrane channels and/or Ca^{2+} release from intracellular stores include neurotransmitters, such as glutamate and acetylcholine; neurotrophic factors, including bFGF, NGF, and BDNF; neuropeptides, such as thyrotropin releasing hormone, and electrical activity itself (Miller et al., 1989; Levitan and Kramer, 1990; Cheng et al., 1993; Ghosh and Greenberg, 1995). Several different regulatory mechanisms exist in neurons that are involved in rapidly restoring cytoplasmic free calcium concentration to the (low) rest level, which is typically in the range of 50–100 n*M* (Fig. 6). These include the Ca^{2+}-ATPases of the plasma membrane and smooth endoplasmic reticulum, the plasma membrane Na^+/Ca^{2+} exchanger, and cytosolic Ca^{2+}-binding proteins (e.g., calbindin and parvalbumin). Also of critical importance for proper regulation of $[Ca^{2+}]_i$ are ion channels and pumps that control membrane potential; these include voltage-dependent Na^+ channels, K^+ channels, and the Na^+/K^+-ATPase. By controlling the membrane potential, the latter ion-regulating systems determine whether Ca^{2+} influx occurs through voltage-dependent channels and ligand-gated channels such as the NMDA receptor.

The potential cytotoxicity of Ca^{2+} probably placed evolutionary pressures on neurons that resulted in the development of the kinds of regulatory systems described above. Neurons also appear to have evolved Ca^{2+}-responsive signal transduction pathways that, when activated, initiate an array of transcription-dependent and transcription-independent responses that lead to enhanced calcium homeostasis (i.e., the ability to reduce calcium influx and enhance calcium extrusion or buffering). This concept has gained considerable

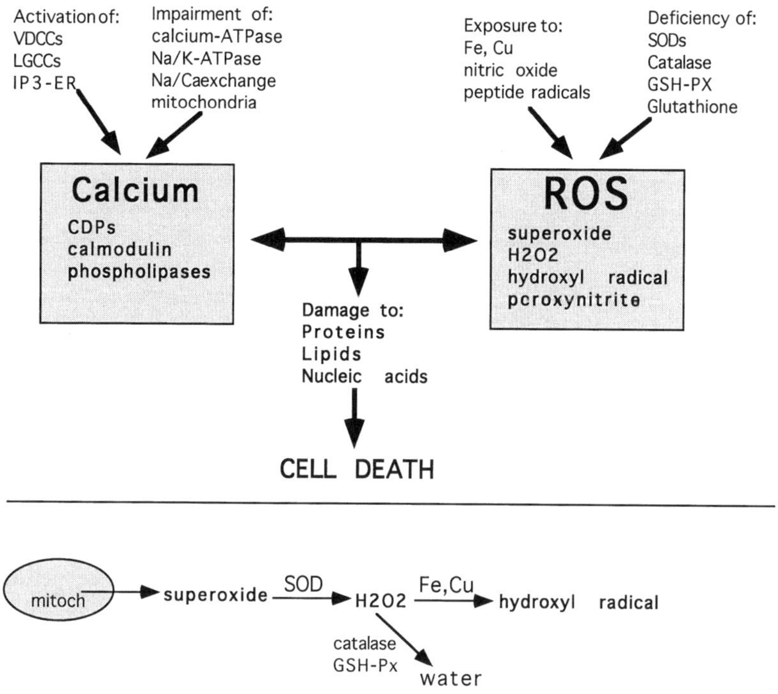

Fig. 6. Interactions of cellular systems involved in the accumulation and removal of Ca^{2+} and ROS in neurons. Elevation of $[Ca^{2+}]_i$ occurs as the result of influx through plasma membrane VDCC and ligand-gated Ca^{2+} channels (LGCC), and via IP_3-induced release from endoplasmic reticulum (ER) stores. Impairment of plasma membrane ion-motive ATPases (Ca^{2+}-ATPase and Na/K-ATPase), reversal of the Na^+/Ca^{2+} exchanger, and impairment of mitochondrial calcium sequestration mechanisms can make very important contributions to elevation of $[Ca^{2+}]_i$ under conditions of metabolic impairment and oxidative stress. Sustained elevation of $[Ca^{2+}]_i$ activates calcium-dependent proteases (CDPs), Ca^{2+} effector proteins such as calmodulin, and phospholipases, all of which can promote damage to cell proteins and membranes. In addition, Ca^{2+} promotes production of various ROSs, including superoxide anion radical and NO. Various oxidative insults induce ROS accumulation in neurons, including Fe^{2+} and Cu^{2+} via the Fenton reaction, and peptide radicals (e.g., amyloid β-peptide) via propagation to the cell membrane. Deficiencies in antioxidant enzymes, such as superoxide dismutases (Cu–Zn- or Mn-SOD), catalase, glutathione peroxidase (GSH-Px), or the thiol antioxidant peptide glutathione, can also promote ROS accumulation and neuronal injury. A major source of ROS is the mitochondria, which produce large amounts of superoxide during the process of respiration. SODs convert superoxide to hydrogen peroxide (H_2O_2). H_2O_2 is normally detoxified by catalase and GSH-Px, which convert it to water. In the presence of Fe^{2+} and Cu^{2+}, however, H_2O_2 is converted to hydroxyl radical, which is highly reactive and promotes membrane lipid peroxidation. Both calcium and ROS are believed to contribute greatly to neuronal injury and death in many, if not all, neurodegenerative conditions.

support from a variety of studies performed during the past 5 yr. Greenberg and coworkers have shown that calcium influx, induced by glutamate or BDNF, can activate MAP kinases and the cAMP response element binding protein (CREB), leading to transcription of genes responsive to this factor (Bading et al., 1993; Ghosh et al., 1994). Activation of MAP kinase pathways is implicated in the trophic and neuroprotective

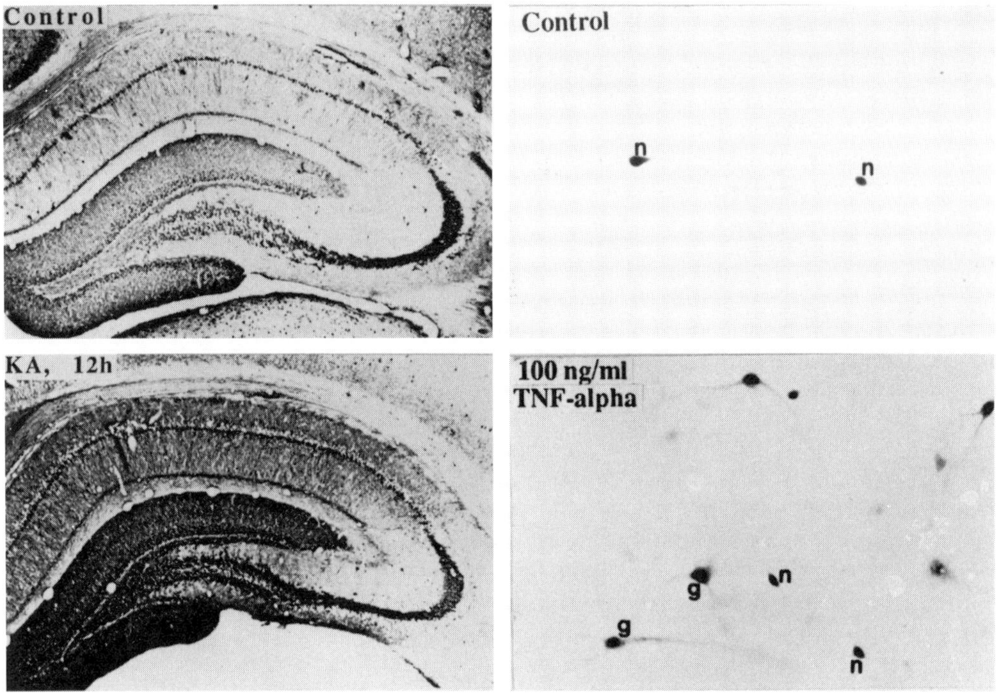

Fig. 7. Induction of the cytoprotective Ca^{2+}-binding protein calbindin D28k by the cytokine TNF, and by neuronal activity. (**Left**) Rats were administered either vehicle (Control) or a convulsant dose of kainic acid (KA; 10 mg/kg intraperitoneally) and were killed 12 h later. Coronal brain sections were immunostained with calbindin antibody, and photographs of the hippocampus were taken under bright-field optics. Note that calbindin immunoreactivity was greatly increased in hippocampal dentate granule and CA1 neurons, and in dendritic fields of those cells, in the rat that received kainic acid. (**Right**) Embryonic rat hippocampal cell cultures were exposed for 2 d to either vehicle (Control) or 100 ng/mL TNFα. The cultures were then immunostained with anticalbindin, and the micrographs show representative bright-field photographs. Note the greater number of calbindin-immunoreactive neurons in the TNFα-treated culture.

actions of several different growth factors, including NGF, BDNF, and bFGF (Lloyd and Wooten, 1992; Nishida and Gotoh, 1993). Trophic factors that stimulate this pathway have been shown to alter levels of expression of glutamate receptor proteins (Mattson et al., 1993a) and Ca^{2+}-binding proteins (Fig. 7) in ways consistent with enhanced Ca^{2+} homeostasis and increased resistance to excitotoxicity and other calcium-mediated insults. We found that bFGF and NGF (Cheng et al., 1993) and TNFα (Furukawa and Mattson, submitted) increase currents through L-type voltage-dependent channels in cultured hippocampal neurons. Paradoxically, these same growth factors protect the same cells against metabolic and oxidative insults (Cheng and Mattson, 1991; Cheng et al., 1994c). When cultured hippocampal neurons are deprived of glucose, their $[Ca^{2+}]_i$ changes over time, so that there is an initial reduction in $[Ca^{2+}]_i$ during the first few hours, followed by a slow progressive rise in $[Ca^{2+}]_i$ during the next 12–14 h, at which time a very large and dramatic elevation of $[Ca^{2+}]_i$ occurs, which immediately precedes neuronal degeneration and death. Pretreatment with bFGF or

NGF for 6–12 h altered the profile of $[Ca^{2+}]_i$ changes following glucose withdrawal, so that the initial suppression of $[Ca^{2+}]_i$ did not occur, and the subsequent elevations of $[Ca^{2+}]_i$ and cell death were greatly delayed (Cheng et al., 1993).

NGF prevents death of cultured sympathetic neurons, and KCl and cAMP prevent their death following NGF withdrawal (Deckwerth and Johnson, 1993). It was originally proposed that NGF prevents cell death by keeping $[Ca^{2+}]_i$ at a high level, and that the cells die when $[Ca^{2+}]_i$ becomes too low (Koike and Tanaka, 1991). However, this initial interpretation of the data may not be correct, because it appears that, as with other paradigms of neuronal cell death described here, the death of sympathetic neurons following NGF withdrawal also results from generation of reactive oxygen species, loss of ion homeostasis, and elevation of $[Ca^{2+}]_i$. Thus, KCl and 8-(chlorophenylthio) cAMP induced a large decrease in the density of voltage-dependent calcium channels in sympathetic neurons, and their $[Ca^{2+}]_i$ remained unchanged for several days following NGF withdrawal in high K^+ medium (Murrell and Tolkovsky, 1993). Moreover, inhibitors of calcium-activated endonucleases (Batistatou and Greene, 1993) prevented death of sympathetic neurons and/or PC12 cells following withdrawal of trophic support. Calcium-dependent proteases (Squier et al., 1994) and other cysteine proteases, such as members of the interleukin-1β-converting enzyme family (Milligan et al., 1995), are implicated in apoptotic cell death in many different neuronal and nonneuronal cell systems.

Excessive elevations of $[Ca^{2+}]_i$ can surely damage and kill cells, but controlled elevations of $[Ca^{2+}]_i$ may activate neuroprotective pathways. Evidence that calcium can activate neuroprotective pathways comes from studies showing that exposure of neurons to a level of NMDA that is subtoxic, but sufficient to elevate $[Ca^{2+}]_i$, can protect neurons against subsequent exposure to an otherwise toxic level of NMDA (Marini and Paul, 1992). The well-known ability of KCl to promote survival of neurons in cell culture may also be explained by the moderate elevation of $[Ca^{2+}]_i$ induced by KCl. In addition, several neurotrophic factors (e.g., bFGF, NGF, and BDNF) are known to induce small elevations of $[Ca^{2+}]_i$ in neurons in which they promote long-term survival and protect against excitotoxic, metabolic, and oxidative insults (Cheng et al., 1993; Levine et al., 1995). We have found that conditions that increase neuronal activity in vivo (e.g., exposure to the seizure-inducing agent kainic acid) induce the expression of the neuroprotective Ca^{2+}-binding protein calbindin in adult rats (Fig. 7). Neurotrophic factors (Collazo et al., 1992) and TNF (Fig. 7; Cheng et al., 1994c) induce calbindin expression in cultured hippocampal neurons. Expression of calbindin in neurons and astrocytes increases their resistance to both apoptosis and necrosis induced by a variety of metabolic and oxidative insults (Mattson et al., 1991, 1995b).

5. ROLES FOR FREE RADICALS IN NEURONAL INJURY AND NEUROPROTECTION

Reactive oxygen species (ROS) certainly play important causal roles in neurodegenerative processes that occur in an array of disorders ranging from stroke (Watson and Ginsberg, 1989) to AD (Benzi and Moretti, 1995) to amyotrophic lateral sclerosis (Wiedau-Pazos et al., 1996). Less appreciated are the important roles ROS play in regulating cell signaling pathways, including those involved in neurotoxicity and neuroprotection. Before describing some of the recently discovered signaling functions of ROS, it is important to briefly describe some of the mechanisms whereby ROS are generated in neurons (Fig. 6). A major source of ROS is mitochondria, which generate superoxide anion radical as a byproduct of respiratory cascades. Superoxide is detoxified by conversion to hydrogen peroxide, a reaction catalyzed by superoxide dismutases. Hydrogen peroxide, though not particularly damaging by itself, can be readily converted to hydroxyl radical in the metal (Fe^{2+} and Cu^{2+}) catalyzed Fenton reaction; hydroxyl radical is extremely reactive and can initiate lipid peroxidation cascades that rapidly disrupt membranes and lead to cell death. Hydroxyl radical formation is normally prevented by the enzymes catalase and glutathione peroxidase, which convert hydrogen peroxide to water. Additional pathways leading to generation of potentially damaging ROS include: calcium-induced disruption of membrane potential across the inner mitochondrial membrane, resulting in an increase in superoxide production; Ca^{2+}-mediated activation of NOS, which generates nitric oxide radical from L-arginine (Dawson et al., 1993); and phospholipid hydrolysis resulting in release of arachidonic acid, which is then acted on by oxygenases (Kuzuya et al., 1993). The most damaging consequences of NO production may result from interaction of nitric oxide with superoxide anion radical, which generates peroxynitrite (Beckman and Crow, 1993). By inducing protein nitration on tyrosine residues, peroxynitrate may prevent phosphorylation of those residues and could, in that way, disrupt the many neurotrophic factor signal transduction pathways that employ tyrosine phosphorylation cascades. These are among the ways in which ROS contribute to Ca^{2+}-mediated cell damage that occurs as the result of exposure to excitotoxic and metabolic insults.

The complexity of ROS formation and detoxification provides an equally complex array of potential neuroprotective targets. Some examples are: NOS inhibitors, such as *N*-methyl-arginine, or NO scavengers, such as hemoglobin (Nathan and Xie, 1994); cyclooxygenase or lipoxygenase inhibitors, such as indomethacin and nordihydroguaiaretic acid, respectively (Goodman et al., 1994); and glutathione and agents such as *N*-acetyl-cysteine, which prevent glutathione depletion (Hamers et al., 1993; Shibata et al., 1995). Some of the most widely effective neuroprotective agents in cell culture and, in some cases, in vivo models of neurodegenerative conditions are compounds that directly neutralize the ultimate agents of destruction, which are oxyradicals and calcium. The tripeptide glutathione is a particularly important endogenous antioxidative agent. It serves to maintain high levels of glutathione peroxidase levels, and, by virtue of its reactive thiol groups, interacts with potentially damaging products of lipid peroxidation, such as 4-hydroxynonenal (Spitz et al., 1990). Glutathione (ethyl ester) protected cultured hippocampal neurons against Aβ toxicity and the toxicity of 4-hydroxynonenal, an aldehydic product of lipid peroxidation believed to play a major role in disruption of ion homeostases effected by oxidative insults (Mark et al., 1996). Glutathione protected PC12 cells against apoptosis induced by serum withdrawal, and against both apoptosis and necrosis induced by 4-hydroxynonenal (Kruman et al., 1996). Other endogenous antioxidants that are effective in protecting cultured neurons against excitotoxic, metabolic, and oxidative insults include vitamin E, which protected cultured hippocampal neurons against Aβ toxicity (Behl et al., 1994; Goodman and Mattson, 1994a), and ubiquinone, which protected cultured neurons against excitotoxicity (Favit et al., 1992).

Neurotrophic factors and cytokines can induce the expression of antioxidant enzymes, but the signaling mechanisms that control the levels of glutathione and other nonenzyme antioxidants are not well understood. A major limiting factor in glutathione synthesis is availability of the precursor cystine. Glutamate can compete with (and thereby reduce) cystine uptake into neurons, and glutamate can kill some cell types by a mechanism involving glutathione depletion (Murphy et al., 1989; Kato et al., 1992). Oxidized glutathione can modulate NMDA receptor function and alter $[Ca^{2+}]_i$ responses to NMDA and depolarization (Sucher and Lipton, 1991), suggesting that factors that regulate redox cycling of glutathione may influence neuronal vulnerability to excitotoxicity.

6. TOO MUCH OF A GOOD THING: NEUROPROTECTIVE SIGNALING PATHWAYS CAN BE OVERACTIVATED

Throughout the signal transduction literature, there are many cases where a low or moderate level of activation of a pathway mediates adaptive responses to environmental demands; overactivation of the same pathway results in cell injury and death. Perhaps the most widely studied such signaling pathway is that of glutamate, the major excitatory neurotransmitter in the mammalian CNS. Low and intermittent levels of activation of glutamate receptors are involved in the regulation of neurite outgrowth, synaptogenesis, and synaptic plasticity, including learning and memory processes (Barger and Mattson, 1995b; Fagnou and Tuchek, 1995). On the other hand, excessive activation of glutamate receptors can induce dendritic degeneration and cell death; this excitotoxic mechanism of neuronal degeneration is implicated in many different neurodegenerative conditions, ranging from stroke to epilepsy to AD (Farooqui and Horrocks, 1994; Mattson et al., 1993b). Data from studies of the sphingomyelin–ceramide–NFκB pathway suggest tightly regulated and complex roles for this pathway in mechanisms of neuronal injury and death. For example, it was shown in cultured mesencephalic cells that high levels of C2-ceramide (10–50 μM) were cytotoxic (Brugg et al., 1996), but lower levels of C2-ceramide were reported to protect cultured embryonic rat hippocampal neurons against the toxicities of glutamate, Aβ, and $FeSO_4$ (Goodman and Mattson, 1996a). Dobrowski et al. (1994) reported that NGF can activate the sphingomyelin–ceramide–NFκB via the low-affinity receptor, and provided evidence that this pathway mediates trophic responses to NGF.

It was reported that BDNF and NT-3 enhance excitotoxicity in cerebellar cultures (Fernandez-Sanchez and Novelli, 1993), and that BDNF, NT-3, and NT-4/5 can potentiate necrotic neuronal cell death induced by glutamate in rat hippocampal cultures (Koh et al., 1995). The latter results appear to be in direct contrast to many other reports that BDNF, NT-3, and NT-4/5 can protect neurons against various excitotoxic, metabolic, and oxidative insults, both in cell culture (Shimohama et al., 1993; Skaper et al., 1993; Cheng and Mattson, 1994; Cheng et al., 1994b; Kokaia et al., 1994) and in vivo (Beck et al., 1994; Zheng et al., 1995). Possible explanations include the timing of exposure of neurons to the neurotrophic factors in relation to the time of the insult, and particular vagaries of the insult paradigms to demonstrate injury exacerbation by neurotrophic factors. Essentially all data available from a variety of adult rat brain injury paradigms relevant to stroke, Huntington's disease, Parkinson's disease, and AD indicate that neurotrophic factors are either neuroprotective or without effect (e.g., Gage and Chen, 1992; Frim et al., 1993a,b; Nozaki et al., 1993; Gash et al., 1996). Therefore, the in vivo relevance of the few reports of injury enhancement in cell culture systems are unclear.

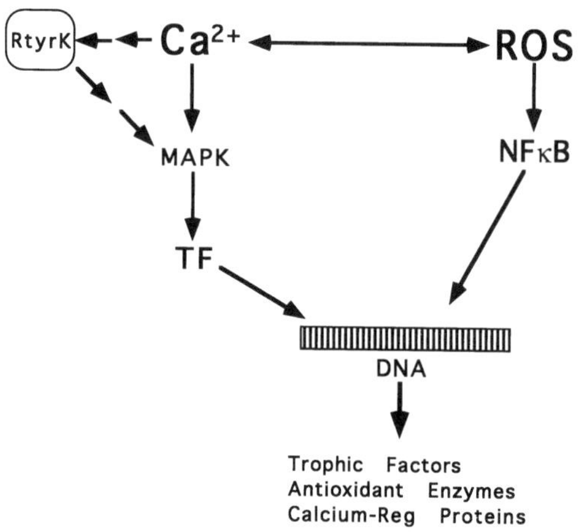

Fig. 8. Ca^{2+} and ROS induce neuroprotective genetic programs. Many brain insults result in elevation of intraneuronal Ca^{2+} and ROS levels. Ca^{2+} can induce neuroprotective pathways, possibly by activating neurotrophic factor signaling pathways that involve receptor tyrosine kinases (RtyrK), mitogen-activated protein kinases (MAPK), and transcription factors (TF). ROS can induce a neuroprotective pathway involving the transcription factor NFκB. Neuroprotective gene products induced by these pathways include trophic factors, antioxidant enzymes, and proteins involved in regulation of Ca^{2+} homeostasis (e.g., calcium-binding proteins).

7. HOMEOPATHIC NEUROPROTECTIVE SIGNALING MECHANISMS

Those who are developing approaches to reduce neurodegenerative processes take great interest in increasing data showing that subtoxic activation of potentially neurotoxic signaling pathways can protect neurons against subsequent exposure to (what would otherwise be) a neurotoxic insult. For example, it is now established that a moderate conditioning ischemia can confer resistance to brain injury induced by a subsequent severe ischemic insult (e.g., Liu et al., 1992). Calcium and ROS, "demons of neuronal damage," are two types of signals that can activate neuroprotective pathways (Fig. 8). Ca^{2+} activates several different signaling pathways in neurons that can contribute to cytoprotection, some of which were described above. For example, Ca^{2+} activates transcription factors that act at AP-1 sites in the promoters of genes. Ca^{2+} may activate neurotrophic factor signal transduction pathways in the absence of ligand. Rosen and Greenberg (1996) showed that activation of voltage-dependent Ca^{2+} channels in PC12 cells results in tyrosine phosphorylation of Shc and its association with Grb2, an adapter protein bound to the guanine nucleotide exchange factor Sos1. Ca^{2+} influx caused tyrosine phosphorylation of the epidermal growth factor receptor and association of Shc, Grb2, and Sos1 with that receptor protein. The work of Rosen et al. (1994) supports a requirement for activation of such neurotrophic factor signaling cascades in the survival- and outgrowth-promoting actions of depolarization-induced Ca^{2+} influx; they showed that Ca^{2+} influx in PC12 cells activates MEK1 and MAP kinases, and that the expression of a dominant negative form of Ras blocks the ability of Ca^{2+} influx to activate this pathway. Some examples of the ability of moderate increases in levels of calcium and ROS to activate neuroprotective pathways were presented above.

When one considers the cellular processes activated by calcium from the evolutionary perspective, it becomes obvious that many of the events induced by Ca^{2+} promote maintenance and plasticity of neuronal circuits. For example, Ca^{2+} induces production and release of neurotrophic factors (e.g., BDNF and sAPPs) from neurons (Ghosh et al., 1994; Nitsch et al., 1993). The production of such factors would be expected to promote the survival, not only of that neuron, but neurons synaptically connected to that neuron, as well.

Cyclic GMP may serve in homeopathic signaling pathways. For example, glutamate induces cGMP production via the NO–soluble guanylyl cyclase pathway (Bredt and Snyder, 1989). As described in Section 2.2., cGMP can protect neurons against excitotoxicity. Therefore, at the same time that glutamate induces increases in potentially neurotoxic levels of Ca^{2+} and ROS, it also activates neuroprotective pathways involving cGMP. Activation of NMDA receptors also increases cAMP levels in hippocampal neurons (Chetkovich and Sweatt, 1993), and cAMP has been shown to protect neurons against excitotoxicity (Mattson and Kate, 1988).

8. GENE THERAPY-BASED NEUROPROTECTIVE APPROACHES

Technologies for inducing or suppressing expression of specific gene products in neural cells are rapidly advancing, and suggest potential for manipulating expression of proteins involved in neuroprotective or neurodegenerative processes. Antisense oligodeoxynucleotides (AODs) that target mRNAs encoding proteins that promote neuronal degeneration have proven effective in cell culture and in vivo paradigms of excitotoxic and ischemic brain injury. For example, administration of AODs to an NMDA-receptor protein (NMDARP-71) to hippocampal cell cultures protected neurons against glutamate toxicity (Mattson et al., 1993a). Intraventricular infusion of AODs directed against the NR1 subunit of the NMDA receptor protected cortical neurons against ischemic injury in adult rats (Wahlstedt et al., 1993).

Overexpression of neuroprotective gene products using a variety of vector types has proven to be an effective strategy for protecting neurons against injury in cell culture and in vivo. Several laboratories have expressed neurotrophic factors in fibroblasts or myoblasts, and then transplanted those cells into the brain and examined whether they would protect neurons against injury. Studies have shown that transplantation of NGF-producing fibroblasts into the septal area protected cholinergic neurons from axotomy-induced degeneration (Dekker et al., 1992; Kawaja et al., 1992). Frim et al. (1993a, b) reported that degeneration of striatal neurons induced by excitotoxins or mitochondrial toxins was attenuated in rats with implants of NGF-producing fibroblasts. Because they are postmitotic, neurons are not readily transfected using approaches such as liposome-mediated DNA transfer or electroporation. On the other hand, viral vectors can be used to express genes in postmitotic cells, including neurons. For example, retroviral vectors were used to produce an NGF-secreting cell line from immortalized CNS-derived neuron progenitor cells (Martinez-Serrano et al., 1995). Transplantation of the transduced cells into the basal forebrain of cognitively impaired aged rats induced a reversal of the spatial learning impairment. Progress is being made with other viral vectors, including adenovirus, adeno-associated virus, and herpes virus (e.g., Lawrence et al., 1996).

In addition to neurotrophic factors, neuroprotective proteins that act intracellularly can be overexpressed. For example, C6 glioma cells stably transfected to express high levels of calbindin were resistant to apoptosis induced by exposure to calcium ionophore or Aβ, compared to control cells. Overexpression of heat-shock proteins in cultured neurons protected them against excitotoxicity (Lowenstein et al., 1991). Studies have shown that expression of Mn-SOD, Cu–Zn-SOD, or glutathione peroxidase in various types of cultured cells can protect them from being killed by different insults. Transgenic mice overexpressing Cu–Zn-SOD exhibited reduced vulnerability of cortical neurons to focal ischemic injury (Yang et al., 1994). Expression of the antiapoptotic gene *bcl-2* in culture and in vivo protected neurons in a variety of cell death paradigms by mechanisms that may involve suppression of ROS accumulation and/or stabilization of Ca homeostasis (Kane et al., 1993; Zhong et al., 1993; Lam et al., 1994; Shimazaki et al., 1994). In *bcl-2* transgenic mice, motoneurons were protected against axotomy-induced death (Dubois-Dauphin et al., 1994), and cortical neurons were resistant to ischemic injury (Martinou et al., 1994).

9. TECHNICAL CONSIDERATIONS IN STUDIES OF MECHANISMS OF NEURONAL INJURY AND NEUROPROTECTION

When the end point of one's studies is cell death (which can be defined and assessed in different ways), and the methods used to measure and manipulate cellular processes are ill-defined (or underemployed), interpretation of the results is often unclear. Many examples can illustrate this point; here are four with clear relevance to the different chapters in this book: the role of production of "death genes" in neuronal degeneration; the role of protein kinases in cell death and neuroprotection; Ca^{2+} and cell death; mitochondrial function and cell survival.

One of the criteria for determining whether neurons die by a programmed cell death mechanism that has become accepted by some neuroscientists is that cycloheximide, a protein synthesis inhibitor, can prevent cell death (Martin et al., 1988; Johnson and Deckwerth, 1993). However, accumulating data suggest that the interpretation of the previous data in studies that employed cycloheximide may be incorrect. We have found that cycloheximide can induce expression of immediate early gene products (i.e., protein) induced by antiapoptotic neurotrophic factors, including c-Fos (Furukawa et al., 1997). For example, we found that low concentrations of cycloheximide (20–200 n*M*) increased levels of *bcl-2* mRNA and protein in embryonic rat hippocampal neurons (Fig. 9). Remarkably, these data show that, within a certain concentration range, the protein synthesis inhibitor cycloheximide stimulates protein synthesis. The concentrations of cycloheximide that induced c-Fos protein protected neurons against excitotoxicity and Aβ toxicity (Furukawa et al., 1997). A biphasic effect of increasing concentrations of cycloheximide on neuronal survival is seen in many systems, with low concentrations promoting survival and higher concentrations killing cells or preventing their rescue by neurotrophic factors. For example, Tortosa et al. (1994) showed that low concentrations of cycloheximide reduced delayed (apoptotic) neuronal death following cerebral ischemia in gerbils, but higher concentrations exacerbated neuronal death. Beyond the lack of convincing evidence for active death programs underlying apoptotic neuronal death in the injured brain, evolutionary considerations argue against mechanisms for actively killing neurons, particularly in the adult brain. In contrast to developmental neuronal death, it seems unlikely that there would be a survival advantage in eliminating postmitotic cells that are unable to replace themselves. We therefore propose that the mechanisms whereby trophic factors and cycloheximide prevent

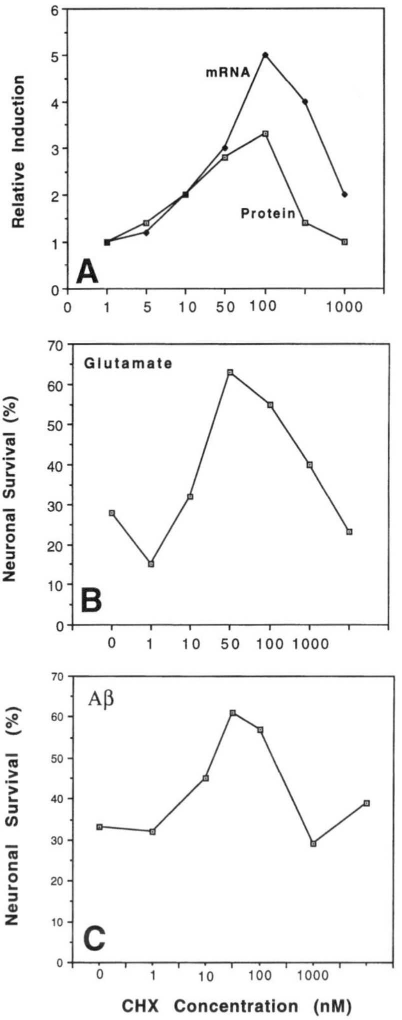

Fig. 9. Cycloheximide induces an increase in *bcl-2* protein levels and protects neurons against excitotoxic and oxidative insults. **(A)** Rat hippocampal cell cultures were exposed for 30 min (mRNA level studies) or 4 h (protein level studies) to the indicated concentrations of cycloheximide (CHX). Levels of *bcl-2* mRNA and protein were quantified by RT-PCR and immunocytochemical methods, respectively. Values represent the mean of three separate cultures. **(B, C)** Cultures were preincubated for 4 h with vehicle or the indicated concentrations of CHX and then exposed to 10 μ*M* glutamate (B) or 5 μ*M* Aβ (C). Neuronal survival was quantified 20 h later. Values represent the mean and SEM of four separate cultures.

apoptosis converge on a common pathway involving activation of life gene products (Fig. 10). In any event, it certainly seems important to re-evaluate the mechanism underlying neuroprotective actions of cycloheximide that have been reported in many paradigms of neuronal death.

Another illustration involves protein kinases, which mediate both rapid and delayed responses to many different environmental signals that affect neuronal survival. For example, glutamate activates metabotropic receptors linked to inositol phospholipid

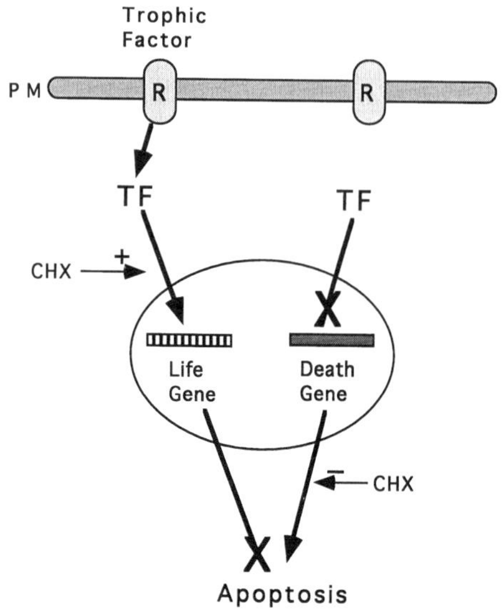

Fig. 10. Hypothetical programmed cell life mechanism for the protection of neurons against various insults by trophic factors and cycloheximide. Trophic factors activate plasma membrane (PM) receptors (R) linked to signaling cascades that induce transcription factors (TF), which induce expression of neuroprotective life genes (left). There are at least two (not mutually exclusive) mechanisms for apoptosis that results when trophic factor is withdrawn. One is that life genes are no longer activated, and the other is that death genes are induced. Although it had previously been assumed that cycloheximide prevents apoptosis by suppressing the expression of death genes, recent findings suggest that cycloheximide may protect neurons by inducing the expression of life genes.

hydrolysis and activation of PKC, and neurotrophic factors activate tyrosine kinase cascades (*see* Section 2.1.). One way in which investigators have attempted to establish cause–effect roles for specific kinases in various cell death paradigms is to use kinase inhibitors. Unfortunately, many, if not all, such kinase inhibitors are nonspecific, because they affect multiple types of kinases. Moreover, several bacterial alkaloid protein kinase inhibitors, including staurosporine and K252a, have quite complex activities, so that within a certain concentration range they can stimulate protein tyrosine phosphorylation (Cheng et al., 1994a; Smith-Swintosky et al., 1996a).

A third example of the importance of technical aspects of studies of mechanisms on neuronal death comes from studies of the role of calcium in various death paradigms. Many studies of nonneuronal cells have emphasized the importance of elevation of $[Ca^{2+}]_i$ in apoptosis (*see* Orrenius and Nicotera, 1994, for review). However, studies of trophic factor deprivation-induced cell death in cultured peripheral neurons, in which NGF increased $[Ca^{2+}]_i$ and the $[Ca^{2+}]_i$ decreased following NGF withdrawal, were interpreted to indicate that the reduction in $[Ca^{2+}]_i$ was causally involved in the cell death (Koike and Tanaka, 1991). Unfortunately, the $[Ca^{2+}]_i$ was not monitored during the entire period following NGF withdrawal until cell death, and it was therefore not established that $[Ca^{2+}]_i$ remained low during the death process. Indeed, recent data indicate that increases of $[Ca^{2+}]_i$, causally linked to cell death, do occur prior to cell death in such trophic factor withdrawal models. A similar pattern of changes of $[Ca^{2+}]_i$ occur in cultured hippocampal cells deprived of glucose (Cheng et al., 1993). During the early period of glucose deprivation, there is a decrease of $[Ca^{2+}]_i$, followed by a period of 10–12 h during which $[Ca^{2+}]_i$ is maintained at a low level. Subsequently, there is a large and relatively rapid elevation of $[Ca^{2+}]_i$ that immediately precedes cell death.

Recently, many investigators have begun to use the dye MTT (3-[4,5-dimethylthiazol-2-yl]-2,5-diphenyl-tetrazolium bromide) as a measure of cell viability (e.g., Helson et al., 1995). In fact, MTT is a compound that is reduced by mitochondrial metabolites and is therefore an indicator of mitochondrial function. Although mitochondrial function decreases when cells die, it is also true that mitochondrial metabolism can be (greatly) decreased without cell death. An example comes from studies of Shearman et al. (1995), which clearly showed that subtoxic levels of amyloid β-peptide cause a pronounced decrease in MTT reduction in cultured PC12 cells. These findings show that, perhaps not surprisingly, some conditions can affect mitochondrial metabolism without affecting cell survival. These kinds of observations suggest that it is important to know exactly what one is measuring in a cell survival assay.

10. PERSPECTIVES

The history of neurotrophic factor discovery and characterization of neurotrophic factor signal transduction pathways demonstrates the importance of cell culture paradigms of trophic factor action. Repeatedly, it has been shown that neurotrophic factors that promote long-term survival and outgrowth of developing neurons in culture can also protect those neurons (in vitro and in vivo) against excitotoxic, metabolic, and oxidative insults (*see* Mattson and Scheff, 1994, for review). Mechanisms elucidated in cell cultures of embryonic neurons are proving to be operative in the adult nervous system, and, importantly, may provide effective treatments for neurodegenerative conditions. For example, the initial discoveries in hippocampal cell cultures, showing that bFGF can promote long-term neuronal survival (Walicke et al., 1986; Morrison et al., 1988), and can protect neurons against excitotoxic (Mattson et al., 1989) and metabolic (Cheng and Mattson, 1991) insults, led to successful trials of bFGF in animal models of ischemic stroke (Nozaki et al., 1993; Koketsu et al., 1994; Fisher et al., 1995), and to an ongoing trial of bFGF in human stroke patients (*see* Chapter 6). Identification of individual molecular components of cellular signaling pathways that transduce environmental inputs into neuroprotective responses is rapidly progressing. Each component identified provides a novel target for the development of neuroprotective drugs.

ACKNOWLEDGMENTS

I thank G. Sears for editorial assistance, and S. W. Barger, A. J. Bruce, B. Cheng, K. Furukawa, R. J. Mark, and V. L. Smith-Swintosky for their critical contributions to the research from my laboratory. Supported by the NIH, the Alzheimer's Association, the Kentucky Spinal Cord and Head Injury Trust, and the Metropolitan Life Foundation.

REFERENCES

Aniksztejn, L., Bregestovski, P., and Ben-Ari, Y. (1991) Selective activation of quisqualate metabotropic receptor potentiates NMDA but not AMPA responses. *Eur. J. Pharmacol.* **205,** 327–328.

Bading, H., Ginty, D., and Greenberg, M. (1993) Regulation of gene expression in hippocampal neurons by distinct calcium signaling pathways. *Science* **260,** 181–186.
Barger, S. W. and Mattson, M. P. (1995a) Secreted form of the Alzheimer's amyloid precursor protein stimulates a membrane-associated guanylate cyclase. *Biochem. J.* **311,** 45–47.
Barger, S. W. and Mattson, M. P. (1995b) Excitatory amino acids and growth factors: biological and molecular interactions regulating neuronal survival and plasticity, in *CNS Neurotransmitters and Neuromodulators (II)* (Stone, T. W., ed.) CRC, Boca Raton, pp. 273–294.
Barger, S. W., Fiscus, R. R., Ruth, P., Hofmann, F., and Mattson, M. P. (1995a) Role of cyclic GMP in the regulation of neuronal calcium and survival by secreted forms of β amyloid precursor. *J. Neurochem.* **64,** 2087–2096.
Barger, S. W., Horster, D., Furukawa, K., Goodman, Y., Krieglstein, J., and Mattson, M. P. (1995b) Tumor necrosis factors α and β protect neurons against amyloid β-peptide toxicity: evidence for involvement of a κB-binding factor and attenuation of peroxide and Ca^{2+} accumulation. *Proc. Natl. Acad. Sci. USA* **92,** 9328–9332.
Barger, S. W. and Mattson, M. P. (1996) Induction of neuroprotective κB-dependent transcription by secreted forms of the Alzheimer's β-amyloid precursor. *Mol. Brain Res.* **40,** 116–226.
Baskin, D. S., Widmayer, M. A., Browning, J. L., Heizer, M. L., and Schmidt, W. K. (1994) Evaluation of delayed treatment of focal cerebral ischemia with three selective kappa-opioid agonists in cats. *Stroke* **25,** 2047–2053.
Batistatou, A. and Greene, L. A. (1993) Internucleosomal DNA cleavage and neuronal cell survival/death. *J. Cell Biol.* **122,** 523–532.
Beck, T., Lindholm, D., Castren, E., and Wree, A. (1994) Brain-derived neurotrophic factor protects against ischemic cell damage in rat hippocampus. *J. Cereb. Blood Flow Metab.* **14,** 689–692.
Beckman, J. S. and Crow, J. P. (1993) Pathological implications of nitric oxide, superoxide and peroxynitrite formation. *Biochem. Soc. Trans.* **21,** 330–334.
Behl, C., Davis, J., Lesley, R., and Schubert, D. (1994) Hydrogen peroxide mediates amyloid β protein toxicity. *Cell* **77,** 817–827.
Behl, C., Widmann, M., Trapp, T., and Holsboer, F. (1995) 17-β estradiol protects neurons from oxidative stress- induced cell death in vitro. *Biochem. Biophys. Res. Commun.* **216,** 473–482.
Benzi, G. and Moretti, A. (1995) Are reactive oxygen species involved in Alzheimer's disease? *Neurobiol. Aging* **16,** 661–674.
Bernstein, B. W. and Bamburg, J. R. (1985) Reorganization of actin in depolarized synaptosomes. *J. Neurosci.* **5,** 2565–2569.
Berry, M. D., Juorio, A. V., and Paterson, I. A. (1994) Possible mechanisms of action of (–) deprenyl and other MAO-B inhibitors in some neurologic and psychiatric disorders. *Prog. Neurobiol.* **44,** 141–161.
Bowling, A. C., Schulz, J. B., Brown R. H., and Beal, M. F. (1993) Superoxide dismutase activity, oxidative damage, and mitochondrial energy metabolism in familial and sporadic amyotrophic lateral sclerosis. *J. Neurochem.* **61,** 2322–2325.
Bredt, D. S. and Snyder, S. H. (1989) Nitric oxide mediates glutamate-linked enhancement of cGMP levels in the cerebellum. *Proc. Natl. Acad. Sci. USA* **86,** 9030–9033.
Bruce, A. J., Boling, W., Kindy, M. S., Peschon, J., Kraemer, P. J., Carpenter, M. K., Holtsberg, F. W., and Mattson, M. P. (1996) Altered neuronal and microglial responses to brain injury in mice lacking TNF receptors. *Nature Med.* **2,** 788–794.
Brugg, B., Michel, P. P., Agid, Y., and Ruberg, M. (1996) Ceramide induces apoptosis in cultured mesencephalic neurons. *J. Neurochem.* **66,** 733–739.
Butt, T. R. and Karathanasis, S. K. (1995) Transcription factors as drug targets, opportunities for therapeutic selectivity. *Gene Exp.* **4,** 319–336.
Cheng, B. and Mattson, M. P. (1991) NGF and bFGF protect rat and human central neurons against hypoglycemic damage by stabilizing calcium homeostasis. *Neuron* **7,** 1031–1041.

Cheng, B., McMahon, D., and Mattson, M. P. (1993) Modulation of calcium current, intracellular calcium levels and cell survival by hypoglycemia, NGF and bFGF in cultured hippocampal neurons. *Brain Res.* **607,** 275–285.

Cheng, B. and Mattson, M. P. (1994) NT-3 and BDNF protect hippocampal, septal, and cortical neurons against metabolic compromise. *Brain Res.* **640,** 56–67.

Cheng, B., Barger, S. W., and Mattson, M. P. (1994a) Staurosporine, K-252a and K-252b stabilize calcium homeostasis and promote survival of CNS neurons in the absence of glucose. *J. Neurochem.* **62,** 1319–1329.

Cheng, B., Goodman, Y., Begley, J. G., and Mattson, M. P. (1994b) Neurotrophin 4/5 protects hippocampal and cortical neurons against energy deprivation- and excitatory amino acid-induced injury. *Brain Res.* **650,** 331–335.

Cheng, B., Christakos, S., and Mattson, M. P. (1994c) Tumor necrosis factors protect neurons against excitotoxic/metabolic insults and promote maintenance of calcium homeostasis. *Neuron* **12,** 139–153.

Cheng, B. and Mattson, M. P. (1995) PDGFs protect hippocampal neurons against energy deprivation and oxidative injury: evidence for induction of antioxidant pathways. *J. Neurosci.* **15,** 7095–7104.

Chetkovich, D. M. and Sweatt, J. D. (1993) NMDA receptor activation increases cyclic AMP in area CA1 of the hippocampus via calcium/calmodulin stimulation of adenylyl cyclase. *J. Neurochem.* **61,** 1933–1942.

Chiou, G. C. and Hong, S. J. (1995) Prevention and reduction of neural damage in ischemic strokes by w-(N,N′-diethylamino)-n-alkyl-3,4,5-trimethoxybenzoate compounds. *J. Pharmacol. Exp. Ther.* **275,** 474–478.

Collazo, D., Takahashi, H., and McKay, R. D. G. (1992) Cellular targets and trophic functions of neurotrophin-3 in the developing rat hippocampus. *Neuron* **9,** 643–656.

Cunningham, D. D., Pulliam, L., and Vaughan, P. J. (1993) Protease nexin-1 and thrombin: injury-related processes in the brain. *Thromb. Haemost.* **70,** 168–171.

Dawson, V., Dawson, T., Bartley, D., Uhl, G., and Snyder, S. (1993) Mechanisms of nitric oxide-mediated neurotoxicity in primary brain cultures. *J. Neurosci.* **13,** 2651–2661.

Deckwerth, T. L. and Johnson Jr., E. M. (1993) Temporal analysis of events associated with programmed cell death (apoptosis) of sympathetic neurons deprived of nerve growth factor. *J. Cell Biol.* **123,** 1207–1222.

Dekker, A. J., Gage, F. H., and Thal, L. J. (1992) Delayed treatment with nerve growth factor improves acquisition of a spatial task in rats with lesions of the nucleus basalis magnocellularis, Evaluation of the involvement of different neurotransmitter systems. *Neuroscience* **48,** 111–119.

Dobrowsky, R. T., Werner, M. H., Castellino, A. M., Chao, M. V., and Hannun, Y. A. (1994) Activation of the sphingomyelin cycle through the low-affinity neurotrophin receptor. *Science* **265,** 1596–1599.

Dubois-Dauphin, M., Frankowski, H., Tsujimoto, Y., Huarte, J., and Martinou, J. C. (1994) Neonatal motoneurons overexpressing the *bcl*-2 protooncogene in transgenic mice are protected from axotomy-induced cell death. *Proc. Natl. Acad. Sci. USA* **91,** 3309–3313.

Fagnou, D. D. and Tuchek, J. M. (1995) The biochemistry of learning and memory. *Mol. Cell. Biochem.* **149–150,** 279–286.

Farinelli, S. E., Park, D. S., and Greene, L. A. (1996) Nitric oxide delays the death of trophic factor-deprived PC12 cells and sympathetic neurons by a cGMP-mediated mechanism. *J. Neurosci.* **16,** 2325–2334.

Farooqui, A. A. and Horrocks, L. A. (1994) Excitotoxicity and neurological disorders: involvement of membrane phospholipids. *Int. Rev. Neurobiol.* **36,** 267–323.

Favaron, M., Manev, H., Simian, R., Bertolino, M, Szekely, A. M., DeErausquin, G., Guidotti, A., and Costa, E. (1990) Down-regulation of protein kinase c protects cerebellar granule neurons in primary culture from glutamate-induced neuronal death. *Proc. Natl. Acad. Sci. USA* **87,** 1983–1987.

Favit, A., Nicoletti, F., Scapagnini, U., and Canonico, L. P. (1992) Ubiquinone protects cultured neurons against spontaneous and excitotoxin-induced degeneration. *J. Cereb. Blood Flow Metab.* **12,** 638–645.

Fernandez-Sanchez, M. T. and Novelli, A. (1993) Basic fibroblast growth factor protects cerebellar neurons in primary culture from NMDA and non-NMDA receptor mediated neurotoxicity. *FEBS Lett.* **335,** 124–131.

Fisher, M., Meadows, M. E., Do, T., Weise, J., Trubetskoy, V., Charette, M., and Finklestein, S. P. (1995) Delayed treatment with intravenous basic fibroblast growth factor reduces infarct size following permanent focal cerebral ischemia in rats. *J. Cereb. Blood Flow Metab.* **15,** 953–959.

Forloni, G., Lucca, E., Angeretti, N., Chiesa, R., and Vezzani, A. (1996) Neuroprotective effect of somatostatin on non-apoptotic NMDA-induced neuronal death: role of cGMP. *J. Neurochem.* **68,** 319–327.

Frandsen, A. and Schousboe, A. (1993) Excitatory amino acid-mediated cytotoxicity and calcium homeostasis in cultured neurons. *J. Neurochem.* **60,** 1202–1211.

Fraser, A. and Evan, G. (1996) A license to kill. *Cell* **85,** 781–784.

Frim, M. D., Short, P. M., Rosenberg, S. W., Simpson, J., Xandra, A. B., Breakfield, O., and Isacson, O. (1993a) Local protective effects of nerve growth factor-secreting fibroblasts against excitotoxic lesions in the rat striatum. *J. Neurosurg.* **78,** 267–273.

Frim, D. M., Simpson, J., Uhler, T. A., Short, M. P., Bossi, S. R., Breakefield, X. O., and Isacson, O. (1993b) Striatal degeneration induced by mitochondrial blockade is prevented by biologically delivered NGF. *J. Neurosci. Res.* **35,** 452–458.

Furukawa, K., Smith-Swintosky, V. L., and Mattson, M. P. (1995) Evidence that actin depolymerization protects hippocampal neurons against excitotoxicity by stabilizing $[Ca^{2+}]_i$. *Exp. Neurol.* **133,** 153–163.

Furukawa, K. and Mattson, M. P. (1995a) Cytochalasins protect hippocampal neurons against amyloid β-peptide toxicity: evidence that actin depolymerization suppresses Ca^{2+} influx. *J. Neurochem.* **65,** 1061–1068.

Furukawa, K. and Mattson, M. P. (1995b) Taxol stabilizes $[Ca^{2+}]_i$ and protects hippocampal neurons against excitotoxicity. *Brain Res.* **689,** 141–146.

Furukawa, K., Barger, S. W., Blalock, E., and Mattson, M. P. (1996) Activation of K^+ channels and suppression of neuronal activity by secreted β-amyloid precursor protein. *Nature* **379,** 74–78.

Furukawa, K., Estus, S., Fu, W., and Mattson, M. P. (1997) Neuroprotective action of cycloheximide involves induction of bcl-2 and antioxidant pathways. *J. Cell Biol.* **136,** 1137–1150.

Gage, F. H. and Chen, K. S. (1992) Neural transplants: prospects for Alzheimer's disease. *Curr. Opinion Neurol. Neurosurg.* **5,** 94–99.

Garthwaite, G. and Garthwaite, J. (1988) Cyclic GMP and cell death in rat cerebellar slices. *Neuroscience* **26,** 321–326.

Gash, D. M., Zhang, Z., Ovadia, A., Cass, W. A., Yi, A., Simmerman, L., et al. (1996) Functional recovery in parkinsonian monkeys treated with GDNF. *Nature* **380,** 252–255.

Geddes, J. W., Schwab, C., Craddock, S., Wilson, J., and Pettigrew, L. C. (1994) Alterations in tau immunostaining in the rat hippocampus following transient cerebral ischemia. *J. Cereb. Blood Flow Metab.* **14,** 554–564.

Gerber, U., Sim, J. A., and Gahwiler, B. H. (1992) Reduction of potassium conductances mediated by metabotropic glutamate receptors in rat CA3 pyramidal cells does not require protein kinase C or protein kinase A. *Eur. J. Neurosci.* **4,** 792–797.

Gereau, R. W. and Conn, P. J. (1994) Potentiation of cAMP responses by metabotropic glutamate receptors depresses excitatory synaptic transmission by a kinase-independent mechanism. *Neuron* **12,** 1121–1129.

Ghosh, A., Carnahan, J., and Greenberg, M. E. (1994) Requirement for BDNF in activity-dependent survival of cortical neurons. *Science* **263,** 1618–1623.

Ghosh, A. and Greenberg, M. E. (1995) Calcium signaling in neurons, molecular mechanisms and cellular consequences. *Science* **268,** 239–247.

Goodman, Y. and Mattson, M. P. (1994a) Secreted forms of β-amyloid precursor protein protect hippocampal neurons against amyloid β-peptide-induced oxidative injury. *Exp. Neurol.* **128,** 1–12.
Goodman, Y. and Mattson, M. P. (1994b) Staurosporine and K-252 compounds protect hippocampal neurons against amyloid β-peptide toxicity and oxidative injury. *Brain Res.* **650,** 170–174.
Goodman, Y., Steiner, M. R., Steiner, S. M., and Mattson, M. P. (1994) Nordihydroguaiaretic acid protects hippocampal neurons against amyloid β-peptide toxicity, and attenuates free radical and calcium accumulation. *Brain Res.* **654,** 171–176.
Goodman, Y. and Mattson, M. P. (1996a) Ceramide protects hippocampal neurons against excitotoxic and oxidative insults, and amyloid β-peptide toxicity. *J. Neurochem.* **66,** 869–872.
Goodman, Y. and Mattson, M. P. (1996b) K^+ channel openers protect hippocampal neurons against oxidative injury and amyloid β-peptide toxicity. *Brain Res.* **706,** 328–332.
Goodman, Y., Bruce, A. J., Cheng, B., and Mattson, M. P. (1996) Estrogens attenuate and corticosterone exacerbates excitotoxicity, oxidative injury and amyloid β-peptide toxicity in hippocampal neurons. *J. Neurochem.* **66,** 1836–1844.
Hamers, F. P., Brakkee, J. H., Cavalletti, E., Tedeschi, M., Marmonti, L., Pezzoni, G., Neijt, J. P., and Gispen, W. H. (1993) Reduced glutathione protects against cisplatin-induced neurotoxicity in rats. *Cancer Res.* **53,** 544–549.
Hara, H., Onodera, H., Yoshidomi, M., Matsuda, Y., and Kogure, K. (1990) Staurosporine, a novel protein kinase C inhibitor, prevents post-ischemic neuronal damage in the gerbil and rat. *J. Cereb. Blood Flow Metab.* **10,** 646–653.
Hayes, R. L., Jenkins, L. W., and Lyeth, B. G. (1992) Neurotransmitter-mediated mechanisms of traumatic brain injury, acetylcholine and excitatory amino acids. *J. Neurotrauma* **9,** S173–S187.
Helson, L., Helson, C., and Ainsworth, S. (1995) Camptothecin cytotoxic effects in vitro: dependency on exposure duration and dose. *Anticancer Drugs* **6,** 612–614.
Holland, J. P., Sydserff, S. G., Taylor, W. A., and Bell, B. A. (1994) Calcitonin gene-related peptide reduces brain injury in a rat model of focal cerebral ischemia. *Stroke* **25,** 2055–2058.
Hu, B. R. and Wieloch, T. (1994) Tyrosine phosphorylation and activation of mitogen-activated protein kinase in the rat brain following transient cerebral ischemia. *J. Neurochem.* **62,** 1357–1367.
Huang, H. M., and Gibson, G. E. (1993) Altered beta-adrenergic receptor-stimulated cAMP formation in cultured skin fibroblasts from Alzheimer donors. *J. Biol. Chem.* **268,** 14,616–14,621.
Huang, H. M., Lin, T. A., Sun, G. Y., and Gibson, G. E. (1995) Increased inositol 1,4,5-trisphosphate accumulation correlates with an up-regulation of bradykinin receptors in Alzheimer's disease. *J. Neurochem.* **64,** 761–766.
Jin, L. W. and Saitoh, T. (1995) Changes in protein kinases in brain aging and Alzheimer's disease. Implications for drug therapy. *Drugs Aging* **6,** 136–149.
Johnson, B. and Byerly, L. (1993) A cytoskeletal mechanism for Ca^{2+} channel metabolic dependence and inactivation by intracellular Ca^{2+}. *Neuron* **10,** 797–804.
Johnson, E. M. and Deckwerth, T. L. (1993) Molecular mechanisms of developmental neuronal death. *Annu. Rev. Neurosci.* **16,** 31–46.
Kane, D. J., Sarafian, T. A., Anton, R., Hahn, H., Gralla, E. B., Valentine, J. S., Ord, T., and Bredesen, D. E. (1993) Bcl-2 inhibition of neural death, decreased generation of reactive oxygen species. *Science* **262,** 1274–1277.
Kato, S., Negishi, K., Mawatari, K., and Kuo, C. H. (1992) A mechanism for glutamate toxicity in the C6 glioma cells involving inhibition of cystine uptake leading to glutathione depletion. *Neuroscience* **48,** 903–914.
Kato, H., Araki, T., Itoyama, Y., and Kogure, K. (1995) Rolipram, a cyclic AMP-selective phosphodiesterase inhibitor, reduces neuronal damage following cerebral ischemia in the gerbil. *Eur. J. Pharmacol.* **272,** 107–110.

Kawaja, D. M., Rosenberg, B. M., Yoshida, K., and Gage, H. F. (1992) Somatic gene transfer of nerve growth factor promotes the survival of axotomized septal neurons and the regeneration of their axons in adult rats. *J. Neurosci.* **12,** 2849–2864.

Keaney Jr., J. F., Shwaery, G. T., Xu, A., Nicolosi, R. J., Loscalzo, J., Foxall, T. L., and Vita, J. A. (1994) 17β-estradiol preserves endothelial vasodilator function and limits low-density lipoprotein oxidation in hypercholesterolemic swine. *Circulation* **89,** 2251–2259.

Keegan, K. and Halegoua, S. (1993) Signal transduction pathways in neuronal differentiation. *Curr. Opinion Neurobiol.* **3,** 14–19.

Kindy, M. S. (1993) Inhibition of tyrosine phosphorylation prevents delayed neuronal death following cerebral ischemia. *J. Cereb. Blood Flow Metab.* **13,** 372–377.

Kochhar, A., Zivin, J. A., Lyden, P. D., and Mazzarella, V. (1988) Glutamate antagonist therapy reduces neurologic deficits produced by focal central nervous system ischemia. *Arch. Neurol.* **45,** 148–153.

Koh, J.-Y., Palmer, E., and Cotman, C. W. (1991) Activation of the metabotropic glutamate receptor attenuates N-methyl-D-aspartate neurotoxicity in cortical cultures. *Proc. Natl. Acad. Sci. USA* **88,** 9431–9435.

Koh, J. Y., Gwag, B. J., Lobner, D., and Choi, D. W. (1995) Potentiated necrosis of cultured cortical neurons by neurotrophins. *Science* **268,** 573–575.

Koike, T. and Tanaka, S. (1991) Evidence that nerve growth factor dependence of sympathetic neurons for survival in vitro may be determined by levels of cytoplasmic free Ca (2+). *Proc. Natl. Acad. Sci. USA* **88,** 3892–3896.

Kokaia, Z., Othberg, A., Kokaia, M., and Lindvall, O. (1994) BDNF makes cultured dentate granule cells more resistant to hypoglycaemic damage. *Neuroreport* **5,** 1241–1244.

Koketsu, N., Berlove, D. J., Moskowitz, M. A., Kowall, N. W., Caday, C. G., and Finklestein, S. P. (1994) Pretreatment with intraventricular basic fibroblast growth factor decreases infarct size following focal cerebral ischemia in rats. *Ann. Neurol.* **35,** 451–457.

Kolesnick, R. and Golde, D. W. (1994) The sphingomyelin pathway in tumor necrosis factor and interleukin-1 signaling. *Cell* **77,** 325–328.

Kruman, I., Bruce-Keller, A. J., Bredesen, D. E., Waeg, G., and Mattson, M. P. (1997) Evidence that 4-hydroxynonenal mediates oxidative stress-induced neuronal apoptosis. *J. Neurosci.* **17,** 5089–5100.

Kuzuya, T., Hoshida, S., Kim, Y., Oe, H., Hori, M., Kamada, T., and Tada, M. (1993) Free radical generation coupled with arachidonate lipoxygenase reaction related to reoxygenation induced myocardial cell injury. *Cardiovasc. Res.* **27,** 1056–1060.

Lam, M., Dubyak, G., Chen, L., Nunez, G., Miesfeld, R., and Distelhorst, C. (1994) Evidence that BCL-2 represses apoptosis by regulating endoplasmic reticulum-associated Ca^{2+} fluxes. *Proc. Natl. Acad. Sci. USA* **91,** 6569–6573.

Lawrence, M. S., Ho, D. Y., Sun, G. H., Steinberg, G. K., and Sapolsky, R. M. (1996) Overexpression of Bcl-2 with herpes simplex virus vectors protects CNS neurons against neurological insults in vitro and in vivo. *J. Neurosci.* **16,** 486–496.

Levine, E. S., Dreyfus, C. F., Black, I. B., and Plummer, M. R. (1995) Differential effects of NGF and BDNF on voltage-gated calcium currents in embryonic basal forebrain neurons. *J. Neurosci.* **15,** 3084–3091.

Levitan, E. S. and Kramer, R. H. (1990) Neuropeptide modulation of single calcium and potassium channels detected with a new patch clamp configuration. *Nature* **348,** 545–547.

Levitzki, A. and Gazit, A. (1995) Tyrosine kinase inhibition: an approach to drug development. *Science* **267,** 1782–1788.

Li, H. and Buchan, A. M. (1993) Treatment with an AMPA antagonist 12 hours following severe normothermic forebrain ischemia prevents CA1 neuronal injury. *J. Cereb. Blood Flow Metab.* **13,** 933–939.

Lindvall, O., Kokaia, Z., Bengzon, J., Elmer, E., and Kokaia, M. (1994) Neurotrophins and brain insults. *Trends Neurosci.* **17,** 490–496.

Liou, H. C. and Baltimore, D. (1993) Regulation of the NF-kappa B/rel transcription factor and I kappa B inhibitor system. *Curr. Opinion Cell Biol.* **5,** 477–487.
Liu, Y., Kato, H., Nakata, N., and Kogure, K. (1992) Protection of rat hippocampus against ischemic neuronal damage by pretreatment with sublethal ischemia. *Brain Res.* **586,** 121–124.
Lloyd, E. D. and Wooten, M. W. (1992) pp42/44 MAP kinase is a component of the neurogenic pathway utilized by nerve growth factor in PC12 cells. *J. Neurochem.* **59,** 1099–1109.
Lowenstein, D. H., Chan, P. H., and Miles, M. F. (1991) The stress protein response in cultured neurons: characterization and evidence for a protective role in excitotoxicity. *Neuron* **7,** 1053–1060.
Macdonald, R. L. and Kelly, K. M. (1993) Antiepileptic drug mechanisms of action. *Epilepsia* **34,** S1–S8.
Marini, A. M. and Paul, S. M. (1992) N-methyl-D-aspartate receptor-mediated neuroprotection in cerebellar granule cells requires new RNA and protein synthesis. *Proc. Natl. Acad. Sci. USA* **89,** 6555–6559.
Mark, R. J., Hensley, K., Butterfield, D. A., and Mattson, M. P. (1995) Amyloid β-peptide impairs ion-motive ATPase activities: evidence for a role in loss of neuronal Ca^{2+} homeostasis and cell death. *J. Neurosci.* **15,** 6239–6249.
Mark, R. J., Lovell, M. A., Markesbery, W. R., Uchida, K., and Mattson, M. P. (1997) A role for 4-hydroxynonenal in disruption of ion homeostasis and neuronal death induced by amyloid β-peptide. *J. Neurochem.* **68,** 255–264.
Martin, D. P., Schmidt, R. E., DiStefano, P. S., Lowry, O. H., Carter, J. G., and Johnson, E. M. (1988) Inhibitors of protein synthesis and RNA synthesis prevent neuronal death caused by nerve growth factor deprivation. *J. Cell Biol.* **106,** 829–844.
Martinez-Serrano, A., Fischer, W., and Bjorklund, A. (1995) Reversal of age-dependent cognitive impairments and cholinergic neuron atrophy by NGF-secreting neural progenitors grafted to the basal forebrain. *Neuron* **15,** 473–484.
Martinou, J. C., Dubois-Dauphin, M., Staple, J. K., Rodriguez, I., Frankowski, H., Missotten, M., et al. (1994) Overexpression of bcl-2 in transgenic mice protects neurons from naturally occurring cell death and experimental ischemia. *Neuron* **13,** 1017–1030.
Mattson, M. P. and Kater, S. B. (1988) Intracellular messengers in the generation and degeneration of hippocampal neuroarchitecture. *J. Neurosci. Res.* **21,** 447–464.
Mattson, M. P. (1989) Acetylcholine potentiates glutamate-induced neurodegeneration in cultured hippocampal neurons. *Brain Res.* **497,** 402–406.
Mattson, M. P., Murrain, M., Guthrie, P. B., and Kater, S. B. (1989) Fibroblast growth factor and glutamate: opposing actions in the generation and degeneration of hippocampal neuroarchitecture. *J. Neurosci.* **9,** 3728–3740.
Mattson, M. P. (1991) Evidence for the involvement of protein kinase C in neurodegenerative changes in cultured human cortical neurons. *Exp. Neurol.* **112,** 95–103.
Mattson, M. P. (1992) Calcium as sculptor and destroyer of neural circuitry. *Exp. Gerontol.* **27,** 29–49.
Mattson, M. P., Rychlik, B., Chu, C., and Christakos, S. (1991) Evidence for calcium-reducing and excitoprotective roles for the calcium binding protein (calbindin-D28k) in cultured hippocampal neurons. *Neuron* **6,** 41–51.
Mattson, M. P., Kumar, K., Cheng, B., Wang, H., Michaelis, E. K. (1993a) Basic FGF regulates the expression of a functional 71 kDa NMDA receptor protein that mediates calcium influx and neurotoxicity in cultured hippocampal neurons. *J. Neurosci.* **13,** 4575–4588.
Mattson, M. P., Barger, S. W., Cheng, B., Lieberburg, I., Smith-Swintosky, V. L., and Rydel, R. E. (1993b) β-amyloid precursor protein metabolites and loss of neuronal calcium homeostasis in Alzheimer's disease. *Trends Neurosci.* **16,** 409–415.
Mattson, M. P., Lovell, M. A., Furukawa, K., and Markesbery, W. (1995a) Neurotrophic factors attenuate glutamate-induced accumulation of peroxides, elevation of $[Ca^{2+}]_i$ and neurotoxicity,

and increase antioxidant enzyme activities in hippocampal neurons. *J. Neurochem.* **65,** 1740–1751.

Mattson, M. P., Cheng, B., Baldwin, S., Smith-Swintosky, V. L., Keller, J., Geddes, J. W., Scheff, S. W., and Christakos, S. (1995b) Brain injury and tumor necrosis factors induce expression of calbindin D-28k in astrocytes: a cytoprotective response. *J. Neurosci. Res.* **42,** 357–370.

Mattson, M. P. and Scheff, S. W. (1994) Endogenous neuroprotection factors and traumatic brain injury, mechanisms of action and implications for therapies. *J. Neurotrauma* **11,** 3–33.

Mattson, M. P. and Furukawa, K. (1996) Programmed cell life: anti-apoptotic signaling and therapeutic strategies in neurodegenerative disorders. *Restorative Neurol. Neurosci.* **9,** 191–205.

Mattson, M. P. and Begley, J. G. (1996) Amyloid β-peptide alters thrombin-induced calcium responses in cultured human neural cells. *Amyloid* **3,** 28–40.

Mattson, M. P., Borges, S. W., Furukawa, K., Bruce, A., Wyss-Coray, T., Mark, R. J., and Mucke, L. (1997) Cellular signaling roles of TGFβ, TNFα, and βAPP in brain injury responses and Alzheimer's disease. *Brain Res. Rev.* **23,** 47–61.

Merlio, J. P., Ernfors, P., Kokaia, Z., Middlemas, D. S., Bengzon, J., Kokaia, M., et al. (1993) Increased production of the TrkB protein tyrosine kinase receptor after brain insults. *Neuron* **10,** 151–164.

Miller, R. J., Murphy, S. N., and Glaum, S. R. (1989) Neuronal Ca^{2+} channels and their regulation by excitatory amino acids. *Ann. NY Acad. Sci.* **568,** 149–158.

Milligan, C. E., Prevette, D., Yaginuma, H., Homma, S., Cardwell, C., Fritz, L. C., et al. (1995) Peptide inhibitors of the ICE protease family arrest programmed cell death of motoneurons in vivo and in vitro. *Neuron* **15,** 385–393.

Morrison, R. S., Keating, R. F., and Moskal, J. R. (1988) Basic fibroblast growth factor and epidermal growth factor exert differential trophic effects on CNS neurons. *J. Neurosci. Res.* **21,** 71–79.

Murphy, T. H., Miyamoto, M., Sastre, A., Schnaar, R. L., and Coyle, J. T. (1989) Glutamate toxicity in a neuronal cell line involves inhibition of cystine transport and oxidative stress. *Neuron* **2,** 1547–1558.

Murray, J. K. and England, J. P. (1992) Inhibitors of cyclic nucleotide phosphodiesterases as therapeutic agents. *Biochem. Soc. Trans.* **20,** 460–464.

Murrell, R. D. and Tolkovsky, A. M. (1993) Role of voltage-gated Ca^{2+} channels and intracellular Ca^{2+} in rat sympathetic neuron survival and function promoted by high K^+ and cyclic AMP in the presence or absence of NGF. *Eur. J. Neurosci.* **5,** 1261–1272.

Nabeshima, T., Ogawa, S. I., Nishimura, H., Fuji, K., Kameyama, T., and Sasaki, Y. (1991) Staurosporine facilitates recovery from the basal forebrain lesion-induced impairment of learning an deficit of cholinergic neuron in rats. *J. Pharmacol. Exp. Ther.* **257,** 562–566.

Nabeshima, T., Nitta, A., Fuji, K., Kameyama, T., and Hasegawa, T. (1994) Oral administration of NGF synthesis stimulators recovers reduced brain NGF content in aged rats and cognitive dysfunction in basal forebrain lesioned rats. *Gerontology* **40,** 46–56.

Nishida, E. and Gotoh, Y. (1993) The MAP kinase cascade is essential for diverse signal transduction pathways. *TIBS* **18,** 128–131.

Nishizuka, Y. (1995) Protein kinase C and lipid signaling for sustained cellular responses. *FASEB J.* **9,** 484–496.

Nitsch, R. M., Farber, S. A., Growdon, J. H., and Wurtman, R. J. (1993) Release of amyloid beta-protein precursor derivatives by electrical depolarization of rat hippocampal slices. *Proc. Natl. Acad. Sci. USA* **90,** 5191–5193.

Nobes, C. D. and Tolkovsky, A. M. (1995) Neutralizing anti-p21ras Fabs suppress rat sympathetic neuron survival induced by NGF, LIF, CNTF and cAMP. *Eur. J. Neurosci.* **7,** 344–350.

Nozaki, K., Finklestein, S. P., and Beal, M. F. (1993) Basic fibroblast growth factor protects against hypoxia-ischemia and NMDA neurotoxicity in neonatal rats. *J. Cereb. Blood Flow Metab.* **13,** 221–228.

O'Connor, J. J., Rowan, M. J., and Anwyl, R. (1994) Long-lasting enhancement of NMDA receptor-mediated synaptic transmission by metabotropic glutamate receptor activation. *Nature* **367,** 557–559.
Ohno, M., Yamamoto, T., and Watanabe, S. (1991) Effect of staurosporine, a protein kinase inhibitor, on impairment of working memory in rats exposed to cerebral ischemia. *Eur. J. Pharmacol.* **204,** 113–116.
Ohno, M., Yamamoto, T., Ueki, S., and Watanabe, S. (1992) Protection by N-methyl-D-asparate receptor antagonists against impairment of working memory in rats following transient cerebral ischemia. *Neurosci. Lett.* **138,** 1–4.
Olney, J. W. (1994) New mechanisms of excitatory transmitter neurotoxicity. *J. Neural Transm.* **43 Suppl,** 47–51.
O'Neill, C., Fowler, C. J., Winblad, B., and Cowburn, R. F. (1994) G-protein coupled signal transduction systems in the Alzheimer's disease brain. *Biochem. Soc. Trans.* **22,** 167–171.
Orrenius, S. and Nicotera, P. (1994) The calcium ion and cell death. *J. Neural. Transm.* **43 Suppl.,** 1–11.
Quayle, J. M., Bonev, A. D., Brayden, J. E., and Nelson, M. T. (1994) Calcitonin gene-related peptide activated ATP-sensitive K^+ currents in rabbit arterial smooth muscle via protein kinase A. *J. Physiol.* (London) **475,** 9–13.
Roberts-Lewis, J. M. and Siman, R. (1993) Spectrin proteolysis in the hippocampus: a biochemical marker for neuronal injury and neuroprotection. *Ann. NY Acad. Sci.* **679,** 78–86.
Rosen, L. B., Ginty, D. D., Weber, M. J., and Greenberg, M. E. (1994) Membrane depolarization and calcium influx stimulate MEK and MAP kinase via activation of Ras. *Neuron* **12,** 1207–1221.
Rosen, L. B. and Greenberg, M. E. (1996) Stimulation of growth factor receptor signal transduction by activation of voltage-sensitive calcium channels. *Proc. Natl. Acad. Sci. USA* **93,** 1113–1118.
Rosenmund, C. and Westbrook, G. (1993) Calcium-induced actin depolymerization reduces NMDA channel activity. *Neuron* **10,** 805–814.
Rydel, R. E. and Greene, L. A. (1988) cAMP analogs promote survival and neurite outgrowth in cultures of rat sympathetic and sensory neurons independently of nerve growth factor. *Proc. Natl. Acad. Sci. USA* **85,** 1257–1261.
Sapolsky, R. M. (1994) The physiological relevance of glucocorticoid endangerment of the hippocampus. *Ann. NY Acad. Sci.* **746,** 294–304.
Savolainen, K., Hirvonen, M. R., and Naarala, J. (1994) Phosphoinositide second messengers in cholinergic excitotoxicity. *Neurotoxicology* **15,** 493–502.
Schoepp, D. D. and Conn, P. J. (1993) Metabotropic glutamate receptors in brain function and pathology. *Trends Pharmacol. Sci.* **14,** 13–20.
Schreck, R., Rieber, P., and Baeuerle, P. A. (1991) Reactive oxygen intermediates as apparently widely used messengers in the activation of the NF-κB transcription factor and HIV-1. *EMBO J.* **10,** 2247–2258.
Seger, R. and Krebs, E. G. (1995) The MAPK signaling cascade. *FASEB J.* **9,** 726–735.
Selkoe, D. J. (1991) The molecular pathology of Alzheimer's disease. *Neuron* **6,** 487–498.
Semkova, I., Wolz, P., Schilling, M., and Krieglstein, J. (1996) Selegiline enhances NGF synthesis and protects CNS neurons from excitotoxic and ischemic damage. *Eur. J. Pharmacol.* **315,** 19–30.
Shearman, M. S., Hawtin, S. R., and Tailor, V. J. (1995) The intracellular component of cellular 3-(4,5-dimethylthiazol-2-yl)-2,5-diphenyltetrazolium bromide (MTT) reduction is specifically inhibited by β-amyloid peptides. *J. Neurochem.* **65,** 218–227.
Sheterline, P., Clayton, J., and Sparrow, J. (1995) Actin. *Protein Profile* **2,** 1–103.
Shibata, S., Tominaga, K., and Watanabe, S. (1995) Glutathione protects against hypoxic/hypoglycemic decreases in 2-deoxyglucose uptake and presynaptic spikes in hippocampal slices. *Eur. J. Pharmacol.* **273,** 191–195.
Shimazaki, K., Ishida, A., and Kawai, N. (1994) Increase in bcl-2 oncoprotein and the tolerance to ischemia-induced neuronal death in the gerbil hippocampus. *Neurosci. Res.* **20,** 95–99.

Shimohama, S., Tamura, Y., Akaike, A., Tsukahara, T., Ohara, O., Watanabe, S., and Kimura, J. (1993) Brain-derived neurotrophic factor pretreatment exerts a partially protective effect against glutamate-induced neurotoxicity in cultured rat cortical neurons. *Neurosci. Lett.* **164,** 55–58.

Siman, R. and Noszek, J. C. (1988) Excitatory amino acids activate calpain I and induce structural protein breakdown in vivo. *Neuron* **1,** 279–287.

Skaper, S. D., Negro, A., Facci, L., and Dal Toso, R. (1993) Brain-derived neurotrophic factor selectively rescues mesencephalic dopaminergic neurons from 2,4,5-trihydroxyphenyl-alanine-induced injury. *J. Neurosci. Res.* **34,** 478–487.

Smith-Swintosky, V. L., Zimmer, S., Fenton, J. W., and Mattson, M. P. (1995) Protease nexin-I and thrombin modulate neuronal Ca^{2+} homeostasis and sensitivity to glucose deprivation-induced injury. *J. Neurosci.* **15,** 5840–5850.

Smith-Swintosky, V. L., Kraemer, P. J., McCants, N., Maki, A., Brown, R. W., Keller, J., Goodman, Y., and Mattson, M. P. (1996a) Bacterial alkaloids mitigate seizure-induced hippocampal damage and memory deficits. *Exp. Neurol.* **141,** 287–296.

Smith-Swintosky, V. L., Pettigrew, L. C., Sapolsky, R. M., Phares, C., Craddock, S. D., Brooke, S. M., and Mattson, M. P. (1996b) Metyrapone, an inhibitor of glucocorticoid production, reduces brain injury induced by focal and global ischemia and seizures. *J. Cereb. Blood Flow Metab.* **16,** 585–598.

Spitz, D. R., Malcolm, R. R., and Roberts, R. J. (1990) Cytotoxicity and metabolism of 4-hydroxy-2-nonenal and 2-nonenal in H_2O_2-resistant cells lines. Do aldehyde by-products of lipid peroxidation contribute to oxidative stress? *Biochem. J.* **267,** 453–459.

Squier, M. K., Miller, A. C., Malkinson, A. M., and Cohen, J. J. (1994) Calpain activation in apoptosis. *J. Cell Physiol.* **159,** 229–237.

Stein-Behrens, B., Mattson, M. P., Chang, I., Yeh, M., and Sapolsky, R. M. (1994) Stress exacerbates neuron loss and cytoskeletal pathology in the hippocampus. *J. Neurosci.* **14,** 5373–5380.

Strijbos, P. J., Relton, J. K., and Rothwell, N. J. (1994) Corticotrophin-releasing factor antagonist inhibits neuronal damage induced by focal cerebral ischaemia or activation of NMDA receptors in the rat brain. *Brain Res.* **656,** 405–408.

Sucher, N. J. and Lipton, S. A. (1991) Redox modulatory site of the NMDA receptor-channel complex, Regulation by oxidized glutathione. *J. Neurosci. Res.* **30,** 582–591.

Takami, K., Kiyota, Y., Iwane, M., Miyamoto, M., Tsukuda, R., Igarashi, K., et al. (1993) Upregulation of fibroblast growth factor-receptor messenger RNA expression in rat brain following transient forebrain ischemia. *Exp. Brain Res.* **97,** 185–194.

Tanaka, C., Nishino, N., Hashimoto, T., Kitamura, N., Yoshihara, C., and Saito, N. (1993) Second messenger systems in brains of patients with Parkinson's or Huntington's disease. *Adv. Neurol.* **60,** 175–180.

Tortosa, A., Rivera, R., and Ferrer, I. (1994) Dose-related effects of cycloheximide on delayed neuronal death in the gerbil hippocampus after bilateral transitory forebrain ischemia. *J. Neurol. Sci.* **121,** 10–17.

Trojanowski, J. Q. and Lee, V. M. (1995) Phosphorylation of paired helical filament tau in Alzheimer's disease neurofibrillary lesions: focusing on phosphatases. *FASEB J.* **9,** 1570–1576.

Wahlestedt, C., Golanov, E., Yumamoto, S., Yee, F., Ericson, H., Yoo, H., Inturrisi, C. E., and Reis, D. J. (1993) Antisense oligodeoxinucleotides to NMDA-R1 receptor channel protect cortical neurons from excitotoxicity and reduce focal ischaemial infarctions. *Nature* **363,** 260–263.

Walicke, P., Cowan, W. M., Ueno, N., Baird, A., and Guillemin, R. (1986) Fibroblast growth factor promotes survival of dissociated hippocampal neurons and promotes neurite extension. *Proc. Natl. Acad. Sci. USA* **83,** 3012–3016.

Watson, B. D. and Ginsberg, M. D. (1989) Ischemic injury in the brain. Role of oxygen radical-mediated processes. *Ann. NY Acad. Sci.* **559,** 269–281.

Wiedau-Pazos, M., Goto, J. J., Rabizadeh, S., Gralia, E. B., Roe, J. A., Lee, M. K., Valentine, J. S., and Bredesen, D. E. (1996) Altered reactivity of superoxide dismutase in familial amyotrophic lateral sclerosis. *Science* **271,** 515–518.

Wong, G. H. (1995) Protective roles of cytokines against radiation: induction of mitochondrial MnSOD. *Biochim. Biophys. Acta* **1271,** 205–209.

Xie, G. X., Meng, F., Mansour, A., Thompson, R. C., Hoversten, M. T., Goldstein, A., Watson, S. J., and Akil, H. (1994) Primary structure and functional expression of a guinea pig kappa opioid (dynorphin) receptor. *Proc. Natl. Acad. Sci. USA* **91,** 3779–3783.

Yang, G., Chan, P. H., Chen, J., Carlson, E., Chen, S. F., Weinstein, P., Epstein, C. J., and Kamii, H. (1994) Human copper-zinc superoxide dismutase transgenic mice are highly resistant to reperfusion injury after focal cerebral ischemia. *Stroke* **25,** 165–170.

Zheng, J. L., Stewart, R. R., and Gao, W. Q. (1995) Neurotrophin-4/5, brain-derived neurotrophic factor, and neurotrophin-3 promote survival of cultured vestibular ganglion neurons and protect them against neurotoxicity of ototoxins. *J. Neurobiol.* **28,** 330–340.

Zhong, L. T., Kane, D. J., and Bredesen, D. E. (1993) BCL-2 blocks glutamate toxicity in neural cell lines. *Mol. Brain Res.* **19,** 353–355.

Index

O

P